ERRATA

Chemistry of Plant Hormones

Edited by Nobutaka Takahashi

Please note the following corrections for Chapter 1.

Page 4, Figure 3: Helminthosporlo should be spelled Helminthosporol. Cotylenib A should be spelled Cotylenol A.

Page 5: Figure 4 should be Figure 2. Continued.

Page 7: The first 3 structures of Figure 5 (AH-toxin, AK-toxin, and HMT-toxin) should be Figure 4. Structures of host-specific phytotoxins.

Chemistry of Plant Hormones

Editor

Nobutaka Takahashi, Ph.D.
Professor
Department of Agricultural Chemistry
Faculty of Agriculture
The University of Tokyo
Bunkyo-ku, Tokyo, Japan

CRC Press, Inc.
Boca Raton, Florida

Library of Congress Cataloging-in-Publication Data
Main entry under title:

Chemistry of plant hormones.

    Bibliography: p.
    Includes index.
    1. Plant hormones. I. Takahashi, Nobutaka,
1930-
QK731.C516   1986   581.19'27   85-17475
ISBN 0-8493-5470-6

This book represents information obtained from authentic and highly regarded sources. Reprinted material is quoted with permission, and sources are indicated. A wide variety of references are listed. Every reasonable effort has been made to give reliable data and information, but the author and the publisher cannot assume responsibility for the validity of all materials or for the consequences of their use.

All rights reserved. This book, or any parts thereof, may not be reproduced in any form without written consent from the publisher.

Direct all inquiries to CRC Press, Inc., 2000 Corporate Blvd., N.W., Boca Raton, Florida, 33431.

© 1986 by CRC Press, Inc.
Second Printing, 1986

International Standard Book Number 0-8493-5470-6

Library of Congress Card Number 85-17475
Printed in the United States

PREFACE

Plants have the photosynthetic ability to produce sugars from carbon dioxide and water by use of sunshine energy, while most living organisms other than plants have no such ability. Thus, their lives in nature really depend upon this unique ability of plants.

It is one of the most important research subjects in biology, biochemistry, and chemistry to clarify the mechanism regulating the life cycles of higher plants. It is an attractive and challenging target not only in an academic sense, but also from the viewpoint of application. This kind of research has been contributing to the development of new techniques for increasing the supply of foods for humankind and feed for animals.

About 50 years have passed since indole-3-acetic acid, which was later recognized as real auxin, was isolated as a heteroauxin from human urine; and 46 years since the isolation of gibberellin as the metabolite of a plant pathogen, *Gibberella fujikuroi*. In these years the number of the principal plant hormone groups has increased to five, namely, auxins, gibberellins, cytokinins, abscisic acid, and ethylene; and many plant growth regulators other than plant hormones have been isolated and characterized. Information on their physiological activity and chemistry should make a tremendous contribution to the understanding of the regulatory mechanism of higher plant life cycles and to the development of new technology in the cultivation of higher plants.

Many excellent books have been published which cover the physiology of plant growth regulators, but rather few on their chemistry. Thus, this book deals mainly with the chemistry of the principal plant hormones and not so much with their physiology. Further, due to limited space, plant growth regulators other than the principal plant hormones are described briefly only in the Introduction.

N. T.

THE EDITOR

Nobutaka Takahashi, Ph.D., is Professor of Pesticide Chemistry in the Faculty of Agriculture at The University of Tokyo. He graduated from The University of Tokyo in 1952 and obtained a Ph.D. He assumed his present position after serving as Assistant Professor from 1964 to 1969. In addition to his university post, Professor Takahashi has since 1977, held the position of Director of Pesticide Chemistry Laboratory III in Japan's prestigious Institute of Physical and Chemical Research.

Dr. Takahashi is Vice President of the Japanese Society of Plant Growth Regulation, a council member of the International Plant Growth Substances Association, the Agricultural Chemical Society of Japan, the Pesticide Science Society of Japan, and the Japanese Society of Plant Physiologists. He is also a member of the Chemical Society of America, Japan and London, and the American Society of Plant Physiologists.

Dr. Takahashi is the author of more than 200 papers and review articles and 7 books. He is noted for his research into the chemistry and physiology of biologically active natural products, including insect sex pheromones, naturally occurring insecticides, and plant growth regulators such as gibberellins, cytokinins, abscisic acid, and brassinosteroids.

CONTRIBUTORS

Nobuhiro Hirai, Ph.D.
Research Associate
Department of Food Science and
 Technology
Faculty of Agriculture
Kyoto University
Kyoto, Japan

Hidemasa Imaseki, Ph.D.
Professor
Research Institute for Biochemical
 Regulation
Faculty of Agriculture
Nagoya University
Chikusa, Nagoya, Japan

Hajime Iwamura, Ph.D.
Associate Professor
Department of Agricultural Chemistry
Faculty of Agriculture
Kyoto University
Kyoto, Japan

Koichi Koshimizu, Ph.D.
Professor
Department of Food Science and
 Technology
Faculty of Agriculture
Kyoto University
Kyoto, Japan

Shingo Marumo, Ph.D.
Professor
Department of Agricultural Chemistry
Faculty of Agriculture
Nagoya University
Chikusa, Nagoya, Japan

Nobutaka Takahashi, Ph.D.
Professor
Department of Agricultural Chemistry
Faculty of Agriculture
The University of Tokyo
Bunkyo-ku, Tokyo, Japan

Isomaro Yamaguchi, Ph.D.
Department of Agricultural Chemistry
Faculty of Agriculture
The University of Tokyo
Bunkyo-ku, Tokyo, Japan

Hisakazu Yamane, Ph.D.
Department of Agricultural Chemistry
Faculty of Agriculture
The University of Tokyo
Bunkyo-ku, Tokyo, Japan

TABLE OF CONTENTS

Chapter 1
Introduction 1
Nobutaka Takahashi

Chapter 2
Auxins 9
Shingo Marumo

Chapter 3
Gibberellins 57
Nobutaka Takahashi, Isomaro Yamaguchi, and Hisakazu Yamane

Chapter 4
Cytokinins 153
Koichi Koshimizu and Hajime Iwamura

Chapter 5
Abscisic Acid 201
Nobuhiro Hirai

Chapter 6
Ethylene 249
Hidemasa Imaseki

Index 265

Chapter 1

INTRODUCTION

Nobutaka Takahashi

The life cycle of higher plants is regular, though complex. Each stage, together with the shift to the next stage, is controlled by endogenous plant growth regulators. The situation is further complicated by the need for the life cycle to accomodate to environmental conditions such as light intensity, daylength, humidity, and nutritional conditions. Such responses to environmental change are believed to be due to the quantity and availability of endogenous plant hormones and other bioactive substances. In this sense the life cycle and its sensitive adjustment to outside stimuli may be due to the mediating role of such compounds, which thus fulfill a most important role.

At this stage we know of many kinds of plant growth regulators associated with a wide variety of physiological functions in higher plants. In many cases these have been isolated and chemically characterized. Among the most important of these are the so-called plant hormones (phytohormones).

The concept of plant hormones differs substantially from that of hormones in animals and insects, because the differentiation of organ tissues in plants is less extensive than in animals. Plant hormones can be broadly defined as follows: (1) they must be chemically characterized and shown to be biosynthesized in some plant organ, (2) they must be broadly distributed within the plant kingdom, (3) they must show specific biological activity in very low concentration and must be shown to play a fundamental role in regulating physiological phenomena in vivo, and (4) they are usually translocated within the plant from a biosynthesis site to an action site.

At present, five groups of plant growth regulators — auxins, gibberellins, cytokinins, abscisic acid, and ethylene — are regarded as plant hormones; however, the distinction between plant hormones and plant growth regulators other than plant hormones is not always easily seen. At this stage one new group of compounds that regulate plant growth, brassinolide and related compounds, must be considered as true plant hormones on account of their wide distribution in the plant kingdom and their unique biological activity. A less clear cut case is that of "florigen", the hypothetical flower-inducing hormone. Logical as the hypothesis may be, the evidence to support the existence of such a hormone is incomplete and it has never been isolated. Despite this, it is logical to expect that various new plant hormones will be isolated and chemically characterized in the future; the wide variety of physiological function required to maintain the complicated life cycle of the higher plants requires that a considerable complexity of chemical control should exist.

Quite apart from these compounds that might be considered plant hormones, there are many natural products that show interesting physiological activity in the higher plants. In general these compounds are of limited distribution in the plant kingdom and show a rather restricted range of activities. They have been isolated not only from plant tissues but also as metabolites of microorganisms. They can be grouped according to their origin (plant, microorganism) and physiological activity as follows:

•Plant growth regulators of plant origin
1. Plant growth promotors (Figure 1)
    Brassinolide and related compounds: brassinolide, castasterone, dolicholide, 6-deoxocastasterone, 2-deoxocastasterone, etc.
    Strigol (germination stimulant from witchweed)
    Phaseolic acid

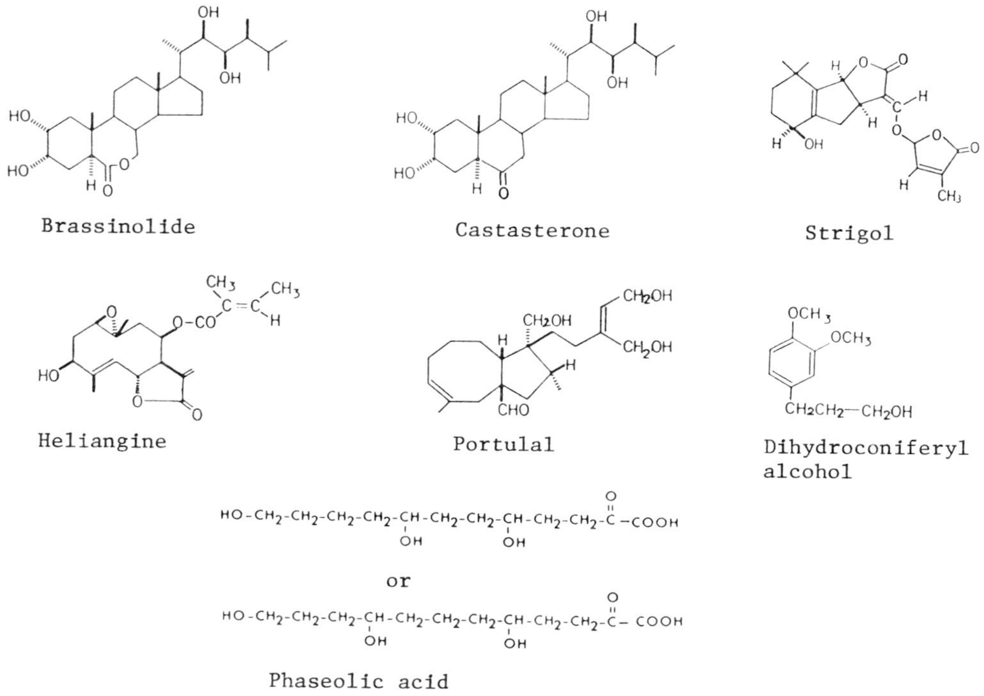

Chemistry of Plant Hormones

Brassinolide

Castasterone

Strigol

Heliangine

Portulal

Dihydroconiferyl alcohol

HO-CH$_2$-CH$_2$-CH$_2$-CH$_2$-CH-CH$_2$-CH$_2$-CH-CH$_2$-CH$_2$-C-COOH
                                   OH   OH

or

HO-CH$_2$-CH$_2$-CH$_2$-CH-CH$_2$-CH$_2$-CH$_2$-CH-CH$_2$-CH$_2$-C-COOH
                            OH   OH

Phaseolic acid

FIGURE 1. Structures of plant growth regulators of plant origin.

    Dihydroconiferyl alcohol (synergist of gibberellin)
2. Plant growth inhibitors (Figure 2)
    Benzoic acid and related compounds
    Cinnamic acid and related compounds
    Phenolic compounds including flavonoids
    Unsaturated lactones: scoporetin, parasorbic acid, psoraen, seselin, etc.
    Growth inhibitors isolated from *Podocarpus*: podolactones A—E, inumakilactones A—D, ponalactone A and its glucoside, hallactones A, B, sellowins A—C, nagilactones A—G, podolide, etc.
    Momilactone and related compounds: momilactones A, B, annonalide
    Jasmonic acid and related compounds: jasmonic acid and its methyl ester, cucurbic acid and its glucoside
    Asparagusic acid and related compounds: asparagusic acid and its sulfoxide, dihydroasparagusic acid, S-acetyldihydroasparagusic acid, etc.
    Poly-yne-ene compounds: matricariaester, 2-(Z)-dehydromatricariaester, methyl 2-(Z)-decene-4,6-diynoate, etc.
    Batatacins: batatacins I, II, III
    Rooting inhibitors from *Eucalyptus*: G-regulators, G1, G2, G3, grandinol
    Growth inhibitors in liverwort and algae: lunularic acid and related compounds
    Growth inhibitors from bulbs of *Lycoris radiata*: lycoricidinol and lycoricidine
    Others: harrintonolide, methyl pheoborides, juglone, 3-acetyl-6-methylbenzaldehyde, lignans with germination inhibitory activity, 4,8,13-duvatriene-1,3-diol

•Plant Growth Regulators of Microbial Origin
1. Plant growth promotors (Figure 3)
    Helminthosporol and related compounds from *Helminthosporium sativum*: helmintho-

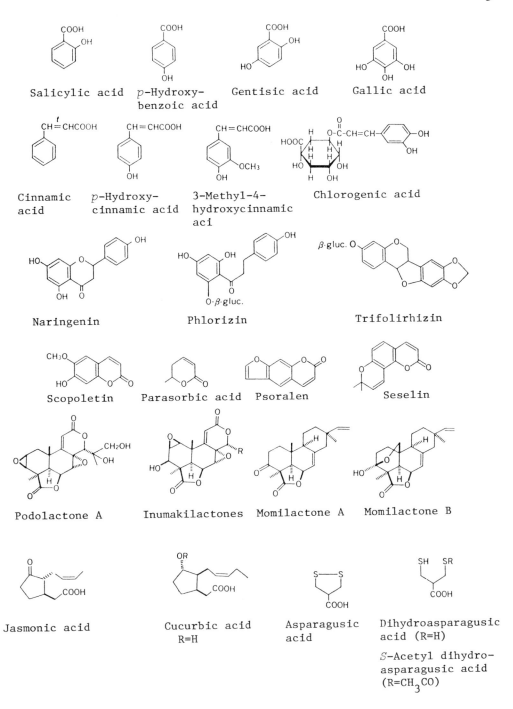

FIGURE 2. Structures of plant growth inhibitors.

sporol and *cis*-sativendiol, etc.
Sclerin and related compounds from *Sclerotinia* spp.: sclerin, sclerotinins A, B
Malformins from *Aspergillus* spp. (malformation inducing substances): malformins A_1, A_2, B_1, B_2, C
Cotylenol and related compounds from *Clodosporium* sp.: cotylenol, cotylenins A—F

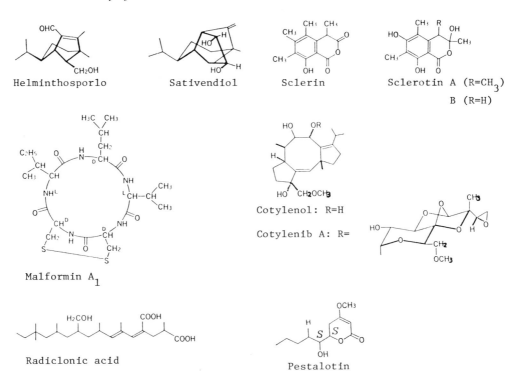

FIGURE 3. Structures of plant growth promotors of microbial origin.

Radiclonic acid from *penicillium* sp.
Synergist to gibberellins from *Pestalotia crytmeraecola* and other unidentified fungi: pestalotin and related compounds

2. Plant growth inhibitors and phytotoxins

Host-specific toxins (Figure 4) have been isolated mainly from *Helminthosporium* and *Alternaria* spp. and have been shown to be used for host recognition by plant pathogens: AM-toxins I—III (*Alternaria mali*), AK-toxin (*A. kikuchiana*), HMT-toxins (*Helminthosporium madys* race T), HC-toxin (*H. carbonum*), toxin from *A. alternata* F. sp. *cycopersici*.

Nonspecific toxins (Figure 5) involve compounds with a wide variety of structures and cause very divergent symptoms on host plants. Some examples follow.

a. Terpenoids: diacetoxysciprenol (*Fusarium equiseti*), aphidicolin (*Harziella entomophilla*), ophiobolins (*Ophiobolus* and *Helminthosporium* spp.), Fusicoccins (*Fusicoccum amygdali*).
b. Other carbocyclic compounds: phyllosinol (epoxidon) and related compounds (*Phyllosticta* sp.).
c. Aromatic and heteroaromatic compounds: piryculol (*Pyricularia oryzae*), fusaric acid (*Gibberella fujikuroi*, *Fusarium*, and *Nectria* spp.), α-picolic acid (*Pyricularia oryzae*), tenuazoic acid (*P. oryzae*, *Alternaria longipes*).
d. Amino acids and peptides: rhizobitoxine and dihydro derivative (*Rhizobium japonicum*), coronatine (*Pseudomonas coronafaciens*), lycomarasmins (*Aspergillus flavus* and *oryzae*), tentoxin (*Alternaria* spp.), phaseotoxin (*Pseudomonas phaseolicola*), tabtoxin (*Pseudomonas tabaci*).

CH$_3$-CH=CH-C≡C-C≡C-CH=CH-COOCH$_3$ (Z)

Matricaria ester; 8Z, 8E

CH$_3$-C≡C-C≡C-C≡C-CH=CH-COOCH$_3$ (Z)

2-Z-Dehydromatricaria ester

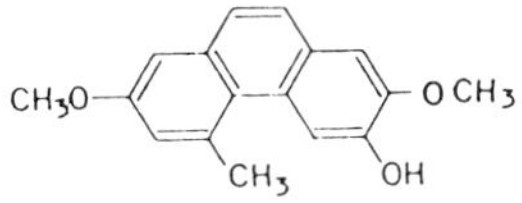

Batatasin I

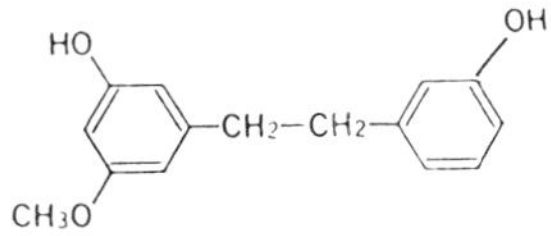

Batatasin II

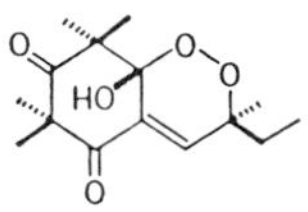

G-Regulator G-I 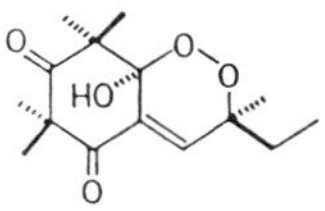   G-II 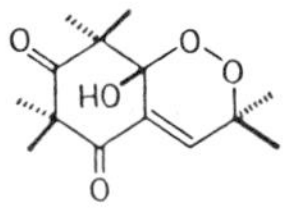   G-III 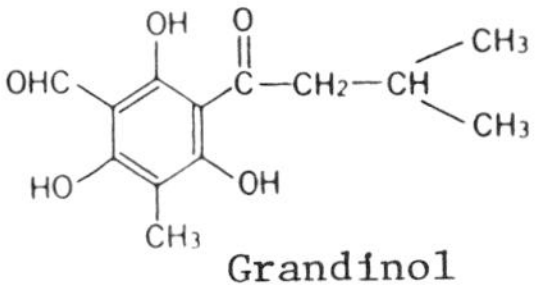   Grandinol

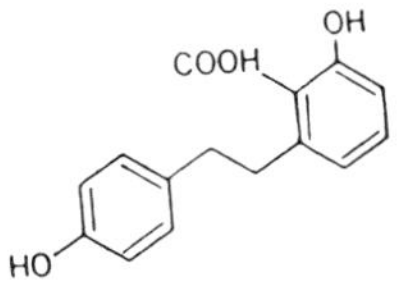

Lunularic acid 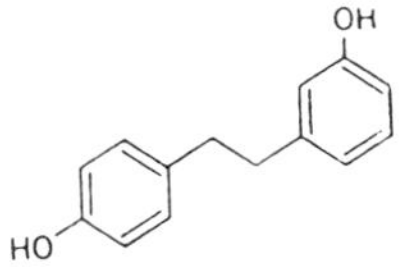   Lunularin

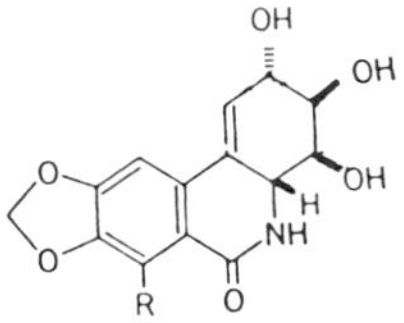

Lycoricidinol (R=OH)
Lycoricidine (R=H)

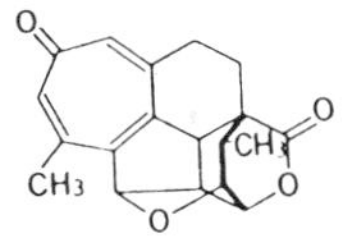

Harringtonolide

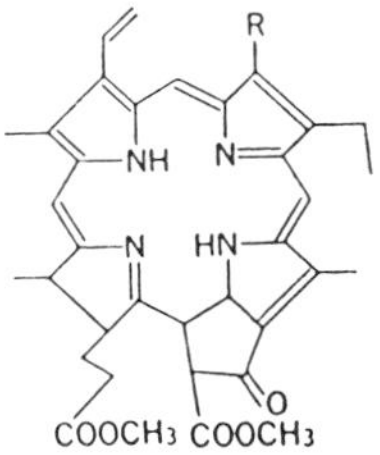

Methyl pheophorbide a (R=CH$_3$)
                    b (R=CHO)

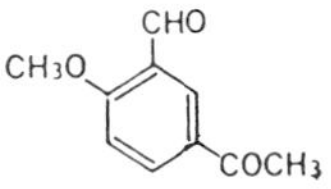

3-Acetyl-6-methoxy-benzaldehyde

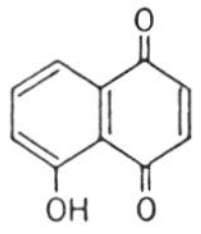

Juglone

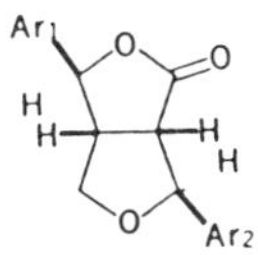

4,8,13-duvatrien-1,3-diol

Lignans from *Aegilops ovata*

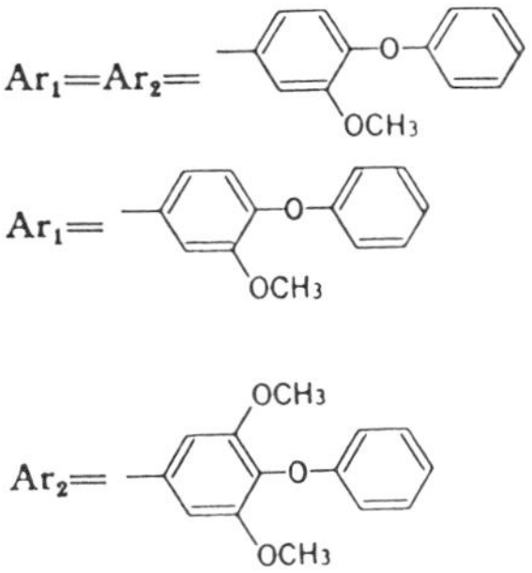

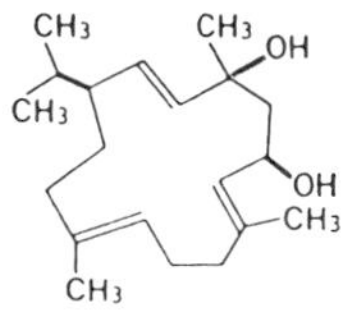

FIGURE 4. Structures of host-specific phytotoxins.

Clarification of the mechanism for the regulation of the life cycle in higher plants has been one of the most important areas of research in plant physiology. An ideal approach to such research should include all the following aspects: (1) isolation and characterization of endogenous compounds responsible for the physiological phenomenon; (2) exogenous ap-

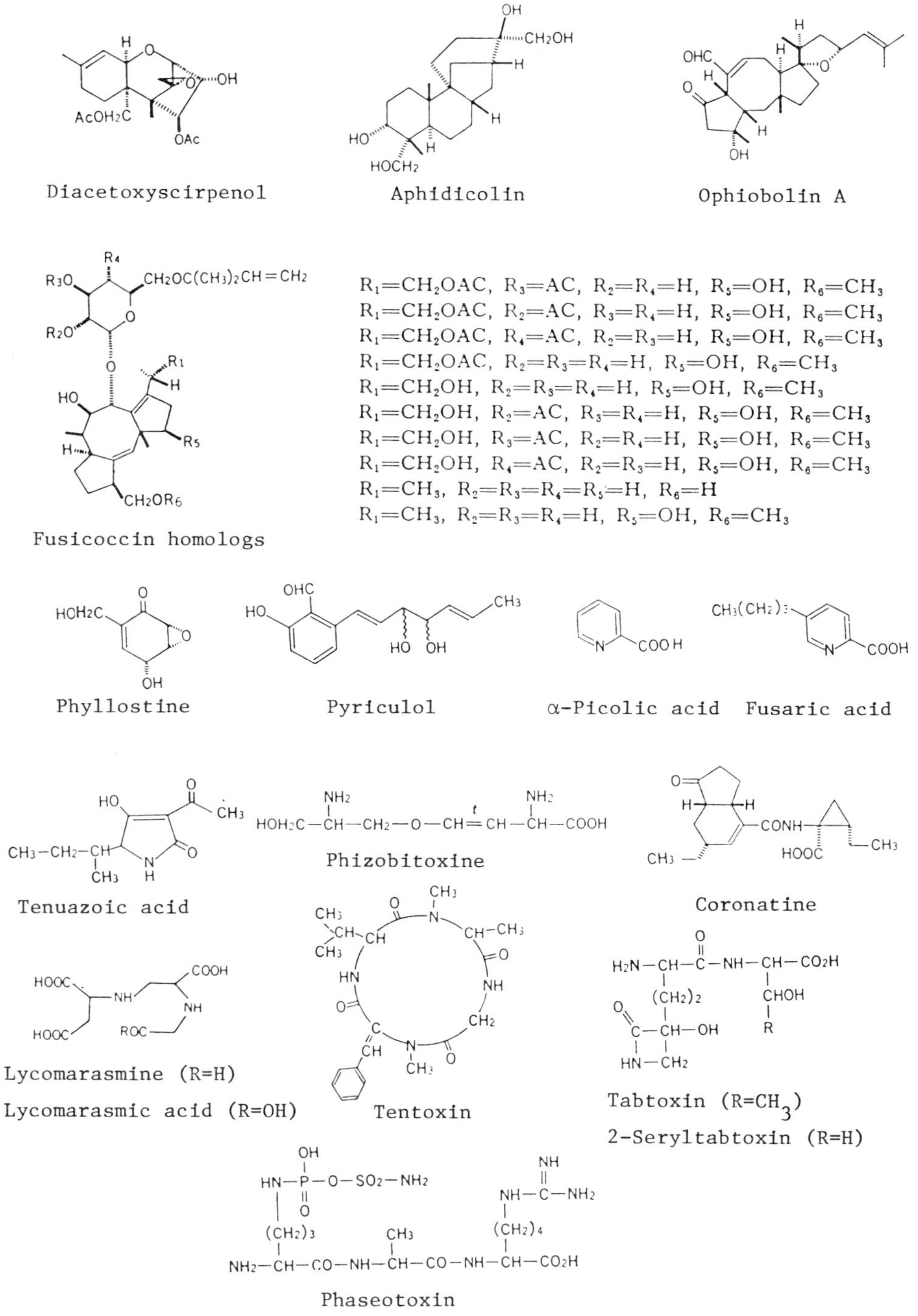

FIGURE 5. Structures of non-specific phytotoxins.

plication to other species to check for the appropriate physiological response (in this way we can hope to establish the generality of the response); (3) examination of the fluctuation in the level of the endogenous regulator in the course of the life cycle of higher plants. This permits correlation between the level of the regulator and the growth and differentiation of

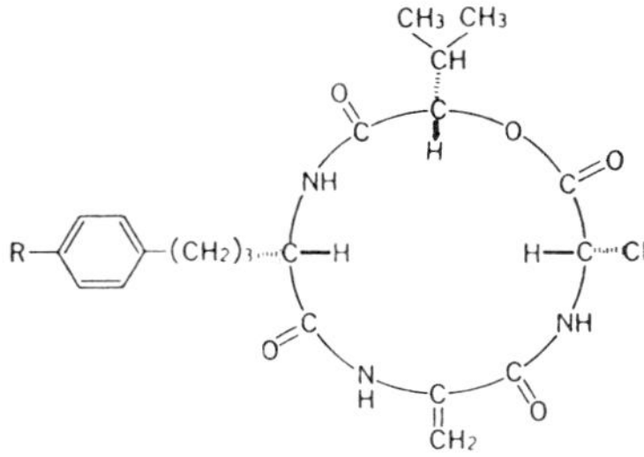

AH-toxin I (R=OCH$_3$)
II (R=H)
III (R=OH)

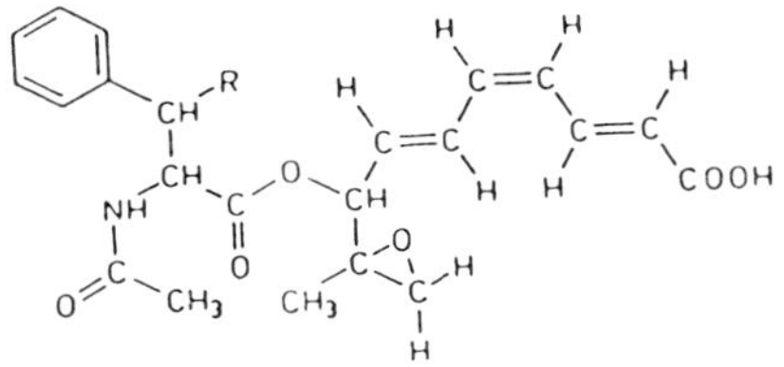

AK-toxin

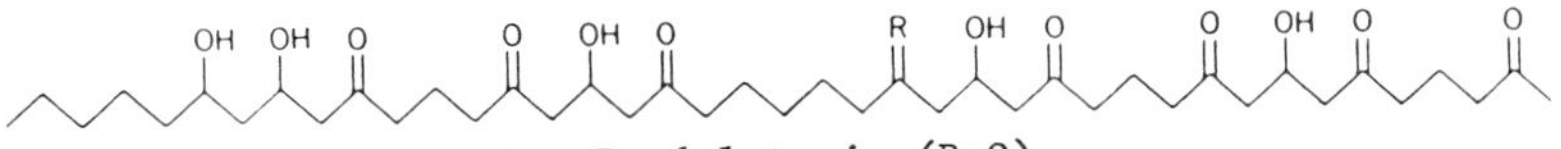

HMT-toxin: Band 1 toxin (R=O)
Band 2 toxin (R=H, OH)

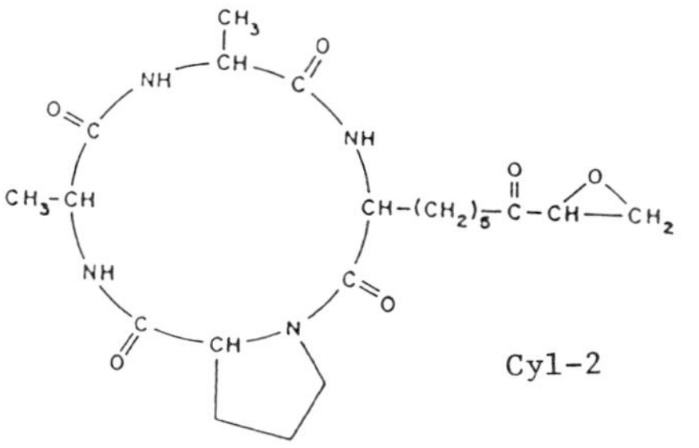

Cyl-2

FIGURE 5. Continued.

the plant, as well as revealing environmental responses. In the course of such studies, the biosynthetic and metabolic pathways should be clarified. An approach such as this can only be brought to fruition by the integration of biological and chemical methodology.

Due to space limitation, this book describes only the chemistry of plant hormones and the reader is referred to other excellent books[1-3] for information on growth regulators other than the recognized plant hormones.

ACKNOWLEDGMENT

The author wishes to express his thanks to Professor Crow of The Australian National University, Canberra, for critical comments.

REFERENCES

1. **Letham, D. S.**, Natural-occurring plant growth regulators other than the prinicpal hormones of higher plants, in *Phytohormones and Related Compounds — A Comprehensive Treatise*, Vol. 1, Lethan, D. S., Goodwin, P. B., and Higgins, T. J. V., Eds., Elsevier/North Holland, Amsterdam, 1978, 349.
2. **Bearder, J. R.**, Plant hormones and other growth substances — their background, structures and occurrence, in *Hormonal Regulation of Development I, Encyclopedia of Plant Physiology*, Vol. 9, MacMillan, J. Ed., Springer-Verlag, Berlin, 1980, 56.
3. **Wood, R. K. S., Ballio, A., and Graniti, A.**, Eds., *Phytotoxins in Plant Disease*, Academic Press, New York, 1972.

Chapter 2

AUXINS

Shingo Marumo

TABLE OF CONTENTS

I.	Definition of a Plant Hormone and Auxin	10
II.	The History of Chemical Research on Auxins	10
III.	The Chemistry of Natural Auxins	16
IV.	Anti-Auxins	19
	A. Synthetic Anti-Auxins	23
	B. Natural Anti-Auxins	26
V.	The Analysis of Auxins	27
	A. The GC and Combined GC/MS of Auxins	27
	B. HPLC of IAA	33
	C. Radioimmunoassay of IAA	38
VI.	The Synthesis of Auxins	39
	A. IAA	39
	B. Isotopically Labeled IAA	41
	C. 4-Chloro-Indole-3-Acetic Acid	41
VII.	Biosynthesis and Metabolism of Auxins	41
	A. Biosynthesis of IAA	41
	B. Metabolism of IAA	45
VIII.	The Biological Activity of Auxins	47
Acknowledgments		47
References		48

I. DEFINITION OF A PLANT HORMONE AND AUXIN

A Committee of the American Society of Plant Physiologists published the nomenclature of chemical plant regulators[1] in 1954, in which plant hormones and auxins are defined as follows:

"(Plant) Hormones (Synonym: Phytohormones) are regulators produced by plants, which in low concentrations regulate plant physiological processes. Hormones usually move within the plant from a site of production to a cite of action."

Auxin is a generic term for compounds characterized by their capacity to induce elongation in shoot cells. They resemble indole-3-acetic acid (IAA)(**1**) in physiological action. Auxins may, and generally do, affect other processes besides elongation, but elongation is considered critical. Auxins are generally acids with an unsaturated cyclic nucleus or their derivatives.

Auxin precursors are also defined as compounds which, in the plant, can be converted into auxins. Anti-auxins are defined as compounds which competetively inhibit the action of auxins.

As seen from this definition, the term "auxin", as a type of plant hormone, does not refer to IAA alone; rather, it represents a group of compounds with similar physiological activities, such as elongation of the *Avena* coleoptile.

As natural auxins, IAA (**1**) and 4-chloroindole-3-acetic acid (**2**) have been isolated from higher plants. 5-Hydroxyindole-3-acetic acid (**3**) was suggested to be present in tomatoes, and 1-methoxy- and 4-methoxyindole-3-acetonitrile have been isolated from the diseased clubroots of Chinese cabbage, *Brassica pekinensis*. Many other related metabolites of natural auxins have also been identified in higher plants (see Section III).

Synthetic auxins, which were found from the screening of a large number of synthetic compounds, include 1-naphthalene-acetic acid (NAA) (**4**), phenylacetic acid (**5**), *cis*-cinnamic acid (**6**), and 2,4-dichlorophenoxyacetic acid (2,4-D)(**7**). Two new synthetic auxins, α-chloro-β-(3-chloro-*o*-tolyl) propionitrile (**8**)[2] and 1,2-benzisothiazol-3-ylacetic acid (BIA)(**9**),[3] were reported. Among them, 2,4-D (**8**), in particular, is an important synthetic auxin that is used extensively in place of IAA in plant physiology studies. Phenylacetic acid (**5**) has been isolated from the etiolated seedlings of *Phaseolus mungo*[4] and stem extracts of tomato and sunflower,[5] and now is recognized as a natural auxin. A new synthetic auxin (**8**) exerted growth-promoting activity in general auxin bioassays, such as elongation of mung bean hypocotyl and *Avena* coleoptile segments. BIA (**9**) showed obvious auxin-like activity in the split pea internode curvature test.

Anti-auxins have been discovered throughout the biological investigation of synthetic auxins and related compounds. Natural compounds with anti-auxin activity have been isolated from higher plants and fungi. Synthetic and natural anti-auxins are described in Section IV.

II. THE HISTORY OF CHEMICAL RESEARCH ON AUXINS

An auxin was the first plant hormone to be chemically identified, and auxins played important roles in early research on plant hormones because they were the sole type of hormone available for plant physiologists to use in their physiological studies until gibberellin (isolated from a fungus) was recognized as a second type of plant hormone. The early history of chemical research on auxins is recorded in *Phytohormones* (Went and Thimann, 1937).[6]

The first substantial evidence that auxin was present in plants was reported by Darwin and Darwin in 1880 in their publication, *The Power of Movement in Plants*. They illuminated seedlings of *Phalaris canariensis* horizontally and showed that the effect of this light was perceived by the tip of a seedling and that the effect was transmitted from the tip to a lower part of the seedling's coleoptile, causing the latter to bend toward the light. This transmittance of a light stimulus downward in the phototropism of *Phalaris* was confirmed by Boysen Jensen[7] in 1913 in his experiments with *Avena sativa*. He cut off the

(1) indole-3-CH₂COOH structure

(2) 4-Cl-indole-3-CH₂COOH structure

(3) 5-HO-indole-3-CH₂COOH structure

(4) naphthalene-1-CH₂COOH structure

(5) phenyl-CH₂COOH structure

(6) trans-cinnamic acid structure (C=C with H, H, COOH)

(7) 2,4-dichlorophenoxyacetic acid (OCH₂COOH with 2,4-Cl)

(8) 2-chloro-3-methylphenyl-CH₂CH(Cl)-C≡N structure

(9) benzisothiazole-CH₂COOH structure

tip of the *Avena* coleoptile and replaced it on the decapitated coleoptile and the wound was covered with cocoa-butter. Light then was spotted on the tip only. It induced curvature both of the tip and of the coleoptile, evidence that the light stimulus was transmitted from the tip across the gap to the coleoptile. From these results Jensen claimed that the existence of a growth substance in the *Avena* coleoptile during phototropic curvature was proved.

Boysen Jensen's experiment was extended by Paál in 1914 and 1919.[8] He showed that even without light similar curvature could be induced when the excised tip was placed on one side of the stump, evidence that some diffusible substance emanating from the tip accelerated the growth of the coleoptile.

In 1921, Stark[9] introduced the method of applying a small block of agar on one side of the decapitated coleoptile, the agar being mixed with various tissue extracts. He could obtain no promoting activity, only inhibitory activity was found. Then in 1925, Seubert[10] was able to prove that an agar block containing saliva, diastase, and malt extract promoted the coleoptile's growth. This was the first evidence that growth-promoting substances exist outside of plants.

Finally, in 1928, Went[11] succeeded in extracting the active substance from the *Avena* coleoptile by combining Paal's experiment on the curvature of the coleoptile and Stark's agar block method. He placed the excised tip of the coleoptile on an agar block for a brief period, after which he placed the block on one side of the stump of the decapitated coleoptile. The coleoptile curved away from the site of the agar block (negative curvature). He next measured the angle of the curve made by the coleoptile, the degree of which was proportional to the concentration of the active substance contained in the agar block. He called this the "*Avena* curvature test", and it has been used extensively since; today, it is recognized as

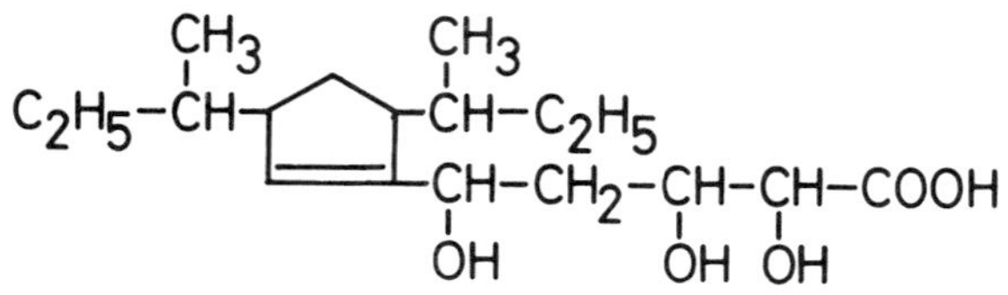

(10)

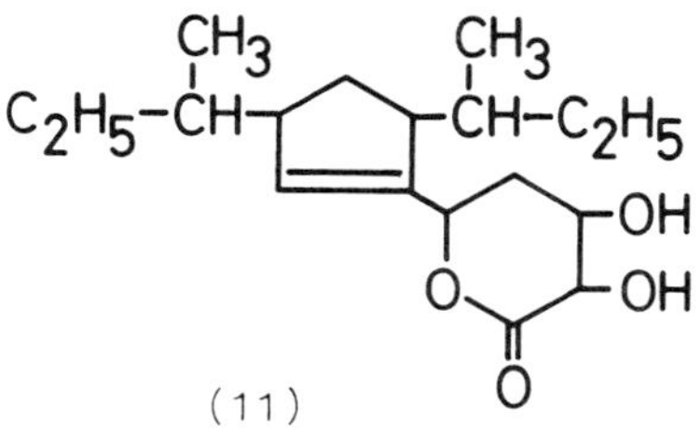

(11)

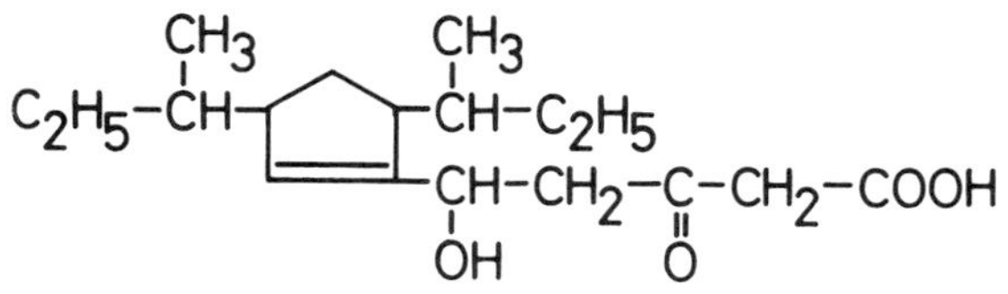

(12)

the standard bioassay for auxins. With this test Went also showed that the active substance in the *Avena* coleoptile is stable on boiling as well as on exposure to light.

In 1928, Nielsen[12] found that two pathogenic fungi, *Rhyzopus suinus* and *Absidia ramosa*, produced an active substance in their culture media when tested with the *Avena* curvature method and that this active substance could be extracted with diethyl ether. Dolk and Thimann[13] later confirmed (1932) that diethyl ether extracted this active substance only from an acidified culture medium, evidence that the substance itself was acidic.

The first isolation of a natural product active in the *Avena* curvature test was reported by Kögl et al.[14] in 1933. They found that human urine contained a large amount of an active substance. Human urine, 150 ℓ, was fractionated sequentially through solvent extraction, the removal of precipitates by lead salt formation, and the heating of the ether-soluble acidic fraction with 1.5% HCl in methanol, until (in the final step) vacuum distillation gave a crude crystalline mixture of the active substance. On recrystallization from alcohol-ligroin or from aqueous acetone the mixture gave two substances: auxin a, m.p. 196°C and auxin a lactone, m.p. 173°C. This latter lactone was considered an artifact of auxin a that resulted from lactonization due to the heating in acidic methanol. Both compounds were reported to have almost the same activity. In 1934, the same group isolated another active substance, which they named auxin b, from malt and corn germ oil.[15] It showed the same activity as the auxin a extracted from human urine.

The chemical structures of auxin a, auxin a lactone, and auxin b were postulated by Kögl et al.,[16,17] based on chemical degradation, as shown in **(10)**, **(11)**, and **(12)**.

The same year that auxin b was isolated, Kögl et al.[18] also succeeded in isolating a fourth active substance from human urine when they used charcoal instead of diethyl ether to adsorb the active substances in urine. The active fraction they separated through a charcoal column was easily purified, and the active compound was identified as indole-3-acetic acid. At that time, this compound had already been isolated from human urine by Salkowski in research unrelated to auxin activity; he had identified the structure by total synthesis. Kögl and associates confirmed that synthetic IAA had the same activity as a natural sample isolated from human urine, their activities in the *Avena* curvature test being almost comparable to the activities of auxins a and b. The term ''hetero-auxin'' was applied to IAA when it was first isolated by Kögl's group, but today plant physiologists know it to be a true, endogenous auxin distributed widely, if not ubiquitously, in the plant kingdom.

After Kögl's report of the isolation and structures of auxin a and b, their isolation from human urine was attempted by other scientists. Bennet-Clark et al.[19] examined an ether

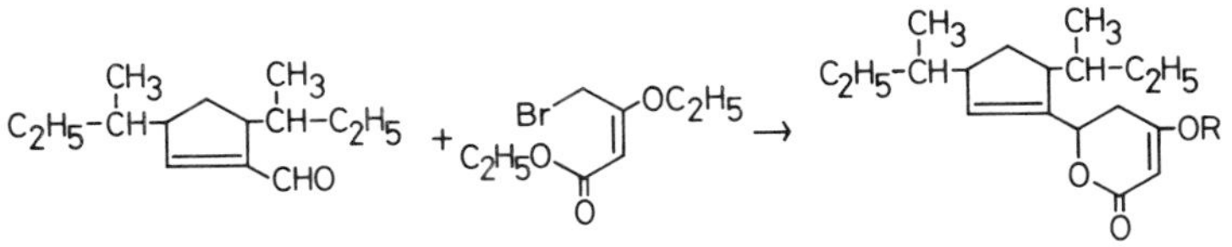

FIGURE 1. The synthesis of stereoisomeric mixture of auxin b lactone (the final step is shown).[23,24]

extract from human urine, but found only IAA. Wieland et al.[20] identified IAA and its methyl ester in human urine by an isolation procedure similar to Kögl's for auxins a and b. Munakata, who repeated Kögl's research in 1966, isolated only IAA and *o*- and *p*-hydroxyphenylacetic acid from human urine; no other auxin-like substance was detected.[21]

In 1952, Luckwill[22] was the first to use paper chromatography to analyze plant auxins. He found that two distinct auxins were contained in extracts from the young leaves of broccoli and several other plants. One of his two auxins had an R_f value identical to that of IAA; the other had many of the properties described for auxin b. His data, however, did not provide enough evidence for the positive identification of auxin b.

The total synthesis of a stereoisomeric mixture of auxin b lactone, (Figure 1) was done by Matsui and Hwang in 1966.[23,24] The key step in their synthesis was the Reformasky reaction of 3,5-di-*sec*-butyl-1-cyclopentenealdehyde (13) with ethyl γ-bromo-β-ethoxycrotonate (14), which afforded 4-ethoxy-6-(3,5-di-*sec*-butyl-1-cyclopentene-1-yl)-5,6-dihydro-2-pyrone (15, R = C_2H_5). Upon hydrolysis with formic acid this compound (15) gave auxin b lactone (16), R = H) in a stereoisomeric mixture. The synthetic auxin b lactone formed was an unstable oil which showed only weak promoting activity when bioassayed by the *Avena* coleoptile test, the *Avena* root growth test, and the pea stem section test. The results of these tests, measured by Nakamura et al.[25] differed entirely form those reported for auxin b and for IAA.

Because of this, Vliegenthart and Vliegenthart,[26] who took over Kögl's laboratory, decided to reinvestigate the nature of auxins a and b. They used very small samples (1 to 10 mg) of the authentic products that remained in that laboratory (necessary information about coded samples was obtained from the original laboratory notebooks and from published results). They analyzed Kögl's auxin a by its mass spectrum, and their accurate mass measurement showed it to be a C_{24} compound with the molecular formula, $C_{24}H_{40}O_5$, not the C_{18} compound established as auxin a. Comparison of its mass spectrum with that of cholic acid proved Kögl's auxin a to be identical to cholic acid. Furthermore, the mass spectra of its methyl ester (sample no. 1719) and of the methyl ester of cholic acid were identical.

As auxin a lactone, Kögl's samples no. 1503 and 8083, and a preparation dated 7-6-38 were examined. Mass spectral analyses showed that these samples had intense peaks at m/z 110 which were due to the $C_6H_6O_2$ ion. Based on this empirical formula and m.p. of 172 to 173°C for the samples, it was decided that hydroquinone had been analyzed. Also, the UV absorption of authentic hydroquinone was superimposable on the absorption data for auxin a lactone reported by Kögl.

In addition, samples no. 6853 and 7344 of auxin b were analyzed. According to the laboratory notebooks these samples had been isolated from malt. Mass spectral analysis showed that the samples contained sulfur, and accurate mass measurement proved auxin b to be thiosemicarbazide, CH_5N_3S. From results of similar mass spectral analyses, auxinglutaric acid was identified as phthalic acid, and the 3,5-dinitrobenzoate of auxin b as ethyl α-cyanophenylpyruvate.

Vliegenthart and Vliegenthart concluded that none of the samples investigated had the compositions originally proposed. No relation existed among the structures of auxin a, auxin a lactone, and auxin b. Thus, Kögl's auxins a and b actually were nonexistent. Results of the reinvestigation offered no explanation for the originally claimed biological activity of auxins a and b.

Thus, the first proposed natural plant growth-promoting substances reported by Kögl and associates were removed from the family of plant hormones. It should be emphasized, however, that IAA, one of the compounds Kögl's group isolated during their auxin research, is now accepted as one of the most important hormones present in plants.

Kögl and Kostermans[27] isolated IAA from yeast in 1934. The active substance reported to be present in the fungus *Rhyzopus suinus* (named Rhyzopin by Nielsen) was isolated and identified as IAA by Thimann[28] in 1935. Thus, all of the auxins that had been found in human urine, yeast, and fungus were identified as IAA. These results indicated that the active substance contained in higher plants also might be IAA.

The actual isolation of IAA from higher plant material was done by Haagen-Smit et al.[29] Haagen-Smit had engaged in auxin research as a member of Kögl's group in the Netherlands. Later he moved to the U.S. and continued his work there. His group used 270 kg of yellow cornmeal that had undergone alkaline hydrolysis at pH 10.5 for 50 hr. The alkaline hydrolyzate formed was fractionated by the isolation procedure originally established by Kögl et al.[14] to isolate auxins a and b.

Haagen-Smit's group did not succeed in isolating the active substance by this method, therefore, the isolation procedure was modified. The acidic fraction, obtained from the alkaline hydrolyzate after several purification steps, was esterified with 1.5% HCl in methanol, and the esterified mixture was distilled under a high vacuum (0.01 mmHg). The distillate was hydrolyzed with $3 N$ KOH, and the resulting acidic substance was kept in a xylene solution for 3 months, after which it gave a crude crystalline sample of IAA (yield 7 mg from 45 kg cornmeal).

Subsequently, Haagen-Smit and associates[30] were able to isolate a free form of IAA from fresh corn kernels (100 kg) in the milk stage, about 15 days after fertilization. The isolation procedure used included neither alkaline hydrolysis nor esterification. They concluded that the free auxin detected by *Avena* bioassay test in immature corn kernels was chiefly, or perhaps totally, IAA.

In addition to Haagen-Smit's work with cornmeal, Berger and Avery[31] proved that an auxin precursor could be extracted from dormant maize kernels which, upon alkaline hydrolysis, liberated free IAA. Redemann et al.[32] were interested in a fruit-setting factor which had been obtained from an ethanol extract of immature kernels of sweet corn (*Zea mays rugosa* va. Golden Cross). The factor was identified as the ethyl ester of IAA. This ethyl ester was approximately 100 times more effective than its free acid for inducing parthenocarpy in the tomato; it was not clear whether the ethyl ester itself was a true, endogenous auxin or an artifact derived from the free acid during ethanol treatment of the kernels.

Thus, IAA was proved to be distributed in several higher plant species. Based on these studies and its important physiological activity in higher plants, the compound was recognized as the first true plant hormone.

In 1951, a new technique, paper chromatography, was introduced for the study of plant growth regulators by Bennet-Clark et al.,[19] who described a method for paper chromatographic separation of auxins at a Society of Experimental Biology conference. In their method macerated plant tissue was treated with alcohol. The alcohol extract was concentrated, and its aqueous residue was treated with diethyl ether at pH 3. The ether-soluble acidic substances obtained were used for paper chromatography, the chromatograms being developed with a solvent mixture of isopropanol/H_2O) (10:1) with ammonia in the base of the tank (the solvent was later modified to isopropanol/ammonia/H_2O [8:1:1] or *n*-butanol/ammonia/H_2O [4:1:1]).

The developed chromatograms were tested by the *Avena* coleoptile bioassay. Using this method Bennet-Clark and Kefford[33] examined a variety of etiolated seedlings and roots: pea

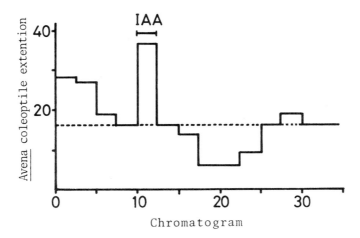

FIGURE 2. Activity in *Avena* coleoptile extension, shown as a histogram of squares cut from a chromatogram of etiolated broad bean shoot extract.

(Pisum sativum) shoots and roots, broad bean *(Vicia faba)* shoots and roots, sunflower *(Helianthus annus)* shoots, maize *(Zea mays)* roots, potato tubers and shoots, and *Aegopodium* rhizomes.

All the chromatograms of their plant extracts showed three active areas, as shown in Figure 2. The central active area (R_f approximately 0.3) corresponds to IAA, which was detected as the characteristic colored spot with Ehrlich's and Salkowski's reagents. The name "accelerator α" was given to the active area between R_f 0 and 0.15, and "inhibitor β" to the inhibitory area with R_f values greater than IAA. No information on the chemical nature of these α and β substances was given at that time. Later, various auxin complexes conjugated with amino acids or sugars were isolated and identified chemically from the R_f area corresponding to accelerator α, as described in the following chapter. Inhibitor β was correlated with the induction of dormancy in *Betula pubescens* by Eagles and Wareing,[34] and finally was identified as abscisic acid.

After Bennet-Clark and Kefford's publication, paper chromatography came to be used extensively to analyze the natural auxins present in plant extracts. As a result, IAA is recognized as a widespread, perhaps universal, auxin in plants. Chromatography studies have amassed evidence based mainly on the presence (on paper chromatograms of plant extracts) of an acidic substance with R_f value similar to that of synthetic IAA. This substance promotes cell extension in coleoptile sections and gives characteristic, although not highly specific, color reactions with Ehrlich's and Salkowski's reagents.

During paper chromatography of natural auxins, extracts of auxins were obtained from some higher plants that did not show the characteristic properties of IAA. For example, Luckwill and Powell[35] found that the auxin present in the seeds, pollen, fruit, and leaves of the apple had an R_f value approximately 10% higher than that of IAA on paper chromatograms developed with butanol/ammonia/water and that it did not give the characteristic color reaction with Ehrlich's reagent even when present (according to its biological activity) in sufficient amount. This auxin was named *Malus* auxin, but its chemical structure has yet to be determined.

The auxin in sunflower was investigated by Guttenberg et al.[36] in 1954. They reported that the active substance present in sunflower hypocotyl sections was not IAA, but a new substance stable on treatment with acid or pea enzyme. They named it "acid fast auxin". Abe and Marumo[37] reinvestigated the active substances in immature sunflower seeds and found at least two active substances which induced elongation of *Avena* coleoptile sections in methanol extracts. The major substance was identified as the ethyl ester of chlorogenic acid and the minor one as IAA. THe paper chromatographic behavior of ethyl chlorogenate as well as its resistance to acid hydrolysis and pea enzyme closely resembled the data for the acid fast auxin reported by Guttenberg and co-workers. Therefore, ethyl chlorogenate and acid fast auxin were considered to be identical.

Scientists from the Citrus Research Center at Riverside, Cal. claimed in a series of publications[38-41] that a new, natural growth-promoting substance had been found in young citrus fruits. They investigated the nature of this new compound by thin layer chromatography (TLC), paper electrophoresis, and spectrometric analysis and showed that the compound could not be IAA, rather it might be nonindolic. They called it "citrus auxin", and considered it to have physiological significance in citrus fruits.

In 1964, Gandar and Nitsch[42] reported the isolation of a new plant growth substance from immature seeds of *Pisum sativum*. The substance, called "substance F", was described as having a molecular weight of 296 as determined by mass spectrometry (MS), and it was active as an auxin on the mesocotyl in the *Avena* curvature and Jerusalem artichoke tests. This compound was neutral and showed UV absorption similar to that of methyl indole-3-acetate, but elementary analysis showed it contained only a trace (0.6%) of nitrogen, evidence that it is nonindolic.

The complete identification of these "new" active substances, whose presence in plants was suggested mainly by the paper chromatographic method described above, was accomplished by the use of the modern techniques of natural product chemistry: isolation of the active substance in its pure form; determination of its structure by chemical and spectrometric IR, UV, ^{1}H and ^{13}C NMR, and mass) analyses and confirmation of the proposed structure by total synthesis. For example, citrus auxin was reinvestigated by Jakahashi et al. and substance F by Marumo et al. Their studies as well as other recent chemical studies on natural auxins are described in Section III.

III. THE CHEMISTRY OF NATURAL AUXINS

As stated in the previous chapter, IAA was first isolated from human urine in 1934.[18] The isolation of IAA from higher plants was achieved in 1941 when IAA was liberated from yellow cornmeal after its alkaline hydrolysis.[29] Later (1946), free IAA was isolated from immature corn kernels.[30] Since then, the presence of IAA in higher plants has been reported by Berger and Avery[31] and Redemann et al.[32] in corn kernels, by Okamoto et al.[4] in etiolated seedlings of *Phaseolus mungo*, by Nomoto and Tamura[43] in diseased clubroots of Chinese cabbage infected with *Plasmodiophora brassicae*, and by Igoshi et al.[44] in the young fruit of *Citrus unshiu*. Thus, IAA has been proven to be distributed widely in higher plants and is recognized as the first true plant hormone.

The presence of "new" natural auxins other than IAA have been reported in several higher plant species based mainly on the results of paper chromatographic methods. Some have been reinvestigated by the modern techniques of natural product chemistry. As a result, several natural auxins have been isolated and their structures determined.

In 1968, Marumo et al.[45,46] investigated the active substance contained in immature seeds of *Pisum sativum* that had been named substance F before by Gandar and Nitsch[42] and was claimed to be nonindolic. The major compound (yield 25 mg) was suspected to be the methyl ester of chlorine-containing IAA from the results of spectrometric (UV, IR, ^{1}H NMR, and mass) analyses. Its structure was shown to be the methyl ester of 4-chloroindole-3-acetic acid (4-Cl-IAA-Me) (**18**) by synthesis with Fox and Bullock's[47] method. The minor active compound (yield 3.8 mg) was identified as free 4-chloroindole-3-acetic acid (4-Cl-IAA) (**2**) by the comparison of its IR spectrum with that of the synthetic compound. In another experiment, Gandar and Nitsch[48] independently isolated the same compound from *Pisum sativum* and found it to be actually the methyl ester of chlorine-containing IAA.

4-Cl-IAA was the second natural auxin found in higher plants. Interestingly, 4-Cl-IAA and its methyl ester gave only faint coloration with Ehrlich's reagent; the minimum amount needed for their detection was 20 μg, much larger than the amount (0.5 μg) needed for the detection of IAA.[49] Thus, these compounds were not readily detectable on paper chromatograms with Ehrlich's reagent. In addition 4-Cl-IAA has an R_f value (0.45) very similar to that of IAA (0.42) in isopropanol/ammonia/water (8:1:1), the commonly used solvent mixture for paper chromatography of auxins. Therefore, the chromatographic behaviors of these auxins are apt to be confused when the compounds are in a crude mixture of plant extracts. Recently, Engvild[50] recommended the use of Ehmann's spray reagent[51] to color 4-Cl-IAA and other chloro-indolic compounds. Use of this reagent produced a stable blue color with 4-Cl-IAA-Me and both IAA and methyl indole-3-acetate (IAA-Me) **(17)** showed stable dark blue colors.

The presence of 4-Cl-IAA-Me both in immature seeds of *Pisum sativum* and other species in Vicieae has been demonstrated by Engvild's[52] work. He examined 15 species of higher plants that had been cultured in perlite/vermiculite containing ^{36}Cl as NaCl. Immature seeds were harvested from these cultures and then treated with butanol. Autoradiograms of two-dimensional thin layer chromatograms of the butanol extracts showed that at least 11 of 15 species incorporated radioactive chloride ion into their organic compounds. In *Pisum sativum*, two of these compounds showed R_f values corresponding to 4-Cl-IAA and 4-Cl-IAA-Me, but in the other species the radioactive compounds were not identical to the auxins of *Pisum sativum*.

Engvild et al.[53] then developed a method for the quantitative determination of 4-Cl-IAA-Me by gas chromatography/mass spectrometry (GC/MS) that used deuterium-labeled 4-Cl-IAA-Me as the internal standard. With this method they showed that *Pisum sativum* contains large amounts (5 mg/kg fresh weight) of 4-Cl-IAA-Me and that *Vicia faba* and *Lathyrus latifolium* contain smaller amounts of it. Subsequently, they[54] identified 4-Cl-IAA-Me in 5 species of immature seeds of the Vicieae; *Lens cultinaris*, *Lathyrus maritimus*, *Lathyrus odoratus*, *Vicia sativa*, and *Lathyrus sativus*. The content of 4-Cl-IAA-Me varies from 1 mg/kg of fresh weight in *Lathyrus sativus* to 0.02 mg/kg in *Lens cultinaris*. *Lathyrus maritimus* also contains IAA-Me (0.34 mg/kg) as well as 4-Cl-IAA-Me. Hofinger and Böttger[55] also identified 4-Cl-IAA and 4-Cl-IAA-Me in immature seeds of *Vicia faba* by the GC/MS method.

West[56] surveyed indole derivatives in tomatoes. He could detect no IAA, but 5-hydroxytryptamine which was present in the young plant as well as in the fruit. He suggested 5-hydroxytryptamine was one of the hormones of growth in the tomatoes, but it is clear that the compound becomes active in the plant after being converted into 5-hydroxyindole-3-acetic acid **(3)**.

In 1970, Nomoto and Tamura[43] investigated active substances contained in the roots of Chinese cabbage, *Brassica pekinensis* Rupr. infected with *Plasmodiophora brassicae* Woronin. The fungal infection produced enlarged malformations on roots which formed "club", and the malformed clubroots had already been shown to contain two unknown auxin-like compounds. Nomoto and Tamura first examined the auxin activity of the acid fraction obtained from the diseased clubroots, but no appreciable activity was found. Remarkable activity, however, was detected in the neutral fraction, and two new compounds, A (yield 10 mg) and B (5 mg), were isolated from the neutral ethyl acetate-soluble fraction of the clubroots (30 kg). In addition, the known auxin-related compounds, indole-3-acetonitrile (IAN) **(21)** (20 mg), IAA-Me **(17)** (1 mg), and indole-3-acetamide (IAM) **(22)** (1 mg) were isolated from this neutral fraction.

The structure of compound-A was determined to be 1-methoxy-indole-3-acetonitrile **(19)** on the basis of spectrometric analysis. Evidence that compound B has a methoxyl group at the C-4 of its indole nucleus indicated a comparison of its ^1H NMR spectrum with the spectra of the regioisomeric 4-, 5-, 6-, and 7-chloroindole-3-acetic acid methyl esters. 4-

Methoxyindole-3-acetonitirile (**20**) was synthesized by the method of Govindachari et al.,[57] the synthetic compound being identical to natural compound B on the basis of their UV, IR, ^{1}H NMR, and mass spectra.

The biological activity of these new methoxyindole derivatives was assayed by the *Avena* coleoptile straight growth test. The results were unexpected: 1-methoxyindole-3-acetonitrile showed slight activity, but the 4-methoxy isomer was almost inactive. This methoxy-substituted indole-3-acetonitrile is assumed to be the metabolite of the diseased roots, rather than of the causal fungus, because 1-methoxyglucobrassicin (neoglucobrassicin) (**39**) has been isolated by Gmelin and Virtanen[58] as a constituent characteristic of the *Brassica* family.

In 1971, citrus auxin, whose presence in the fruit and seeds of *Citrus* species had been suggested by Khalifah et al.,[38-41] was investigated by Takahashi and his associates.[44] His group harvested the young fruit (10 kg) of *Citrus unshiu* one week after full bloom. Auxin-like substances were extracted from this fruit with ether, and the biological activity was bioassayed by the *Avena* curvature test. The fruit contained 0.5 to 1 mg/kg (approximately 0.2 μg per fruit) of IAA and 5 mg/kg (approximately 1 μg per fruit) of IAM (**22**); no other active substances were detected in any of the fractionated portions. Auxin activity in fruit harvested 2 months after full bloom was also examined; in this case no IAA, its amide, or citrus auxin were detected, the content of IAA being estimated as less than 1 μg/kg (approximately 0.002 μg per fruit).

Takahashi and co-workers concluded that the absence of nonindolic promotors such as citrus auxin throughout the fruit development stage cast doubt upon its physiological significance in "Citrus" species. Goren et al.[59,60] reached a similar conclusion in a separate experiment. Both groups published a single paper[61] stating that the concept of a citrus auxin is not justified and should be avoided in the future.

In 1970, Hofinger et al.[62] found indole-3-acrylic acid (IACRA) (**23**) in the ether and ethyl acetate extracts of lentil *(Lens culinaris)* seedlings and identified it by a comparison of its chromatographic behavior with that of authentic IACRA. On paper chromatograms in isopropanol/ammonia/water, IACRA has an R_f (0.29) very close to that of IAA (0.33). Isobutanol/methanol/water, however, gives a reasonable resolution of the two compounds: IACRA, R_f 0.68, IAA, 0.51. Ehrlich's reagent gives a green color with IACRA, but IAA gives violet. The reaction of dimethylaminocinnamaldehyde with IACRA gives a brown color and with IAA, violet-blue.

Further confirmation[63] of the presence of IACRA in lentil seedlings was made by a GC comparison (retention time) of the methyl ester of a natural sample (prepared with diazomethane) with an authentic compound. Hofinger and associates claimed that the presence of IACRA in relatively large quantities in lentil seedlings and its high auxin activity show that this new natural compound is the main auxin that controls lentil root growth. This conclusion, however, has yet to be confirmed by other chemical evidence such as the isolation of IACRA in its pure form and unambiguous identification by spectrometric analysis.

Etiolated seedlings of *Phaseolus mungo* were investigated by Isogai et al.[64,65] These seedlings contained relatively large quantities of IAA and such related metabolites as IAM (**22**), IAN (**21**), and indole-3-carboxyaldehyde (**24**), all of which have been isolated and identified chemically. Indole-3-ethanol IEt) (**25**) was isolated by Rayle and Purves[66] from cucumber *(Cucumis sativus* L.) seedlings and identified by MS, TLC and GC, and UV and visible spectroscopies, as well as by its physiological characteristics.

The presence of IEt (**25**) in *Helianthus annuus* was also demonstrated by Rajagopal[67] from its R_f value, color reaction, and UV spectrum. IAN (**21**) was isolated in pure form from cabbage heads by Jones et al.,[68] who subsequently isolated indole-3-aldehyde (**24**) and indole-3-carboxylic acid(**26**) from cell-free aqueous extracts of cabbage.[69] In their experiments, the yield of indole-3-aldehyde was 6.5 mg from maturing cabbage (var. January King; 1.9 kg) and 67 mg from fully grown cabbage (var. Early Offenham 9.1 kg). From the latter variety, 3.5 mg of indole-3-carboxylic acid was also isolated.

Auxin complexes conjugated with amino acids or sugars have been isolated, or their presence in higher plants indicated by chromatographic techniques. Row et al.[70] showed that the ethanol extracts of 30-day-old tomato seedlings contained an Ehrlich-positive substance which was identical on paper chromatography with synthetic indole-3-acetyl-DL-aspartic acid (IAA-aspartate) (28). This compound was stable under chromatography (ammonia was used) and turned purple with Ehrlich's reagent. Because IAA-aspartate had been identified as a product of the application of exogenous IAA to such a plant as *Pisum sativum*, their research was the first to demonstrate that IAA-aspartate is present in plant extracts. Klämbt[71] also showed IAA-aspartate is present in *Magnolia* sp., *Gleditschia triacanthos*, *Lupinus angustrifolus*, and *Papaver simneferum* from their R_f values at 0.08 in isopropanol/ammonia/water (80:15:1) and their dark blue coloration with Ehrlich's reagent and red-violet color with Salkowsli's reagent. He also found that N-malonyl-tryptophan (29) was present in the plant extracts, as ascertained by their R_f values at 0.095 and the gray-blue and yellow-blue coloration imparted by the above two spray reagents.

4-Chloroindole-3-acetyl-L-aspartic acid monomethyl ester (30) was isolated by Hattori and Marumo[72] from immature seeds of *Pisum sativum*. The yield was 2.2 mg from 241 kg of seeds. This compound induced hypocotyl swelling in *Phaseolus mungo*, a phenomenon which has been used as a bioassay for the isolation of 4-Cl-IAA. Interestingly, Marumo and Hattori[73] isolated two kinds of 4-chlorotryptophan derivatives from the same methanol extract of immature pea seeds (241 kg). Their structures, deduced from spectrometric data, were verified by synthesis as those of the malonyl amide derivatives of 4-chlorotryptophan; the methyl ester (31) (yield 3.9 mg) and its ethyl ester (32) (1.5 mg). (The D configuration assigned to these compounds must be reinvestigated because this is not definitive.) This is the first finding of natural 4-chlorotryptophan.

The biological activity of a mixture of (31) and (32) was assayed by the *Phaseolus mungo* test. This mixture induced hypocotyl swelling and root shortening at 1.0 μg/mℓ. The activity is slightly less than that of racemic 4-chlorotryptophan and is comparable to the activity caused by 0.05 μg/mℓ of 4-Cl-IAA. It is notable that the response appears 2.5 days after the application, or about 1 day later compared with the case of 4-Cl-IAA and 4-chlorotryptophan, evidence that the compounds must be metabolized to 4-Cl-IAA through 4-chlorotryptophan before becoming active.

Bandurski and his co-workers[74-80] extensively surveyed the conjugates of IAA contained in the kernels of corn *(Zea mays)*, and identified all of their structures. Bandurski[81] summarized their experimental results and grouped the conjugates into three classes of esters: the IAA-*myo*-inositol group (2-O-IAA-*myo*-inositol (33) and its DL isomer) which constitutes about 15% of the total; the IAA-*myo*-inositol glycosides group (5-O-β-L-arabinopyranosyl-2-O-IAA-*myo*-inositol (34) and its DL isomer and 5-O-L-galactopyranosyl-2-O-IAA-*myo*-inositol (35) which constitutes about 25% of the total; and high molecular weight IAA β-(1 → 4) glucan (37) which makes up about 50% of the total IAA in the kernels. Free IAA, its 2-O, 4-O, and 6-O-IAA-glucose esters and its (IAA)$_n$ inositol ester (36) constitute the remainder.

Glucobrassicin[82] (38) and neoglucobrassicin[58,83] (39) sulfur-containing glucoside derivatives, and ascorbigen A[84,85] (41) have been isolated from several species of cabbage, and 1-sulfoglucobrassicin[86] (40) has been isolated from *Isatis tinctoria*.

The structures and origins of the natural auxins and related metabolites found in nature are summarized in Table 1.

IV. ANTI-AUXINS

Anti-auxins, which competetively inhibit the actions of auxins, have been discovered through biological studies on synthetic auxins and related compounds. Natural anti-auxins also have been isolated from higher plants and fungi.

Table 1
NATURAL AUXINS AND RELATED METABOLITES IDENTIFIED IN HIGHER PLANTS AND OTHER NATURAL SOURCES

Structures	Origins	Ref.
R=H: Indole-3-acetic acid (IAA)(**1**) R=CH$_3$: Methyl indole-3-acetate (IAA Me)(**17**)	Human urine Yeast Fungus (*Rhyzopus suinus*) Etiolated seedlings of *Phaseolus mungo* Diseased roots of *Brassica pekinensis* Young fruit of *Citrus unshiu* Yellow cornmeal Corn kernels	18,20 27 28 4 43 44 29,30 32
R=H: 4-Chloroindole-3-acetic acid (4-Cl-IAA)(**2**) R=CH$_3$: Methyl 4-chloroindole-3-acetate (4-Cl-IAA-Me) (**18**)	Immature seeds of *Pisum sativum*, *Vicia fava*, *Lathyrus*, *latifolium*, *Lens cultinaris*, *Lathyrus maritimus*, *L. odoratus*, *L. Sativus*, and *Vicia sativa*	45,46,48 52—55
5-Hydroxyindole-3-acetic acid (5-OH-IAA) (**3**)	Tomatoes*	56
1-Methoxyindole-3-acetonitrile (**19**)	Diseased roots of *Brassica pekinensis*	43
4-Methoxyindole-3-acetonitrile (**20**)	Diseased roots of *Brassica pekinensis*	43,57
Indole-3-acetonitrile (IAN) (**21**)	Cabbage (*Brassica oleracea*) Etiolated seedlings of *Phaseolus mungo* Diseased roots of *Brassica pekinensis*	68 64,65 43
Indole-3-acetamide (IAM) (**22**)	Etiolated seedlings of *Phaseolus mungo* Young fruit of *Citrus unshiu*	64,65 44
Indole-3-acrylic acid (IACRA) (**23**)	Lentil seedlings (*Lens cultinaris*)*	62,63

Table 1 (continued)
NATURAL AUXINS AND RELATED METABOLITES IDENTIFIED IN HIGHER PLANTS AND OTHER NATURAL SOURCES

Structures	Origins	Ref.
Indole-3-carboxyaldehyde (Indole-3-CHO) (**24**)	Cabbage (*Brassica leracea*) Etiolated seedlings of *Phaseolus mungo*	69 64,65
Indole-3-ethanol (IEt) (**25**)	Cucumber (*Cucumis sativus*) *Helianthus annuus*	66 67
Indole-3-carboxylic acid (**26**)	Cabbage (*Brassica oleracea*)	69
Indole-3-acetaldehyde (IAald) (**27**)	*Pisum sativum** *Helianthus annuus** Insect galls of chestnut (*Catanea* sp.) (as a dimethylacetal form)	67 67 87
Indole-3-acetyl-L-aspartic acid (IAA-aspartate) (**28**)	Tomato seedlings* *Magnolia* sp.* *Gleditschia triacanthos** *Lupinus augustifolus** *Papaver somniferum**	70 71
N-Malonyl-tryptophan (**29**)	The same plants as those with IAA-aspartate, except tomato seedlings*	71
Monomethyl 4-chloroindole-3-acetyl-L-aspartic acid (**30**) (R=H, R′=CH₃; or R=CH₃, R′=H)	Immature seeds of *Pisum sativum*	72

Table 1 (continued)
NATURAL AUXINS AND RELATED METABOLITES IDENTIFIED IN HIGHER PLANTS AND OTHER NATURAL SOURCES

Structures	Origins	Ref.
N-malonyl derivatives of 4-chlorotryptophan R=CH$_3$: (31), R=CH$_2$CH$_3$: (32)	Immature seeds of *Pisum sativum*	73
2-O-(indole-3-acetyl)-*myo*-inositol (33) 1-DL-(indole-3-acetyl)-*myo*-inositol	Corn kernels (*Zea mays*)	74—81
5-O-β-L-arabinopyranosyl-2-O-(indole-3-acetyl)-*myo*-inositol (34) 5-O-β-L-arabinopyranosyl-DL-(indole-3-acetyl)-*myo*-inositol	Corn kernels (*Zea mays*)	74—81
5-O-L-Galactopyranosyl-2-O-(indole-3-acetyl)-*myo*-inositol (35)	Corn kernels (*Zea mays*)	74—81
General structure of indole-3-acetyl-*myo*-inositol (36)	Corn kernels (*Zea mays*)	74—81
4-O-(indole-3-acetyl)-D-glucopyranose (37) 6-O-(indole-3-acetyl-D-glucopyranose (indole-3-acetyl)-glucan β-(1→4)-Cellulosic glucan with 7 to 50 glucose units/IAA	Corn kernels (*Zea mays*)	74—81

Table 1 (continued)
NATURAL AUXINS AND RELATED METABOLITES IDENTIFIED IN HIGHER PLANTS AND OTHER NATURAL SOURCES

Structures	Origins	Ref.
Glucobrassicin (38)	Cabbage (*Brassica oleracea*)	82,83
Neoglucobrassicin (39)	Cabbage (*Brassica oleraceae*)	58,83
1-Sulfoglucobrassicin (40)	*Isatis tinctoria (I. Japonica)*	86
Ascorbigen A (41)	Cabbage (*Brassica oleracea*)	84,85

Notes: (1) *These compounds have not been isolated in a pure form, but their presence in higher plants has been indicated by chromatographic methods.

(2) In addition to those listed in Table 1, indole-3-propionic acid and indole-3-butyric acid (as well as IAA and indole-3-COOH) have been detected in *Nicotiana glauca* and *N. langsdorffii* by gas-liquid chromatography. An unusually high level of indole-3-butyric acid was observed in tumor-prone hybrids of these plants.[111]

(3) Tryptamine and 5-hydroxytryptamine (as well as IAA, indole-3-CHO, and malonyltryptophan) have been found in barley and tomato shoots. Tomato shoots also contained indole-3-lactic acid and tryptophol. Barley shoots contained 3-aminomethylindole, 3-methylaminomethylindole, gramine, *N*-methyltryptamine, and 5-hydroxy-*N*-methyltryptamine.[112]

A. Synthetic Anti-Auxins

In 1953, MacRae and Bonner[88] made a comprehensive survey of anti-auxins reported in the literature and clarified the relationship between chemical structure and anti-auxin activity. Bonner[89] showed that 2,4-dichloroanisole (**42**), as an analogue of 2,4-dichlorophenoxyacetic acid (2,4-D) (**7**) inhibited the growth of *Avena* coleoptile induced by IAA and 2,4-D. *Trans*-cinnamic acid was found by van Overbeck et al.[90] to antagonize IAA, 2,4-D, and NAA (**4**) although its *cis* isomer (**6**) was active as an auxin. Maleic hydrazide (**47**) was reported by Currier et al.[91] and Leopord and Klein[92] to possess the anti-auxin activity. Smith et al.[93] showed that the (+) isomer of α-(2-naphthoxy)-propionic acid was active in the *Avena* coleoptile section test but this positive activity could be antagonized by its (−) isomer (**64**): thus, the latter (**64**) is an anti-auxin. The optical isomerism in synthetic auxins and anti-auxins was reviewed by Matell,[94] Åberg,[95] and Mitsui and Fujita,[96] and the relationship between the absolute stereochemistry and biological activity of these compounds has been

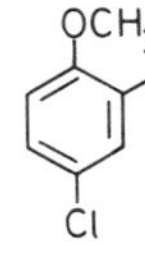
(42)

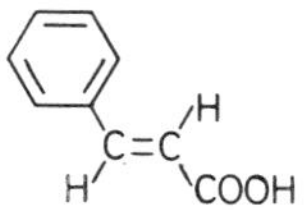
(43)

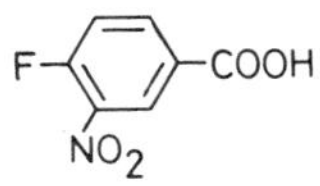
(44)

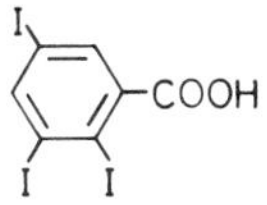
(45)

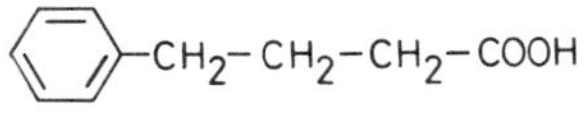
(46)

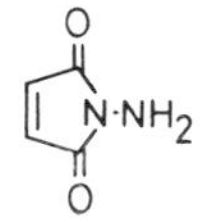
(47)

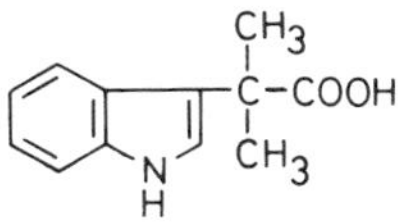
(48)

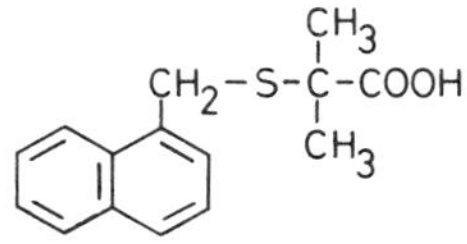
(49)

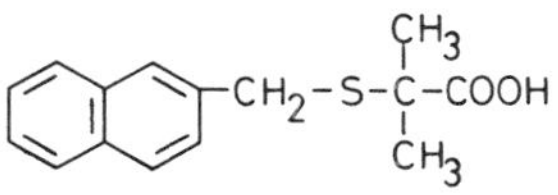
(50)

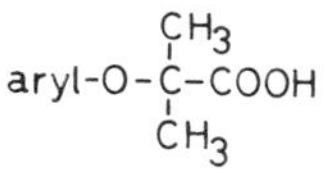

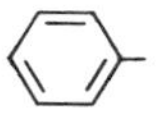
(51)

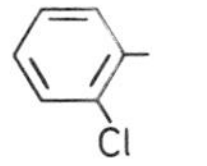
(52)

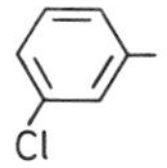
(53)

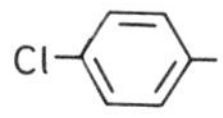
(54)

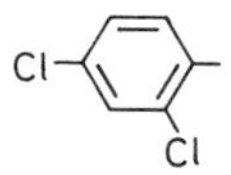
(55)

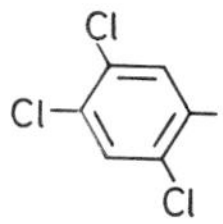
(56)

FIGURE 3. The structures of synthetic anti-auxins.

determined. 2,6-Dichloro- and 2,4,6-trichlorophenoxyacetic acids (**66** and **67**) were found by MacRae and Bonner[97] to inhibit the auxin-induced *Avena* coleoptile growth.

The structures of synthetic anti-auxins summarized by MacRae and Bonner[88] are listed in Figure 3. MacRae and Bonner indicated that most (if not all) of anti-auxins listed in Figure 3 could be derived from the active auxins by eliminating one of the structural features essential to the auxin activity, based on ''the two point attachment concept'' proposed by Bonner et al.[98] and Muir and Hansch.[99] This was attained in three ways: (1) elimination of the essential carboxy group, e.g., 2,4-dichloroanisole **(42)** from 2,4-D **(7)**; (2) eliminationof the essential ortho substituent from the ring system, e.g., phenoxyacetic acid **(65)** from 2,4-D; (3) elimination of proper spatial relationship between the ring system and the polar group

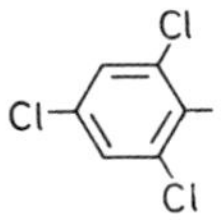

(57)

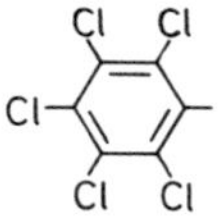

(58)

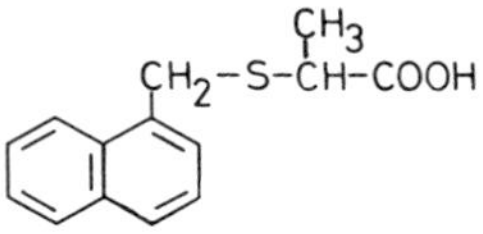

(59)

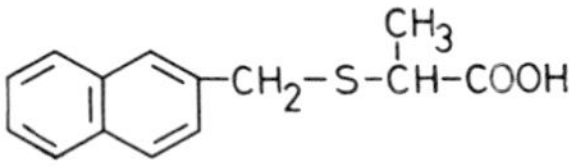

(60)

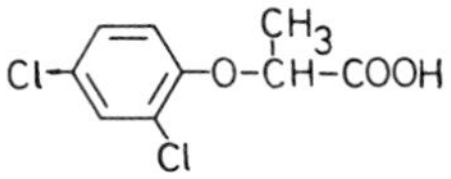

(61)

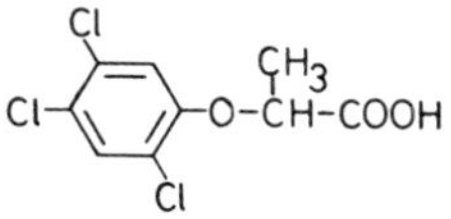

(62)

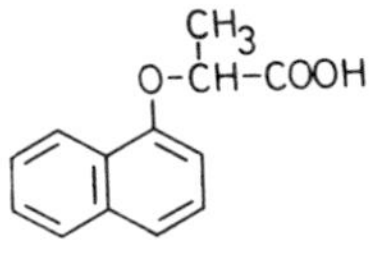

(63)

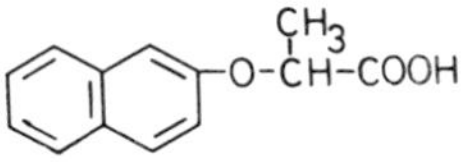

(64)

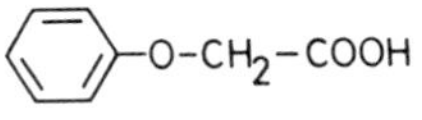

(65)

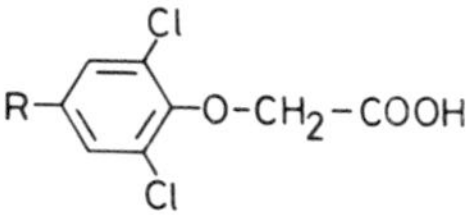

(66) R=H (67) R=Cl

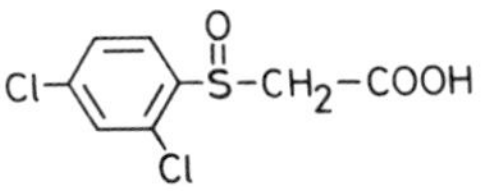

(68)

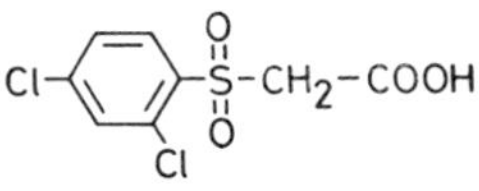

(69)

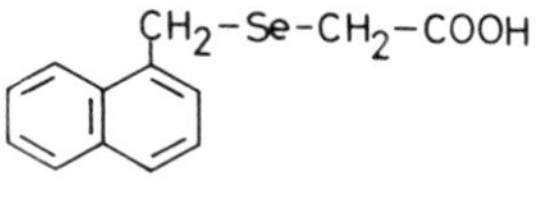

(70)

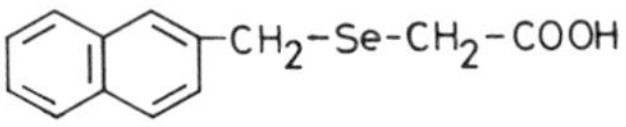

(71)

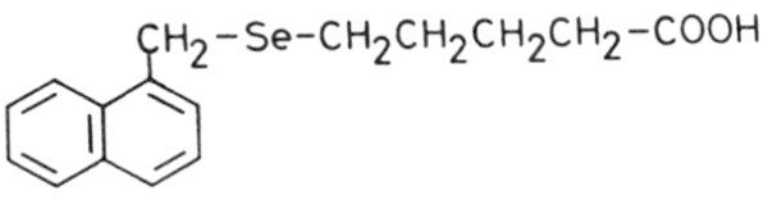

(72)

FIGURE 3. Continued.

by introducing a bulky group in the side chain, e.g., 2,4-dichlorophenoxyisobutyric acid **(55)** from 2,4-D.

Relationships between chemical structure and biological activity also have been reviewed by Veldstra,[100] including auxins, anti-auxins, and auxin synergists.

4-Chlorophenoxyisobutyric acid (PCIB) **(54)** was recently shown by Moloney and Pilet[101] to interact strongly with an auxin-specific binding site extracted from maize roots. The effect of PCIB on the ripening in Bartlett pear *(Pyres communis)*[102] and also on the formation of meristematic centers during regeneration of isolated tissue framents of *Riella helicophylla*[103] were investigated.

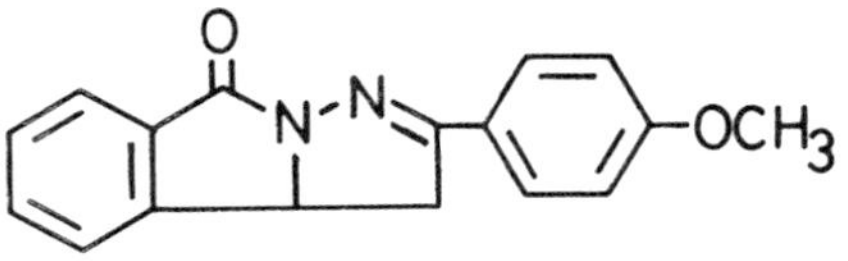

(73)

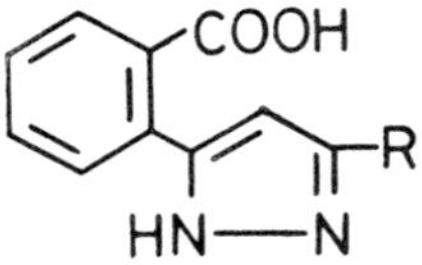

(74) 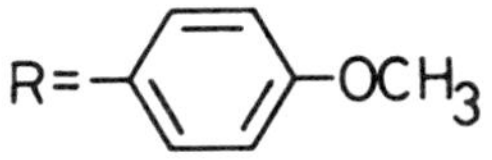 R= —C₆H₄—OCH₃

(75) 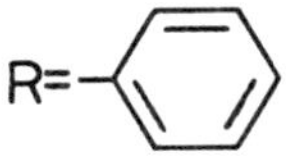 R= —C₆H₅

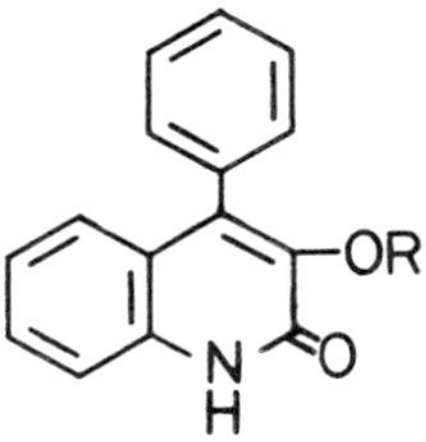

(76) R=H
(77) R=CH$_3$
(78) R=CH$_2$COOH

Two new auxin transport inhibitors were synthesized by Beyer et al.: DPX-1840 (2-(4-methoxyphenyl)-33a-hydro-*8H*-pyrazolo [5,1-a] isoindol-8-one)[104] (**73**) and its dehydrogenation and ring-cleavage product, 2-(3-aryl-5-pyrazoyl) benzoic acid (**74** and **75**).[105]

These growth regulators inhibit auxin transport by reduction of basipetal auxin transport capacity in bean when incorporated into receiver agar cylinder or applied foliarly to intact plants. Their morphological effects include growth retardation, epinasty, breaking of apical dominance, loss of geo- and phototropic responsiveness, synergism with ethylene in abscission, and induction of parthenocarpy.

DPX-1840 (**73**) was biologically converted, when applied to etiolated bean hypocotyl hooks, into its ring-opened pyrazoylbenzoic acid (**74**), the latter compound was considered to be a biologically active form of DPX-1840.

B. Natural Anti-Auxins

Viridicatin (**76**), an antibacterial metabolite of several *Penicillium* strains, including *P. crustosum*,[106] was chemically converted by Taniguchi and Satomura[107] into plant growth regulators possessing anti-auxin activity. The methylation or carboxymethylation of viridicatin lost the antibacterial activity, while they promoted the root elongation of rice seedlings, just as 2,4,6-T (**57**) did. 3-*O*-Carboxymethylviridicatin (**78**) also competitively inhibited IAA-induced elongation of *Avena* coleoptile section. A lamina inclination of rice seedlings was also inhibited by this compound and the inhibition was competitively reversed by IAA.

Based on these data, Taniguchi and Satomura suggested that a functional group at the 3-position might be important for anti-auxin activity, and the amido-hydrogen was another necessary group because the methylation of this group completely abolished the plant growth activity. Taniguchi et al.[108] found that 3-*O*-Carboxymethylviridicatin (**78**) enhanced the formation of IAA oxidase, which resulted in the exhibition of the anti-auxin activity.

(79)

(80) : R^1=OH, R^2~R^5=H, R^6=OH, R^7=CH$_3$
(81) : R^1=R^2=OH, R^3~R^5=H, R^6=OH, R^7=CH$_3$
(82) : R^1=R^2=H, R^3,R^4=>O, R^5=R^6=OH, R^7=CH$_3$
(83) : R^1=R^2=H, R^3,R^4=>O, R^5=OH, R^6=R^7=H

(84) : R^1=R^2=H, R^3,R^4=>O, R^5=R^6=OH, R^7=H, R^8=CH$_3$
(85) : R^1~R^4=H, R^5=OH, R^6=H, R^7=R^8=CH$_3$
(86) : R^1=R^2=H, R^3,R^4=>O, R^5=OH, R^6=H, R^7,R^8=>CH$_2$

2-Pyruvoylaminobenzamide[109] (79) was isolated as an anti-auxin from the culture broth of a fungus, *Colletotrichum lagenarium,* a cause of anthracnose disease in cucurbitaceous plants. The compound at 30 ppm did not affect the straight growth of *Avena* coleoptile segments; however, when 10 ppm were applied in combination with 1 ppm of IAA, it interfered with a response of the segments to IAA.

Six nagilactones, A (80), B (81), C (82), D (83), E (85), and F (86) and inumakilactone A (84), which were nor- or bisnorditerpenoids dilactones isolated from *Podocarpus nagi,* were found by Hayashi and Sakan[110] to be effective plant growth regulators with anti-auxin-like activity.

All of these nagilactones and inumakilactone A strongly inhibited elongation of the *Avena* coleoptile sections at 10^{-4} to 10^{-5} M concentration. Notably, four of them (80—83), which have the α-pyrone structure, showed dual activity — inhibitory or promotive — depending upon their concentrations, while nagilactone E and F were only inhibitory. Interaction between nagilactone B (and E) and IAA was examined on elongation of *Avena* segments; this showed a clear competitive inhibitory effect of the former compound(s) on the action of IAA.

V. THE ANALYSIS OF AUXINS

The qualitative and quantitative analysis of natural auxins has been established by the use of paper chromatography,[113,114,117] TLC,[115-117] and partition and adsorption chromatographies[117] on various adsorbents. These separation techniques have been effectively applied to the analysis and identification of natural auxins and related indole compounds in plant extracts. Most of these studies are reviewed in Section III. This section examines the analysis of auxins and related compounds in plant extracts by GC, combined GC/MS, and high performance liquid chromatography (HPLC). A radioimmunoassay of IAA is also described.

A. The GC and Combined GC/MS of Auxins

GC was first applied to the indole auxins by Powell[118] and Stowe and Schilke.[119] Powell devised the preliminary purification procedure of indole compounds occurring in plant ex-

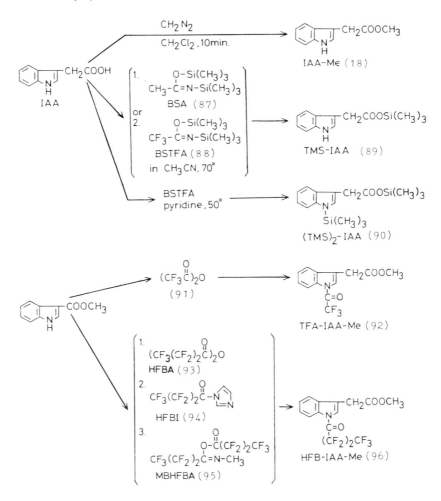

FIGURE 4. Preparation of the methyl ester, TMSi, TFA, and HFB derivatives of IAA for GC analysis.

tracts so that they could be processed by GC and spectrophotofluorometry. He then confirmed the presence of IAA in maize seeds and in cabbage by the GC method.

Grunwald et al.[120] examined the methyl esterification of indole acids and the GC separation of the indole methyl esters. They showed that the esterification with diazomethane in methylene chloride was complete within 10 min and the recovery exceeded 95%. They prepared the methyl ester of 10 indole acids, e.g., methyl indole-3-acetate (IAA-Me) (18) in Figure 4, and selected, by use, three suitable GC stationary phases (Versamid 900, Hi-Eff-8-BP, and SE-30) that permitted the complete separation of all ten esters, while SE-52 gave an unsatisfactory result. Dedio and Zalik[121] also found that indole acids were successfully esterified with diazomethane and that the use of silicon phase, SE-30 and SE-52, was effective for separation of the indole methyl esters.

Grunwald and Lockard[122] resolved nine of ten closely related indole acids as their methyl ester or trimethylsilyl (TMSi) derivatives separated on an OV-101 column. They employed two procedures for the preparation of TMSi-indole derivatives, e.g., TMSi-IAA (89) in Figure 4: (1) an indole acid was dissolved in acetonitrile and bis(trimethylsilyl) acetamide (BSA) (87) was added under anhydrous conditions and kept at 70°C as described by Klebe et al.,[123] (2) a TMSi derivative was prepared by adding bis(trimethylsilyl)trifluoroacetamide

Table 2
COMPARISON OF RELATIVE RETENTION (r), EFFECTIVE PLATE VALUE (N), RESOLUTION (R) OF VARIOUS INDOLE TMS DERIVATIVES, & INDOLE METHYL ESTERS ON COLUMN SUBSTRATE OV–101

Indole, TMS derivative	r	N	R	Indole, methyl ester	r	N	R
Indole-2-carboxylate	0.27	935		Indole-2-carboxylate	0.42	318	
			1.91				3.92
Indole-1-propionate	1.09	848		Indole-3-acetate	0.80	1063	
			2.30				0.18
Indole-3-acetate	1.49	905		Indole-5-carboxylate	0.82	653	
			1.69				0.52
Indole-3-carboxylate	1.85	1095		Indole-3-carboxylate	0.88	1290	
			1.98				0.30
Indole-3-propionate	2.32	1272		Indole-1-propionate	0.91	1373	
			0.20				1.87
Indole-5-carboxylate	2.37	1278		Indole-3-propionate	1.12	1110	
			3.92				3.28
Indole-3-butyrate	3.52	1765		Indole-3-butyrate	1.59	1745	
			0.61				0.38
Indole-3-lactate	3.73	1214		Indole-3-lactate	1.65	1628	
			2.19				4.73
Indole-3-glyoxylate	4.59	2490		Indole-3-glyoxylate	2.61	1834	
			3.79				1.42
Indole-3-acrylate	6.31	2190		Indole-3-acrylate	2.97	1695	

Note: Column characteristics: column temperature 200°C; detector temperature 250°C; carrier gas helium; column 5% OV-101 on Anakrom ABS 80/90 mesh; retention time ethyl indole-3-acetate 4.8 min.

(BSTFA) (88) instead of BSA. Silylation of the carboxyl group was accomplished in less than 1 min, however, silylation of the indole nitrogen atom was more difficult. Indole acids with the short ($< C_4$) aliphatic carboxyl group were not difficult to be silylated with BSA, except for indole-2-carboxylic acid, which was not completely silylated even after 120 min; however, BSTFA completely silylated within 10 min. All other tested indole acids were completely silylated in less time. The action of BSTFA as a silylating agent was similar to BSA. BSTFA was more effective as a silyl donor than BSA, therefore, it was recommended for the formation of TMSi-indole acids.

A flame ionization detector (FID) was usually used for detection of the methyl ester and TMSi derivatives of indole acids. An alkali flame ionization detector (AFID) was employed by Swartz and Powell.[124] This detector was selective for nitrogen compounds and could detect 10^{-2} to 10^{-1} μg IAA-Me and TMSi-IAA.

The GC results obtained by Grunwald and Lockard of the TMS and methyl ester derivatives of ten indole acids most likely found in biological materials are presented in Table 2. The relative retention time values (r) are given with respect to ethyl indole-3-acetate. They concluded that, using TMSi and methyl ester derivatives and the stationary phase OV-101, all indole-acids could be resolved at the 99% level except indole-3-butyrate from indole-3-lactate.

DeYoe and Zaerr[125] analyzed IAA in shoots of Douglas fir and its seasonal variation. The shoots, collected in December and June, were extracted with methanol. The methanol extracts were purified by solvent partitioning and through Sephadex® LH-20. IAA in the purified plant extracts was derivatized to IAA-Me, TMSi-IAA, and TMSi-IAA-Me, all of which were analyzed with the polar XE-60 and moderately polar Hi-Eff-8-BP stationary phases. IAA was thus identified in Douglas fir and its amount was shown to vary seasonably.

Free IAA and conjugated IAA, from the latter of which IAA was liberated by alkaline hydrolysis, was quantified by Bandurski and Schulze[126] by GLC analysis in the *Avena* and *Zea* seedlings. The ether extract of the seedlings was preliminarily purified through DEAE-cellulose and Sephadex® LH-20 columns. The purified sample was then silylated with BSTFA and pyridine at 50°C[127] to produce Bis-TMSi-IAA (90), in which both carboxyl and indolic nitrogen were silylated. The derivative was analyzed by GLC on 5% SP-2401 on 100/120 Supelcoport. The upper limit of free IAA in *Avena* was approximately 16 μg/kg and in *Zea*, approximately 24 μg. The amount of 1N alkali-labile IAA in *Zea* was approximately 330 μg/kg and there was very little 1 N alkali-labile IAA in *Avena*.

Brook et al.[128] demonstrated that trifluoroacetylation of IAA-Me offered a very sensitive method with an electron capture detector (ECD). IAA-Me, prepared with ethereal diazomethane by Williams, procedure,[129] was acylated with trifluoroacetic anhydride (TFAA) (91) to N-trifluoroacetyl (TFA)-IAA-Me (92).

Seeley and Powell[130] showed that heptafluorobutyryl (HFB)-IAA-Me (96) and TFA-IAA-Me had excellent chromatographic properties for qualitative and quantitative analysis in the nanogram to picogram level. HFB-IAA-Me was prepared with heptafluorobutyrylimidazol (HFBI) (94) or heptafluorobutyric anhydride (HFBA) (93).[130] The limit of detection (5 times the noise level) of HFB-IAA-Me with ECD was below 10 pg.

The method was applied to IAA in apple seeds, which was converted into TFA-IAA-Me. Thus, the amount of IAA present in 26 g of apple seeds was approximately 15 μg.

Hofinger[131] developed a new derivatization reagent, N-methylbis(heptafluorobutyramide) (MBHFBA) (95). The reagent was less powerful than HFBA for acylation property, but the reaction proceeded in the neutral conditions. IAA-Me was acylated with MBHFBA and dimethylaminopyridine at 120°C for 2 hr to HFB-IAA-Me that was quantified in the picogram range by high performance (HP) GC-ECD method. The HPGC-ECD results of some synthetic N-HFB-indole acid methyl esters are shown in Figure 5.

IAA in the leaves of *Bryonia dioica* was analyzed by the HPGC-ECD method and the IAA content was estimated at 62 ng/g of fresh weight.

The indolo-α-pyrone spectrofluorometric method has been reported by Stessl and Venis[132] and Knegt and Bruinsma.[133] This method quantified IAA selectively discriminated from other indole compounds and included the condensation of IAA with acetic anhydride in the presence of trifluoroacetic acid. The product was 2-methylindolo-α-pyrone (2-MIP) (97), which showed specific fluorescence at 490 nm (excitation at 440 nm). Very small amounts of IAA in crude plant extracts could be selectively determined by this method. The time required was in the order of 2 hr and the amount of fresh plant material for analysis was enough in 0.05 to 5 g. Kamisaka and Larsen[134] improved the method by modifying the reaction solvent.

Although the 2-MIP method was specific for IAA, the co-existence of a new auxin, 4-Cl-IAA (and 5-OH-IAA), interfered with a quantitation of IAA,[135] because 4-Cl-IAA exhibited similar fluorescence activity and their separate determination was not possible by this method. The method was recently modified by the use of HPLC that could separate 2-MIP (IAA) from its corresponding chloro (and hydroxyl) derivative(s) (see V.B, high-performance liquid chromatography of auxins).

Recently, combined GC/MS has been used for the identification of endogenous auxins. GC/MS is summarized by Yokota et al.[117] to be an analytical method whereby a mixture is separated by GC and the separated compounds are led directly into the ion source of a mass spectrometer. Identification of the components is based both upon their GLC retention times and their mass spectra. When GC/MS is operated with the technique of selected-ion monitoring (SIM) or mass fragmentography (MF), the GC/SIM method is very effective not only for identification of a trace of compound, but also for its quantitative analysis. In the GC/SIM method, GC is conducted by monitoring the ion current from a specific ion (such as a molecular ion) and characteristic fragment ions of the compound.

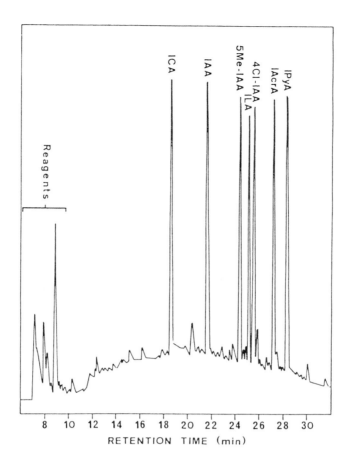

FIGURE 5. HPGC-ECD of some synthetic N-HFB-indolic acid methyl esters: indolyl-carboxylic (ICA), -acetic (IAA), -lactic (ILA), -acrylic (IAcrA), -pyruvic (IPyA), 5-methylindolylacetic (5-Me-IAA), and 4-chloroindoleacetic (4-Cl-IAA) acids (ca tng each). 1.2 bar, program temperature, 130—235°C, 4°/min, attenuation.

Bridges et al.[136] identified IAA in roots of the seedlings of *Zea mays* by GC/MS. Crude IAA, extracted from 400 g of the fresh seedlings, was bis-trimethylsilylated to (TMSi)$_2$-IAA (**90**), which was purified by GC on SE-33 and then analyzed with a single beam MS spectrometer. Over 50 ng of IAA/g of fresh weight was found in the steles, and 29 ng in the tips. In this method, less than 0.1 μg of IAA was required to measure a complete spectrum.

IAA in the phloem and xylem transport system of *Ricinus communis* was identified by Hall and Medlow[137] using bis-TMSi-IAA. White et al.[138] analyzed IAA in apical portions of the stem of *Phaseolus vulgaris* by GC/MS of TMSi-IAA.

4-Cl-IAA-Me, previously isolated from immature seeds of *Pisum sativum*, has been identified by Engvild et al.[53,54] by the GC/MS method in seven species of the Vicieae plants.

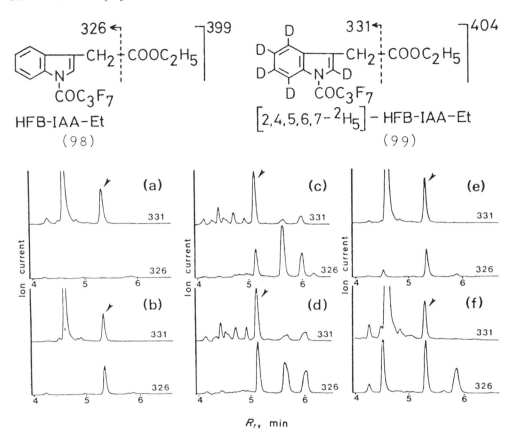

FIGURE 6. Selected ion chromatograms from extracts: (a) standard of 250 pmol $^2H_5$-IAA only; (b) standard of 250 pmol IAA and 250 pmol $^2H_5$-IAA; (c) *Zea* coleoptile, (d) *Zea* mesocotyl; (e) dark-grown *Pisum* epicotyl; (f) light-grown *Pisum* epicotyl. Arrows indicate the peaks from $^2H_5$-IAA-HFB-Et.

Hofinger and Böttger[55] also identified 4-Cl-IAA and its methyl ester in immature seeds of *Vicia faba* by GC/MS.

Allen et al.[139] applied the GC/MS method for the analysis of IAA in pea and maize seedlings. They converted a crude sample of IAA extracted from 0.5 to 2.5 g of the seedlings into HFB-IAA-Et (**98**), which was analyzed using [^2H]-HFB-IAA-Et (**99**) as an internal standard. For quantitative analysis, m/z 326.1 ($M^+ - COOC_2H_5$) and 399.0 (M^+) of HFB-IAA-Et and the corresponding 331.1 and 404.0 of [$^2H_5$]-HFB-IAA-Et were monitored simultaneously. The selected ion chromatograms from the extracts of *Zea* coleoptile and mesocotyl and *Pisum* epicotyl are shown in Figure 6.

Based on these selected ion chromatograms, the IAA levels of 30 to 90 pmol/g of fresh weight were obtained for dark-grown *Pisum sativum* epicotyl and 71 to 199 pmol/g of fresh weight for dark-grown *Zea mays* seedlings.

Caruso et al.[140] determined the IAA content in Douglas fir by GC/SIM using methylene-d_2IAA as an internal standard. IAA-Me was derivatized to TFA-IAA-Me (**92**) and the molecular ion (m/z 285) and its fragmented ion (m/z 226) were used in selected ion monitoring, compared with the corresponding m/z 287 and 228 of d_2-IAA. Thus, endogenous IAA of Douglas fir was determined to be 1.6 μg/g of fresh weight in the shoot and 2.9 μg in the seedlings, respectively.

River and Pilet[141] analyzed IAA in maize root tips by GC/SIM using 5-methyl-IAA as an internal standard. Crude IAA in the plant extract was derivatized to HFB-IAA-Me (**96**).

The molecular ion (m/z 385) and the $M^+ - COOCH_3$ ion (m/z 326) were monitored, compared with the corresponding ions (m/z 399 and 340) of 5-Me-IAA.

Little et al.[142] identified IAA in the cambial region of *Picea sitchensis* by combined GC/MS with single-ion monitoring. The plant extracts containing IAA were reacted with BSA to produce $(TMSi)_2$-IAA, and the most intense peak at m/z 202 in the mass spectrum was monitored. The diffusible, free, and bound forms of IAA throughout 1 year were 0.06 to 0.30, 0.46 to 3.85, and 0.04 to 0.20 µg/g of oven-dried weight, respectively.

B. HPLC of IAA

HPLC has recently been used for the analysis of IAA and related indole compounds in plant extracts. Various types of HPLC, such as anion exchange, silica gel adsorption, and reversed-phase (ion suppression or ion-pair) partition, have been reported. Several different types of detectors have also been reported.

In 1978, Sweetser and Swartzfager[143] developed a HPLC method for the analysis of IAA in plant extracts. They examined three types of HPLC columns: (1) an anion exchange column, Partisil-10-SAX with a mobile phase of 0.01 M KH_2PO_4 and 0.05 M $NaClO_4$; (2) two reversed-phase columns with a bonded phase of octadecyl (ODS) silane on totally porous particles, Zorbax-ODS with a mobile phase of 25% acetonitrile, 75.9% H_2O and 0.1% formic acid, and Partisil-10-ODS with a mobile phase of 30% methanol and 70% (0.01 M KH_2PO_4 + 0.05 M $NaClO_4$); (3) an adsorption column, Zorbax-SIL with a mobile phase of either 1.5% methanol, 0.1% formic acid in CH_2Cl_2 (0.5% H_2O saturated), or 20% ethyl acetate-80% heptane (0.1% in formic acid and 0.005% H_2O). Two selective detectors, a fluorescence detector and an electrochemical, carbon paste amperometric (EC) detector, were used.

Among these columns examined, Partisil-SAX retained IAA on the column a longer time than the other two types of columns; IAA has a large k' of 5.0 to 5.6 that facilitated the separation of IAA from most other components in the plant extracts. This column also separated IAA well from *cis, trans-* and *trans, trans-* abscisic acid.

The IAA levels of six plant species have been determined with the Partisil-SAX column: pinto bean *(Phaseolus vulgaris)*, soybean *(Glycine max* cv Wye and cv Kent), cotton (Gossypium hirsutum cv Stoneville 213), wheat *(Tritricum aestivum* cv Selkirk), and corn *(Zea mays* cv Pioneer brand 3331). The highest concentration of IAA in various plant tissues was found in immature seeds. The range of IAA concentrations in leaf, petiole, and stem tissues was mainly due to the age of the tissue; the younger tissue generally showed higher IAA levels. In most species, the stem and petiole had an IAA concentration higher than even younger leaves. This should indicate that the stem was either a significant storage site for IAA or an important source of IAA synthesis.

Both the reversed-phase columns, Zorbax-ODS and Partisil-10-ODS, and the adsorption column, Zorbax-SIL, could be used as pre-columns for collecting the IAA fraction from crude plant extracts prior to injection to Partisil-SAX. These columns were also used directly for the quantitation of IAA in plant extracts with a fluorescent detector. These columns separated IAA from abscisic acid.

The EC detector gave a linear response for IAA from the sub-nanogram level to the microgram level (detection limit, approximately 0.05 ng), and the fluorescent detector gave a similar response from 0.5 ng to over 400 ng (detection limit, 0.5 to 1 ng). The latter detector was at least 13 times more sensitive than the UV (254 nm) detector. Although the fluorescent detector was less sensitive than the EC detector, it was more convenient and had good long-term stability. Only milligram amounts of plant tissues were required for the analysis of IAA by this HPLC method.

Crozier et al.[144] extended Sweetser and Swartzfager's HPLC method into the picogram level analysis by selecting a fluorimeter. They used Hypersil-ODS (5 µm) as a reversed-phase column and a mobile phase of an increasing gradient of ethanol in ammonium acetate

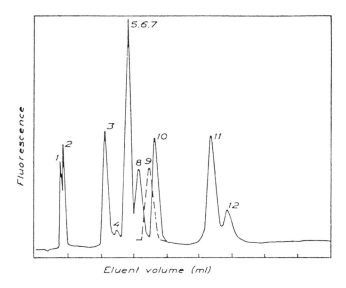

FIGURE 7. HPLC separation of indolecarboxylic acids with ion-suppression chromatography on a Nucleosil C_{18} column. Eluent: 22.0% methanol in 0.01 M phosphate buffer (pH 3.0). Compounds: tryptophan (1), 5-OH-IAA (2), 3-indolelacetic acid (3), 3-indolepyruvic acid (4), 3-indoleacrylic acid (5), 3-indoleglyoxylic acid (6), 5-indolecarboxylic acid (7), IAA (8), 4-Cl-IAA (9), 3-indolepropionic acid (10), 2-indolecarboxylic acid (11), and 3-indolebutyric acid (12). Peak 4 is uncertain because of the rapid degradation of 3-indolepyruvic aid in aqueous solution. Because of the small amounts available, 4-Cl-IAA (peak 9) was chromatographed separately.

buffer (pH 3.5, 20 mM). They found that a sensitivity of fluorimetry was practically instrument dependent, and an assessment of several commercial HPLC fluorescence monitors confirmed that the best results were obtained with a Perkin-Elmer 650-10 LC spectrofluorimeter; the limit of detection was 1 pg and this limit was approximately 10^3 times lower than that of an UV monitor at 280 nm. They have then analyzed endogenous IAA in elongating shoots, xylem sap, and callus of Douglas fir.

Sandberg and Anderson[145] isolated IAA from Scotch pine, *Pinus sylvestris* L. by reversed-phase ion-pair HPLC with a spectrofluorimetric detector (excitation 285 ± 5 nm, and emission 360 ± 10 nm). They used a Nucleosil C_{18} (10 μm) column with tetrabutylammonium (TBA) ion as a counter ion. First, they examined two types (ion-suppression and ion-pair) of chromatography (ISC and IPC) that were generally used for reversed-phase HPLC of acidic indoles. The IPC included the addition of a lipophilic counter ion, commonly a TBA ion, to the effluent, which increased the selectivity for negatively charged compounds.

ISC and IPC were compared in the separation of 12 kinds of indole acids that are reported to occur in plants. The chromatographic results are shown in Figures 7 and 8, respectively. The superiority of the IPC system was obtained. On the contrary, the ISC system failed to separate IAA from 4-Cl-IAA that was a potent natural auxin. The optimal concentration of TBA is 0.01 M. They analyzed, with this HPLC method, the IAA content of single pine seedlings, the amount of which needed was only 400 mg or less.

Průkryl and Vančura[146] applied reversed-phase HPLC to the analysis of IAA and related 26 indole compounds that might occur in the metabolic pathway of IAA. As ODS-modified silica gel columns, MicroPak CH (10 μm) and LiChrosorb RP-18 (5 μm) were selected for analytical and preparative separation.

Satisfactory separation on the MicroPak CH column with the solvent mixture of ethanol (20) and 1% acetic acid (80) was achieved with several series of indole compounds: (a)

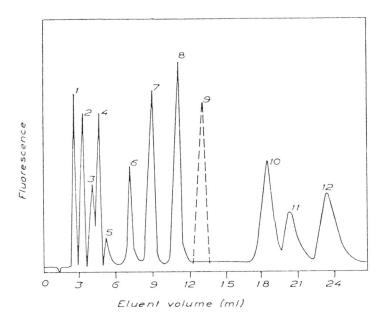

FIGURE 8. HPLC separation of indolecarboxylic acids with ion-pair chromatography on a Nucleosil C_{18} column. Eluent: 27.5% methanol in 0.01 M phosphate buffer (pH 6.5) with 0.01 M TBA. Compounds: tryptophan (1), 5-OH IAA-(2), 3-indolelacetic acid (3), 3-indoleacrylic acid (4), 3-indolepyruvic acid (5), 3-indoleglyoxylic acid (6), 5-indolecarboxylic acid (7), IAA (8), 4-Cl-IAA (9), 3-indolepropionic acid (10), 2-indolecarboxylic acid (11), and 3-indolebutylic acid (12). Peak 5 is uncertain because of the rapid degradation of 3-indolepyruvic acid in aqueous solution. Because of the small amounts available, 4-Cl-IAA (peak 9) was chromatographed separately.

indole-3-acetic, -propionic, and -butyric acid; (b) some derivatives of IAA which have different functional groups, e.g., 5-OH-IAA, IAM, IAN, indole-ethyl acetate, -glycolic acid, and -glyoxalic acid; some derivatives of indole-3-propionic acid, viz., indole-3-aminopropionic acid, -propionitrile, -lactic acid; some derivatives of indole-3-butyric acid, viz., indole-3-butyramide; (c) some substances with the same functional group in different positions on the indole ring, viz., indole-5-, -3-, and -2-carboxylic acid. The retention volume of the isomer with the 2-carboxyl group was substantially higher compared with the 3- and 5-isomers.

The LiChrosorb RP-18 column with a mobile phase of ethanol (30) and 1% acetic acid (70) exhibited a fourfold higher efficiency for the separation of IAA and the sequence of indole compounds eluted from the column was similar to that obtained with MicroPak CH. In particular, satisfactory separation was obtained with some biologically interesting substances such as indole-3-aminopropionic acid, -ethylamine, -triethylamine, -acetonitrile, -pyruvic acid, and -propionic acid. On the other hand, the separation of indole-3-acetic acid, -butyramide, -glyoxalic acid, -aldehyde, -ethanol, and -methanol became less satisfactory.

A lower limit of determination of three indole compounds tested with the UV detector at 280 nm was 17.5 ng for IAA, 15.6 ng for IAN, and 5.0 ng for indole-3-aminopropionic acid.

Sjut[147] applied reversed-phase HPLC to the separation of IAA, 4-Cl-IAA, and 5-OH-IAA; the former two compounds have been identified as natural auxins and the last one was suggested to be present in tomatoes. Prepacked columns of Spherisorb-ODS (10 μm) and LiChrosorb RP-8 (7 μm) were used with a fixed wavelength UV monitor (254 nm). Spherisorb-ODS is a spherical, totally porous silica with a C_{18} bonded stationary phase and

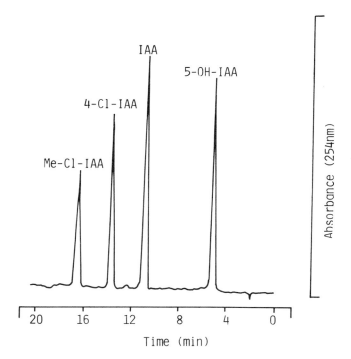

FIGURE 9. Separation of IAA and substituted IAA derivatives by the 7-μm LiChrosorb RP-8 column (250 × 4.6 mm i.d.). Flow rate: 1.5 mℓ/min. Mobile phase: linear gradient, 20% 0.1N acetic acid to 70% 0.01N acetic acid in methanol over 15 min. Sample injected: 5-OH-IAA 250 ng; others 500 ng.

LiChrosorb RP-8 is a similar but irregularly shaped silica with a C_8 bonded phase. The solvent systems were binary mixtures of acidic buffers in water (0.1 N acetic acid) and methanol (0.1 N acetic acid in methanol).

Both reversed-phase columns were suitable for the separation of these three important indole acids plus 4-Cl-IAA Me. The results obtained with the analytical LiChrosorb RP-8 column are shown in Figure 9. This column could also be used to recover IAA from 4-Cl-IAA and 5-OH-IAA for the selective quantitation of IAA by the indolo-α-pyrone method that was interfered with by the presence of these chloro- and hydroxyl-substituted IAA derivatives.

Blakesley et al.[148] reinvestigated the 2-methylindolo-α-pyrone (2-MIP) method for selective quantitation of IAA co-occurring with 4-Cl-IAA and 5-OH-IAA. The latter two compounds interfered with the quantitation of IAA when they were present together, due to their similar fluorescence activity. Hypersil-ODS (5 μm) allowed the simultaneous analysis of IAA and detection of 4-Cl-IAA and 5-OH-IAA.

They found that two fluorescence peaks were separated with the Hypersil-ODS column following derivatization in purified tissues of both *Pinus sylvestris* and *Chamaecyparis lawsoniana*. The first peak had an identical fluorescence emission spectrum (480 nm) following excitation at 445 nm and the retention time in HPLC to the 2-MIP of IAA. The second peak was unidentified.

Reversed-phase ion-pair chromatography was performed by Monsdale[149] for the analysis of IAA in plant extracts with a stainless steel column packed with Partisil-10-ODS.

Since IAA has a carboxyl group, reversed-phase chromatography can be performed either with a mobile phase of low pH using the ion-suppressing principle or with a mobile phase of a pH greater than 5 (when the acid is in an ionic dissociated form). The mobile phases were (for the ion-suppression chromatography) 5% acetic acid (pH 3.0)-methanol (70:30 and 60:40); (for the ion-pair chromatography) 0.01 M tetramethylammonium phosphate (TMA) or tetrabutylammonium hydrogen sulfate (TBA) (in 0.001 M K_2HPO_4/KH_2PO_4, pH 6.6)-methanol (90:10 and 60:40). The column effluent was monitored with the UV detector (278 nm).

The ion-pairing of the ionized IAA with 0.001 M TMA or TBA in the phosphate buffer solution increased the retention time of IAA in the chromatograms compared with that under the ion-suppressed conditions. Thus, the application of this reversed-phase ion-pair chromatography to partially purified apple and tobacco extracts gave an excellent separation of IAA from other UV-absorbing peaks.

Jensen[150] examined the effects of both pH of the mobile phase and alkyl chain length of the stationary phase on the elution pattern of indole acids with reversed-phase Rad-Pak C_{18} (5 μm), Rad-Pak C_8 (10 μm), and Hypersil-ODS (5 μm).

IAA was eluted much more rapidly at pH 7.0 (10 mM ammonium acetate) than pH 3.5 (10 mM acetic acid). It might be explained that the ionized IAA at pH 7.0 was distributed preferentially into the more polar aqueous mobile phase. The more polar Rad-Pak C_8 with a shorter alkyl chain possessed some distinct properties compared with its C_{18} equivalents with a longer alkyl chain.

Reversed-phase HPLC was applied by Nonhebel et al.[151] to the metabolic study of IAA using a Shandon column packed with Hypersil-ODS (5 μm) and the mobile phase of a 30 min gradient of 10 to 60% methanol in ammonium acetate buffer (pH 3.5, 20 mM). They analyzed radioactivity in the methanolic extract of *Zea mays* L. cv Fronica root segments which had been incubated with [^{14}C]-IAA, and provided a support for Reinecke and Bandurski's[152] previous finding that IAA was metabolized to oxindole-3-acetic acid.

Hardin and Stutte[153] established the one chromatographic separation method of some of the major plant hormones. For this purpose, they examined two columns: μ-Bondapak C_{18} (with the solvent system of a gradient increase of methanol in 0.67% acetic acid) and a Radial Compression Module (RCM-100) equipped with a RadialPak A cartridge (0.5% acetic acid-acetonitrile, 80:20, and methanol-acetonitrile, 58:42). A mixture of IAA, zeatin, zeatin-riboside, and *cis,cis*- and *cis,trans*-abscisic acid was used. The chromatographic results of the hormone mixture on two columns are shown in Figure 10.

The baseline separation of all components of the hormone mixture was achieved with the RCM-100 column which was superior for their separation to the μ-Bondapak C_{18} column. This technique was then applied to the analysis of endogenous hormone in the crude solvent extracts obtained from 1 g of soybean tissue in which IAA, *cis,trans*-abscisic acid, and *trans*-zeatin riboside were identified.

Durley et al.[154] analyzed IAA, abscisic acid, and phaseic acid in Sorghum bicolor leaves with reversed-phase chromatography. The methanol extract of the plant was initially purified by treatment with ammonia and then by chromatography on short polyvinyl-pyrrolidone column. This procedure is very rapid and highly effective in reducing a dry weight of the sample. The sample thus purified was sufficiently pure for the analysis of IAA by HPLC on the Beckman Ultrasphere ODS column (5 μm) with a mobile phase of water-acetonitrile-acetic acid (71:28:1) and a fluorescence detector with excitation at 254 nm and emission at 360 nm. The detection limit was approximately 100 pg. Five grams (5 g) of the fresh leaves were sufficient for this analysis of IAA.

Twenty-five amino acid conjugates of IAA were separated by Hollenberg et al.[155] with two reversed-phase HPLC, Varian MCH-10 and Partisil PXS 10/25 ODS-2, using the solvent of 1% acetic acid or acetonitrile. Each conjugate could be separated from all of the others on both columns, and the observed separation was more distinct than those previously obtained with TLC and paper chromatographic systems.

They then incubated cucumber seedlings with [^{3}H]-IAA and the tritiated metabolites were

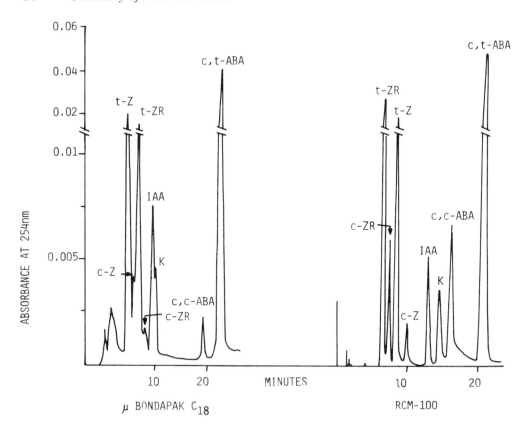

FIGURE 10. Separation of plant hormones on μ-Bondapak C_{18} and RCM-100. Flow rate: 2 mℓ/min. t-ZR = *trans*-zeatin riboside; c-ZR = *cis*-zeatin riboside; t-Z = *trans*-zeatin; c-Z = *cis*-zeatin; IAA = indole-3-acetic acid; K = kinetin; c,c-ABA = *cis,cis*-abscisic acid; c,t-ABA = *cis, trans*-abscisic acid.

extracted from the cucumber and separated by HPLC. Two peaks of major of tritium radioactivity coincided with those of authentic IAA-[^{14}C]-Asp and IAA-[^{14}C]-Glu.

C. Radioimmunoassay of IAA

A specific radioimmunoassay (RIA) has been developed by Pengelly and Mems[156] for the analysis of nanogram quantities of IAA in plant extracts. Antibody specific for IAA in the 0.2 to 12 ng range obtained from rabbits immunized with IAA bound to bovine serum albumin (BSA). IAA was coupled to BSA by the Ranadine and Sehon method[157] for antibodies to serotonin, which formed the covalent bond between IAA and BSA via the indole nitrogen with formaldehyde treatment.

RIA depends upon the competition of IAA and [^{3}H]-IAA for binding sites on rabbit anti-IAA antibody molecules. After binding equilibrium has been reached, the [^{3}H]-IAA-antibody complex is separated from free [^{3}H]-IAA by precipitation with ammonium sulfate. The amount of unlabeled IAA in the sample is then calculated by comparing the fraction of total [^{3}H]-IAA in the bound form with a standard curve.

The specificity and sensitivity of the RIA method depend on the quality of antibody used. In assays with the antibody Pengelly has prepared, 2,4-dichlorophenoxyacetic acid and indoles structurally related to IAA reacted from 300- to 3000-fold less than did IAA itself. However, α- and β-naphthaleneacetic acid reacted significantly and hence interfered with the assay.

Pengelly chose the crown-gall tumor and the pith tissue of tobacco *(Nicotiana tabacum*

L.) for the RIA of IAA, since the former gall was auxin-autotrophic and the latter pith required exogenous auxin for growth in culture. The crown-gall tissue gave a high IAA value (26.7 ng/g of fresh weight) by RIA, whereas the pith tissue gave little (< 0.5 ng) or no IAA. These data were in agreement with the findings of Kulescha and Gautheret who measured the auxin content in these tissues by bioassay.

In the RIA method, the critical factor is antibody specificity. Fuchs and co-workers[158] previously immunized rabbits with IAA coupled via the carboxyl end of the side chain to a carrier. Their antibody has a low affinity for IAA, cross-reacted extensively with other indoles. Pengelly and Mems attached IAA to BSA with the indole nitrogen that left the side chain on the critical 3-position free. Thus, the prepared anti-IAA antibody was high-specific and approximately 1000 times as sensitive as the preparations obtained by Fuchs and associates.

The RIA for IAA has been validiated by Pengelly[159] in Bandurski's group[159] by direct comparison with a physicochemical assay utilizing GC/SIM and [$^2H_4$]-IAA as an internal standard. the IAA content in the acetone extract of 4 cm shoot tips from 4-day-old, dark-grown corn seedlings was analyzed to be 0.40 ng/g of dry weight using GC/SIM and 0.56 ng using RIA. The two methods showed good agreement when estimating the free IAA content. However, the poor agreement was obtained between the two methods when the alkaline hydrolyzed extract was used to the analysis of IAA. The inhibitors contained in the hydrolyzates were removed with purification by DEAE- and LH-20 Sephadex® chromatography. Following such purification, the RIA estimate (2.11 ng/g of dry weight) of the total (free + bound) IAA content agreed with that (2.14 ng) of the GC/SIM method within 2% on the average.

VI. THE SYNTHESIS OF AUXINS

The synthetic methods for IAA and its isotopically labeled compounds are described herein. These are very useful tools for studying the biosynthesis and metabolism of IAA. Synthesis of 4-chloroindole-3-acetic acid (4-Cl-IAA), which is a new auxin isolated from *Pisum sativum,* is also described.

A. IAA

The synthetic methods of IAA can be classified into three groups: (1) synthesis via Fisher indole synthesis, (2) condensation of indole with electrophiles, and (3) others. These three methods are illustrated in Table 3.

Method 1, which was originally developed by Ellinger[160] in 1904, consists in the phenylhydrazone formation from aldehydic esters or ketoesters and subsequent cyclization of the phenylhydrazone leads to indole derivatives. Ellinger[160] achieved the first synthesis of IAA by cyclization of the phenylhydrazone of methyl β-formylpropionate and alkaline hydrolysis.King and L'Ecuyer[161] similarly synthesized IAA from the phenylhydrazone of diethyl α-acetylglutarate. Tanaka[162] synthesized IAA from β-cyanopropionaldehyde. Yoshimura et al.[163] have shown that IAA can be prepared from the phenylhydrazone of alkyl β-formylpropionate in good yield (91%).

In method 2, the key reaction is the condensation of indole with an electrophile: (a) Majima and Hoshimo[164] synthesized IAN by treating indole-magnesium-iodide with chloroacetonitrile. Alkaline hydrolysis of IAN gave IAA. (b) IAA also was synthesized via a Mannich reaction of indoles;[165] indole was reacted with formaldehyde and *N*-methylaniline to give a Mannich base which afforded IAN after treatment with NaCN. Snyder and Pilgrim[166] similarly synthesized IAN by the reaction of gramine with NaCN. Cyanomethylation of indole with diethylaminoacetonitrile also gave IAN.[167] (c) IAA has been obtained by condensation of indole with α-hydroxy acid[168,169] or (d) with α-halo acid[170-172] under alkaline conditions in very high yields (92 to 99%). (e) Reaction of indole with oxalyl chloride gave indoleglyoxylic acid, which was converted to IAA by Wolf-Kishner reduction.[173]

Table 3
SYNTHETIC METHODS OF IAA

Synthetic reactions	Ref.

Method 1: Via Fisher indole synthesis — 160—163,182[a]

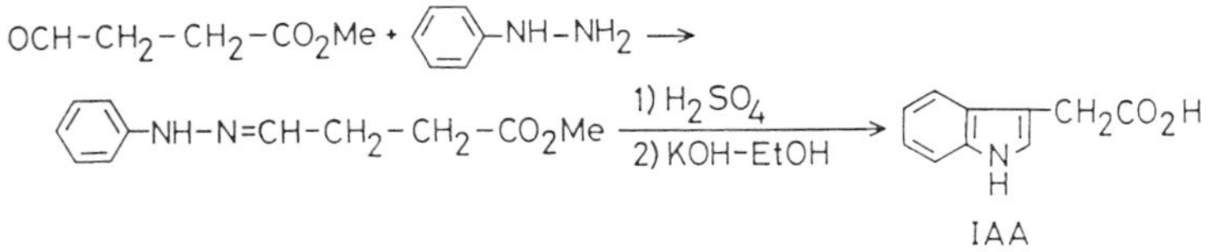

$OCH-CH_2-CH_2-CO_2Me$ + Ph-NH-NH$_2$ → Ph-NH-N=CH-CH$_2$-CH$_2$-CO$_2$Me $\xrightarrow[2) KOH-EtOH]{1) H_2SO_4}$ indole-CH$_2$CO$_2$H (IAA)

Method 2: Condensation of indole with electrophiles

(a) Mazima's synthesis — 164

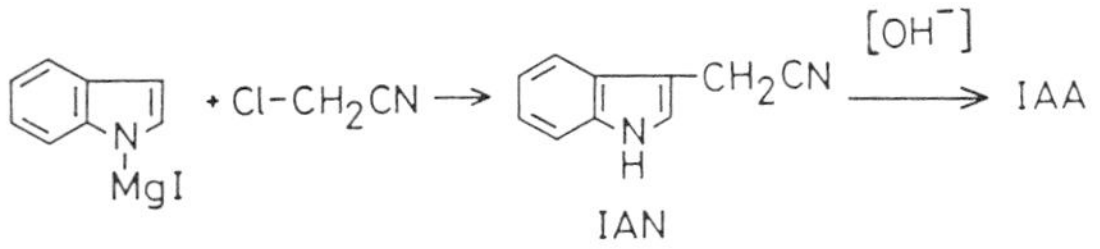

indole-MgI + Cl-CH$_2$CN → indole-CH$_2$CN (IAN) $\xrightarrow{[OH^-]}$ IAA

(b) Synthesis via Mannich reaction — 165—167,183[a]

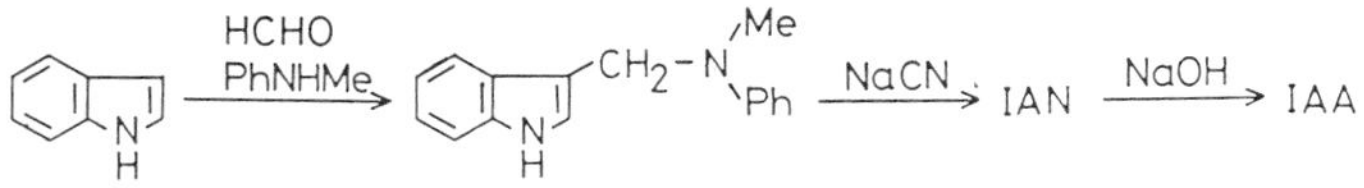

indole $\xrightarrow[PhNHMe]{HCHO}$ indole-CH$_2$-N(Me)(Ph) $\xrightarrow{NaCN}$ IAN $\xrightarrow{NaOH}$ IAA

(c) Condensation of indole with α-hydroxy acids — 168,169

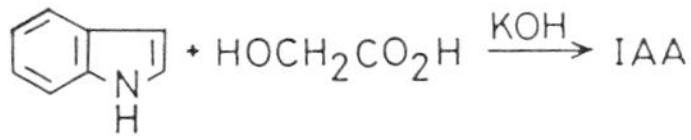

indole + HOCH$_2$CO$_2$H $\xrightarrow{KOH}$ IAA

(d) Condensation of indole with α-halo acids — 170—172

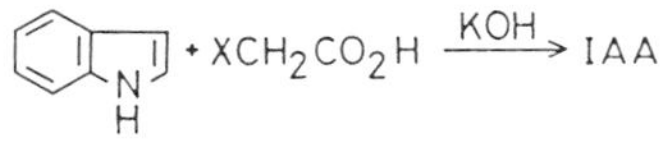

indole + XCH$_2$CO$_2$H $\xrightarrow{KOH}$ IAA

(e) Condensation of indole with oxalyl chloride — 173

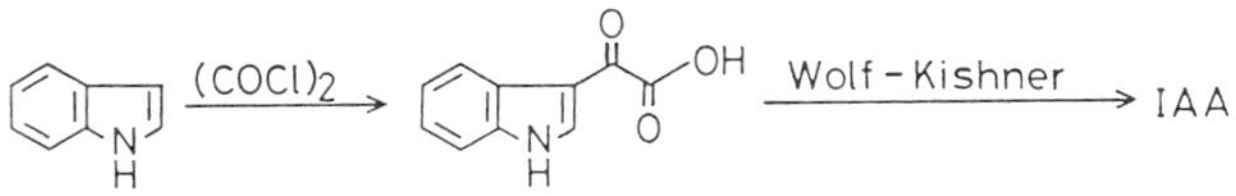

indole $\xrightarrow{(COCl)_2}$ indole-CO-CO-OH $\xrightarrow{Wolf-Kishner}$ IAA

Method 3: Others

(a) — 174

3-hydroxy-1-acetylindole $\xrightarrow[AcONH_4]{NCCH_2CO_2H,}$ IAN $\xrightarrow{NaOH}$ IAA

(b) — 175

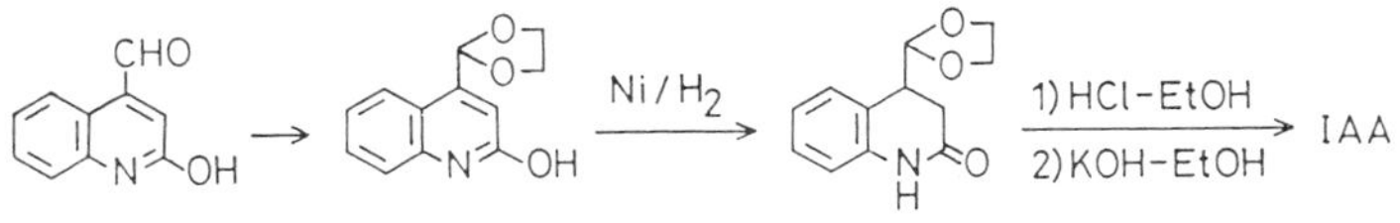

2-hydroxycinchoninaldehyde → (dioxolane derivative) $\xrightarrow{Ni/H_2}$ (reduced) $\xrightarrow[2) KOH-EtOH]{1) HCl-EtOH}$ IAA

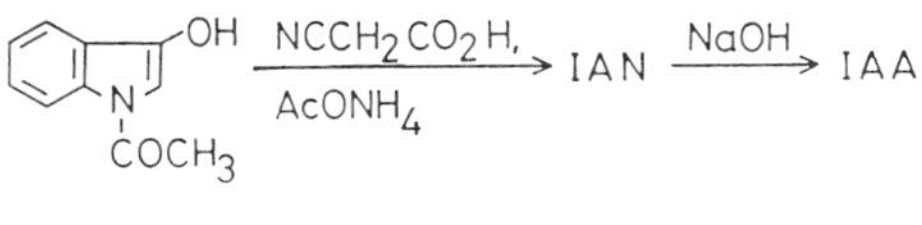

(c) — 176

indole + N$_2$CHCO$_2$Et $\xrightarrow{CuCl}$ indole-CH$_2$CO$_2$Et $\xrightarrow{[OH^-]}$ IAA

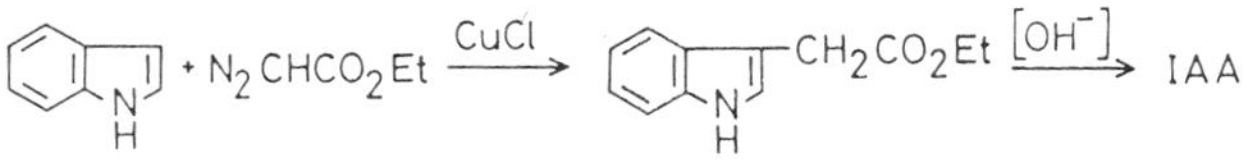

Table 3 (continued)
SYNTHETIC METHODS OF IAA

| Synthetic reactions | Ref. |

(d) [indole-CHO] → [indole-CH=C(NH-CO-NH-CO) hydantoin adduct] $\xrightarrow{H_2SO_4}$ [indole-CH_2-CO-COOH] $\xrightarrow{H_2O_2-NaOH}$ IAA   177

[a] References for 4-Cl-IAA.

Four syntheses of IAA in method 3 which are different from the above two methods have been reported. (a) 1-Acetyl-indoxyl was treated with cyanoacetic acid in the presence of ammonium acetate to give IAN.[174] (b) The second synthesis used 2-hydroxycinchoninaldehyde as the starting material.[175] (c) IAA also was prepared by the addition of ethyl diazoacetate to indole.[176] (d) Bentley et al.[177] synthesized indole-3-pyruvic acid from the condensation of indole-3-aldehyde with hydantoin, which was then converted to IAA by oxidative decarboxylation.

B. Isotopically Labeled IAA

Experimental details on the synthesis of [^{14}C]-labeled IAA have been reported by Pichat et al.[178] [2-^{14}C]-IAA was synthesized from [2-^{14}C]-indole according to the procedure reported by Majima and Hoshimo.[164] Robinson and Good[179] reported on the details of a semimicro synthesis of [^{14}C]-labeled IAA in the benzene ring. They synthesized the labeled IAA from α-ketoglutaric acid and [^{14}C]-phenylhydrazine. Ramamurthy and Viswanathan[180] synthesized [1-^{14}C]-IAA by the condensation of indole and [1-^{14}C]-glycolic acid in the presence of glyoxylic acid.

Recently, combined GC/SIM has been used widely for plant hormone research and the sensitivity of this technique was increased with introduction of the use of an internal standard such as deuterium-labeled IAA (see Section V). Magnus et al.[181] synthesized 4,5,6,7- and 2,4,5,6,7-deuterium-labeled IAA for use in MS assays; both d_4- and d_5-IAA were synthesized from α-keto-γ-cyanobutyric acid ethyl ester and pentadeuterophenylhydrazone as shown in Figure 11.

C. 4-Chloro-Indole-3-Acetic Acid

4-Cl-IAA is the second natural auxin found in higher plants (see Section III). Fox and Bullock[182] prepared a number of IAA with different substitutents at different positions of the benzene ring. Among them, 4-Cl-IAA was synthesized by Mazima's method (2 (a)) in 9 to 31% yields. Marumo synthesized 4-Cl-IAA by this method for identification of the auxin isolated from *Pisum sativum*. A better procedure (40% yield) for the synthesis of 4-Cl-IAA was reported by Engvild[183] (method 2 (b)).

VII. BIOSYNTHESIS AND METABOLISM OF AUXINS

A. Biosynthesis of IAA

IAA has been established to be biosynthesized from L-tryptophan, one of the amino acids. Evidence that IAA was produced via four different pathways in higher plants has been obtained from many biosynthetic studies. The biosynthetic pathwasy of IAA are shown in Figure 12.

The first pathway in the biosynthesis of IAA is that tryptophan $\xrightarrow{1}$ indole-3-pyruvic acid $\xrightarrow{2}$ indole-3-acetaldehyde $\xrightarrow{3}$ IAA. Yamaki and Nakamura[184] and Stowe and Thimann[185]

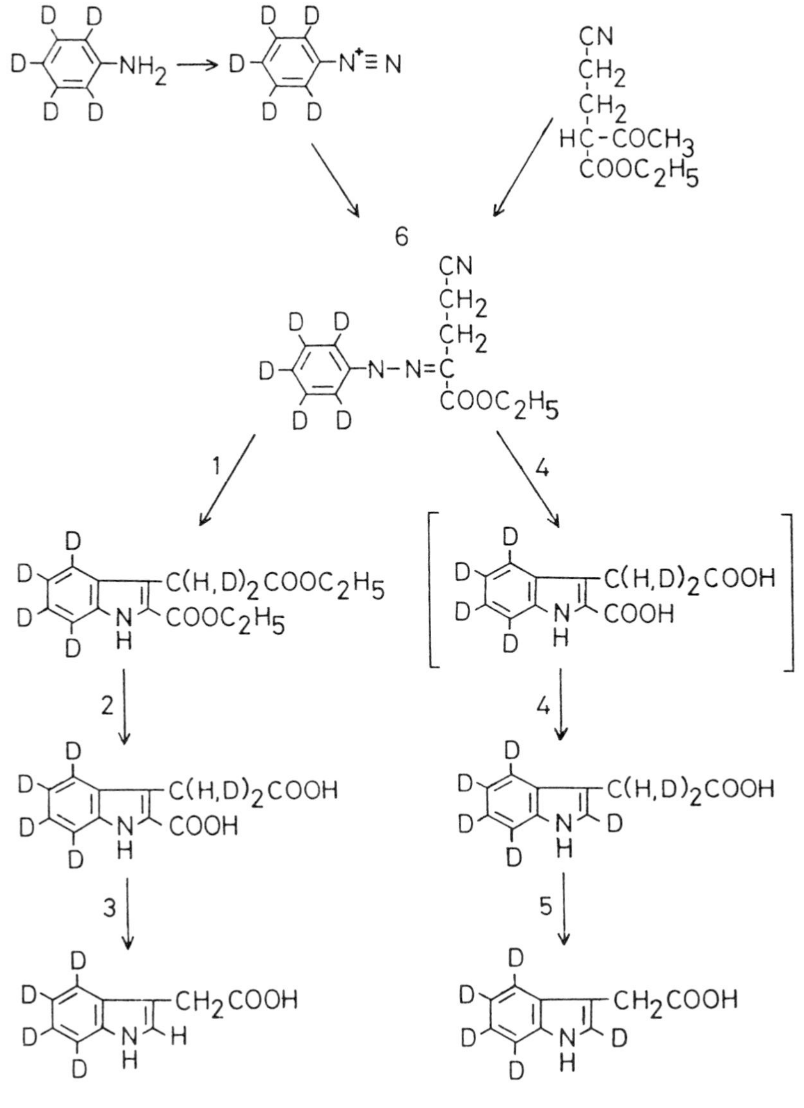

FIGURE 11. Syntheses of IAA-4,5,6,7-d_4 and IAA-2,4,5,6,7-d_5. Conditions used for the indicated steps were as follows: (1) DCl in C_2H_5OD/D_2) at reflux; (2) KOH/C_2H_5OH, at room temperature; (3) 1 equivalent KOH at 200°C; (4) DCl/D_2O)-D_3PO_4-pyridine at reflux; (5) 14% aqueous KOH at 120°C; (6) NaOH-50% aqueous ethanol at −15°C.

suggested that indole-3-pyruvic acid could be a native metabolite in vegetative tissues. This suggestion was confirmed by Winter's[186] experiment using corn. Rajagopal[187] then revealed that indole-3-acetaldehyde was present in etiolated shoots of *Pisum sativum* and *Helianthus annuus*. Gibson et al.[188] confirmed the biosynthetic route of IAA via indole-3-pyruvic acid by the metabolic experiments with 3-[^{14}C]-tryptophan that was fed to excised shoots of both tomato and barley. Radioactive indole-3-pyruvic acid, indole-3-acetaldehyde, and IAA were extracted from the excised shoots after the feeding experiments.

Reaction 1 in the first pathway involves a transamination of tryptophan to indole-3-pyruvic acid, which is catalyzed by tryptophan aminotransferase. This enzyme was extracted from bushbean seedlings by Forest and Wightman[189] and some properties of the purified enzyme

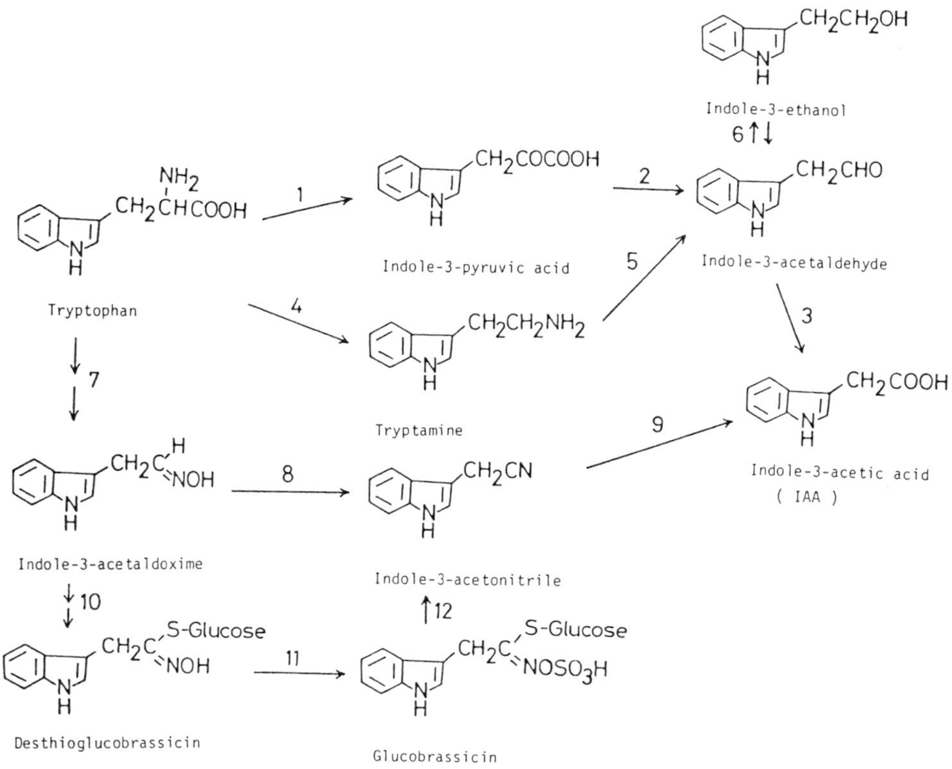

FIGURE 12. Biosynthesis of IAA.

were also examined. Truelson[190] has shown that the enzyme was distributed in 29 species (16 families) of higher plants.

Reaction 2, by which indole-3-pyruvic acid is converted to indole-3-acetaldehyde, is catalyzed by indolepyruvic acid decarboxylase. Gibson et al.[188] has found this enzyme in the crude supernatant from tomato shoots.

Reaction 3, which converts indole-3-acetaldehyde to IAA, is catalyzed by two enzymes, dehydrogenase and oxidase. The presence of an NAD-dependent indoleacetaldehyde dehydrogenase in mung bean seedlings has been shown by Wightman and Cohen.[191] Meanwhile, Rajagopal[192] reported indoleacetaldehyde oxidase of *Avena* coleoptile that converted indole-3-acetaldehyde to IAA. Miyata et al.[193] partially purified this enzyme from epicotyl of *Pisum sativum* and found it to be active only in the presence of molecular oxygen.

The second pathway in the biosynthesis of IAA is that tryptophan $\xrightarrow{4}$ tryptamine $\xrightarrow{5}$ indole-3-acetaldehyde $\xrightarrow{3}$ IAA. In 1937, Skoog[194] observed that tryptamine was active in the *Avena* curvature test. White[195] first isolated tryptamine from *Acacia,* and it has since been detected in several plants such as watermelon,[196] tobacco,[197] and *Coleus.*[198]

Reaction 4, which converts tryptophan to tryptamine, is catalyzed by tryptophan decarboxylase. Phelps and Sequeira[199] found this enzyme in the tobacco plant. Sherwin[200] demonstrated that the formation of tryptamine from tryptophan with the extract of cucumber hypocotyl was mediated by tryptophan decarboxylase. The tryptamine-forming L-decarboxylase from tomato shoots has been partially purified and charcterized by Gibson et al.[201]

Reaction 5, which converts tryptamine to indole-3-acetaldehyde, is mediated by tryptamine oxidase. Sherwin and Purves[202] have shown that [^{14}C]-tryptamine, when fed to cucumber seedlings, converted to [^{14}C]-IAA. Evidence that an amine oxidase participated in this conversion was obtained by using amine oxidase inhibitors.

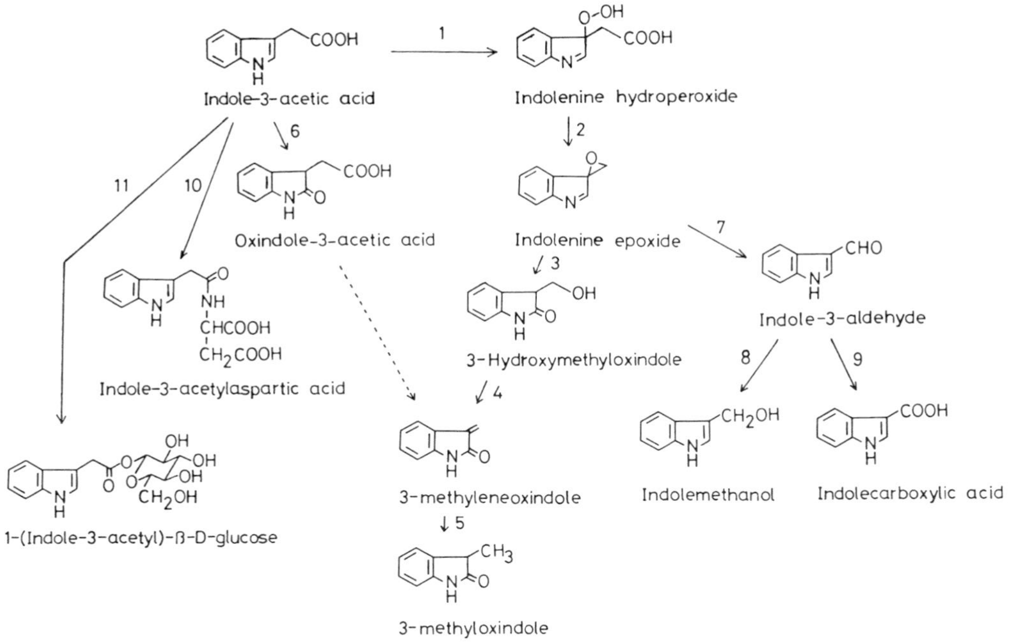

44   *Chemistry of Plant Hormones*

FIGURE 13.   Metabolism of IAA.

The third pathway with which the biosynthesis of IAA is concerned is indole-3-ethanol $\stackrel{6}{\rightleftharpoons}$ indole-3-acetaldehyde $\stackrel{3}{\rightarrow}$ IAA. Indole-3-ethanol has been isolated from green shoots of cucumber seedlings by Rayle and Purves,[66] and also was found as a natural constituent of tomato.[203] Indole-3-ethanol showed auxin activity against tomato, pea, radish, beet, and cucumber.[203,204] In the in vitro tracer experiment, Rayle and Purves[205] have shown that [^{14}C]-indole-3-ethanol was converted to [^{14}C]-IAA in cucumber seedling shoots.

Reaction 6, which mediates conversion of indole-3-ethanol to indole-3-acetaldehyde, is catalyzed by indole-3-ethanol oxidase. The enzyme was purified by Vickery and Purves[206] from cucumber seedlings having more than 3000-fold activity.

The fourth pathway in the biosynthesis of IAA is that tryptophan $\stackrel{7}{\rightarrow}$ indole-3-acetaldoxime $\stackrel{8}{\rightarrow}$ indoleacetonitrile $\stackrel{9}{\rightarrow}$ IAA. The pathway includes another side route in which indole-3-acetaldoxime $\stackrel{10}{\rightarrow}$ desthioglucobrassicin $\stackrel{11}{\rightarrow}$ glucobrassicin $\stackrel{12}{\rightarrow}$ indoleacetonitrile $\stackrel{9}{\rightarrow}$ IAA. These pathways have been detected mainly in the Brassicaceae, but glucobrassicin itself was found in the Tovariaceae, Capparaidaceae, and Resedaceae,[207] providing evidence that the pathway via indole-3-acetaldoxime was present in these plants. Kindl[208] demonstrated the occurrence of indole-3-acetaldoxime in cabbage by direct isolation. It showed auxin activity in the wheat cylinder, pea segment, and pea curvature tests.[209] Indoleacetonitrile was more active than IAA in the *Avena* straight growth test[210] and showed auxin activity in radish, cabbage, and turnip,[211] and in sunflower, cucumber, and balsamine;[212] however, it had little activity against pea, tomato, and broad bean.[213-215] The latter inactive plants might not convert indoleacetonitrile to IAA. Glucobrassicin, which occurred in large amounts in *Brassica* genus, showed growth-promoting activity in the wheat coleoptile test by Kutáček et al.,[216] but very limited responses to savoy cabbage.[217]

Radioactive tryptophan has been shown by Kutáček et al.[218,219] and other research groups[208,215] to be converted to indole-3-acetaldoxime, glucobrassicin, and indoleacetonitrile in cabbage. Wightman[215] has shown that [^{14}C]-indoleacetonitrile was converted to [^{14}C]-IAA in cabbage. Metabolic experiments with leaves of *Isatis tinctoria* L. showed that tritiated indole-3-acetaldoxime was efficiently converted to desthioglucobrassicin and glucobrassicin by Mahadevan and Stowe.[220] In the in vivo experiment with horseradish peroxidase tryptophan was possibly converted to indole-3-acetaldoxime.[208]

Reaction 8, which converts indole-3-acetaldoxime to indole-acetonitrile, is catalyzed by indoleacetaldoxime hydrolyase found in banana leaves.[221] The enzyme has been partially purified from *Gibberella fujikuroi*.[222] The purified enzyme was activated by dehydroascorbic acid, ascorbic acid, and pyridoxal phosphate.

Nitrilase is an enzyme that mediates reaction 9. This reaction converts indoleacetonitrile to IAA. Nitrilase was found in the families Gramineae (grasses), Cruciferae (cabbage and radish), and Musaceae (banana).[223]

Reaction 12, which converts glucobrassicin to indoleacetonitrile, was catalyzed by mirosinase (found in cruciferous plants). The role of this enzyme was considered to be rather limited.[224]

B. Metabolism of IAA

IAA has been shown to be metabolized mainly in two pathways; the oxidative metabolism with IAA oxidase or peroxidase and the formation of bound auxins with amino acids or sugars. In the oxidative metabolism, the degradation of IAA in plant tissues is that IAA $\xrightarrow{1}$ indolenine hydroperoxide $\xrightarrow{2}$ indolenine epoxide $\xrightarrow{3}$ 3-hydroxymethyloxindole $\xrightarrow{4}$ 3-methyleneoxindole $\xrightarrow{5}$ 3-methyloxindole (Figure 13).

IAA was enzymatically oxidized with peroxidase to give indolenine hydroperoxide. The unstable hydroperoxide was previously considered to degrade chemically to 3-methyloxindole. Tuli and Moyed,[225] however, reported that methyleneoxindole was enzymatically produced from IAA. Basu and Tuli[226] prepared extracts from wheat seedlings and wheat germ which catalyzed conversion of 3-hydroxymethyloxindole to 3-methyleneoxindole. A specific 3-methyleneoxindole reductase which reduced 3-methyleneoxindole to 3-methyloxindole has been purified from peas.[227,228] Hager and Schmidt[229] have prepared crude cell-free extracts from corn coleoptile which converted IAA to 3-methyleneoxindole.

Tuli and Moyed[230] hypothesized that 3-methyleneoxindole is tenfold more active than IAA itself in growth promotion of pea and mung bean stem segments, i.e., IAA acts as a growth promotor only after oxidative conversion to 3-methyleneoxindole. They discussed the role of chlorogenic acid and 2,4-dichlorophenoxyacetic acid in relation to metabolic conversion of IAA to 3-methyleneoxindole. Evans and Ray,[231] however, demonstrated that synthetic 3-methyleneoxindole, prepared from IAA and *N*-bromosuccinimide, lacked growth-promoting activity in coleoptile and pea stem segments. They considered that the activity of 3-methyleneoxindole previously reported by Tuli and Moyed was due to contamination with IAA. Marumo et al.[232] also demonstrated that both 4-chloro-3-methyleneoxindole prepared from 4-Cl-IAA and 3-methyleneoxindole has no appreciable auxin activity in six different bioassays. Higher biological activity of 4-Cl-IAA than IAA may be due to its stronger resistance to peroxidase-catalyzed oxidation.

Recently, Reinecke and Bandurski[152] identified 1-[^{14}C]-oxindole-3-acetic acid as a catabolic product of [1-^{14}C]-IAA in corn kernels. This is the first indication that corn kernels have a major oxidative pathway (without decarboxylation) of IAA that produced oxindole-3-acetic acid.

Indole-3-aldehyde is one of the major oxidation products of IAA in higher plants. The aldehyde has been shown to be a native compound in pea seedlings,[233] barley, and tomato shoots.[2033] Hinman and Lang[234] suggested that it might arise from indolenine epoxide (reaction 7). A metabolic conversion of [^{14}C]-IAA to [^{14}C]-indole-3-acetaldehyde was shown in pea seedlings[235] and bean stem.[236] The enzyme that catalyzed conversion of IAA to indole-3-acetaldehyde was found in pea,[237] Japanese radish,[238] and corn.[239] Magnus et al.[233] demonstrated that indole-3-aldehyde was reversibly reduced to indolemethanol (reaction 8) and

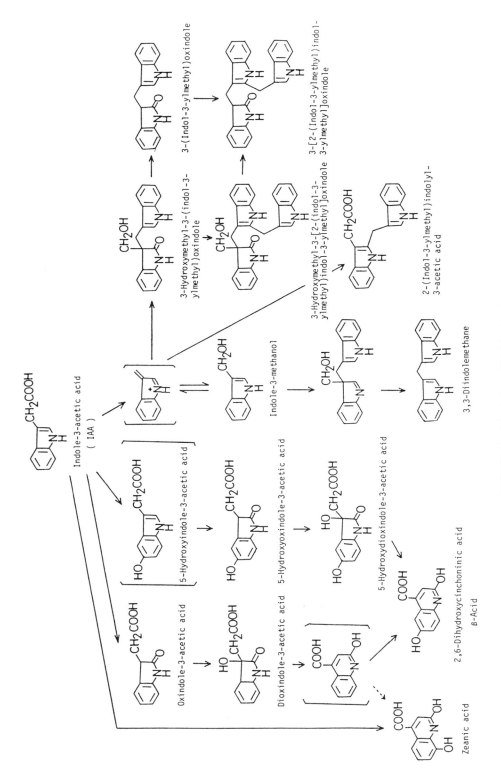

FIGURE 14. Metabolism of IAA in rice plant.

irreversibly oxidized to indolecarboxylic acid (reaction 9). Indolecarboxylic acid was identified in several *Nicotiana* species[240] and corn.[241]

Many metabolites in the bound form of IAA were detected in plants. Andreae and Good[242] first discovered indole-3-acetylaspartic acid in pea seedlings (reaction 10). Zenk has isolated 1-(indole-3-acetyl-β-D-glucose from leaves of *Colchicum neapolitanum* Ten. and Davies[236] also isolated this compound from pea and bean stems (reaction 11).

Cohen and Bandurski,[244] have reviewed chemistry and physiology of bound auxins in detail, and in this review they have summarized that the roles of bound auxins are played for the transport of IAA, for the storage and subsequent re-use of IAA, for protection of IAA from enzymatic destruction, and for the homeostatic control of the concentration of IAA in the plant.

Several reviews have been published on biosynthesis and metabolism of IAA in plants.[215,145-148]

Recently, Kawarada and associates reported the isolation of oxindole- and dioxindole-3-acetic acid derivatives from rice bran as the oxidation metabolites of IAA.[249,250] Suzuki and Kawarada[251] have isolated many indole dimers and trimers as enzymatic oxidation products of IAA catalyzed by horseradish peroxidase. These indole compounds isolated by Kawarada's group are shown in Figure 14.

The figure indicates that there are new metabolic pathways of IAA in rice plants besides those listed in Figure 13 which produced various oxidized or polymerized indole compounds. Among them, 5-hydroxyindole-3-acetic acid and 5-hydroxydioxindole-3-acetic acid were isolated as the synergistic substances with auxin in the auxin-induced ethylene production by etiolated mung bean hypocotyl segments. Matsushima et al.[252] have isolated zeanic acid as a plant growth stimulator. Zeanic acid and 2,6-dihydroxycinchoninic acid (β-acid) might be nonenzymatically derivatized from indole-type metabolites. The biological activity of indole dimers and trimers have not yet been assayed.

VIII. THE BIOLOGICAL ACTIVITY OF AUXINS

Although the critical activity of auxins is defined to stimulate elongation of excised stem sections prepared from the elongation zone of plants, biological activities of applied auxins are so diverse that compiling a complete list of the activities is quite difficult, if not impossible. The diversity of the biological activities results from the nature of plant tissues to which auxins are applied and the concentration of auxins. Different tissues respond differently to the same concentration and the same species of auxin. Even the anatomically identical tissues give different responses depending upon their age and other physiological states.

Thus, auxins have been shown to induce or stimulate stem and root cell elongation, cell division, xylem and phloem element differentiation, adventitious root formation, parthenocarpy, and many other reactions. Biochemically, auxins have been known to activate certain enzymes and stimulate the biosynthesis of others. Since a detailed review of the biological activities of auxins is not within the scope of this book, readers should refer to other reference books, including proceedings of the International Conference of Plant Growth Substances and other reviews.[253-260]

ACKNOWLEDGMENTS

The author should like to acknowledge his debt to Professor H. Imaseki, Associate Professor K. Wada, and Dr. M. Katayama in the Faculty of Agriculture, Nagoya University, for their invaluable help in preparing this chapter.

REFERENCES

1. Nomenclature of chemical plant regulators, *Plant Physiol.*, 29, 307, 1954.
2. **Vendrig, J. C.**, α-Chloro-β-(3-chloro-*o*-tolyl)propionitrile, a new synthetic auxin, *Nature (London)*, 234, 557, 1971.
3. **Branca, C. and Gaetani, E.**, 1,2-Benzisothiazol-3-ylacetic acid, a new synthetic auxin, *Atti Acad. Naz. Lincei, Cl. Sci. Fis., Mat. Nat. Rend.*, 54, 275, 1973.
4. **Okamoto, T., Isogai, Y., and Koizumi, T.**, Studies on plant growth regulators. II. Isolation of indole-3-acetic acid, phenylacetic acid and several plant growth inhibitors from etiolated seedlings of *Phaseolus*, *Chem. Pharm. Bull.* 15, 159, 1967.
5. **Schneider, E. A. and Wightman, F.**, Metabolism of auxin in higher plants, *Ann. Rev. Plant Physiol.*, 25, 487, 1974.
6. **Went, F. W. and Thimann, K. V.**, *Phytohormones*, Macmillan, New York, 1937.
7. **Boysen Jensen, P.**, Über die Leitung des phototropischen Reizes in der *Avena*- koleoptile, *Ber. Bot. Ges.*, 31, 559, 1913.
8. **Paäl, A.**, Über phototropische Reizleitungen, *Ber. Bot. Ges.*, 32, 499, 1914; Über phototropische Reizleitung, *Jahrb. Wiss. Bot.*, 58, 406, 1919.
9. **Stark, P.**, Studien über traumatotrope und haptotrope Reizleitungsvorgänge mit besonderer Berücksichtigung der Reizüber tragung auf fremde Arten und Gattungen, *Jahrb. Wiss. Bot.*, 60, 67, 1921.
10. **Seubert, E.**, Über Wachstumsregulatoren in der Koleoptile von *Avena*, *Z. Bot.*, 17, 49, 1925.
11. **Went, F. W.** Wuchstoff und Wachstum, *Recl. Trav. Bot. Neerl.*, 25, 1, 1928.
12. **Nielsen, N.**, Untersuchungen über Stoffe, die das Wachstum der *Avena*-coleoptile beschleunigen, *Planta*, 6, 376, 1928.
13. **Dolk, H. E. and Thimann, K. V.**, Studies on the growth hormone of plants. I., *Proc. Natl. Acad. Sci. U.S.A.*, 18, 30, 1932.
14. **Kögl, F., Haagen Smit, A. J., and Erxleben, H.**, Über ein Phytohormon der Zellstreckung. Reindarstellung des Auxins aus menschlichem Harn. IV. Mitteilung, *Z. Physiol. Chem.*, 214, 241, 1933.
15. **Kögl, F., Erxleben, H., and Haagen Smit, A. J.**, Über die Isolierung der Auxine a and b aus pflanzlichem Materialen. IX. Mitteilung, *Z. Physiol. Chem.*, 225, 215, 1934.
16. **Kögl. F., Erxleben, H., and Haagen Smit, A. J.**, Über ein Phytohormon der Zellstreckung. Zur Chemie des krystallisierten Auxins. V. Mitteilung, *Z. Physiol. Chem.*, 216, 31, 1933.
17. **Kögl, F. and Erxleben, H.**, Über die Konstitution der Auxine a und b. X. Mitteilung über pflanzliche Wachstumsstoffe, *Z. Physiol. Chem.*, 227, 51, 1934; Synthese der "Auxin-glutarsäure" und einiger Isomerer. XV. Mitteilung, *Z. Physiol. Chem.*, 235, 181, 1935.
18. **Kögl, F., Haagen Smit, A. J., and Erxleben, H.**, Über ein neues Auxin ("Heteroauxin") aus Harn. XI. Mitteilung, *Z. Physiol. Chem.*, 228, 90, 1934.
19. **Bennet-Clark, T. A., Tambiah, M. S., and Kefford, N. P.**, Estimation of plant growth substances by partition chromatography, *Nature (London)*, 169, 452, 1952.
20. **Wieland, O. P., DeRopp R. S., and Avener, J.**, Identity of auxin in normal urine, *Nature (London)*, 173, 776, 1954.
21. **Fukushima, I. and Munakata, K.**, The Annual Meeting of the Japan Agric. Chem. Soc., 1966.
22. **Luckwill, L. C.**, Application of paper chromatography to the separation and identification of auxins and growth-inhibitors, *Nature (London)*, 169, 375, 1952.
23. **Matsui, M. and Hwang, Y.-S.**, The synthesis of the proposed structure for auxin b lactone, *Proc. Jpn. Acad.*, 42, 488, 1966.
24. **Hwang, Y.-S. and Matsui, M.**, Synthesis of the stereoisomeric mixture of the compound having the proposed structure for auxin b lactone, *Agric. Biol. Chem.*, 32, 81, 1968.
25. **Nakamura, T., Takahashi, N., Matsui, M., and Hwang, Y.-S.**, Activity of synthesized auxin b lactone as the plant growth regulator of auxin type, *Plant Cell Physiol.*, 7, 693, 1966.
26. **Vliegenthart, J. A. and Vliegenthart, J. F. G.**, Reinvestigation of authentic samples of auxins a and b, and related products by mass spectrometry, *Rec. Trav. Chim. Pays-Bas.*, 85, 1266, 1966.
27. **Kögl, F. and Kostermans, D. G. F. K.**, Hetero-auxin als Stoffwechselproduct niederer pflanzlicher Organismen. Isolierung aus Hefe. XIII. Mitteilung, *Z. Physiol. Chem.*, 228, 113, 1934.
28. **Thimann, K. V.**, On the plant growth hormone produced by *Rhyzopus sinus*, *J. Biol. Chem.*, 109, 279, 1935.
29. **Haagen-Smit, A. J., Leech, W. D., and Bergern, W. R.**, Estimation, isolation and identification of auxins in plant material, *Science*, 93, 624, 1941.
30. **Haagen-Smit, A. J., Dandliker, W. B., Wittwer, S. H., and Murneed, A. E.**, Isolation of 3-indoleacetic acid from immature corn kernels, *Am. J. Bot.*, 33, 118, 1946.
31. **Berger, J. and Avery, G. S., Jr.**, Isolation of an auxin precursor and an auxin (indoleacetic acid) from maize, *Am. J. Bot.*, 31, 199, 1944.

32. **Redemann, W. R., Wittwer, S. H., and Sell, H. M.,** The fruit-setting factor from the ethanol extracts of immature corn kernels, *Arch. Biochem. Biophys.,* 32, 80, 1951.
33. **Bennet-Clark, T. A. and Kefford, N. P.,** Chromatography of the growth substances in plant extracts, *Nature (London),* 171, 645, 1953.
34. **Eagles, C. F. and Wareing, P. F.,** Dormancy regulators in woody plants, *Nature (London),* 199, 874, 1963.
35. **Luckwill, L. C. and Powell, L. E., Jr.,** Absence of indole-acetic acid in the apple, *Science,* 123, 226, 1956.
36. **Guttenberg, H. V., Nehring, G., and Blanke, I.,** Papier-chromatographische Untersuchung und Molekulargewichtsbestimmung des "säurefesten Wuchsstoffes", *Naturwissenschaften,* 41, 334, 1954.
37. **Abe, H. and Marumo, S.,** Identification of auxin-active substances as ethyl chlorogenate and indole-3-acetic acid in immature seeds of *Helianthus annuus, Agric. Biol. Chem.,* 36, 42, 1972.
38. **Khalifah, R. A., Lewis, L. N., and Coggins, C. W., Jr.,** New natural growth promoting substance in young citrus fruit, *Science,* 142, 399, 1963.
39. **Khalifah, R. A. and Radick, P. C.,** Fluorometric, chromatographic, and spectronic evidence of the nonindolic nature *Citrus* auxin, *J. Exp. Bot.,* 16, 511, 1965.
40. **Lewis, L. N., Khalifah, R. A., and Coggins, C. W., Jr.,** Seasonal changes in citrus auxin and 2 auxin antagonists as related of fruit development, *Plant Physiol.,* 40, 500, 1965.
41. **Khalifah, R. A., Lewis, L. N., and Coggins, C. W., Jr.,** Differentiation between indoleacetic acid and the *Citrus* auxin by column chromatography, *Plant Physiol.,* 41, 208, 1966.
42. **Gandar, J. C. and Nitsch, J. P.,** Extraction dúe substance de croissance ápartir des jeunes graines de *Pisum sativum, Colloq. Int. C. N. R. S.,* 123, 169, 1964.
43. **Nomoto, M. and Tamura, S.,** Isolation and identification of indole derivatives in clubroots of Chinese cabbage, *Agric. Biol. Chem.,* 34, 1590, 1970.
44. **Igoshi, M., Yamaguchi, I., Takahashi, N., and Horise, K.,** Plant growth substances in the young fruit of *Citrus unshiu, Agric. Biol. Chem.,* 35, 629, 1971.
45. **Marumo, S., Abe, H., Hattori, H., and Munakata, K.,** Isolation of a novel auxin, methyl 4-chloroindoleacetate from immature seeds of *Pisum sativum, Agric. Biol. Chem.,* 32, 117, 1968.
46. **Marumo, S., Hattori, H., Abe, H., and Munakata, K.,** Isolation of 4-chloroindolyl-3-acetic acid from immature seeds of *Pisum sativum, Nature (London),* 219, 959, 1968.
47. **Fox, S. W. and Bullock, M. W.,** Synthesis of indole-3-acetic acids and 2-carboxyindole-3-acetic acids with substituents in the benzene ring, *J. Am. Chem. Soc.,* 73, 2756, 1951.
48. **Gandar, J. C. and Nitsch, C.,** Isolement de l'ester méthylique d'un acide chloro-3-indolelylacétique á partir de grains immatures de pois, *Pisum sativum, C. R. Acad. Sci. Ser. D:,* 265, 1795, 1967.
49. **Marumo, S., Hattori, H., and Abe, H.,** Chromatography of a new natural auxin, 4-chloroindolyl-3-acetic acid and related chloro derivatives, *Anal. Biochem.,* 40, 488, 1971.
50. **Engvild, K. C.,** Simple identification of the neutral chlorinated auxin in pea by thin layer chromatography, *Physiol. Plant.,* 48, 435, 1980.
51. **Ehmann, A.,** The van Urk-Salkowski reagent — a sensitive and specific chromogenic reagent for silica gel thin-layer chromatographic detection and identification of indole derivatives, *J. Chromatogr.,* 132, 267, 1977.
52. **Engvild, K. C.,** Natural chlorinated auxins labeled with radioactive chloride in immature seeds, *Physiol. Plant.,* 34, 286, 1975.
53. **Engvild, K. C., and Egsgaard, H., and Larsen, E.,** Determination of 4-chloroindoleacetic acid methyl ester in *Vicia* species by gas chromatography-mass spectrometry, *Physiol. Plant.,* 53, 79, 1981.
54. **Engvild, K. C., Egsgaard, H., and Larsen, E.,** Determination of 4-chloroindoleacetic acid methyl ester in *Vicia* species by gas chromatography-mass spectrometry, *Physiol. Plant.,* 53, 79, 1981.
55. **Hofinger, M. and Böttger, M.,** Identification by GC-MS of 4-chloroindole-3-acetic acid and its methyl ester in immature *Vivia faba* seeds, *Phytochemistry,* 18, 653, 1979.
56. **West, G. B.,** Indole derivatives in tomatoes, *J. Pharm. Pharmacol. Suppl.,* 11, 275T, 1959.
57. **Govindachari, T. R., Pillai, P. M., Nagarajan, K., and Viswasnathan, N.,** Synthesis of 5- and 8-methoxy-yobyrines, *Tetrahedron,* 21, 2957, 1965.
58. **Gmelin, R. and Virtanen, A. I.,** Neoglucobrassicin, another thioglucoside in *Brassica* species with an indole group in the molecule, *Suom. Kemistil. B,* 35, 34, 1962 (CA, 57, 16534, 1962).
59. **Goren, R. and Goldschmidt, E. E.,** Regulative systems in the developing citrus fruit. I. The hormonal balance in orange fruit tissues, *Physiol. Plant.,* 23, 937, 1970.
60. **Goldschmidt, E. E., Monselise, S. P., and Goren, R.,** On the identification of native auxins in citrus tissues, *Can. J. Bot.,* 49, 241, 1971.
61. **Goldschmidt, E. E., Goren, R., Monselise, S. P., Takahashi, N., Igoshi, H., Yamaguchi, I., and Hirose, K.,** Auxins in citrus: a reappraisal, *Science,* 174, 1256, 1971.
62. **Hofinger, M., Gaspar, Th., and Darimont, E.,** Occurrence, titration and enzymatic degradation of 3-(3-indolyl)-acrylic acid in *Lens culinaris* Med. extract, *Phytochemistry,* 9, 1757, 1970.

63. **Hofinger, M., Monseur, X., Pais, M., and Jarreau, F. X.,** Further confirmation of the presence of indolylacrylic acid in lentil seedlings and identification of hypaphorine as its precursor, *Phytochemistry,* 14, 375, 1975.
64. **Isogai, Y., Okamoto, T., and Koizumi, T.,** Plant growth regulators. I. Isolation of indole-3-acetamide, 2-phenyl-acetamide, and indole-3-carboxaldehyde, from etiolated seedlings of phaseolus, *Chem. Pharm. Bull.,* 15, 151, 1967.
65. **Okamoto, T., Isogai, T., Koizumi, T., Fujishiro, H. and Sato, Y.,** Studies on plant growth regulators. III. Isolation of indole-3-acetonitrile and methyl indole-3-acetate from the neutral fraction of the moyashi extract, *Chem. Pharm. Bull.,* 15, 163, 1967.
66. **Rayle, D. L. and Purves, W. K.,** Isolation and identification of indole-3-ethanol from cucumber seedlings, *Plant Physiol.,* 42, 520, 1967.
67. **Rajagopal, R.,** Occurrence of indoleacetaldehyde and tryptophol in the extracts of etiolated shoots of *Pisum* and *Helianthus* seedlings, *Physiol. Plant.,* 20, 655, 1967.
68. **Jones, E. R. H., Henbest, H. B., Smith, G. F., and Bentley, J. A.,** 3-Indolylacetonitrile: a naturally occurring plant growth hormone, *Nature (London),* 169, 485, 1952.
69. **Jones, E. R. H. and Taylor, W. C.,** Some indole constituents of cabbage, *Nature (London),* 179, 1138, 1957.
70. **Row, V. V., Sanford, W. W., and Hitchcock, A. E.,** Indole-3-acetyl-D,L-aspartic acid as a naturally-occurring indole compound in tomato seedlings, *Contrib. Boys Thomson Inst.,* 21, 1, 1961.
71. **Klämbt, H. D.,** Indole-3-acetylasparginsaure, ein natürlich vorkommendes Indolderivat, *Naturwissenschaften,* 47, 397, 1960.
72. **Hattori, H. and Marumo, S.,** Monomethyl-4-chloroindolyl-3-acetyl-L-aspartate and absence of indolyl-3-acetic acid in immature seeds of *Pisum sativum, Planta,* 102, 85, 1972.
73. **Marumo, S. and Hattori, H.,** Isolation of D-4-chloro-tryptophan derivatives as auxin-related metabolites from immature seeds of *Pisum sativum, Plant,* 90, 208, 1970.
74. **Labarca, C., Nicholls, P. B., and Bandurski, R. S.,** A partial characterization of indoleacetylinositols from *Zea mays, Biochem. Biophys. Res. Commun.,* 20, 641, 1965.
75. **Ueda, M. and Bandurski, R. S.,** A quantitative estimation of alkali-lable indole-3-acetic acid compounds in dormant and germinating maize kernels, *Plant Physiol.,* 44, 1175, 1969.
76. **Ueda, M., Ehmann, A., and Bandurski, R. S.,** Gas-liquid chromatographic analysis of indole-3-acetic acid myoinositol esters in maize kernels, *Plant Physiol.,* 46, 715, 1970.
77. **Piskornik, Z. and Bandurski, R. S.,** Purification and partial characterization of a glucan containing indole-3-acetic acid, *Plant Physiol* 50, 176, 1972.
78. **Ueda, M. and Bandurski, R. S.,** Structure of indole-3-acetic acid myoinositol esters and pentamethyl-myoinositols, *Phytochemistry,* 13, 243, 1974.
79. **Ehmann, A.,** Identification of 2-*O*-(indole-3-acetyl)-d-glucopyranose, 4-*O*-(indole-3-acetyl)-d-glucopyranose and 6-*O*-(indole-3-acetyl)-d-glucopyranose from kernels of *Zea mays* by gas-liquid chromatography-mass spectrometry, *Carbohydr. Res.,* 34, 99, 1974.
80. **Ehmann, A. and Bandurski, R. S.,** The isolation of di-*O*-(indole-3-acetyl)-myo-inositol and tri-*O*-(indole-3-acetyl)-myo-inositol from mature kernels of *Zea mays, Carbohydr. Res.,* 36, 1, 1974.
81. **Bandurski, R. S.,** Chemistry and physiology of conjugates of indole-3-acetic acid, in *Plant Growth Substances,* No. III, Mandava, N. B., Ed., ACS Symp. Ser., American Chemical Society, Washington, D.C., 1979, 1.
82. **Gmelin, R., Saarlvirta, M., and Virtanen, A. I.,** Glucobrassicin, a precursor of thiocyanate and ascorbigen in the species of *Brassica oleraces, Suom. Kemistil. B,* 33, 172, 1960 (CA, 55, 8384, 1961).
83. **Gmelin, R. and Virtanen, A. I.,** Neogulucobrassicin, ein zweiter SCN⁻-precursor, *Acta Chim. Scand.,* 16, 1378, 1962.
84. **Prochazaka, Z., Sandava, V., and Sorm, F.,** The structure of ascorbigen. Preliminary communication, *Chem. Listy,* 51, 1197, 1957; *Collect. Czech. Chem. Commun.,* 22, 654, 1957 (CA 51, 13843h, 17920e, 1957).
85. **Kiss, G. and Neukon, H.,** Über die Struktur des Ascorbigens, *Helv. Chim. Acta,* 49, 989, 1966.
86. **Elliot, M. C. and Stowe, B. B.,** A novel sulphonated natural indole, *Phytochemistry,* 9, 1629, 1970.
87. **Yokota, T., Okabayashi, M., Takahashi, N., Shimura, I., and Umeya, K.,** in *Plant Growth Substances 1973: Proc. 8th Int. Conf. Plant Growth Substances,* Hirokawa Publishing, Tokyo, 1974, 28.
88. **MacRae, D. H. and Bonner, J.,** Chemical structure and anti-auxin activity, *Physiol. Plant.,* 6, 485, 1953.
89. **Bonner, J.,** Relation of respiration and growth in the *Avena* coleoptile, *Am. J. Bot.,* 36, 429, 1949.
90. **van Overbeck, J., Blondeau, R., and Horne, V.,** *Trans*-cinnamic acid as an anti-auxin, *Am. J. Bot.,* 38, 589, 1951.
91. **Currier, H. B., Day, B. E., and Craft, A. S.,** Some effects of maleic hydrazide on plants, *Bot. Gaz.,* 112, 272, 1951.
92. **Leopord, A. G. and Klein, W. H.,** Maleic hydrazide as anti-auxin in plants, *Science,* 114, 9, 1951.

93. **Smith, M. S., Wain, R. L., and Wightman, F.**, Antagonistic action of certain stereoisomers on the plant growth-regulating activity of their enantiomorphs, *Nature (London)*, 169, 883, 1952.
94. **Matell, M.**, Stereochemical studies on plant growth substances, *K. Lantbrukshoegsk. Ann.*, 20, 205, 1953.
95. **Åberg, B.**, On optically active plant growth regulators, *K. Lantbrukshoegsk. Ann.*, 20, 243, 1953.
96. **Mitsui, T. and Fujita, N.**, Optical isomerism and physiological activity of plant growth substances (in Japanese), *Kagaku-no-Ryoiki*, 12, 2, 1958.
97. **MacRae, D. H. and Bonner, J.**, Diortho substituted phenoxy-acetic acid as anti-auxins, *Plant Physiol.*, 27, 837, 1952.
98. **Foster, R. J., MacRae, D. H., and Bonner, J.**, Auxin induced growth inhibition a natural consequence of two-point attachment, *Proc. Natl. Acad. Sci. U.S.A.*, 38, 1014, 1951.
99. **Muir, R. M. and Hansch, C.**, The relationship of structure and plant-growth activity of substituted benzoic and phenoxyacetic acids, *Plant Physiol.*, 26, 369, 1951.
100. **Veldstra, H.**, The relation of chemical structure to biological activity in growth substances, *Ann. Rev. Plant Physiol.*, 4, 151, 1953.
101. **Moloney, M. M. and Pilet, P. E.**, Auxin binding in roots: a comparison between maize roots and coleoptiles, *Planta*, 153, 447, 1981.
102. **Frenkel, C. F. and Haard, N.**, Initiation of ripening in Bartlett pear with an antiauxin α-(*p*-chlorophenoxy) isobutyric acid, *Plant Physiol.*, 52, 380, 1973.
103. **Groth, R.**, Der Einflutz des Antiauxins *p*-chlorophenoxyisobuttersäure auf die Bildung meristematischer Zeutren bei der Regeneration isolierter Gewebefragmente von *Riella helicophylla* (Bory et Mont.) Mont., *Planta (Berlin)*, 129, 235, 1976.
104. **Beyer, E. M., Jr.**, Auxin transport: a new synthetic inhibitor, *Plant Physiol.*, 50, 322, 1972.
105. **Beyer, E. M., Jr., Johnson, A. L., and Sweetser, P. B.**, A new class of synthetic auxin transport inhibitors, *Plant Physiol.*, 57, 839, 1976.
106. **Taniguchi, M. and Satomura, Y.**, Isolation of viridicatin from *Penicillium crustosum*, and physiological activity of viridicatin and its 30-carboxymethylene derivative on microorganisms and plants, *Agric. Biol. Chem.*, 34, 506, 1970.
107. **Taniguchi, M. and Satomura, Y.**, Structure and physiological activity of carbostyril compounds, *Agric. Biol. Chem.*, 36, 2169, 1972.
108. **Taniguchi, M., Yamaguchi, M. and Satomura, Y.**, Regulation of IAA-metabolizing enzymes in plants by an anti-auxin, 3-(4-phenylcarbostyriloxy)acetic acid, *Agric. Biol. Chem.*, 37, 819, 1973.
109. **Kimura, Y., Inoue, T., and Tamura, S.**, Isolation of 2-pyruvoylaminobenzamide as an antiauxin from *Colletotrichum lagenarium*, *Agric. Biol. Chem.*, 37, 2213, 1973.
110. **Hayashi, Y. and Sakan, T.**, Nagilactones, plant growth regulators with an antiauxin activity, in *Plant Growth Substances 1973: Proc. 8th Int. Conf. Plant Growth Substances*, Horikawa Publishing, Tokyo, 1974, 525.
111. **Bayer, M. H.**, Gas chromatographic analysis of acidic indole auxins in *Nicotiana*, *Plant Physiol.*, 44, 267, 1969.
112. **Schneider, E. A., Gibson, R. A., and Wightman, F.**, Biosynthesis and metabolism of indole-3-acetic acid, *J. Exp. Bot.*, 23, 152, 1972.
113. **Stowe, B. B., and Thimann, K. V.**, The paper chromatography of indole compounds and some indole-containing auxins of plant tissue, *Arch. Biochem. Biophys.*, 51, 499, 1954.
114. **Weller, L. E., Wittwer, S. H., and Sell, H. M.**, The detection of 3-indoleacetic acid in cauliflower heads. Chromatographic behavior of some indole compounds, *J. Am. Chem. Soc.*, 76, 629, 1954.
115. **Byrd, D. J., Kochen, W., Idzko, D., and Knorr, E.**, The analysis of indolic tryptophan metabolisms in human urine, *J. Chromatogr.*, 94, 85, 1974.
116. **Raj, R. K. and Hutzinger, D.**, Indoles and auxins. VIII. Partition chromatography of naturally occurring indoles on cellulose thin layers and Sephadex columns, *Anal. Biochem.*, 33, 471, 1970.
117. **Yokota, T., Murofushi, N., and Takahashi, N.**, Extraction, purification, and identification, in *Hormonal Regulation of Development I*, MacMillan, J., Ed., Springer Verlag, Berlin, 1980.
118. **Powell, L. E.**, Preparation of indole extracts from plants for gas chromatography and spectrophotofluorometry, *Plant Physiol.*, 39, 836, 1964.
119. **Stowe, B. B. and Schilke, J. F.**, Regulateurs Naturls de la Croissance Végétale, *Colloq. Int. C. N. R. S.*, 123, 409, 1964.
120. **Grunwald, C., Vendrell, M., and Stowe, B. B.**, Evaluation of gas and other chromatographic separation of indolic methyl esters, *Anal. Biochem.*, 20, 484, 1967.
121. **Dedio, W. and Zalik, S.**, Gas chromatography of indole auxins, *Anal. Biochem.*, 16, 36, 1966.
122. **Grunwald, C. and Lockard, R. G.**, Analysis of indole acid derivatives by gas chromatography using liquid phase OV-101, *J. Chromatogr.*, 52, 491, 1970.
123. **Klebe, J. F., Finkbeiner, H., and White, D. M.**, Silylation with bis (trimethylsilyl) acetamide, a highly reactive silyl donor, *J. Am. Chem. Soc.*, 88, 3390, 1966.

124. **Swartz, H. J. and Powell, L. E.**, Determination of indole acetic acid from plant samples by an alkali flame ionization detector, *Physiol. Plant*, 47, 25, 1979.
125. **DeYoe, D. R. and Zaerr, J. B.**, Indole-3-acetic acid in Douglas fir, *Plant Physiol.*, 58, 299, 1976.
126. **Bandurski, R. S. and Schulze, A.**, Concentrations of indole-3-acetic acid and its esters in *Avena* and *Zea*, *Plant Physiol.*, 54, 257, 1974.
127. **Ehmann, A. and Bandurski, R. S.**, Purification of indole-3-acetic acid myoinositol esters on polystyrene-divinyl-benzene resins, *J. Chromatogr.*, 72, 61, 1972.
128. **Brook, J. L., Biggs, R. H., St. John, P. A., and Anthony, D. S.**, Gas chromatography of several indole derivatives, *Anal. Biochem.*, 18, 453, 1967.
129. **Williams, C. M.**, Gas chromatography of urinary aromatic acids, *Anal. Biochem.*, 4, 423, 1962.
130. **Seeley, S. D. and Powell, L. E.**, Gas chromatography and detection of microquantities of gibberellins and indole-acetic acid as their fluorinated derivatives, *Anal. Chem.*, 58, 39, 1974.
131. **Hofinger, M.**, A method for the quantitation of indole auxins in the picogram range by high performance gas chromatography of their *N*-heptafluorobutyryl methyl esters, *Phytochemistry*, 19, 219, 1980.
132. **Stessl, A. and Venis, M. A.**, Determination of submicrogram levels of indole-3-acetic acid: a new, highly specific method, *Anal. Biochem.*, 34, 344, 1970.
133. **Knegt, E. and Bruinsma, J.**, A rapid, sensitive and accurate determination of indolyl-3-acetic acid, *Phytochemistry*, 12, 753, 1973.
134. **Kamisaka, S. and Larsen, P.**, Improvement of the indole-pyrone fluorescence method for quantitative determination of endogenous indole-3-acetic acid in lettuce seedlings, *Plant Cell Physiol*, 18, 595, 1977.
135. **Böttger, M., Engvild, K. C., and Kaiser, P.**, Response of substituted indoleacetic acids in the indolo-α-pyrone fluorescence determination, *Physiol. Plant*, 43, 62, 1978.
136. **Bridges, I. G., Hillman, J. R., and Wilkins, M. B.**, Identification and localisation of auxin in primary roots of *Zea mays* by mass spectrometry, *Planta*, 115, 189, 1973.
137. **Hall, S. M. and Medlow, G. C.**, Identification of IAA in Phloem and root pressure saps of *Ricinus communis* L. by mass spectrometry, *Planta*, 119, 257, 1974.
138. **White, J. C., Medlow, G. C., Hillman, J. R., and Wilkins, M. B.**, Correlative inhibition of lateral bud growth in *Phaseolus vulagaris* L. Isolation of indoleacetic acid from the inhibitory resion, *J. Exp. Bot.*, 26, 419, 1975.
139. **Allen, J. R. F., Rivier, L., and Pilet, P.-E.**, Quantification of indol-3-yl acetic acid in pea and maize seedlings by gas-chromatography-mass spectrometry, *Phytochemistry*, 29, 525, 1982.
140. **Caruso, J. L., Smith, R. G., Smith, L. M., Cheng, T. Y., and Daves, G. D., Jr.**, Determination of indole-3-acetic acid in Douglas fir using deuterated analog and selected ion monitoring, *Plant Physiol.*, 62, 841, 1978.
141. **River, L. and Pilet, P.-E.**, Indolyl-3-acetic acid in cap and apex of maize roots: identification and quantification by mass fragmentography, *Planta*, 120, 107, 1974.
142. **Little, C. H. A., Herald, J. K., and Browning, G.**, Identification and measurement of IAA and abscisic acid in the cambial region of *Picea sitchensis* Carr. by combined gas chromatography-mass spectrometry, *Planta*, 139, 133, 1978.
143. **Sweetser, P. B. and Swartzfager, D. G.**, Indole-3-acetic acid levels of plant tissue as determined by a new high performance liquid chromatographic method, *Plant Physiol.*, 61, 254, 1978.
144. **Crozier, A., Loferski, K., Zaerr, J. B., and Morris, R. O.**, Analysis of picogram quantities of indole-3-acetic acid by high performance liquid chromatography-fluorescence procedures, *Planta*, 150, 366, 1980.
145. **Sandberg, G. and Andersson, A.**, Identification of 3-indole-acetic acid in *Pinus sylvestris* L. by gas chromatography-mass spectrometry, and quantitative analysis by ion-pair reversed-phase liquid chromatography with spectrofluorimetric detection, *J. Chromatogr.*, 205, 125, 1981.
146. **Průkryl, M. W. Z. and Vančura, V.**, High-performance liquid chromatography of plant hormones. I. Separation of plant hormones of the indole type, *J. Chromatogr.*, 191, 129, 1980.
147. **Sjut, V.**, Reverse-phase high-performance liquid chromatography of substituted indole acetic acids, *J. Chromatogr.*, 209, 107, 1981.
148. **Blakesley, D., Hall, J. F., Weston, G. D., and Elliott, M. C.**, Simultaneous analysis of indole-3-acetic acid and detection of 4-chloroindole-3-acetic acid and 5-hydroxyindole-3-acetic acid in plant tissues by high-performance liquid chromatography of their 2-methylindole-α-pyrone derivatives, *J. Chromatogr.*, 258, 155, 1983.
149. **Monsdale, D. M. A.**, Reverse-phase ion-pair high-performance liquid chromatography of the plant hormones indolyl-3-acetic acid and abscisic acid, *J. Chromatogr.*, 209, 489, 1981.
150. **Jensen, E.**, Analysis of indole derivatives by reversed-phase high-performance liquid chromatography, *J. Chromatogr.*, 246, 126, 1982.
151. **Nonhebel, H. M., Crozier, A., and Hillman, J. R.**, Analysis of [^{14}C]indole-3-acetic acid metabolites from the primary roots of *Zea mays* seedling using reverse-phase high-performance liquid chromatography, *Physiol. Plant.*, 57, 129, 1983.

152. **Reinecke, D. M. and Bandurski, R. S.**, Metabolic conversion of ^{14}C-indole-3-acetic acid to ^{14}C-oxindole-3-acetic acid, *Biochem. Biophys. Res. Commun.*, 103, 429, 1981.
153. **Hardin, J. M. and Stutte, C. A.**, Analysis of plant hormones using high-performance liquid chromatography, *J. Chromatogr.*, 208, 124, 1981.
154. **Durley, R. C.. Kaunangara, T., and Simpson, G. M.**, Leaf analysis for abscisic acid, phaseic and 3-indolylacetic acids by high-performance liquid chromatography, *J. Chromatogr.*, 236, 181, 1982.
155. **Hollenberg, S. M., Chappell, T. G., and Purves, W. K.**, High-performance liquid chromatography of amino acid conjugates of indole-3-acetic acid, *J. Agric. Food Chem.*, 29, 1173, 1981.
156. **Pengelly, W. and Mems, F., Jr.**, A specific radioimmunoassay for nanogram quantities of the auxin, indole-3-acetic acid, *Planta*, 136, 173, 1977.
157. **Ranadine, N. S. and Sehon, A. H.**, Antibodies to serotonin, *Can. J. Biochem.*, 45, 1701, 1967.
158. **Fuchs, S., Haimovich, J., and Fuchs, Y.**, Immunological studies of plant hormones. Detection and estimation by immunological assays, *Eur. J. Biochem.*, 18, 384, 1971.
159. **Pengelly, W. L., Bandurski, R. S., and Schulze, A.**, Validation of a radioimmunoassay for indole-3-acetic acid using gas chromatography-selected ion monitoring-mass spectrometry, *Plant Physiol*, 68, 96, 1981.
160. **Ellinger, A.**, Über die Constitution der Indol-gruppe im Eiweiss (Synthese der sogen. Skatolcarbonsäure) und die Quelle der Kynuresaure, *Ber. Dtsch. Chem. Ges.*, 37, 1801, 1904.
161. **King, F. E. and L'Ecuyer, P.**, Synthesis of indoleacetic acids, *J. Chem. Soc.*, 1901, 1934.
162. **Tanaka, Z.**, Synthesis of β-indoleacetic acid. II., *J. Pharm. Soc. Jpn.*, 60, 219, 1940.
163. **Yoshimura, I., Sakamoto, H., and Matsunaga, T.**, Japan 6932,780 (Cl. 16E332), 26 Dec 1969, Appl. 16 Dec 1966.
164. **Majima, R. and Hoshino, T.**, Synthetische Versuche in der Indol-Gruppe. VI. Eine neue Synthese von β-Indolyl-alkylaminen, *Ber. Dtsch. Chem. Ges.*, 58, 2042, 1925.
165. **Candiano, L., Alexandru, S., and Eugenia, C.**, Preparation of 3-indolylacetic acid, Romania, 64,006 (Cl. C07D209/12), 31 Mar 1978, Appl. 81,276 28 Jan 1975.
166. **Snyder, H. R. and Pilgrim, F. J.**, Preparation of 3-indole-acetic acid; new synthesis of tryptophol, *J. Am. Chem. Soc.*, 70, 3770, 1948.
167. **Eliel, E. L. and Murphy, N. J.**, Cyanomethylation of indole with diethylaminoacetonitrile, *J. Am. Chem. Soc.*, 75, 3589, 1953.
168. **Johnson, H. E.**, Synthesis of 3-indolealkanoic acids, U.S. 3,047,585, July 31, 1962, Appl. Oct. 28, 1960.
169. **Young, D. W.**, 3-indolealkanoic acids, U.S. 3,256,296 (Cl. 260-319), June 14, 1966.
170. **Avramenko, V. G., Pershin, G. N., Mushulov, P. I., Makeeva, O. O., Eryshev, B. Ya., Shagalov, L. B., and Suvorov, N. N.**, Indole derivatives. V. Synthesis and tuberculostatic activity of indole-3-alkanoic acids, *Khim. Farm Zh.*, 4, 15, 1970.
171. **Naruto, S. and Yonemitsu, O.**, Photo-induced Friedel-Crafts reactions. IV. Indoleacetic acids, *Chem. Pharm. Bull.*, 20, 2163, 1972.
172. **Suvorov, N. N., Avramenko, V. G., Shagalov, L. B., Eryshev, B. Ya., and Mushulov, P. I.**, 3-Indolylacetic acid, *Sint. Geterotsikl. Soedin.*, 47, 1972.
173. **Nenitzescu, C. D. and Răileanu, D.**, A new synthesis of indole-3-acetic acid, *Acad. Repub. Pop. Rom., Stud. Cercet. Chim.*, 7, 243, 1959.
174. **Nenitzescu, C. D. and Răileanu, D.**, Synthesen des Heteroauxins, des Tryptamins und des Serotonins, *Chem. Ber.*, 91, 1141, 1958.
175. **Ochiai, E. and Dodo, T.**, Synthesis of indole derivatives. VI. A new synthesis of indole-3-acetic acid, *Itsuu Kenkyusho Nempo*, 14, 33, 1965.
176. **Nametkin, S. S., Melńikov, N. N., and Bokarev, K. S.**, Preparation of heteroauxin, *Zh. Priklad. Khim. (Moscow)*, 29, 459, 1956.
177. **Bentley, J. A., Farrar, K. R., Housley, S., Smith, G. F., and Taylor, W. C.**, Some chemical and physical properties of 3-indolepyruvic acid, *J. Biol.*, 64, 44, 1956.
178. **Pichat, L., Audinot, M., and Monnet, J.**, Synthèses de [14]C-2 indole et d'acide β-([14]C-2 indolyl) acetique(hétéroauxine), *Bull. Soc. Chim. Fr.*, 85, 1954.
179. **Robinson, J. R. and Good, N. E.**, Synthesis of indoleacetic acids, *Can. J. Chem.*, 35, 1578, 1957.
180. **Ramamurthy, T. V. and Viswanathan, K. V.**, Studies on the condensation between indole and ketoacids using ^{14}C labeled precursors, *Proc. Nucl. Chem. Radiochem. Symp.*, 1980, 261.
181. **Magnus, B., Bandurski, R. S., and Schulze, A.**, Synthesis of 4,5,6,7, and 2,4,5,6,7 deuterium-labeled indole-3-acetic acid for use in mass spectrometric assays, *Plant Physiol.*, 66, 775, 1980.
182. **Fox, S. W. and Bullock, M. W.**, Synthesis of indole-3-acetic acids and 2-carboxyindole-3-acetic acids with substituents in the benzene ring, *J. Am. Chem. Soc.*, 73, 2756, 1951.
183. **Engvild, K. C.**, Preparation of chlorinated 3-indoleacetic acids, *Acta Chem. Scand.*, B31, 338, 1977.
184. **Yamaki, T. and Nakamura, K.**, Formation of indoleacetic acid in maize embryo, *Sci. Papers Coll. Gen. Ed. Univ. Tokyo*, 2, 81, 1952.

185. **Stowe, B. B. and Thimann, K. V.**, The paper chromatography of indole compounds and some indole-containing auxins of plant tissues, *Arch. Biochem. Biophys.*, 51, 499, 1954.
186. **Winter, A.**, Evidence for the occurrence of indolepyruvic acid invivo, *Arch. Biochem. Biophys.*, 106, 131, 1964.
187. **Rajagopal, R.**, Occurrence of indoleacetaldehyde and tryptophol in the extracts of etiolated shoots of *Pisum* and *Helianthus* seedlings, *Physiol. Plant.*, 20, 655, 1967.
188. **Gibson, R. A., Schneider, E. A., and Wightman, F.**, Biosynthesis and metabolism of indole-3yl-acetic acid, *J. Exp. Bot.*, 23, 381, 1972.
189. **Forest, J. C. and Wightman, F.**, Amino acid metabolism in plants. III. Purification and some properties of a multispecific aminotransferase isolated from bushbean seedlings (*Phaseolus vulgaris* L.), *Can. J. Biochem.*, 50, 813, 1972.
190. **Truelson, T. A.**, Indole-3-pyruvic acid as an intermediate in the conversion of tryptophan to indole-3-acetic acid. II. Distribution of tryptophan transaminase activity in plants, *Physiol. Plant.*, 28, 67, 1973.
191. **Wightman, F. and Cohen, D.**, Intermediary steps in the enzymatic conversion of tryptophan to IAA in cell free systems from mung bean seedlings, in *Biochemistry and Physiology of Plant Growth Substances: Proc. 6th Int. Conf. Plant Growth Substances,* Wightman, F. and Setterfield, G., Eds., Runge Press, Ottawa, 1968, 273.
192. **Rajagopal, R.**, Metabolism of indole-3-acetaldehyde. III. Some characteristics of the aldehyde oxidase of *Avena* coleoptiles, *Physiol. Plant.*, 24, 272, 1971.
193. **Miyata, S., Suzuki, Y., Kamisaka, S., and Masuda, Y.**, Indole-3-acetaldehyde oxidase of pea seedlings, *Physiol. Plant.*, 51, 402, 1981.
194. **Skoog, F.**, A deseeded *avena* test method for small amounts of auxin and auxin precursors, *J. Gen. Physiol.*, 20, 311, 1937.
195. **White, E. P.**, Alkaloids of the Leguminosae, *N. J. J. Sci. Technol. B.*, 25, 137, 1944.
196. **Dannenburg, W. N. and Liverman, J. L.**, Conversion of tryptophan-2-C^{14} to indoleacetic acid by watermelon tissue slices, *Plant Physiol.*, 32, 263, 1957.
197. **Sequeira, L.**, Origin of indoleacetic acid in tobacco plants infected by *Pseudomonas solanacearum, Phytopathology*, 55, 1232, 1965.
198. **Valdovinos, J. G. and Perley, J. E.**, Metabolism of tryptophan in petioles of *Coleus, Plant Physiol.*, 41, 1632, 1966.
199. **Phelps, P. H. and Sequeira, L.**, Auxin biosynthesis in a host-parasite complex, in *Biochemistry and Physiology of Plant Growth Substances: Proc. 6th Int. Conf. Plant Growth Substances,* Wightman, F. and Setterfield, G., Eds., Runge Press, Ottawa, 1968, 197.
200. **Sherwin, J. E.**, A tryptophan decarboxylase from cucumber seedlings, *Plant Cell Physiol.*, 11, 865, 1970.
201. **Gibson, R. A., Barrett, G., and Wightman, F.**, Biosynthesis and metabolism of indole-3yl-acetic acid, *J. Exp. Bot.*, 23, 775, 1972.
202. **Sherwin, J. E. and Purves, W. K.**, Tryptophan as an auxin precursor in cucumber seedlings, *Plant Physiol.*, 44, 1303, 1969.
203. **Schneider, E. A., Gibson, R. A., and Wightman, F.**, Biosynthesis and metabolism of indole-3yl-acetic acid, *J. Exp. Bot.*, 23, 152, 1972.
204. **Rayle, D. L. and Purves, W. K.**, Studies on 3-indole ethanol in higher plants, in *Biochemistry and Physiology of Plant Growth Substances: Proc. 6th Int. Conf. Plant Growth Substances,* Wightman, F. and Setterfield, G., Eds., Runge Press, Ottawa, 1968, 153.
205. **Rayle, D. L. and Purves, W. K.**, Conversion of indole-3-ethanol to indole-3-acetic acid in cucumber seedling shoots, *Plant Physiol.*, 42, 1091, 1967.
206. **Vickery, L. E. and Purves, W. K.**, Isolation of indole-3-ethanol oxidase from cucumber seedlings, *Plant Physiol.*, 49, 716, 1972.
207. **Schraudolf, H.**, Zur Verbreitung von Glucobrassicin und Neoglucobrassicin in höheren Pflanzen, *Experientia*, 21, 520, 1965.
208. **Kindl, H.**, Oxydasen und Oxygenasen in höheren Pflanzen. I. Über das Vorkommen von Indolyl-(3)-acetaldehydoxim und seine Bildung aus L-Tryptophan, *Hoppe-Seyler's Z. Physiol. Chem.*, 349, 519, 1968.
209. **Fawcett, C. H.**, Auxin activity of certain oximes, *Nature (London)*, 204, 1200, 1964.
210. **Bentley, J. A. and Housley, S.**, Studies on plant growth hormones. I. Biological activities of 3-indoleacetaldehyde and 3-indoleacetonitrile, *J. Exp. Bot.*, 3, 393, 1952.
211. **Michel, B. E.**, Growth responses of crucifers to indoleacetic acid and indoleacetonitrile, *Plant Physiol.*, 32, 632, 1957.
212. **Ballin, G.**, Untersuchungen über das Vorkommen und die Wirkungsweise von Indole-3-acetonitril in höheren Pflanzen, *Planta*, 58, 261, 1962.
213. **Thimann, K. V.**, Hydrolysis of indoleacetonitrile in plants, *Arch. Biochem.*, 44, 242, 1953.
214. **Wain, R. L. and Wightman, F.**, *The Chemistry and Mode of Action of Plant Growth Substances,* Butterworths, London, 1956, 234.

215. **Wightman, F.**, Metabolism and biosynthesis of 3-indoleacetic acid and related indole compounds in plants, *Can. J. Bot.*, 40, 689, 1962.
216. **Kutáček, M., Bulgakov, R. and Oplištilová, K.**, O růstové aktivitě glucobrassicinu v biologických testech, *Biol. Plant.*, 8, 252, 1966.
217. **Skytt Andersen, A. and Muir, R. M.**, Auxin activity of glucobrassicin, *Physiol. Plant.*, 19, 1038, 1966.
218. **Kutáček, M., Procházka, Ž., Grünberger, D., and Stajkova, R.**, On the bound form of ascorbic acid. XVII. Biogenesis of ascorbigen, 3-indoleacetonitrile and other indole derivatives in *Brassica oleracea* L. from DL-tryptophan-3-^{14}C, *Collect. Czech. Chem. Commun.*, 27, 1278, 1962.
219. **Kutáček, M. and Kefeli, V. I.**, The present knowledge of indole compounds in plants of the Brassicaceae family, in *Biochemistry and Physiology of Plant Growth Substances: Proc. 6th Int. Conf. Plant Growth Substances*, Wightman, F. and Setterfield, G., Eds., Runge Press, Ottawa, 1968, 127.
220. **Mahadevan, S. and Stowe, B. B.**, An intermediate in the synthesis of glucobrassicins from 3-indoleacetaldoxime by woad leaves, *Plant Physiol.*, 50, 43, 1972.
221. **Mahadevan, S.**, Conversion of 3-indoleacetaldoxime to 3-indole-acetonitrile by plants, *Arch. Biochem. Biophys.*, 100, 557, 1963.
222. **Shukla, P. S. and Mahadevan, S.**, Indoleacetaldoxime hydrolyase (4.2.1.29). III. Further studies on the nature and mode of action of the enzyme, *Arch. Biochem. Biophys.*, 137, 166, 1970.
223. **Thimann, K. V. and Mahadevan, S.**, Nitrilase. I. Occurrence, preparation, and general properties of the enzyme, *Arch. Biochem. Biophys.*, 105, 133, 1964.
224. **Schraudolf, H. and Weber, H.**, IAN-Bildung aus Glucobrassicin: pH-Abhängigkeit und wachstumsphysiologische Bedeutung, *Planta*, 88, 136, 1969.
225. **Tuli, V. and Moyed, H. S.**, Inhibitory oxidation products of indole-3-acetic acid: 3-hydroxymethyloxindole and 3-methlene-oxindole as plant metabolites, *Plant Physiol.*, 42, 425, 1967.
226. **Basu, P. S. and Tuli, V.**, Enzymatic dehydration of 3-hydroxy-methyloxindole, *Plant Physiol.*, 50, 503, 1972.
227. **Moyed, H. S. and Williamson, V.**, 3-Methyleneoxindole reductase of peas, *Plant Physiol.*, 42, 510, 1967.
228. **Moyed, H. S. and Williamson, V.**, Multiple 3-methyleneoxindole reductase of peas. Differential inhibition of synthetic auxins, *J. Biol. Chem.*, 242, 1075, 1967.
229. **Hager, A. and Schmidt, R.**, Auxintransport und Phtotropismus. I. Die lichtbedingte Bildung eines Hemmstoffers für den Transport von Wuchsstoffen in Koleoptilen, *Planta*, 83, 347, 1968.
230. **Tuli, V. and Moyed, H. S.**, The role of 3-methyleneoxindole in auxin action, *J. Biol. Chem.*, 244, 4916, 1969.
231. **Evans, M. L. and Ray, P. M.**, Inactivity of 3-methyleneoxindole as mediator of auxin action on cell elongation, *Plant Physiol.*, 52, 186, 1972.
232. **Marumo, S., Hattori, H., and Yamamoto, A.**, Biological activity of 4-chloroindolyl-3-acetic acid, in *Plant Growth Substances, 1973: Proc. 8th Int. Conf. Plant Growth Substances*, Horikawa Publishing, Tokyo, 1974, 419.
233. **Magnus, V., Iskric, S., and Kveder, S.**, Indole-3-methanol, a metabolite of indole-3-acetic acid in pea seedlings, *Planta*, 97, 116, 1971.
234. **Hinman, R. L. and Lang, J.**, Peroxidase-catalyzed oxidation of indole-3-acetic acid, *Biochemistry*, 4, 144, 1965.
235. **Morris, D. A., Briant, R. E., and Thomson, P. G.**, The transport and metabolism of ^{14}C-labeled indoleacetic acid in intact pea seedlings, *Planta*, 89, 178, 1969.
236. **Davies, P. J.**, The fate of exogenously applied indoleacetic acid in light grown stems, *Physiol. Plant.*, 27, 262, 1972.
237. **Racusen, D.**, Formation of indole-3-aldehyde by indoleacetic acid oxidase, *Arch. Biochem. Biophys.*, 58, 508, 1955.
238. **Morita, Y., Kameda, K., and Mizuno, M.**, Studies on phytoperoxidase. XVI. Aerobic destruction of indole-3-acetic acid catalyzed by crystalline Japanese-radish peroxidase a and c, *Agric. Biol. Chem.*, 26, 442, 1962.
239. **BeMiller, J. N. and Colilla, W.**, Mechanism of corn indole-3-acetic acid oxidase in vitro, *Phytochemistry*, 11, 3393, 1972.
240. **Bayer, M. H.**, Gas chromatographic analysis of acidic indole auxins in *Nicotiana*, *Plant Physiol.*, 44, 267, 1969.
241. **Stowe, B. B., Vendrell, M., and Epstein, E.**, Separation and identification of indoles of maize and woad, in *Biochemistry and Physiology of Plant Growth Substances: Proc. 6th Int. Conf. Plant Growth Substances*, Wightman, F. and Setterfield, G., Eds., Runge Press, Ottawa, 1968, 173.
242. **Andreae, W. A. and Good, N. E.**, The formation of indole-acetylaspartic acid in pea seedlings, *Plant Physiol.*, 30, 380, 1955.
243. **Zenk, M. H.**, 1-(Indole-3-acetyl)-β-D-glucose, a new compound in the metabolism of indole-3-acetic acid in plants, *Nature (London)*, 191, 493, 1961.

244. **Cohen, J. D. and Bandurski, R. S.**, Chemistry and physiology of the bound auxins, *Ann. Rev. Plant Physiol.*, 33, 403, 1982.
245. **Hare, R. C.**, Indolacetic acid oxidase, *Bot. Rev.*, 30, 129, 1964.
246. **Shantz, E. M.**, Chemistry of naturally-occurring growth-regulating substances, *Ann. Rev. Plant Physiol.*, 17, 409, 1966.
247. **Scott, T. K.**, Auxins and roots, *Ann. Rev. Plant Physiol.*, 23, 235, 1972.
248. **Schneider, E. A. and Wightman, F.**, Metabolism of auxin in higher plants, *Ann. Rev. Plant Physiol.*, 25, 487, 1974.
249. **Kinashi, H., Suzuki, Y., Takeuchi, S., and Kawarada, A.**, Possible metabolic intermediates from IAA to β-acid in rice bran, *Agric. Biol. Chem.*, 40, 2465, 1976.
250. **Suzuki, Y., Kinashi, H., Takeuchi, S., and Kawarada, A.**, (+)-5-Hydroxy-dioxindole-3-acetic acid, a synergist from rice bran of auxin-induced ethylene production in plant tissue, *Phytochemistry*, 16, 635, 1977.
251. **Suzuki, Y. and Kawarada, A.**, Products of peroxidase catalyzed oxidation of indolyl-3-acetic acid, *Agric. Biol. Chem.*, 42, 1315, 1978.
252. **Matsushima, H., Fukumi, H., and Arima, K.**, Isolation of zeanic acid, a natural plant growth-regulator from corn steep liquor and its chemical structure, *Agric. Biol. Chem.*, 37, 1865, 1973.
253. **Letham, D. S., Goodwin, R. B., and Higgins, T. J. V.**, *Phytohormones and Related Compounds: A Comprehensive Treatise*, Vols. 1 and 2, Elsevier/North-Holland, Amsterdam, 1978.
254. **Thimann, K. V.**, *Hormone Action in the Whole Life of Plants*, University of Massachusetts Press, Amherst, 1977.
255. **Audus, L. J.**, *Plant Growth Substances*, Vol. 1, *Chemistry and Physiology*, Leonard Hill, London, 1972.
256. **Wightman, F. and Setterfield, G., Eds.**, *Biochemistry and Physiology of Plant Growth Substances: Proc. 6th Int. Conf. Plant Growth Substances*, Runge Press, Ottawa, 1968.
257. **Carr, D. J., Ed.**, *Plant Growth Substances 1970: Proc. 7th Int. Conf. Plant Growth Substances*, Springer-Verlag, Berlin, 1972.
258. **Sumiki, Y., Ed.**, *Plant Growth Substances 1973: Proc. 8th Int. Conf. Plant Growth Substances*, Horikawa Publishing, Tokyo, 1974.
259. **Pilet, P.-E., Ed.**, *Plant Growth Regulation: Proc. 9th Int. Conf. Plant Growth Substances*, Springer-Verlag, Berlin, 1977.
260. **Skoog, F., Ed.**, *Plant Growth Substances 1979: Proc. 10th Int. Conf. Plant Growth Substances*, Springer-Verlag, Berlin, 1980.
261. **Wareing, P. F., Ed.**, *Plant Growth Substances 1982: Proc. 11th Int. Conf. Plant Growth Substances*, Academic Press, London, 1982.

Chapter 3

GIBBERELLINS

Nobutaka Takahashi, Isomaro Yamaguchi, and Hisakazu Yamane

TABLE OF CONTENTS

I.	Discovery of Gibberellins and Their Occurrence in Nature 58	
	A. Fungal Gibberellins (GAs) 58	
	B. Plant GAs 59	
	C. Conjugated GAs in Higher Plants 65	
	D. Content of GAs in Tissues of Higher Plants................................ 65	
II.	Chemistry... 68	
	A. Isolation and Characterization 68	
	1. Extraction and Solvent Fractionation............................ 69	
	2. Countercurrent Distribution 71	
	3. Charcoal Adsorption Chromatography 71	
	4. Silicic Acid Adsorption Chromatography 72	
	5. Partition Chromatography.. 73	
	6. Chemical and Enzymatical Cleavage of Conjugated GAs 76	
	7. Paper and Thin Layer Chromatography 77	
	8. HPLC... 77	
	9. GC, GC/MS, and GC/SIM... 80	
	10. Immunoassay 83	
	11. Isolation Examples of Free GAs 88	
	12. Isolation Examples of Conjugated GAs 90	
	B. Structural Determination... 91	
	1. Free GAs 96	
	a. $GA_{1,3}$ by Degradation Studies 96	
	b. Chemical Evidence and Spectrometric Methods 97	
	c. Identification of New GAs by GC/MS Analysis........... 98	
	2. Conjugated GAs 102	
	3. Physiochemical Properties of GAs 103	
	a. UV Spectra.. 103	
	b. IR Spectra.. 103	
	c. NMR Spectra... 104	
	d. Mass Spectra.. 107	
	C. Synthesis 114	
	1. Degradation Products of GAs.................................... 114	
	2. Partial Synthesis of GAs....................................... 115	
	3. Total Synthesis of GAs 117	
	4. GA Conjugates 119	
III.	Biosynthesis and Metabolism 119	
	A. Biosynthesis of GAs in *Gibberella fujikuroi*............................. 121	
	1. Stage A: Conversion of Mevalonic Acid to *ent*-Kaurene......... 121	
	2. Stage B: Conversion of *ent*-Kaurene to GA_{12}-7-Aldehyde 121	
	3. Stage C-1: Formation of C_{19}-GAs from GA_{12}-7-Aldehyde via C_{20}-GAs 124	

		4.	Stage C-2: Biosynthetic Conversion of GAs 125
	B.	Metabolism of Nonfungal GAs by *G. fujikoroi* 126
	C.	Biosynthesis of GAs in Higher Plants 126
		1.	Stages A and B: Formation of GA_{12}-7-Aldehyde from MVA 128
		2.	Stages C: Conversion of GA_{12}-7-Aldehyde to C_{19}-GAs and Interrelationship of GAs 130

IV.	Physiology 132
	A. Plant Growth Promoting Effect 132
	B. Effect on Flowering 137
	C. Effect on Dormancy Break and Germination 138
	D. Effect on Parthenocarpy 138
	E. Effect on Sex Expression 139
	F. Activation of Hydrogenase Activity 139

Acknowledgments 139

References 139

I. DISCOVERY OF GIBBERELLINS AND THEIR OCCURRENCE IN NATURE

A. Fungal Gibberellins (GAs)

In Japan the disease of rice seedling plants called "bakanaebyo" (foolish seedling disease) was known before the advent of modern science. In this disease the height of rice plants is taller than healthy ones and the color of the leaves becomes yellowish green and they are often withered. The disease has been shown to be caused by the fungal infection of a plant pathogen, *Gibberella fujikoroi* (sometimes, *Fusarium moniliforme* is used).

In 1926, Kurosawa[1] reported a very important finding. He cultured the pathogen in a rice seed extract medium. When rice seedlings were cultured with an aqueous medium supplemented by a culture filtrate of the fungus, disease symptoms were observed that were identical to those of infected plants. This indicated that the pathogen produces a "toxin" which is responsible for the symptoms of the disease.

This research led to studies on the isolation of the active principle from culture filtrates of the plant pathogen by the Japanese agricultural chemists, Yabuta and Sumiki, at the University of Tokyo. In 1938, they succeeded in isolating two active components which were named gibberellin A and gibberellin B.[2] This information was reported in a short communication in the *Journal of the Agricultural Chemical Society of Japan* (in Japanese) (Figure 1). In Japan, research studies on GAs were discontinued during World War II. After the war, the research was resumed and the early reports on GAs by Japanese scientists attracted many researchers throughout the world, e.g., the Northern Regional Research Laboratory (NRRL) in the U.S. and the Imperial Chemical Industries, Ltd. (ICI) in Great Britain. In 1954, the ICI group[3] reported the isolation of an active principle from *Gibberella fujikoroi* that had plant growth-promoting activity. The biological activity was similar to that of the GAs isolated by Japanese workers, but its chemical nature was clearly different. The British researchers[3] named their substance gibberellic acid.

In 1955, the NRRL group[4] reported the successful separation of their gibberellin A into two components — one called gibberellin X, the other gibberellin A, which was thought to be identical with the Japanese gibberellin A. In 1955, the University of Tokyo group[5]

稲馬鹿苗病菌の生化學 (續報)
植物を徒長せしむる作用ある物質 Gibberellin の結晶に就いて

農學博士 藪田貞治郎, 農學博士 住木諭介

(東京帝國大學農學部)
昭和13年11月26日受理

稲馬鹿苗病菌 Gibberella fujikuroi をグリセリン, KH_2PO_4, NH_4Cl より成る培養液に培養し培養液を活性炭にて處理して有效成分を吸着せしめ之をメタノール, アンモニヤにて溶出し, 更に鹽基性醋酸鉛等にて處理して得たる粗有效成分の粉末 (Crude Gibberellin と命名, 分解點約 60°) を醋酸エステルリグロイン又はアルコールリグロインの混合溶劑にて處理し, 2 種の有效成分を結晶狀に分離するを得たり. 之を Gibberellin A, Gibberellin B と命名す. Gibberellin A は長柱狀結晶, 融點 194~6°, 稲苗地上部を徒長せしむるのみならず根の發育を促進せしむる作用あり. C% = 75.16, H% = 7.60. M.W = 250 (ラスト法). Gibberellin B は短柱狀結晶, 分解點 245~6°. 稲苗地上部を徒長せしむる作用甚だ強く根の發育は阻害す. C% = 65.05, H% = 7.12, M.W = 386 (ラスト法), 357 (滴定: モノカルボキシリツク酸として), $[\alpha]_D$ = +36.13° (メタノール, C = 4.290). (昭和 13 年 9 月東京支部講演會にて發表)

FIGURE 1. The first paper describing the isolation of GA in crystals by Yabuta and Sumiki.

reinvestigated the purity of their gibberellin A, and reported the separation of three gibberellins, A_1, A_2, and A_3, as their methyl esters, and free gibberellins A_1 and A_2.

Direct comparison among various GAs isolated by the three groups disclosed that gibberellin A_1 (University of Tokyo group) and gibberellin A (NRRL group) were identical, and gibberellic acid (ICI group), gibberellin X (NRRL group), and gibberellin A_3 (the University of Tokyo group) were identical; gibberellin A_2 was a new GA.

In 1957, the University of Tokyo group[6] reported the isolation of a new GA, GA_4 (abbreviation of GA_n for gibberellin A_n is used hereafter).

From 1957 to 1968, the ICI group[7-17] isolated nine new GAs, GA_7 and GA_{9-16}, from the culture filtrates of *G. fujikuroi*. After 1968, MacMillan and his collaborators at Bristol University (the Bristol University group) characterized additional GAs as new: $GA_{24,25,36,41,42,47}$[18-21,24] and the University of Tokyo group characterized GA_{40}[22,23] and GA_{55-57}[26,27] as additional metabolites of *G. fujikuroi*. GA_{20}[25] and GA_{37}[21], originally characterized from higher plants, were also identified from the fungus; actually 26 GAs have been identified from the fungus: $GA_{1-4,7,9-16,20,24,25,36,37,40-42,47,54-57}$ (see Figure 2 for their structures).

Schreiber et al.[28] reported the occurrence of 3-O-acetyl GA_3 in the metabolite of *G. fujikuroi*. It has been believed for a long time that among fungi only *G. fujikuroi* produces GAs, in spite of ambiguous papers reporting the production of GA-like substances by microorganisms other than *G. fujikuroi*.[29-32] In 1979, Rademacher and Graebe[33] reported the isolation of GA_4 from the metabolite of *Sphaceloma manihoticola*, a plant pathogen of cassava. This is the first paper reporting the definitive evidence for the occurrence of GAs in microorganisms other than *G. fujikuroi*. Quite recently, Kawanabe et al.[34] disclosed the occurrence of GA_3 in very low levels in *Neurospora crassa*. This information may suggest a wide distribution of GAs in metabolites of fungal Ascomycetes. A list of the fungal GAs and their origin is given in Table 1.

B. Plant GAs

In 1951, Mitchell et al.[35] reported the occurrence of growth-stimulating substances, later interpreted as GA-like substances, from the extract of immature seeds of bean. In 1956,

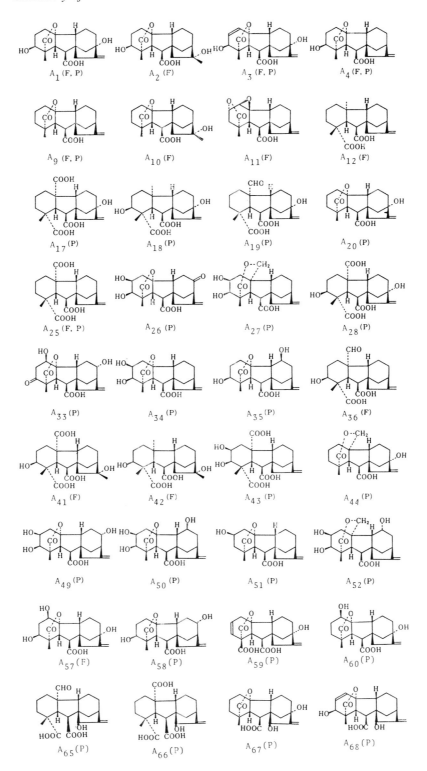

FIGURE 2. Structures of free GAs. P and F in parentheses show origins, plant, and fungus.

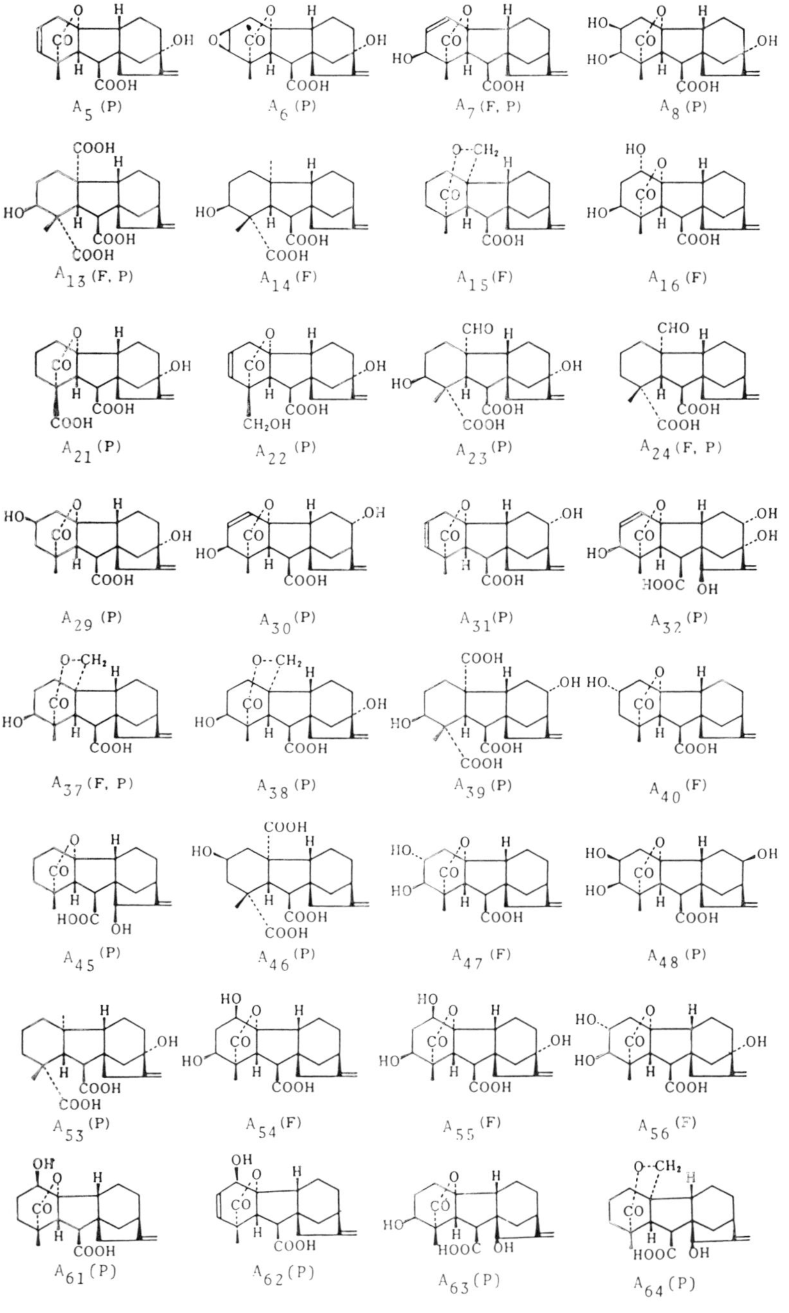

FIGURE 2. Continued.

Radley[36] and West and Phinney[37] published evidence showing the occurrence of GA-like substances in the extracts of higher plants.

In 1958, MacMillan and Suter[38] (in England) reported the isolation and identification of GA_1 from immature seeds of scarlet runner bean, *Phaseolus multiflorus,* (also called *P. coccineus*). In 1959, they also isolated and reported on the structures of three new GAs, $GA_{5,6,8}$, from *P. coccineus*.[39-41] West and Phinney[42] reported the isolation of bean factor-I and -II from immature seeds of *Phaseolus vulgaris* and later they were identified as GA_1

Table 1
ORIGIN OF FUNGAL GAs

GAs	Research group	Ref.,
GAs from Gibberella fujikuroi		
A_1	Univ. of Tokyo	2,5
(= A)	NRRL	4
A_2	Univ. of Tokyo	5
A_3 (Gibberellic acid)	Univ. of Tokyo	5
	ICI	3
(= X)	NRRL	4
3-O-acetyl GA_3	Schreiber et al.	28
A_4	Univ. of Tokyo	6
A_7	ICI	7,9
A_9	ICI	8,9
A_{10}	ICI	10,11
A_{11}	ICI	10,12
A_{12}	ICI	10,13
A_{13}	ICI	14
A_{14}	ICI	15
A_{15}	ICI	16
A_{20}	McInnes et al.	25
A_{24}	Bristol Univ.	18,19
A_{25}	Bristol Univ.	19
A_{36}	Bristol Univ.	20,21
A_{37}	Bristol Univ.	21
A_{40}	Univ. of Tokyo.	22,23
A_{41}	Bristol Univ.	21
A_{42}	Bristol Univ.	21
A_{47}	Bristol Univ.	24
A_{54}	Univ. of Tokyo	26
A_{55}	Univ. of Tokyo	26
A_{56}	Univ. of Tokyo	26
A_{57}	Univ. of Tokyo	27
GAs from microorganisms other than *G. fujikuroi*		
Ephalceoma manthoticola		
GA_4	Rademacher & Graebe	33
Neurospora crassa		
GA_3	Kawanabe et al.	34

and GA_5, respectively. That same year, at the University of Tokyo, Kawarada and Sumiki[43] reported the identification of GA_1 in water sprouts of mandarin orange (*Citrus unshiu*).

These pioneering studies stimulated research by many plant physiologists throughout the world, so that much data appeared in the literature showing the occurrence of GA-like substances in higher plants. The evidence was based primarily on the bioassay of plant extracts eluted from paper chromatography. Thus, plant physiologists obtained evidence for the widespread distribution of GA-like substances in the plant kingdom. They were soon after thought to play an important role in the regulation of growth and differentiation of higher plants. Isolation of additional GAs unique to plants occurred in 1967 when the University of Tokyo group[44,45] succeeded in identifying a new gibberellin, GA_{19}, from young shoots of bamboo (*Phyllostachys edulis*). The yield was low — 14 mg from the water extract of 44,000 kg of bamboo shoots. Koshimizu et al.[46,47] isolated 50 mg of GA_{18} from 142 kg immature pods of yellow lupin (*Lupinus luteus*) at Kyoto University. In 1967, MacMillan[48]

Table 2
CHARACTERIZATION OF NEW GAs FROM HIGHER PLANTS AFTER 1965

Origin	As	Ref.
University of Tokyo group		
Phyllostachys edulis	A_{19}	44,45
Pharbitis nil	A_{20}	49
	$A_{26,27}$	53
	$A_{29}{}^a$	55
Canavalia gladiata	$A_{21,22}$	50,51
Calonyction aculeatum	$A_{30,31,33,34}$	56
Prunus persica	A_{32}	57,58
Cytisus scoparius	A_{35}	59
Phaseolus vulgaris	$A_{37,38}{}^a$	60,61
Kyoto University group		
Lupinus luteus	A_{18}	47
	A_{23}	52
	A_{28}	54
Cucurbita pepo	$A_{39,48,49}$	62,63
Lagenaria leucantha	$A_{50,52}$	68
Bristol University group, collaborating groups		
Phaseolus coccineus	A_{17}	48
Marah macrocarpus	A_{43}	64
	A_{46}	64,67
Pisum sativum	A_{44}	65
	A_{51}	69
Pyrus communis	A_{45}	66
	$A_{62,63}$	74,75
Vicia faba	A_{53}	70
Tricicum aestivum	$A_{54,55}$	71
	$A_{60,61}$	74
Cucurbita maxima	A_{58}	72
Helianthus annus	$A_{64,65,66,67}$	75
Malus sylvestris	A_{68}	75,325

^a Isolated as glucosyl derivatives.

identified GA_{17} in the immature seeds of scarlet runner bean (*P. coccineus*) by combined gas chromatography/mass spectrometry (GC/MS) at Bristol University.

After 1968, many new GAs were isolated from various higher plants by three research groups in particular: the Universities of Tokyo, Kyoto, and Bristol[38,75] (see Table 2).

Some of the new GAs from plants, $GA_{54,60-66}$, were identified without isolation, i.e., by comparison of their GC/MS and GC/MS in selected ion monitoring mode (GC/SIM) with those of authentic GA analogues prepared chemically and biologically from structurally known GAs. Recent progress in analytical methods such as GC/MS and GC/SIM for the identification of GAs in ultramicro quantity has enabled us to conclusively identify many GAs in higher plants. At this time, 56 GAs are recognized to occur in higher plants: $GA_{3-9,13,15-35,37-39,43-46,48-55,58-68}$.

The evidence for the occurrence of GAs in lower plants such as algae, Bryophyta, and Pteridophyta is rare. Marine algae, such as *Fucus* and *Tetraselmis* species have been shown to contain GA-like substances.[76-78] The University of Tokyo group[79] identified GA_{36} in the sporophytes of *Psilotum rudum*. Endo et al.[80] succeeded in the isolation of the antheridial-inducing substance (antheridiogen), An, from prothalli (gametophytes) of *Anemia phylltidis* (Schizaiceae fern). They established its structure[81] (see Figure 3), which is closely related to GAs. The Tokyo group[82] also isolated an antheridiogen from *Lygodium japonicum* and

Gibberellin Glucosyl Ether

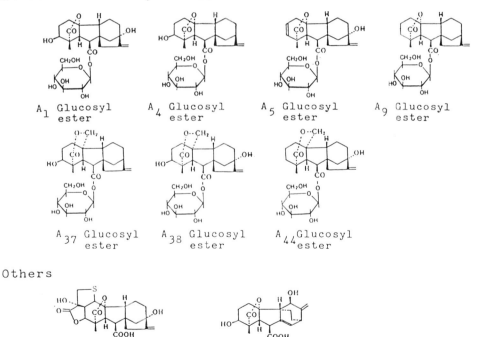

FIGURE 3. Structures of conjugated GAs.

identified it as the methyl ester of GA_9. It is the first evidence for the natural occurrence of a GA methyl ester in plants.

There are currently 68 GAs isolated and characterized from fungi and higher plants, which are called free GAs. Their structures are summarized in Figure 2. In this figure, GAs of

C. Conjugated GAs in Higher Plants

Except for GA_{28} and GA_{32}, free GAs can be extracted with ethyl acetate from acidic aqueous solution. However, studies by several investigators[83-86] showed the occurrence of GA-like substances that were more polar than free GAs. The chemical nature of these polar GA-like substances was clarified independently by two research groups. Schreiber et al.[87,88] reported the isolation of a polar GA conjugate from the mature seeds of scarlet runner bean and characterized it as 2-O-β-glucosyl GA_8. In 1968, the Tokyo group[89] reported the isolation of a GA conjugate from the immature seeds of morning glory (*Pharbitis nil*), and identified it as 3-O-β-glucosyl GA_3. They also identified four glucosyl ether conjugates, 2-O-β-glucosyl ethers of GA_8, GA_{26}, GA_{27}, and GA_{29}.[55] They also reported on 11-O-β-glucosyl GA_{35} from the immature seeds of yellow broom (*Cytisus scoparius*),[59] 3-O-β-glucosyl GA_1 from the immature seeds of *Dolichos lablab*,[90] and 2-O-β-glucosyl GA_8 from the immature seeds of *Phaseolus vulgaris*.[60]

The Tokyo group identified another type of glucosyl conjugate. They found polar neutral GA-like substances in the extracts of mature seeds of bean (*P. vulgaris* cv. Kentucky Wonder), and succeeded in the isolation of four neutral GA conjugates characterized as β-D-glucosyl esters of $GA_{1,4,37,38}$.[60,61] They also characterized glucosyl esters of $GA_{5,44}$ in the immature seeds of *Pharbitis purpurea*.[91] Lorenzi et al.[92] reported the occurrence of GA_9 glucosyl ester in the needles of *Picea sitchensis*. Thus, two types of glucosyl conjugates of GA glucosyl ethers (GAs-GEt) and GA glucosyl esters (GAs-GEs) were shown to be present in higher plant.

A third type of GA conjugate, gibberethione (formerly designated as pharbitic acid), was isolated from the immature seeds of the morning glory in 1974.[93] It was shown to be a conjugate between 3-keto GA_3 and mercaptopyruvic acid. The structures of all known GA conjugates are shown in Figure 3.

The distribution of plant GAs (including conjugates), whose occurrences have been confirmed by isolation, GC/MS, and GC/SIM, etc., is summarized in Table 3.

D. Content of GAs in Tissues of Higher Plants

The amounts of GAs in higher plants vary depending on the kinds of plant tissue and their stages of growth. Immature seeds of Leguminosae plants have been known to be rich sources of endogenous GAs. For example, the immature seeds of *P. coccineus* were shown to contain GA_1 (18.0 mg/kg), GA_5 (1.5), GA_6 (2.5), GA_8 (30), GA_{17} (2.0), GA_{19} (0.5), and GA_{20} (0.5).[96] The immature seeds of *P. vulgaris* were shown to contain GA_1 (1.2 mg/kg), GA_8 (0.7), GA_{38} (0.4), and $GA_{4,5,6,37,44}$ (not quantified),[60] as well as the GA conjugates, GA_8-GEt (0.8 mg/kg) and GA_{38}-GEs (not quantified). While these mature seeds contained only low levels of the free GAs, GA_1 (0.1) and GA_8 (0.01), glucosyl conjugates were present at relatively high levels — GA_8-GEt (2) and GA_1-GEs (0.4), GA_4–GEs (0.1), GA_{37}-GEs (0.1), and GA_{38}-GEs (0.5 mg/kg).[61]

Information on the occurrence and content of endogenous GAs in tissues other than seeds is rare. It can be said generally that vegetative tissues contain very low levels of GAs. The yield of GA_{19} from young shoots of bamboo was reported to be 0.3 μg/kg,[45] although the actual level must be much higher, considering loss in the isolation process and that the GA_{19} level in leaves and shoots of rice plants at the tillering stage is approximately 10 μg/kg.[118] One exception to the low levels is in the water sprout of *Citrus unshiu*, where GA_1 was obtained in a yield of 0.5 mg/kg.[43] Contents of $GA_{3,4,7}$ in the pollen of *Pinces attenuata* have been determined by bioassay.[132] In dormant pollen, GA_3 was present at levels of 10 μg/kg, and GA_4 and GA_7 at levels of 40 μg/kg (GA_3 equivalent) in total.[132] In germinating

Table 3
DISTRIBUTION OF GAs IN THE PLANT KINGDOM

Leguminosae	GAs	Ref.
Phaseolus vulgaris	A_1	42,60,94—96
	A_4	94
	A_5	42,94,96
	A_6	94,96
	A_8	61,94,96
	$A_{17,20,29}$	95
	A_{38}	94
	A_{44}	95
	$A_{1,4,37,38}$-GEs	60,61
	A_8-GEt	61
Phaseolus coccineus	A_1	38,40,96—98
(formerly designated	A_3	96
as *P. multiflorus*)	A_4	97,98
	A_5	39,40,96,98
	A_6	41,96
	$A_{17,19}$	96
	A_{20}	96,98
	$A_{34,48,44}$	99
	A_8-GEt	87,88
Lupinus luteus	A_{18}	46,47
	$A_{19,23}$	52
	A_{28}	54
Wistaria floribunda	$A_{18,23}$	100
Dolichos lablab	$A_{1,8,44}$	101
	A_1-GEt	90
Canavalia gladiata	$A_{21,22}$	50,51
	A_{59}	73
Pisum sativum	$A_{9,17}$	65
	A_{20}	65,102,103
	A_{29}	65,102
	$A_{38,44}$	65
	A_{51}	69
Cytisus scoparius	A_{35}	59
	A_{35}-GEt	59
Vicia faba	A_{53}	70
Vigna unguiculata	$A_{4,6,8,17,20}$	104
Cassia fistula	A_3	105
Leucaena leucocephala	$A_{1,8,17,19,20,23,34,53}$	106
Convolvulaceae		
Pharbitis nil	$A_{3,5}$	49,53
	A_8	53
	$A_{17,19}$	107
	A_{20}	49
	$A_{26,27}$	53
	$A_{29,53}$	107
	$A_{3,8,26,27,29}$-GEt	55,89
Pharbitis purpurea	$A_{3,5,8,17,19,20}$	170
	$A_{26,27,29,34,44,53}$	170
	$A_{5,44}$-GEs	91
Calonyction aculeatum	$A_{8,17,19,27,29,30,31,33,34}$	56
Calystegia soldanella	$A_{3,4,7}$	108
Ipomea batatus	$A_{1,3,19,20,23}$	108
Ipomea pes-caprae	A_{19}	108
Quamoclit pennata	A_3	108
	A_3-GEt	109

Table 3 (continued)
DISTRIBUTION OF GAs IN THE PLANT KINGDOM

Leguminosae	GAs	Ref.
Cucurbitaceae		
Marah macrocarpus (formerly designated as *Echinocystis macrocarpa*)	$A_{4,7,24,25,43}$	64
	A_{46}	64,67
Cucumis sativus	$A_{1,3,4,7}$	110
Cucumis melo	$A_{1,3,5}$	110
Cucurbita pepo	$A_{39,48,49}$	62,63
Cucurbita maxima	A_{43}	111
	A_{58}	72
Lagenaria leucantha	$A_{50,52}$	68
Rosaceae		
Pyrus communis	A_3	112
	$A_{17,25,45}$	66
	$A_{62,63}$	75
Pyrus malus	$A_{4,7,9,62}$	74,114
	A_{63}	75
Prunus persica	$A_{5,32}$	57,58
Prunus armenica	A_{32}	115
Prunus cerasus	A_{32}	116
Malus sylvestris	$A_{4,7,9,12,15,17,20,44,53,67,68}$	325
Gramineae		
Phyllostachys edulis	A_{19}	45
Oryza sativa	A_1	117—119
	A_4	118,119
	A_9	119
	A_{12}	113
	A_{19}	117—119
	$A_{20,24,29,34,44,51,53}$	119
Triticum aestivum	$A_{1,3}$	121
	$A_{4,9}$	120
	$A_{15,17,19,20,24,25}$	123
	$A_{44,54,55}$	114
	$A_{60,61,62}$	74
Secale cereale	$A_{1,3}$	121
	$A_{8,16,24}$	124
Hordeum vulgare	A_1	125
	A_3	125,126
	A_7	126
Avena sativa	A_3	127
Zea mays	$A_{17,19,20,44,53}$	128
Malvaceae		
Altheae rosea	$A_{1,3,8,9}$	129
	A_8-GEt	130
Gossypium hirsutum	$A_{3,4,7,13}$	131
Pinaceae		
Pinus attenata	$A_{3,4,7}$	132
Picea sitchensis	isoA_9	133
	A_9-GEs	92
Rutaceae		
Citrus unshiu	A_1	43
Citrus limon	A_1	134
Citrus sinensis	A_1	134

Table 3 (continued)
DISTRIBUTION OF GAs IN THE PLANT KINGDOM

Leguminosae	GAs	Ref.
Umbelliferae		
Daucus carota	$A_{1,4,7}$	135,136
	$\Delta^{1(10)}A_1$	135,136
Pimpinella anisum	$A_{1,4,7}$	135,136
	$\Delta^{1(10)}A_1$	135,136
Compositae		
Enhydra fluctuans	$A_{9,13}$	137
Helianthus annus	$A_{64,65,66,67}$	75
Others		
Spinacia oleracea	$A_{17,19,20,44,53}$	138
Bryophyllum diageremontianum	A_{20}	139
Corylus avellana	$A_{1,9}$	140
Humulus lupulus	A_{19}	141
Nicotiana tabacum	$A_{1,9,19,20}$	142
Sonneralitia aplata	$A_{1,3}$	143
	Tetrahydro A_3	144
Rhizophora mucranata	$A_{3,5,9}$	143
Bruguiera gymnorlza	A_3	143
Pteridophyte		
Lygodium japonicum	A_9-Me	82
	A_{20}-Me	122
Psilotum nudum	A_{36}	79

pollen, (15 hr after germination) the content of GA_3 has been reported at a level of 150 μg/kg, while GA_4 and GA_7 have been contained in the same level as dormant pollens.[132]

On the basis of GC/MS, cultured cells of tobacco crown gall have been shown to contain very low levels of $GA_{1,9,19,20}$ while much higher levels of similar GAs have been identified from intact plants of tobacco.[142]

II. CHEMISTRY

A. Isolation and Characterization

Currently, 68 free and 16 conjugated GAs are known. The isolation and characterization of each GA reflects the development of technology in organic chemistry of natural products. Originally, large quantities of GAs were required for structural studies. Isolation procedures involved extremely large amounts of materials and extensive purification steps. Several GA bioassay systems have been used for the detection of active fractions from chromatographic purification. Recent development of analytical instruments as well as the accumulation of spectroscopic data on GAs greatly diminish the amounts of material required for structure determination, so that only a few hundred micrograms are sometimes sufficient for structural determination.

The recent development and improvement of chromatographic materials and instrumentation provide high resolution and reproducibility and enable the isolation of minute amounts of GAs. In this chapter, both large and small scale isolation and purification procedures are described. Chromatographic characteristics of GAs and their derivatives are also presented. The theoretical aspects of the methods used for the purification procedures of GAs are discussed elsewhere.[145,146]

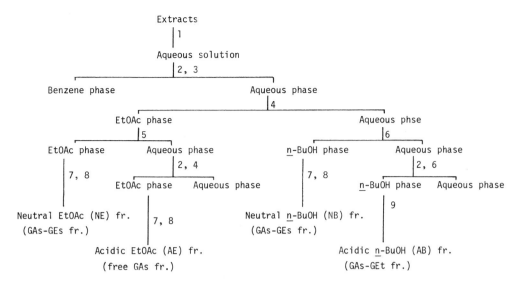

FIGURE 4. General solvent fractionation procedure: (1) evaporation of organic solvent *in vacuo*, (2) pH adjustment to 2.5 to 3.0, (3) extraction with benzene ($\times$ 3), (4) extraction with ethyl acetate ($\times$ 3), (5) extraction with aqueous NaHCO$_3$, (6) extraction with *n*-BuOH, (7) dried over anhydrous Na$_2$SO$_4$, (8) concentration *in vacuo*, (9) pH adjustment to 6.5 and concentration *in vacuo*.

1. Extraction and Solvent Fractionation

Plant materials are usually extracted by homogenization in methanol or 80% aqueous acetone, followed by filtration. This extraction procedure is repeated 3 to 4 times. The organic solvent in the filtrate is removed *in vacuo* and the residual aqueous phase is submitted to solvent fractionation to give an acidic ethyl acetate-soluble (AE) fraction, a neutral ethyl acetate-soluble (NE) fraction, an acidic *n*-butanol soluble (AB) fraction, and a neutral *n*-butanol soluble (NB) fraction (Figure 4).

Most of the free GAs partition into the AE fraction. However, GA_9, GA_{12}, and GA_{25} sometimes appear in the NE fraction; GA_{28} is not completely extracted with ethyl acetate,[54] and GA_{32} is almost nonextractable with ethyl acetate, but extractable with *n*-butanol and partitions into the AB fraction.[57]

All free GAs are unstable in acid (Figure 5);[21,59,147,148] GAs such as GA_3 with a C-1 to C-2 double bond and a 3β-hydroxyl group, and GAs such as GA_1 carrying a 3β-hydroxyl group and a lactone linkage from C-4 to C-10 (γ-lactone) or to C-20 (δ-lactone) are unstable in base. Under basic conditions, isomerization of 3β-hydroxyl group occurs.[148,149] This process is shown in Figure 6; under acidic conditions (pH 2.5 to 3) or basic conditions (pH 8 to 8.5), GAs are usually stable for relatively short periods of time. Fukui and his co-workers[62] avoided the use of base in order to minimize the ismerization of GAs having a 3β-hydroxyl group. They extracted an aqueous solution with ethyl acetate at pH 3. The ethyl acetate extract is concentrated and immediately submitted to charcoal adsorption chromatography for the isolation of $GA_{39,48,49}$.[62]

Conjugated GAs show different behavior than the ethyl acetate-soluble free GAs in a solvent fractionation procedure. Of the conjugated GAs, GAs-GEt cannot be extracted with ethyl acetate but can be extracted with *n*-butanol, and is partitioned into the AB fraction.[53,59,109]

The solvent fractionation procedure used to obtain a GAs-GEs is not yet established. Some of GAs-GEs, such as GA_4-GEs, GA_9-GEs, and GA_{37}-GEs can be extracted with ethyl acetate; GA_1-GEs and GA_{38}-GEs are only partially extracted with this solvent. Complete extraction occurs with the use of *n*-butanol. Using extracts from *Phaseolus vulgaris* seed, Hiraga et al.[61] isolated GA_1-GEs, GA_4-GEs, GA_{37}-GEs, and GA_{38}-GEs from the NE fraction

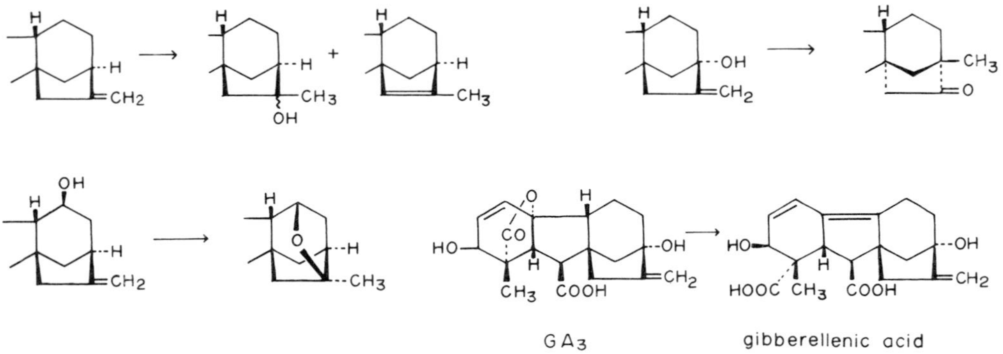

FIGURE 5. Structural change of GAs under acidic condition. The change of GA_3 to gibberellenic acid proceeds even in boiling water.

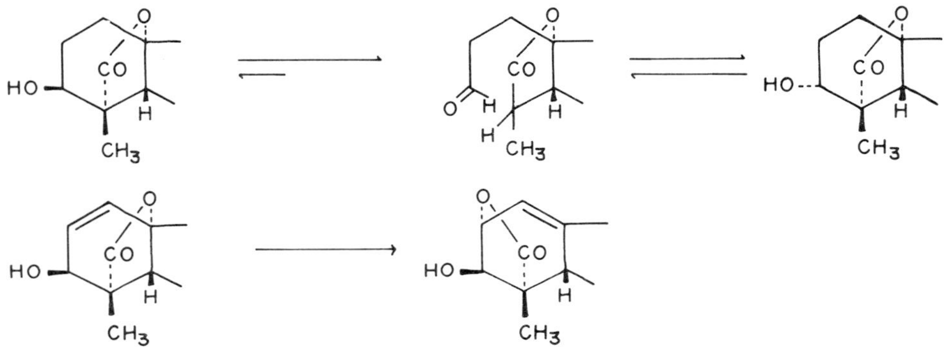

FIGURE 6. Isomerization of GAs under basic condition.

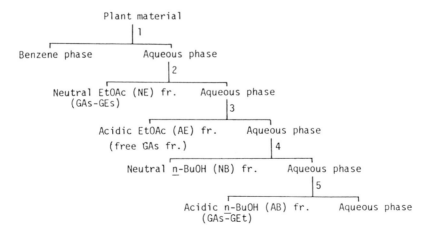

FIGURE 7. Solvent fractionation used in the isolation of GA_1-GEs, GA_4-GEs, GA_{37}-GEs, and GA_{38}-GEs by Hiraga et al.[60] (1) Extraction with benzene at pH 7. (2) Extraction with EtOAc at pH 7. (3) Extraction with EtOAc at pH 2.5. (4) Extraction with n-BuOH at pH 7. (5) Extraction with n-BuOH at pH 2.5.

and GA_1-GEs and GA_{38}-GEs from the NB fraction (Figure 7). In contrast Yamaguchi et al.[91] used the procedure shown in Figure 8 to isolate GA_5-GEs and GA_{44}-GEs from the NB

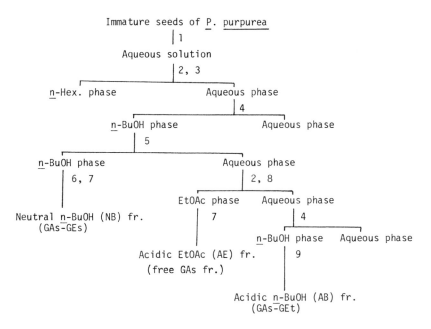

FIGURE 8. Fractionation procedure to obtain an NB fraction by Yamaguchi et al.[91] (1) Extraction with 80% acetone and removal of acetone *in vacuo*. (2) Adjustment of pH to 3. (3) Extraction with *n*-hexane. (4) Extraction with *n*-BuOH. (5) Extraction with aqueous NaHCO$_3$. (6) Washing with sat. saline. (7) Concentration *in vacuo* after dried over anhydrous Na$_2$SO$_4$. (8) Extraction with EtOAc. (9) Concentration *in vacuo* at pH 6.5.

fraction of *Pharbitis purpurea* seeds. Fractions containing GAs-GEs should not be stored in a basic medium, since their aglycones are readily released.

2. Countercurrent Distribution

In the large-scale purification of an AE fraction containing free GAs, the fraction is sometimes submitted to countercurrent distribution conducted between ethyl acetate as the upper phase and a phosphate buffer (1 to 1.5 M, pH 5.4 to 5.5) as a lower phase. In this solvent system, the number of hydroxyl groups in a GA molecule gives a larger effect than the number of carboxyl groups to partition coefficient. After distribution, the lower phase of each fraction is acidified to pH 2.5 to 3, extracted with ethyl acetate 3 to 4 times, combined with the corresponding upper phase, and dried over anhydrous sodium sulfate. The biological activity of each fraction is determined and the active fractions combined are submitted to further purification. Countercurrent distribution is often omitted when the amounts of the AE fraction are low. The distribution coefficients of GAs between an ethyl acetate and an aqueous phase have been examined by Durley and Pharis[150] and are shown in Table 4.

Prior to charcoal chromatography, conjugated GAs and GA$_{32}$, a very polar GA, were purified by Yokota et al.,[53] and Yamaguchi et al.,[57] and Yamane et al.[59] using a countercurrent distribution between 20% *n*-butanol in ethyl acetate and distilled water.

3. Charcoal Adsorption Chromatography

Following countercurrent distribution, charcoal chromatography is often used with or without Celite®. The amount of charcoal required is about 10 to 20 times the weight of the material to be purified. Charcoal is suspended in distilled water and degassed under reduced pressure prior to its packing into a column with distilled water. A sample for isolation of free GAs is suspended in a small amount of aqueous 5 to 10% acetone or methanol with the help of ultra-sonic treatment and charged onto the column. Mixtures of acetone-water

Table 4
PARTITION COEFFICIENTS (Kd = C AQ./C ORG.) OF GAs AND ent-KAURENOIC ACID BETWEEN ETHYL ACETATE AND PHOSPHATE BUFFER[150]

GAs	No. of functional groups OH	No. of functional groups COOH	pH 8.0	pH 6.5	pH 5.0	pH 3.5	pH 2.5
A_1	2	1	∞	∞	1.2	0.17	0.11
A_2	2	1	∞	7.9	0.97	0.19	0.15
A_3	2	1	∞	∞	1.2	0.21	0.17
A_4	1	1	2.2	0.29	0.05	0	0
A_5	1	1	∞	4.8	0.19	0	0
A_6	1	1	∞	5.4	0.49	0.05	0
A_7	1	1	3.2	0.56	0.10	0	0
A_8	3	1	∞	∞	4.9	0.64	0.45
A_9	0	1	0.34	0.06	0	0	0
A_{10}	1	1	11.3	1.6	0.33	0	0
A_{12}	0	2	0.56	0.04	0	0	0
A_{13}	1	3	∞	7.1	0.06	0	0
A_{14}	1	2	∞	0.41	0	0	0
A_{16}	2	1	∞	3.2	0.16	0	0
A_{17}	1	3	∞	∞	0.50	0.04	0
A_{18}	2	2	∞	∞	0.42	0	0
A_{19}	1	2	∞	4.6	0.81	0.10	0
A_{20}	1	1	∞	2.1	0.09	0	0
A_{21}	1	2	∞	∞	9.1	0.89	0.08
A_{22}	2	1	∞	15.1	1.4	0.51	0.19
A_{23}	2	2	∞	∞	19.4	1.0	0.17
A_{24}	0	2	∞	0.83	0	0	0
A_{25}	0	3	13.1	0.66	0	0	0
A_{26}	2	1	∞	∞	3.2	0.44	0.21
A_{27}	2	1	∞	1.6	0.18	0.05	0
A_{28}	2	3	∞	∞	12.7	0.81	0.07
A_{29}	2	1	∞	∞	1.9	0.20	0.15
KA	0	1	0.24	0.04	0	0	0

KA: ent-kaurenoic acid, 0: Kd <0.02, ∞: Kd >20.

or methanol-water are used as eluents by increasing the concentration of organic solvents stepwise. Polar GAs are eluted earlier, less polar ones later; the elution pattern of charcoal chromatography resembles that of reversed phase chromatography. Free GAs are scarcely eluted at concentrations less than 30% acetone or methanol.

In the purification of conjugated GAs and very polar GAs, charcoal chromatography is used widely and effectively. Usually, the amount of charcoal is 3 to 10 times the weight of the concentrates submitted. As an eluent, aqueous acetone or aqueous methanol is used increasing the concentration of organic solvent in water. The elution of conjugated GAs is greatly affected by the ratio of charcoal used and materials submitted. As long as adequate charcoal is used, conjugated GAs are scarcely eluted with acetone concentrations less than 30%. In such cases it is very effective to elute a column with a large volume of 5 to 20% aqueous acetone to remove inorganic salts and other glycosides.[53,57,59]

4. Silicic Acid Adsorption Chromatography

In the isolation of free GAs, silicic acid adsorption chromatography is often used before or after charcoal chromatography. The amount of silicic acid is 20 to 30 times the weight of the material submitted. Silicic acid with or without Celite® is soaked in a packing solvent (usually, *n*-hexane, benzene, or chloroform) and packed into a glass column. It is recommended that the packed column be stabilized overnight before charging a sample. A sample is suspended in a starting eluent and charged onto a column or adsorbed on acid-washed Celite®, dried, and then added to the top of the column. A mixture of ethyl acetate and *n*-hexane, benzene, or chloroform is often used as an eluent with increasing concentrations of ethyl acetate.

Silicic acid adsorption chromatography is also used in the separation of conjugated GAs. For GAs-GEt, a mixture of methanol, chloroform, and acetic acid is used.[61] For GAs-GEs, a mixture of acetone and benzene is used with increasing concentrations of acetone in benzene.[61]

5. Partition Chromatography

Partition chromatography is a very effective method for separating and purifying GAs. The AE fraction is often submitted to partition chromatography without prepurification by countercurrent distribution, charcoal chromatography, and/or silicic acid adsorption chromatography. In partition chromatography, a combination of organic solvents is usually used for the mobile phase and an aqueous buffer for the stationary phase. This partition system is called normal phase partition chromatography. Silicic acid, Sephadex® G-25, G-50, or LH-20 is widely used as a support for the stationary phase.

Pitel et al.[151] reported the separation of GA_4, GA_7 and iso-GA_7 and GA_1, and GA_3 and iso-GA_3 by partition chromatography on Sephadex® G-25. They applied the solvent systems originally developed for thin layer chromatography (TLC) by Kagawa et al.[152] and MacMillan and Suter.[153] In the solvent system of benzene-petroleum ether-acetic acid-water (6:2:5:3), GA_4, GA_7, and iso-GA_7 are clearly separated. GA_1, GA_3, and iso-GA_3 are separated in the solvent system of benzene-ethyl acetate-acetic acid-water (55:25:30:50). In both two-phase solvent systems, the aqueous lower phases are used as the stationary phases and the upper phases as the mobile phases. However, these solvent systems are not suitable for the separation of many GAs.

MacMillan and Wels[154] reported the separation of a wide range of GAs using partition chromatography with Sephadex® LH-20 and three two-phase solvent systems. A variety of GAs are separated in a partition column, termed a wide range column. The solvent system is petroleum ether-ethyl acetate-acetic acid-methanol-water (100:80:5:40:7) (see Figure 9). This column needs careful packing and equilibration. A well-packed column provides a fairly high resolution with theoretical plates 5500/1.5 m for GA_3, although the overlappings of some GAs are still being observed. The condensed region of less polar GAs from GA_{12}-7-aldehyde to GA_{14}-7-aldehyde is expanded by a partition column, termed a narrow range column. The solvent system used is of light petroleum ether-ethyl acetate-acetic acid-methanol-water (50:15:10:10:2). Separation of nonpolar compounds such as *ent*-kaurene, *ent*-kaurenoic acid can be accomplished by using a solvent system consisting of light petroleum ether-acetic acid-methanol (100:1:40).

Silicic acid has been frequently used as a support for the stationary phase in partition chromatography. Usually, a 1.4 to 1.6-fold excess silicic acid over the weight of stationary phase is used for fixation.

Powell and Tautvydas[155] reported the behavior of GA_{1-9} on silicic acid partition column with a solvent system of 0.5 *M* formic acid and increasing ethyl acetate in *n*-hexane. The monobasic GAs are separated, depending on the polarity derived from the number of hydroxyl groups in the molecule as shown in Table 5. Durley et al.[156] examined the behavior of GA_{1-34} (excluding GA_{32}) on Woelm silica gel partition column, using 0.5 *M* formic acid as a stationary phase and continuously increasing ethyl acetate concentration in *n*-hexane as a mobile phase. The elution pattern of GAs are shown in Table 6. After long lasting partition chromatography with formic acid as the ion suppressor, formate formation can occur.

Sephadex partition chromatography is also effective for the purification of conjugated

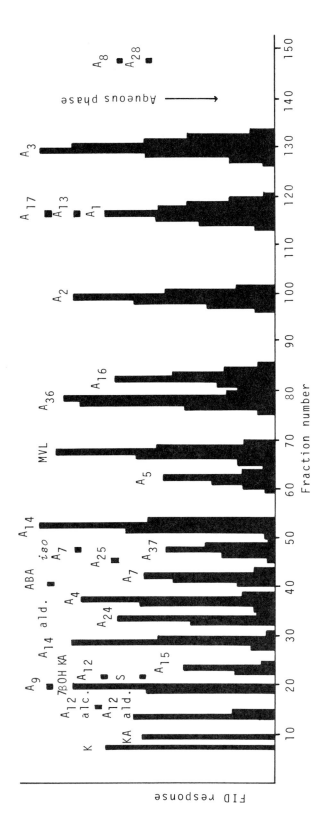

FIGURE 9. Separation of GAs by Sephadex® LH-20 partition chromatography.[154] Column: 145 × 1.5 cm i.d. Stationary phase: aqueous phase of light petroleum ether-ethyl acetate-acetic acid-methanol-water (100:80:5:40:7) mixture. Mobile phase: organic phase of the above mixture. Flow rate: 50 mℓ/hr. Samples: GAs are expressed by A numbers. K: *ent*-kaurene, KA: *ent*-kaurenoic acid, A_{12} Alc.: GA_{12}-7-alcohol, A_{12} ald.: GA_{12}-7-aldehyde, A_{14} ald.: GA_{14}-7-aldehyde, 7 OH KA: *ent*-7-hydroxykaurenoic acid, ABA: abscisic acid. Detection: 150 mℓ fractions collected and analyzed by GC.

Table 5
SEPARATION OF GA_{1-9} BY PARTITION CHROMATOGRAPHY ON SILICA GEL WITH SOLVENT SYSTEM OF 0.5 M FORMIC ACID-ETHYL ACETATE IN n-HEXANE[155]

Ethyl acetate (%)	GAs
0	
1.0	
3.0	
6.0	A_9
9.0	
12.0	
15.0	A_4
16.5	A_4, A_7
18.0	
19.5	A_5
21.0	
22.5	
24.0	A_6
25.5	
27.0	
30.0	
33.0	
36.0	
39.0	
40.5	
42.0	A_1, A_3
43.5	A_1, A_3
45.0	A_2
46.0	A_2 (trace)
48.0	
51.0	
54.0	
57.0	
60.0	
63.0	A_8
66.0	

Column: 8 g silica gel hydrated with 5 mℓ of 0.5 M formic acid.
Eluents: Ethyl acetate-n-hexane saturated with 0.5 M formic acid.
Elution volumes: 25 mℓ for each fraction.

GAs. In the partition chromatography for GAs-GEt, phosphate buffer (1 M, pH 5.4 to 5.5) adsorbed on Sephadex® G-25 or G-50 is often used as a stationary phase and a mixture of n-butanol and ethyl acetate as the mobile phase with increasing concentrations of n-butanol.

Table 6
SEPARATION OF 33 GAs AND ABSCISIC ACID BY PARTITION CHROMATOGRAPHY[156]

Fraction no.	GAs
2	A_9, A_{12}
3	$A_9, A_{11}, A_{14}, A_{24}, A_{31}$
4	$A_4, A_5, A_6, A_7, A_{14}, A_{15}, A_{20}, A_{25}, A_{31}$, ABA
5	A_6, A_{10}, A_{15}
6	A_{10}
8	A_{27}, A_{34}
9	A_{16}, A_{27}, A_{34}
10	$A_{16}, A_{27}, A_{34}, A_{33}$
11	A_{33}
12	A_1, A_3, A_{30}
13	A_1, A_3, A_{30}
14	A_1, A_3, A_{19}
15	A_2, A_{13}, A_{19}
16	A_2, A_{13}, A_{22}
17	$A_{18}, A_{22}, A_{26}, A_{29}$
18	A_{18}, A_{26}, A_{29}
19	A_{17}
20	A_{17}, A_{23}
21	A_{21}, A_{23}
22	A_{21}
23	A_8, A_{28}
24	A_8, A_{28}

Column: Woelm silica gel for partition chromatography (20 g) equilibrated with 0.5 M formic acid-saturated solution of ethyl acetate in n-hexane (gradient solution) (10:90), 20 × 1.3 i.d. cm.

Elution: Varigard gradient system (0.5 M formic acid-saturated solvents are used). Chamber 1, ethyl acetate in n-hexane (65:35, 129 mℓ), Chamber 2, ethyl acetate in n-hexane (20:80, 147 mℓ), Chambers 3 and 4, 100% ethyl acetate (114 mℓ).

Elution volume: 20 mℓ for each fraction.

Yokota et al.[53] used Sephadex® G-50 partition chromatography with a solvent system of phosphate buffer (1 M, pH 5.5)-ethyl acetate-n-butanol, increasing the n-butanol concentration in ethyl acetate, for the purification of GA_3-GEt, GA_8-GEt, GA_{26}-GEt, and GA_{29}-GEt. The solvent system mentioned above is for normal phase partition chromatography and less polar substances are eluted faster than polar ones. Yamaguchi et al.[91,109] used reversed-phase partition chromatography on Sephadex® LH-20 with a mixture of distilled water and n-butanol (100:6) as the eluent for the purification of GA_3-GEt and also GA_5-GEs and GA_{44}-GEs. In this system, polar substances are eluted faster than less polar ones.

6. Chemical and Enzymatical Cleavage of Conjugated GAs

Thus far, the known conjugated GAs are glucosyl ethers and glucosyl esters, except for gibberethione and 3-O-acetyl GA_3. These conjugated GAs are cleaved to afford sugar and aglycon moieties for structural determination. GAs-GEs are easily cleaved in mild basic solution to release sugar and aglycons. Hiraga et al.[61] treated GA_1-GEs, GA_4-GEs, GA_{37}-GEs, and GA_{38}-GEs with 0.1 N aqueous NaOH to obtain free $GA_{1,4,37,38}$ and their C-3 epimers. Yamaguchi et al.[91] obtained the methyl esters of $GA_{5,44}$ by treating their glucosyl esters with 0.01 N NaOCH$_3$. The structure of the aglycon moiety is determined by the procedure described in Sections II. A and B. A sugar moiety which remains in an aqueous solution where the aglycon is removed by solvent extraction is identified by GC after derivation to a trimethylsilyl ether with a mixture of pyridine-N,O-bis(trimethylsilyl)acetamide-trimethylchlorosilane (10:5:1). Chemical cleavage of GAs-GEt is rather complicated. Boiling a GAs-GEt with 0.1 N HCl successfully cleaves an ether linkage and affords glucose, and not simply an aglycon, because the GA itself is very labile under such strong acidic conditions.[54,59] Therefore, to obtain an aglycon without undergoing structural change, enzymatic hydrolysis is used. β-Glucosidase, cellulase, and emulsin are generally used. Yokota et al.[53] used cellulase or emulsin to cleave GA_8-GEt, GA_{26}-GEt, GA_{27}-GEt, and GA_{29}GEt. Yamane et al.[59] also used cellulase to cleave GA_{35}-GEt. Samples are dissolved in 1.0 mℓ of 1 M sodium acetate buffer (pH 4.5) and 10 to 50 mg of enzyme for 10 to 20 mg of GAs-GEt are added and the mixture kept at 37°C for 16 hr. Schliemann[157] reported that the hydrolysis rates of some conjugated GAs by β-glucosidase decreased in the following order: GA_8-O-2-GEt (GA_8-GEt), GA_3-GEs, GA_3-O-3-GEt (GA_3-GEt), GA_5-O-13-GEt, while, GA_1-O-3-GEt (GA_1-GEt) resisted enzymic hydrolysis by β-glucosidase.[90,158]

7. Paper and Thin Layer Chromatography

In the final purification and identification of free and conjugated GAs, both paper and TLC have been used. Paper chromatography is now used only rarely, while TLC is still a very effective and convenient method. In paper chromatography, mixtures of organic solvents and water-containing ions, such as the ammonium cation and the acetate anion, are used as developing solvents. Partition of compounds takes place between the water adsorbed on cellulose and the organic solvents, and compounds are separated based on partition coefficients. The identification by paper chromatography is based on a comparison of the R_f value of an unknown sample with that of an authentic specimen. R_f values in paper chromatography are affected by several experimental conditions such as conditions in the developing chamber and temperature, among others. Therefore, it is recommended to co-chromatograph a sample with an authentic specimen in as many different solvent systems as possible. The disadvantages of paper chromatography are the fairly long time required for development, and the poor resolution compared to TLC or HPLC. It should be remembered GAs possessing the same A ring structure as GA_3 are often partially isomerized when developed in basic solvent systems.[53] The behavior of several free GAs, conjugated GAs, and their derivatives in paper chromatography with several solvent systems is summarized in Tables 7 and 8.[53,158]

TLC is used as both partition and adsorption chromatography. It can be used for both preparative and identification purposes. In TLC higher resolution is achieved, and time required for development is shorter in comparison to paper chromatography. Recently, high performance TLC is available with very high resolution. For TLC of GAs, silica gel is the most commonly used support.

The behavior of an unknown GA and its methyl ester on TLC reflects the number of hydroxyl and carboxyl groups in the molecule. The number of GAs (68 free GAs in 1984) is making "identification" by TLC very difficult. Besides, R_f values are affected by several factors (as with paper chromatography). Nevertheless, TLC is still an effective and important technique, not only as a preliminary test for physicochemical properties, but also as a reliable pre-identification technique when an unknown GA is co-chromatographed with an authentic standard in several different solvent systems. R_f values of free and conjugated GAs and their derivatives are summarized in Tables 9 to 12.[53,55,145,159]

8. HPLC

HPLC is one of the most modern techniques for the separation and identification of natural products. Development of chemically bonded media has greatly expanded the applications of liquid chromatography to the purification of organic compounds. Specifically, the development of extremely small and uniform particles used as packing materials has afforded

Table 7
R_f VALUES OF GAs ON PAPER CHROMATOGRAPHY[53]

GA	Solvent system			
	1	2	3	4
A_1	0.31	0.40	0.52	0.49
A_3	0.29	0.40	0.50	0.48
A_4	0.60	0.57	0.72	0.72
A_5	0.45	0.49	0.59	0.56
A_6	0.33	0.42	0.51	0.51
A_7	0.61	0.57	0.71	0.71
A_8	0.15	0.25	0.40	0.35
A_9	0.73	0.65	0.77	0.78

Solvent systems: (1) n-BuOH:1.5 N NH$_4$OH = 3:1, descending; (2) n-BuOH:trt-amyl alcohol:acetone:ammonia:water = 5:5:5:2:3, descending; (3) 2-PrOH:H$_2$O = 4:1, ascending; (4) 2-PrOH:7 N NH$_4$OH = 5:1, ascending. Paper: Whatman No. 1, temperature: 20° C.

Table 8
R_f VALUES OF GA GLUCOSIDES (GAs-GEt) ON PAPER CHROMATOGRAPHY[158]

GAs-GEt	R_f value
3-O-β-Glucosyl-gibberellin A$_3$ (GA$_3$-GEt)	0.59
3-O-β-Glucosyl-gibberellenic acid	0.22
3-O-β-Glucosyl-iso-gibberellin A$_3$ (iso-GA$_3$-GEt)	0.57
2-O-β-Glucosyl-gibberellin A$_8$ (GA$_8$-GEt)	0.51
2-O-β-Glycosyl-gibberellin A$_{26}$ (GA$_{26}$-GEt)	0.55
2-O-β-Glucosyl-gibberellin A$_{27}$ (GA$_{27}$-GEt)	0.63
2-O-β-Glucosyl-gibberellin A$_{29}$ (GA$_{29}$-GEt)	0.54

Paper: Toyo filter paper No. 51: solvent system: isopropanol:7 N-ammonium hydroxide:water (8:1:1).

extremely high resolution. UV and RI (refractive index) detectors are most commonly used; fluorescence, radioactivity, and conductivity detectors are used on occasion.

Most GAs do not have any characteristic chromophore but only end absorption at 200 to 210 nm due to the carboxyl group and olefinic double bond(s). Therefore, GAs can be detected by UV at 200 to 210 nm or RI levels. However, monitoring eluates by UV at end absorption limits the usable solvent systems, since they must be transparent at 200 to 210 nm. In RI detection, a gradient elution mode cannot be used because the baseline drift is too large, and sensitivity is low.

HPLC of nonderivatized GAs has been examined by many groups. Reeve et al.[160] and Crozier and Reeve[161] showed the separation of radioactive GAs and their biosynthetic precursors by partition chromatography on Partisil 20 with a solvent system of 40% 0.5 M formic acid (stationary phase) and n-hexane containing an increasing concentration of ethyl acetate. The eluate was monitored by both radioactivity and UV (254 nm) detectors. The elution pattern is shown in Figure 10.

Jones et al.[162] reported fractionation of GAs by a combination of preparative and analytical

Table 9
RELATIVE R_f VALUES OF GAs ON TLC[145]

	R_{GA_3} (reproducibility in parentheses)				Color of induced fluorescence[a]
	Solvent system				
GAs	1 (±0.02)	2 (±0.03)	3 (±0.05)	4 (±0.03)	
A_1	1	1	1	1	Blue
A_2	0.92	0.91	1.1	1	Purple
A_3	1	1	1	1	Green-blue
A_4	1.17	1.13	1.48	0	Purple
A_5	1.0	1.12	1.35	0.80	Blue
A_6	1	1.07	0.95	0.95	Blue
A_7	1.17	1.13	1.48	0	Yellow
A_8	0.89	0.84	0.52	1.05	Blue
A_9	1.19	1.25	1.55	0	Purple
Gibberellenic acid	0.50	0.42	0.10	1.03	Green-blue

Note: Solvent systems: (1) Isopropranol:water (4:1), R_f of GA_3; 0.55. (2) Isopropanol:4.5 N-ammonium hydroxide (3:1), R_f of GA_3; 0.55. (3) n-Butanol:4.5 N-ammonium hydroxide (3:1), R_f of GA_3; 0.55. (4) Phosphate buffer (0.1 M pH 6.3) on silica gel impregnated with capryl alcohol, R_f of GA_3; 0.70. Equilibration prior to development was carried out for 16 hr. Impregnation of silica gel with capryl alcohol was accomplished by developing the chromoplate in a 7% solution of capryl alcohol in light petroleum (40 to 60°C) and drying at room temperature.

[a] Fluorescence under UV (365 nm) is induced by heating the plates after spray with 70% sulfuric acid in ethanol or water.

reversed phase columns. A separation pattern on an analytical column of μ-Bondapak C_{18} (300 × 4 mm i.d., Waters Associates) eluted with a linear gradient of methanol in 1% aqueous acetic acid is shown in Table 13. The GAs in each fraction were detected by bioassay and/or GC. By this method they fractionated the extracts of immature seeds of *Pharbitis nil* and identified $GA_{3,5,17,19,20,29,44}$ by GC/MS. They applied this method again to the analysis of endogenous GAs of spinach and their changes in relation to photoperiodism.[138,163,164]

Barendse and Van de Werken[165] separated $GA_{1,3-5,7,9,20}$ by an ODS column with gradient elution modes. This increased the methanol concentration in a mixture of phosphate buffer (0.1 M, pH 2.2)-acetone, or elevated the pH (4.0 to 7.2) of the eluent of 35% methanol in 0.01 M phosphate buffer. A chromatogram is shown in Figure 11. They used this method to identify GA_3 in *P. nil*.

Yamaguchi et al.[166,167] reported the qualitative and semiquantitative analysis of minute amounts of GAs, where a combination of different modes of HPLC is efficiently used as a prepurification technique for GC/SIM analysis. They examined the chromatographic behavior of many GAs and also their recoveries. Behavior of GAs on a gel permeation column of Shodex A-801 (500 × 8 mm i.d., Showa Denko Co., Ltd.) with tetrahydrofuran as eluent, and on a column of Nucleosil $N(CH_3)_2$ (250 × 6 mm i.d.) with methanol containing 0.05% acetic acid as eluent, is shown in Tables 14 and 15. The GPC column with high theoretical plates (15,000/50 cm) is very effective in concentrating GAs in plant extracts. The use of Nucleosil $N(CH_3)_2$ with isocratic elution can classify many GAs depending on their structural characteristics as shown in Figures 12 and 13. These methods were used in the procedure to identify and quantify endogenous GAs in the immature seed and floral organs of *P. nil*.

Yamaguchi et al.[109] used chromatography on C_{18} and C_2 columns for the separation of free GAs and conjugated GAs. They used aqueous ammonium chloride solutions as eluents,

Table 10
R_f VALUES OF GAs ON TLC[145]

	Kieselgel					
	Solvent system					
GAs	1	2	3	4	2	5
A_1	0.06	0.00	0.37	0.95	0.20	0.55
A_2	0.01	0.00	0.19	0.85	0.24	0.67
A_3	0.06	0.00	0.37	0.95	0.13	0.45
A_4	0.20	0.69	0.61	0.80	1.00	1.00
A_5	0.16	0.29	0.59	0.80	0.87	1.00
A_6	0.13	0.16	0.57	0.90	0.82	1.00
A_7	0.19	0.60	0.62	0.80	1.00	1.00
A_8	0.01	0.00	0.24	1.00	0.04	0.20
A_9	0.59	1.00	0.78	0.65	1.00	1.00
A_{10}	0.06	0.36	0.33	0.60	0.91	1.00
A_{11}	0.49	1.00	0.74	0.80	1.00	0.88
A_{12}	0.67	1.00	0.78	0.70	1.00	1.00
A_{13}	0.14	0.11	0.46	0.90	0.39	1.00
A_{14}	0.26	0.75	0.63	0.80	0.86	0.72
A_{15}	0.44	1.00	0.78	0.57	1.00	1.00
A_{18}	0.04	0.02	0.35	0.95	0.34	1.00
A_{19}	0.08	0.00	0.42	0.95	0.12	0.82

Solvent systems: (1) di-isopropyl ether:acetic acid (95:5); (2) benzene:acetic acid:water (8:3:5); (3) ethyl acetate:chloroform:acetic acid (15:5:1); (4) water; (5) benzene:propionic acid:water (8:3:5). With solvent systems 2 and 5 plates were equilibrated overnight with lower phase, then developed with upper phase.

and applied this procedure to the isolation and identification of GA_3-GEt in *Quamoclit penata*.[109]

Koshioka et al.[168] also reported the separation of free and conjugated GAs by reversed phase columns and applied the technique to the identification of GA_{19} in young shoots of apple, and GA_1 and GA_{19} in meristem tissue of maize. The elution pattern of some conjugated GAs and their aglycons is summarized in Table 16.

To enhance the sensitivity and to expand the variety of usable eluents and packing media, specific chromophores are sometimes introduced into GA molecules prior to their development.

Crozier and Reeve[161] reported analyzing GAs as their benzyl esters on Partisil 10 monitoring the eluate at 254 nm. By this method they separated $GA_{1,3-5,7,9,13-15}$ with identification by MS.

Crozier et al.[169] prepared methoxycoumaryl esters by treating GAs with an equimolar amount of 4-bromomethyl-7-methoxycoumarin, one tenth the equivalent of 18-crown-6-ether and a crystal of K_2CO_3 in 100 µℓ of dry acetonitrile at 60°C for 2 hr. The carboxyl groups of all GAs tested were completely esterified and their coumaryl esters were highly fluorescent (λ ex max 320 nm, λ em max 400 nm) and could be detected at the low picogram level by a spectrophotofluorometer. They separated 13 methoxycoumaryl esters by HPLC on ODS Hypersil with mobile phases of gradient methanol or ethanol in 20 mM ammonium acetate buffer (pH 3.5) as shown in Figure 14.

9. GC, GC/MS, and GC/SIM

GC is partition chromatography between a gas phase and a liquid stationary phase on the surface of a support. Very high performance, reproducibility, and sensitivity are obtained.

Table 11
R_f VALUES OF GAs AND THEIR METHYL ESTERS ON TLC[152]

	Free acid						Methyl ester	
	Silica gel G solvent systems			Kieselguhr G solvent systems			silica gel G solvent systems	
GAs	1	2	3	3	4	5	6	7
A_1	0.20	0.49	0.0	0.28	0.0	0.49	0.31	0.29
A_2	0.17	0.40	0.0	0.37	0.23	0.23	0.23	0.13
A_3	0.19	0.54	0.0	0.18	0.0	0.40	0.35	0.32
A_4	0.63	0.95	0.0	0.90	0.67	1.00	0.73	0.75
A_5	0.53	0.87	0.27	0.85	0.45	0.90	0.60	0.69
A_6	0.59	0.87	0.11	0.86	0.33	0.84	0.66	0.67
A_7	0.60	0.90	0.57	0.85	0.45	0.91	0.71	0.72
A_8	0.04	0.30	0.0	0.06	0.0	0.10	0.17	0.12
A_9	0.87	0.95	1.00	1.00	1.00	1.00	0.98	0.96

Solvent systems: (1) benzene:n-butanol:acetic acid (80:15:6); (2) benzene:n-butanol:acetic acid (70:25:5); (3) carbon tetrachloride:acetic acid:water (8:3:5), lower phase plus 10% ethyl acetate; (4) carbon tetrachloride: acetic acid:water (8:3:5), lower phase; (5) carbon tetrachloride:acetic acid:water (8:3:5), lower phase plus 20% ethyl acetate; (6) ethyl acetate:benzene (4:1); (7) ethyl ether:petroleum ether (4:1), developed twice. With solvent systems 3, 4, and 5, plates are equilibrated overnight with upper phase, then developed with lower phase or lower phase plus ethyl acetate.

Table 12
R_f VALUES OF CONJUGATED GAs ON TLC[55,145]

	Solvent systems			
Conjugated GAs	a	b	c	d
GA_3-GEt	0.28	0.37		
Gibberellenic acid-GEt	0.20	0.4		
iso-GA_3-GEt	0.24	0.47		
GA_8-GEt	0.20	0.46		
GA_{26}-GEt	0.33	0.51		
GA_{27}-GEt	0.36	0.55		
GA_{29}-GEt	0.24	0.43		
GA_1-GEs			0.47	0.19
GA_3-GEs			0.47	0.17
GA_4-GEs			0.62	0.33
GA_{37}-GEs			0.62	0.35
GA_{38}-GEs			0.47	0.12

Solvent systems: (a) chloroform:methanol:acetic acid:water (40:15:3:2); (b) acetone:acetic acid (97:3); (c) chloroform:methanol (3:1); (d) benzene:acetone (1:5).

In GC of GAs, samples are derivatized to methyl esters, methyl ester-trimethylsilyl ethers, or trimethylsilyl ester-trimethylsilyl ethers to increase volatility. Methyl esterification of GAs

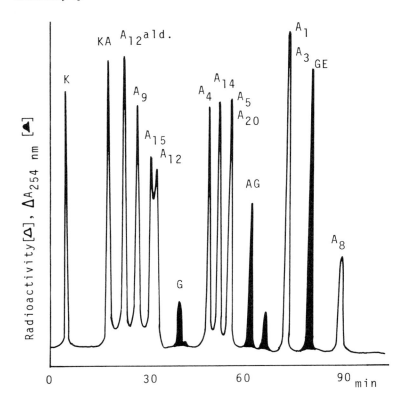

FIGURE 10. HPLC of radioactive GAs and GA precursors with UV-absorbing internal markers.[161] Column: 450 × 10 mm i.d., Partisil 20. Stationary phase: 40% 0.5 M formic acid. Mobile phase: 2 hr gradient 0 to 100% ethyl acetate in hexane. Flow rate: 5 mℓ/hr. Samples are expressed by A numbers: 50,000 dpm (^{14}C)-GA$_3$, (^{3}H)-GA$_5$, (^{14}C)-GA$_{12}$, (^{14}C)-GA$_{19}$, and (^{3}H)-GA$_{20}$; 100,000 dpm (^{3}H)-GA$_1$, (^{3}H)-GA$_4$, (^{3}H)-GA$_8$, (^{3}H)-GA$_9$, and (^{3}H)-GA$_{14}$. K: ent-($_{14}$C)-kaurene (24,000 dpm), KA: ent-(^{3}H)-kaurenoic acid (100,000 dpm), GA$_{12}$ald.: GA$_{12}$-7-aldehyde (100,000 dpm), G: gibberic acid (uncalibrated amount), AG: allgibberic acid (uncalibrated), GE: gibberellenic acid (uncalibrated). Detectors: homogenous radioactivity monitor, 1800 cpm full-scale deflection, absorbance monitor A$_{254}$.

is usually accomplished by treating a sample in methanol with etherial diazomethane. A variety of reagents and methods are known for trimethylsilylation: N,O-bis(trimethylsilyl)acetaminde (BSA), N,O-bis(trimethylsilyl)trifluoroacetamide (BSTFA), hexamethyldisilazane, and trimethylsilylimidazole; these are usually used with or without trimethylchlorosilane (TMCS) in acetonitrile or pyridine solution. A flame ionization detector (FID) is most commonly used for the analysis of GAs and it enables the detection of 10 ng levels of GAs. Identification by GC is accomplished by the comparison of the retention time (t_R) of a sample peak with that of an authentic specimen on several different stationary phases. Behaviors of free and conjugated GA derivatives on GC are summarized in Tables 17 to 19.

A mass spectrometer is also used as a GC detector and the system is called combined GC/MS. In a usual GC/MS analysis full mass spectra of compounds separated by GC are measured and its detection level is a few nanograms in the case of GAs. In a GC/MS analysis, information on t_R on GC and a mass spectrum gives very reliable and confirmatory evidence for identification. Computerized on-line accumulation and analysis of GC/MS data enable the output of the data in the form of mass chromatograms, in which fluctuation in intensity of specific ions selected from full spectral data is traced against t_R on GC. In mass chro-

Table 13
ELUTION PATTERN OF GAs ON REVERSED PHASE HPLC[162]

GAs	Fraction no.
GA_1	10,11
GA_{20}	16,17
GA_{44}	17
GA_{19}	18
GA_{13}	18
GA_{36}	18
GA_{17}	19,20
GA_{37}	20
GA_7	21
GA_9	24
GA_{25}	24
GA_{12}	28

Column: Bondapak C_{18} (300 × 4 mm i.d.); elution: linear gradient of methanol (30 to 100 %) in 1% aqueous acetic acid, run in 30 min at 2 mℓ/min (fractions collected every minute). Underlining indicates fraction in which most of GA eluted.

matography, identification and quantification are determined in the same way as combined GC/SIM, which is described below. In the case of GC/SIM, 1 to 5 ions characteristic of an expected compound are selected for monitoring. The identity is determined by comparing the t_R and relative intensity of the peaks of the selected ions to those of an authentic specimen. Quantification is carried out by measuring a peak area or peak height of one of the selected ions and comparing the values to the calibration curve drawn by using an authentic specimen. In GC/SIM 100 pg of GAs are sufficient for qualitative and quantitative analyses. Usually much less prepurification is needed in GC/SIM than in GC/MS.

A mass chromatogram identifying $GA_{17,44,53}$ in *Pharbitis purpurea* seeds is shown in Figure 15.[170] A GC/SIM profile identifying GA_{19} in spring bleeding sap of *Juglans regia* L. is shown in Figure 16.[171]

Laurent et al.[172] examined methods for GC/MS analysis of conjugated GAs as well as free GAs, and found the permethylated derivatives could be analyzed by GC/MS. They successfully identified GA_8-GEt and tentatively identified GA_1-GEt and GA_{34}-GEt in *Phaseolus coccineus* pods by this method.

10. Immunoassay

In general, immunoassays are highly sensitive and usually specific. Therefore, samples can be assayed without complicated purification. Immunoassay of GAs was first reported by Fuchs et al.[173] in 1971, though its sensitivity was not high. The radioimmunoassay of GAs developed by Weiler and Wieczorek[174] is, however, highly sensitive and specific. The enzyme immunoassay reported by Atzorn and Weiler[175] in 1983 shows even higher sensitivity

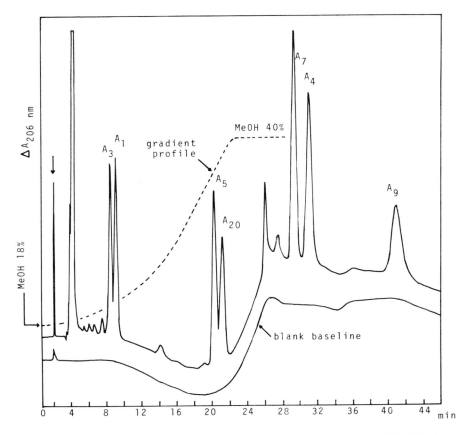

FIGURE 11. Reversed phase HPLC of GA_1, GA_3, GA_4, GA_5, GA_7, GA_9, and GA_{20}.[165] Mobile phase: gradient from 18% MeOH containing 0.01 M H_3PO_4 and acetone (10 mℓ/ℓ) to 40% MeOH, 0.01 M H_3PO_4 at pH 2.2 in 20 min. Flow rate: 1.8 mℓ/min. Detection at 206 nm.

Table 14
RETENTION TIME (t_R) OF GAs AND ABA ON SHODEX® A-801[166]

No. of OH	C_{19}-GAs & ABA	t_R (min)	C_{20}-GAs	t_R (min)
None	GA_9	13.7	GA_{24}	13.0
Mono	GA_5	12.9	GA_{13}	12.0
	GA_{20}	12.9	GA_{14}	12.2
	ABA	12.2	GA_{19}	12.2
			GA_{36}	12.3
Di	GA_1	12.2	GA_{27}	12.8
	GA_3	12.3	GA_{41}	11.8
Tri	GA_8	11.9		

Column size: 500 × 8 mm; solvent: tetrahydrofuran, 1.0 mℓ/min; detection: UV (254 nm) and RI.

than radioimmunoassay. The basic principle of immunoassay is the formation of an antigen (immunogen)-antibody complex. Since GA is too small a molecule to be an immunogen by itself, it must be conjugated to a large molecule such as bovine serum albumin (BSA) or human serum albumin (HSA) to become an immunogen. The immunogen is emulsified in a mixture of saline and Freund's complete adjuvant, and hypodermically injected into rabbits.

Table 15
RETENTION TIME (MIN) OF GAs ON NUCLEOSIL N(CH$_3$)$_2$[166]

No. of OH	Type of GAs									
	C-10 CH$_3$		δ-lactone		C-10 CHO		C-10 COOH		γ-lactone	
None	GA$_{12}$	10.7			GA$_{24}$	32.4			GA$_9$ 27.0	
Mono	GA$_{14}$	11.4	GA$_{37}$	13.8	GA$_{19}$	41.6	GA$_{13}$	28.0	GA$_4$ 27.8	GA$_5$ 38.2
					GA$_{36}$	31.4	GA$_{17}$	21.4	GA$_7$ 27.8	GA$_{20}$ 30.0
									GA$_{31}$ 37.6	GA$_{40}$ 28.0
Di	GA$_{18}$	11.7	GA$_{27}$	15.6					GA$_1$ 22.0	GA$_3$ 28.9
									GA$_{16}$ 25.6	GA$_{22}$ 49.0
									GA$_{26}$ 48.2	GA$_{30}$ 28.6
Tri									GA$_8$ 26.0	

Column size: 250 × 6 mm; particle size: 10 μm; column temperature: 50°C; solvent: 0.05% acetic acid in methanol, 2 mℓ/min; detection: UV (205 nm).

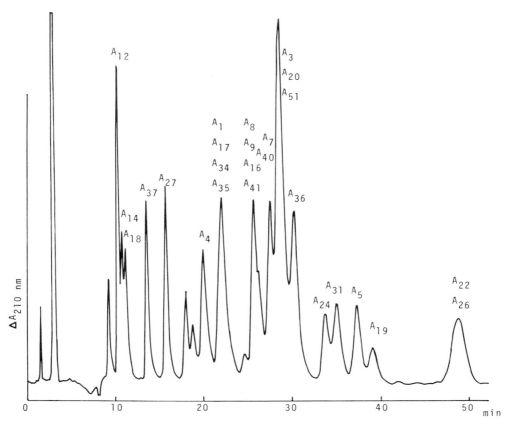

FIGURE 12. HPLC of GAs on Nucleosil 10 N(CH$_3$)$_2$ (250 × 6 mm i.d.).[166,167] Mobile phase: MeOH containing 0.05% AcOH. Flow rate: 2 mℓ/min. Column temp.: 50°C. Detection: 210 nm.

Blood is collected after several injections with appropriate intervals. The antisera raised against such a immunogen reacts with GA itself or its methyl ester to form a complex.

The outline of the radioimmunoassay procedure is as follows. In the solution containing a sample, a tracer, and antisera, the complexes of antibody-tracer and antibody-GA in a

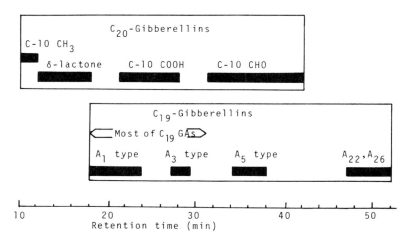

FIGURE 13. Diagram of the elution pattern of GAs on Nucleosid 10 N(CH$_3$)$_2$. Chromatographic conditions are in Figure 12.

Table 16
REVERSED PHASE HPLC
SEPARATION OF CONJUGATED GAs
AND THEIR AGLYCONS[168]

Conjugated GAs	t_R (min)	t_R of aglycon (min)
GA$_8$-O(2)-GEt	11—12	12—13
GA$_3$-O(13)-GEt	20—21	25—26
GA$_1$-O(13)-GEt	21—22	25—26
GA$_3$-O(3)-GEt	22—23	25—26
GA$_1$-O(3)-GEt	23—24	25—26
GA$_1$-GEt	24—25	25—26
GA$_3$-GEs	25	25—26
GA$_5$-O(13)-GEt	33—34	37—38
GA$_4$-GEs	33—34	37—38
GA$_4$-O(3)-GEt	36—37	37—38

Column: Bondapak C$_{18}$ (300 × 3.9 mm i.d.).
Elution: (Pump A) 10% methanol in 1% aqueous acetic acid. (Pump B) 100% methanol linear gradient program; 0—10 min (Pump A, 100%), 10—40 min (Pump B, 0—70%), 40—50 min (Pump B, 70%), 50—80 min (Pump B, 100%).
Flow rate: 2 mℓ/min.

sample are formed. The complexes are separated as precipitates by salting out, and the radioactivity of the antibody-tracer complex in the precipitates is measured. Quantification is carried out by comparing the radioactivity with the calibration curve made at the same time using standard solutions.

In the immunoassay of GAs, the cross-reactivity must be carefully examined, for the GA family is composed of 68 species (in 1984) and some of them are very similar in structure.

Weiler and Weiczorek[173] prepared an immunogen by coupling GA$_3$ to BSA via a GA$_3$-

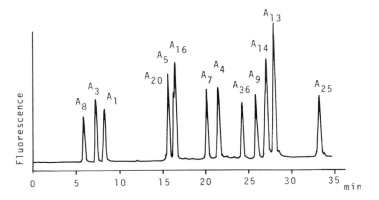

FIGURE 14. Reversed phase HPLC of GA methoxycoumaryl esters on ODS-Hypersil (250 × 5 mm i.d.).[169] Mobile phase: 30 min gradient, 40 to 80% ethanol in 20 mM ammonium acetate buffer (pH 3.5). Flow rate: 1 mℓ/min. Samples: methoxycoumaryl esters of GAs, approximately 9.0 ng mono, 4.5 ng bis- and 3.0 ng triesters. Detector: spectrofluorimeter (λex. 320 nm, λem. 400 nm).

Table 17
RETENTION TIME (MIN) OF GAs ON GC[145]

GAs	GA methyl ester		GA methyl ester-TMSi ether	
	(1) 2% QF-1	(2) 2% SE-33	(3) 2% QF-1	(4) 2% SE-33
A_1	19.7	14.7	16.3	16.8
A_2	20.7	15.1	23.1	8.2
A_3	20.6	17.6	19.1	18.3
A_4	8.4	7.4	11.2	8.9
A_5	10.4	6.7	11.3	8.1
A_6	17.1	9.4	19.1	11.4
A_7	9.1	7.9	12.8	9.5
A_8	38.6	30.7	20.7	29.3
A_9	4.3	3.9	7.4[a]	4.5[a]
A_{10}	10.4	6.9	16.0	10.8
A_{11}	6.9	5.1	12.3[a]	6.0[a]
A_{12}	2.0	4.0	3.5[a]	4.9[a]
A_{13}	6.2	11.9	6.2	12.2
A_{14}	4.8	8.4	4.5	8.7
A_{15}	14.6	9.9	24.9[a]	12.4[a]
A_{18}	10.7	16.5	6.9	13.3
A_{19}	8.7	9.9	9.8	12.9

Column size, 5 ft × 0.25 in., column temperature and carrier gas (N_2) flow rate; (1) 201°C, 60 mℓ/min, (2) 190°C, 80 mℓ/min, (3) 179°C, 84 mℓ/min, (4) 187°C, 75 mℓ/min.

[a] Methyl ester.

isobutyloxycarbonate mixed anhydride, and ^{125}I-labeled tracer (Figure 17). The antisera prepared by Weiler and Wieczorek was sensitive in detecting 10 to 200 fmol (4 to 80 pg) of GA_3 and the cross-reactivity is shown in Table 20. Atzorn and Weiler[175,176] also prepared immunogens of $GA_{1,3,4,7,9}$ by a similar method and used antisera against these immunogens for radio- and enzyme immunoassay.

Nakagawa et al.[177] prepared an immunogen of $GA_{5,20}$, in which they used β-alanine as a spacer and the active ester method is applied instead of a mixed anhydride method as

Table 18
RETENTION TIMES (t_R, MIN) OF GA TMSi ESTER TMSi ETHERS[170]

GAs	t_R	Relative t_R
GA_1	7.26	0.93
GA_3	7.84	1.00
GA_4	4.54	0.58
GA_5	4.32	0.55
GA_7	4.44	0.57
GA_8	7.00 (220°C)	—
GA_9	2.78	0.36
GA_{12}	3.14	0.40
GA_{13}	6.92	0.88
GA_{14}	4.86	0.62
GA_{15}	5.76	0.74
GA_{16}	6.80	0.87
GA_{17}	6.76	0.86
GA_{18}	6.96	0.89
GA_{19}	6.26	0.80
GA_{20}	4.36	0.56
GA_{22}	7.68	0.98
GA_{24}	4.52	0.58
GA_{26}	12.20	1.56
GA_{27}	13.72	1.75
GA_{29}	7.66	0.98
GA_{30}	9.82	1.25
GA_{31}	5.38	0.69
GA_{34}	7.56	0.96
GA_{35}	7.00	0.89
GA_{36}	6.20	0.79
GA_{37}	9.50	1.21
GA_{40}	4.86	0.62
GA_{51}	4.70	0.60
GA_{53}	4.82	0.62

Column, 2% OV-1 on Chromosorb W (1 m × 3 mm i.d.); column temp., 210°C except for GA_8 (220°C); He flow, 50 mℓ/min.

shown in Figure 18. The antisera raised against these immunogens showed cross-reactivity as shown in Tables 21 and 22.

11. Isolation Examples of Free GAs

Many isolation procedures have been reported. Each procedure has its own characteristics suitable for the separation of certain GAs. In this section, three different procedures used for the isolation of free GAs from plant materials are discussed.

Yokota et al.[49,53] isolated $GA_{3,5,8,20,26,27}$ from immature seeds of *Pharbitis nil*. A portion of their work follows. They extracted 108 kg of immature seeds with methanol and obtained an AE fraction (57 g) by the procedure described in Section II. A. 1 and in Figure 4. The AE fraction was purified by the procedure shown in Figure 19.

Murofushi et al.[45] isolated 14 mg of GA_{19} from the boiling water extracts of 44,000 kg of bamboo shoots. The AE fraction obtained was successively purified by 10 transfer countercurrent distribution, repetitive charcoal chromatography, silicic acid adsorption chromatography, partition chromatography, and a recrystallization as shown in Figure 20.

Table 19
GC RETENTION TIMES (t_R) OF CONJUGATED GAs DERIVATIVES[145]

Conjugated GAs	TMSi ester-TMSi ether[a]	Me ester-TMSi ether[a]	TMSi ether	
GA_1-O(3)-GEt	15.9	15.6		
GA_1-O(13)-GEt	15.4	15.1		
GA_3-O(3)-GEt	12.8	14.0		
GA_3-O(13)-GEt	10.0	11.3		
GA_8-O(2)-GEt	12.1	12.8		
Allogibberic acid-O(13)-GEt	3.25	3.20		
GA_1-GEs			14.8[b]	18.8[c]
GA_3-GEs			16.5	20.7
GA_4-GEs			13.1	14.7
GA_{37}-GEs			22.0	23.0
GA_{38}-GEs			25.7	28.0

[a] Column, 3% QF-1 on Gas-Chrom Q (1.5 m × 4 mm i.d.), column temp, 245°C, N_2 flow; 175 mℓ/min.
[b] Column, 2% QF-1 of Chromosorb W (1 m × 3 mm i.d.), column temp. 224°C, N_2 flow; 34 mℓ/min.
[b] Column, 2% OV-1 on Chromosorb W (1 m × 3 mm i.d.), column temp, 243°C, N_2 flow, 33 mℓ/min.

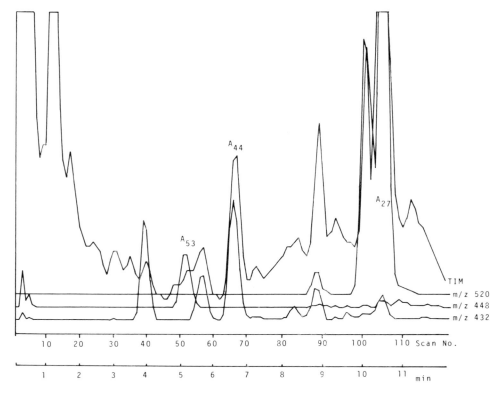

FIGURE 15. Mass chromatogram identifying GA_{27}, GA_{44}, and GA_{53} as their methyl ester-TMSi ethers from *P. purpurea seeds*.[170] Each mass spectrum reconstructed from the scan number 52, 67, and 107 respectively identical to that of authentic specimen.

Fukui and co-workers[62] isolated $GA_{39,48,49}$ and *ent*-12β,17-dihydroxykauren-19-oic acid from seeds (50 kg) of *Cucurbita pepo* L. by the procedure shown in Figure 21. An aqueous concentrate obtained by removal of methanol from the extracts of seeds was acidified to pH 3 with 6 N HCl and extracted with ethyl acetate. The ethyl acetate extract was submitted

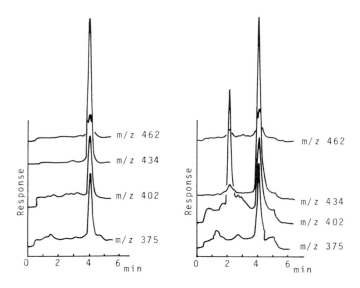

FIGURE 16. GC/SIM profile identifying GA_{19} from bleeding sap of *J. regia* L. as its methyl ester-TMSi ether.[171] (Left) Authentic GA_{19}-Me-TMSi. (Right) Methylated and trimethylsilylated GA_{19} fraction from bleeding sap. Column: 2% OV-1 on Chromosorb W (1 m × 3 mm i.d.), 210°C. He flow: 50 mℓ/min. Ionization: EI 22 eV.

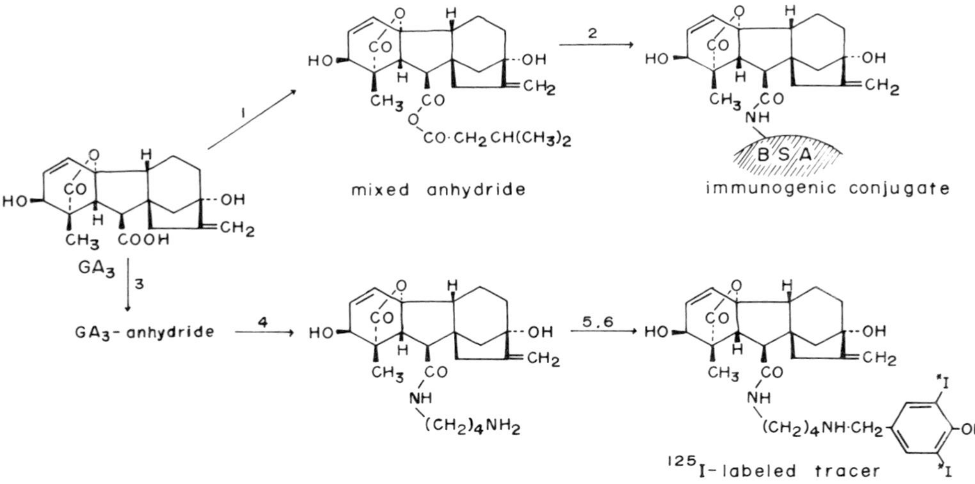

FIGURE 17. Preparation of immunogenic conjugate and ^{125}I-labeled tracer of GA_3.[173] (1) Isobutyloxychlorocarbonate/tri-*n*-butylamine, (2) BSA, (3) dicyclohexylcarbodiimide, (4) putrescine, (5) *p*-hydroxybenzaldehyde/ $NaBCNH_3$, (6) $Na^{125}I$/chloramine T.

to chromatographic purification without solvent fractionation to separate acidic compounds from neutral compounds.

12. Isolation Examples of Conjugated GAs

Yokota et al.[53] isolated GA_3-GEt, GA_8-GEt, GA_{26}-GEt, GA_{27}-GEt, and GA_{29}-GEt from immature seeds of *Pharbitis nil*. The methanol extracts of the immature seeds were submitted to the solvent fractionation as shown in Figure 4 to obtain an AB fraction. The AB fraction

Table 20
CROSS-REACTIVITY OF GAs AND RELATED COMPOUNDS WITH ANTI-GA$_3$-ANTISERUM[173]

Compounds	Cross-reactivity (%)
GA$_3$ Methyl ester	100
GA$_3$ Tetraacetyl-β-D-glucosyl ester	2.9
GA$_7$ Methyl ester	65.5
GA$_1$ Methyl ester	12.9
GA$_4$ Methyl ester	6.3
GA$_9$ Methyl ester	1.9
GA$_8$ Methyl ester	0
GA$_8$-O(2)-β-D-glucoside methyl ester	0
GA$_{13}$ Methyl ester	0
Gibberethion methyl ester	0

Note: All compounds were assayed after diazomethane treatment.

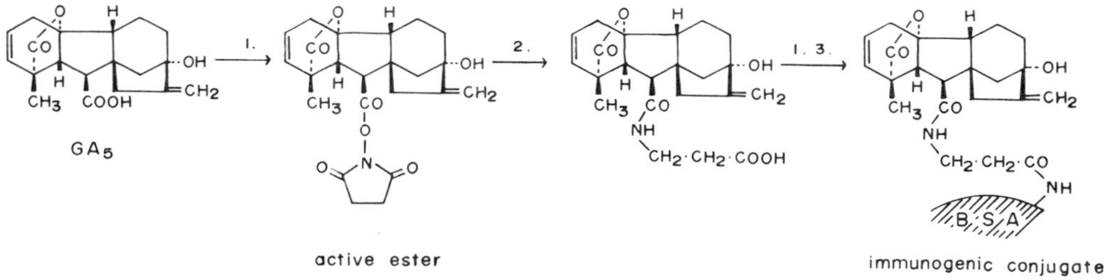

FIGURE 18. Preparation of immunogenic conjugate of GA$_5$.[177] (1) dicyclohexylcarbodiimide, (2) β-alanine, (3) BSA.

was further purified as shown in Figure 22. TLC was very useful for the final separation of each conjugated GA.

Hiraga et al.[61] isolated GA$_1$-GEs, GA$_4$-GEs, GA$_{37}$-GEs, and GA$_{38}$-GEs from *Phaseolus vulgaris* seeds. They extracted the seeds with aqueous acetone. The extract was submitted to the solvent fractionation shown in Figure 5. The NE fraction was purified by the procedure shown in Figure 23 to give GA$_1$-GEs, GA$_4$-GEs, GA$_{37}$-GEs, and GA$_{38}$-GEs. From the NB fraction more GA$_1$-GEs and GA$_{38}$-GEs were isolated through the purification processes shown in Figure 24.

Yamaguchi et al.[91] isolated GA$_5$-GEs and GA$_{44}$-GEs from the immature seeds of *Pharbitis purpurea* by a different purification method. The seeds were extracted with 80% acetone. The extract was submitted to the solvent fractionation procedure shown in Figure 6 to get an NB fraction which was submitted to the purification procedure shown in Figure 25. HPLC played an important role in purifying GA$_5$-GEs and GA$_{44}$-GEs.

B. Structural Determination

GAs have been shown to be related to a diterpene hydrocarbon skeleton, *ent*-gibberellane (**1**). Their nomenclature is based on Rowe's[178] proposal for a new nomenclature for diterpenes

Table 21 CROSS-REACTIVITY OF ANTI-GA$_5$-ANTISERA[177]		Table 22 CROSS-REACTIVITY OF ANTI-GA$_{20}$-ANTISERA[177]	
GAs-Me	Cross-reactivity (%)	GAs-Me	Cross reactivity (%)
GA$_1$	1.90	GA$_1$	0.20
GA$_3$	0.89	GA$_3$	0.05
GA$_4$	0.25	GA$_4$	0.09
GA$_5$	100	GA$_5$	32
GA$_7$	0.11	GA$_7$	0.04
GA$_8$	0.45	GA$_8$	0.12
GA$_9$	24.3	GA$_9$	66
GA$_{12}$	0.01	GA$_{12}$	—
GA$_{13}$	0.01	GA$_{13}$	0.04
GA$_{14}$	0.01	GA$_{14}$	0.04
GA$_{15}$	0.16	GA$_{15}$	0.05
GA$_{17}$	0.01	GA$_{17}$	0.04
GA$_{19}$	0.01	GA$_{19}$	—
GA$_{20}$	24.3	GA$_{20}$	100
GA$_{24}$	0.20	GA$_{24}$	0.25
GA$_{34}$	0.08	GA$_{34}$	0.04
GA$_{36}$	0.03	GA$_{36}$	0.04
GA$_{37}$	0.01	GA$_{37}$	0.04
GA$_{44}$	0.2	GA$_{44}$	0.2
GA$_{53}$	0.01	GA$_{53}$	0.04

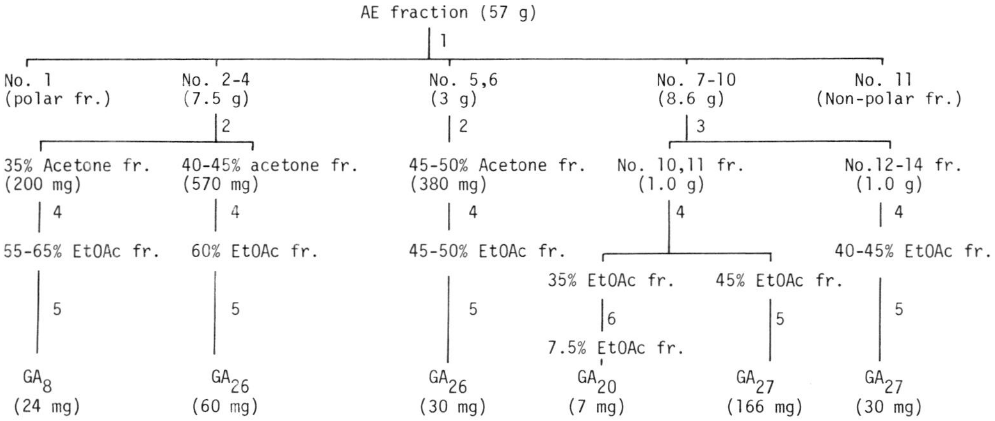

FIGURE 19. Isolation of GA$_{8,20,26,27}$ from *P. nil* seeds.[49,53] (1) Ten transfer countercurrent distribution between EtOAc and phosphate buffer (1*M*, pH 5.5). (2) Charcoal chromatography eluted with water containing an increasing acetone concentration by 5% steps. (3) Charcoal chromatography eluted with water containing continuously increasing acetone concentration. (4) Silicic acid adsorption chromatography eluted with benzene containing an increasing EtOAc concentration by 5% steps. (5) Crystallization in EtOH-EtOAc-*n*-hexane, (6) Partition chromatography between 1 *M* phosphate buffer (pH 5.62) on silicic acid (stationary phase) and benzene containing an increasing EtOAc concentration (mobile phase).

at the IUPAC meeting of 1968; for example, GA$_3$ is expressed as *ent*-3,-10,13-trihydroxy-20-norgibberella-1,16-diene-7,19-dioic-19,10-lactone, although until 1968 a nomenclature based on the common carbon skeleton to gibbane GAs (2) had been used and GA$_3$ had been expressed as 2,4a,7-trihydroxy-1-methyl-8-methylene-gibb-3-ene-1,10-dicarboxylic acid 1 → 4a lactone. However, in general, a convenient trivial nomenclature to allocate A-number to new naturally occurring and fully characterized GAs in the series A$_1$ to A$_n$ has been used.

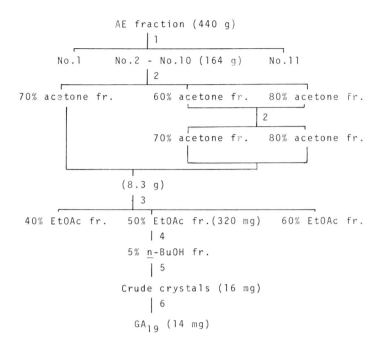

FIGURE 20. Isolation of GA_{19} from bamboo shoots.[45] (1) Countercurrent distribution between EtOAc and phosphate buffer (pH 5.5). (2) Charcoal adsorption chromatography eluted with water containing an increasing acetone concentration. (3) Silicic acid adsorption chromatography eluted with benzene containing an increasing EtOAc concentration, (4) Partition chromatography between phosphate buffer (1.5 M, pH 5.3) on silicic acid (stationary phase) and benzene containing an increasing concentration of n-BuOH (mobile phase). (5) Evaporation of solvent *in vacuo*. (6) Recrystallization in acetone-n-hexane.

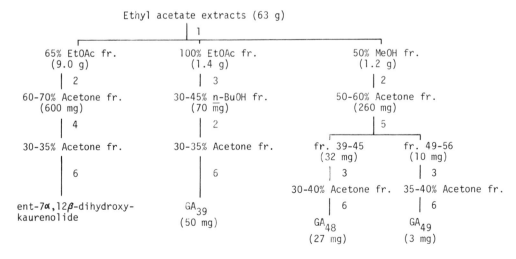

FIGURE 21. Isolation of $GA_{39,48,49}$ and ent-7α,12β-dihydroxy-kaurenolide from *Cucurbita* seeds.[62] (1) Silicic acid-Celite® adsorption chromatography eluted with benzene containing an increasing EtOAc concentration and then with EtOAc containing an increasing MeOH concentration. (2) Charcoal-Celite® adsorption chromatography eluted with water containing an increasing acetone concentration. (3) Partition chromatography between 1 M phosphate buffer (pH 5.5) on Celite® (stationary phase) and benzene containing an increasing n-BuOH concentration by 5% steps. (4) Silicic acid-Celite® adsorption chromatography eluted with benzene containing an increasing acetone concentration. (5) Partition chromatography on Sephadex® LH-20 with a solvent system of petroleum ether-EtOAc-AcOH-MeOH-water (100:80:5:40:7). (6) Crystallization in EtOAc-n-hexane.

94 Chemistry of Plant Hormones

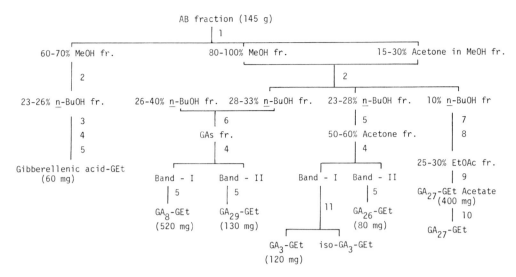

FIGURE 22. Isolation of GA glucosyl ethers from the immature seeds of *P. nil*.[53] (1) Charcoal adsorption chromatography eluted with water containing an increasing MeOH concentration. (2) Partition chromatography between phosphate buffer (1 M, pH 5.5, stationary phase) on Sephadex® G-50 and EtOAc containing an increasing concentration of *n*-BuOH (mobile phase). (3) Paper chromatography developed in a solvent system of 2-PrOH-7N NH$_4$OH-H$_2$O (8:1:1). (4) Silica gel TLC of triplicated development in a solvent system of CHCl$_3$:MeOH:AcOH:H$_2$O (70:20:3:2). (5) Charcoal adsorption chromatography eluted with water containing an increasing acetone concentration. (6) Silicic acid adsorption chromatography eluted with a solvent system of CHCl$_3$:MeOH:AcOH (90:10:0.5). (7) Acetylation with acetic anhydride chromatography eluted with benzene containing an increasing EtOAc concentration. (9) Crystallization in a mixture of EtOAc and *n*-hexane. (10) alkaline hydrolysis. (11). Paper chromatography in a solvent system of 2-PrOH:7N NH$_4$OH:H$_2$O (8:1:1).

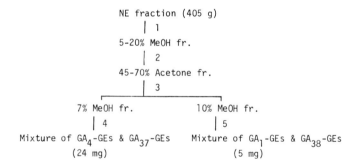

FIGURE 23. Isolation of GA$_4$-GEs and GA$_{37}$-GEs from the NE fraction of *P. vulgaris* seeds.[61] (1) Silica gel adsorption chromatography eluted with EtOAc containing an increasing MeOH concentration. (2) Charcoal adsorption chromatography eluted with water containing an increasing acetone concentration. (3) Silica gel adsorption chromatography eluted with benzene containing an increasing MeOH concentration. (4) Silica gel TLC developed in a solvent system of CHCl$_3$:MeOH (4:1). (5) Silica gel TLC developed in a solvent system of CHCl$_3$:MeOH (3:1).

The allocation of A-number has been suggested by Professors MacMillan (University of Bristol, England) and Takahashi (University of Tokyo) in order to remove the duplication of similar A-numbers for different GAs.[179]

GAs occur naturally in several chemical forms, free GAs, GAs-GEt, and GAs-GEs. Free GAs are classified into C_{19}-GAs and C_{20}-GAs depending upon the total carbon number. The C_{20}-GAs constitute the 26 members, GA$_{12-15,17-19,23-25,27,28,36-39,41-44,46,52,53,64-66}$, with the basic structure (3) varying in the carbon substituent (C-20) at C-10, namely methyl, hydroxylmethyl

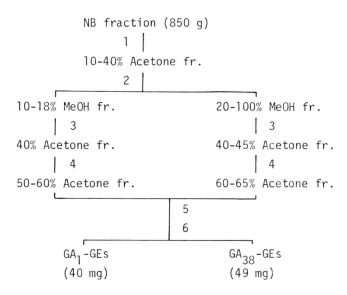

FIGURE 24. Isolation of GA_1-GEs and GA_{38}-GEs from the NB fraction of *P. vulgaris* seeds.[61] (1) Charcoal adsorption chromatography eluted with water containing an increasing acetone concentration. (2) Silica gel adsorption chromatography eluted with $CHCl_3$ containing an increasing MeOH concentration. (3) Charcoal adsorption chromatography eluted with water containing an increasing acetone concentration. (4) Silica gel adsorption chromatography eluted with benzene containing an increasing acetone concentration. (5) Silica gel TLC develpmed in a solvent system of $CHCl_3$:MeOH (3:1). (6) TLC developed in a solvent system of acetone:benzene (4:1).

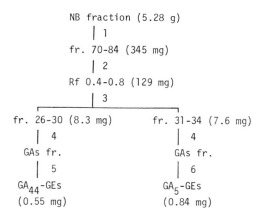

FIGURE 25. Isolation of GA_5-GEs and GA_{44}-GEs from *P. purpurea* seeds.[91] (1) Sephadex® LH-20 (750 × 35 mm i.d.) chromatography eluted with a solvent system of MeOH:$CHCl_3$ (4:1) in 5 mℓ fractions. (2) TLC in a solvent system of $CHCl_3$:MeOH:AcOH (25:10:1). (3) Sephadex® LH-20 (335 × 22 mm i.d.) chromatography eluted with a solvent system of n-BuOH:H_2O (6:100) in 6 mℓ fractions. (4) TLC in a solvent system of $CHCl_3$:MeOH:AcOH (25:5:1), duplicated development. (5) HPLC on Nucleosil $5C_{18}$ eluted with 30% 2-PrOH in H_2O). (6) HPLC on Nucleosil $5C^{18}$ eluted with 50% MeOH in H_2O.

(which forms δ-lactone with a carboxyl group at C-4), an aldehyde, and a carboxyl group are the substituents. Exceptionally, $GA_{41,42}$ contain a hydrated exomethylene at C-16.

The C_{19}-GAs constitute 42 members: $GA_{1-11,16,20-22,26,29-35,40,45,47-51,54-63,67,68}$ containing the

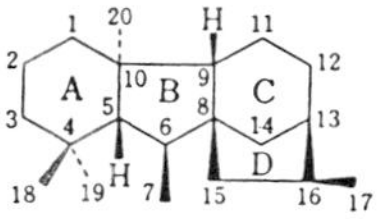

96 *Chemistry of Plant Hormones*

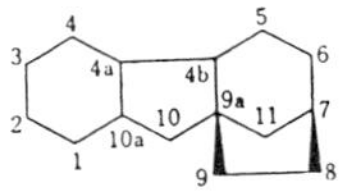

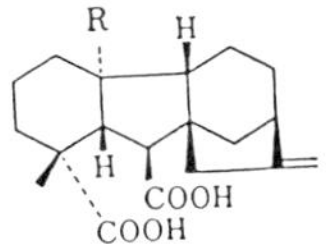

FIGURE 26. Structures 1 to 8 and 11.

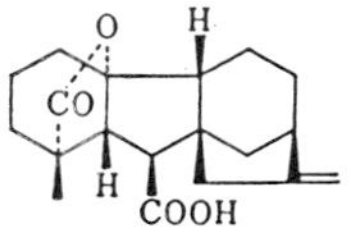

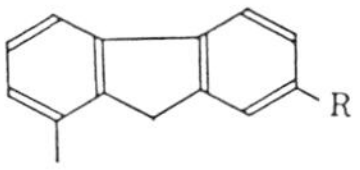

basic structure (4) with hydroxyl, ketone, and epoxide groups and double bonds. Exceptionally, some GAs such as $GA_{21,22,59}$ contain β-oriented one-carbon substituents in oxidized form such as hydroxylmethyl and carboxyl groups, and $GA_{2,10}$ contain a hydrated exomethylene at C-16 (Figure 26).

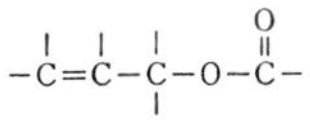

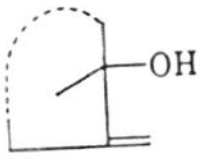

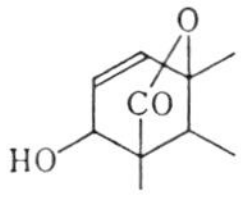

In the early years of GA research, chemical structures of GAs were studied mainly by the University of Tokyo and ICI groups. These two groups carried out extensive degradation studies using the fungal gibberellins GA_1 and GA_3 for the elucidation of structure. After several proposals and corrections, Cross et al.[180] of the ICI group succeeded in presenting a correct chemical structure of GA_3.

At this writing, 68 free GAs, 7 GAs-GEt, 7 GAs-GEs, and gibberethione have been identified from natural sources. An outline of the structural determination and physicochemical properties of GAs are described elsewhere in this chapter.

1. Free GAs

a. $GA_{1,3}$ by Degradation Studies[181-199]

GA_1 has the molecular formula, $C_{19}H_{26}O_6$, and contains two hydroxyls, a carboxyl, γ-lactone, and a double bond, while GA_3, with the molecular formula, $C_{19}H_{24}O_6$, has two hydroxyls, a carboxyl, γ-lactone, and two double bonds. On catalytic hydrogenation under controlled condition, GA_3-methyl ester (GA_3-Me) was converted to GA_1-Me by simple hydrogenation and an acidic compound by hydrogenolysis of the γ-lactone ring. Dehydrogenation of $GA_{1,3}$ yielded 1,7-dimethylfluorene (5), or gibberene. Thus, GA_1 and GA_3 may contain four carbon rings (hydrofluorene and one additional ring). The above chemical evidence suggests that GA_3 must have a structure in which one additional double bond exists relative to the structure of GA_1; easy conversion of GA_3 to an acidic compound by hydrogenolysis indicates that GA_3 has a partial structure (6). Treatment of GA_1 with a strong acid afforded gibberellin C, which is a ketonic compound with the same molecular formula as GA_1, while GA_3 was converted into allogibberic acid via gibberellenic acid under mild acidic conditions. Treatment of allogibberic acid with a strong acid gave gibberic acid.

Allogibberic acid afforded formaldehyde and a ketol by ozonolysis. Oxidation of the ketol with sodium bismuthate yielded a keto-acid. These results suggest that allogibberic acid contains an exocyclic methylene with a tertiary hydroxyl at the allylic position, suggesting the partial structure (7). Selenium dehydrogenation of the keto-acid afforded 1-methyl-7-

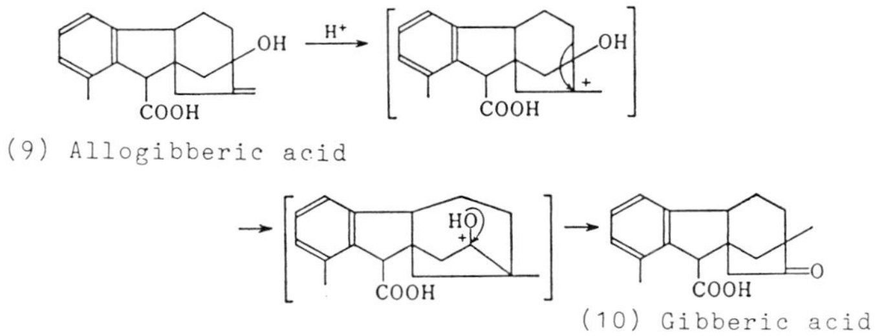

FIGURE 27. Conversion of allogibberic acid to gibberic acid thorough Wagner-Meerwein rearrangement.

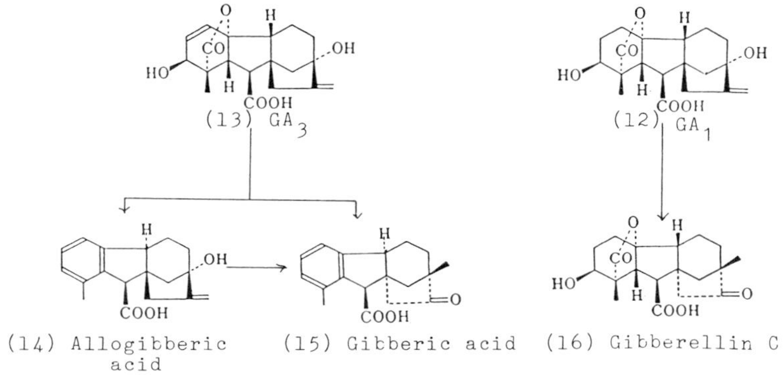

FIGURE 28. Structures of acid degradation products of GA_1 and GA_3.

hydroxyfluorene **(8)**, indicating that the tertiary hydroxyl of allogibberic acid is located at the C-7 position of hydrofluorene. On the other hand, gibberic acid is a keto-acid which has the same molecular formula as allogibberic acid. Considering the results of degradation studies of gibberic acid and the fact that conversion of allogibberic acid into gibberic acid proceeds through Wagner-Meerwein rearrangement as shown in Figure 27, the structures of allogibberic acid and gibberic acid were assigned as **(9)** and **(10)** in Figure 26, respectively. By this assignment the results of all chemical reactions using both compounds can be explained. Later, total synthesis of gibbereic acid was achieved to confirm this structural assignment unambiguously.

$GA_{1,3}$ are considered to have a similar C and D ring structure as that of allogibberic acid. The A ring of GA_3 was converted to an aromatic ring by loss of the γ-lactone and the secondary hydroxyl by acid treatment. The presence of the partial structure **(11)** was indicated by the conversion of GA_3 to an α,β-unsaturated ketone by manganese dioxide oxidation. Based on all the evidence on the structures of A, C, and D rings, GA_1 and GA_3 can be assigned structures **(12)** and **(13)**, respectively.

In 1962, by X-ray analysis McCapra et al.[200,201] confirmed the structure of GA_3 presented by Cross et al.,[180] using a crystal of methyl bromogibberellate. Its absolute configuration was assigned by analysis of CD spectra of GA_3-related compounds. Later, Hartsuck and Lipscomb[202] reported X-ray analysis of di-*p*-bromobenzoate of GA_3. Structural relationships including stereochemistry between GA_3 and the main acid degradation products, gibberellin C **(14)**, allogibberic acid **(15)**, and gibberic acid **(16)**, are summarized in Figure 28.

b. Chemical Evidence and Spectrometric Methods

NMR spectroscopy and MS, which were introduced into GA research in the 1960s in addition to IR, UV, CD, and ORD, have proven to be very useful in the structural elucidation of GAs.

Chemical structures of $GA_{2,4-11,16}$[9,11-13,17,40,204] have been determined by chemical degradation studies, by correlating them chemically to $GA_{1,3}$ (Figure 29) and/or ^{1}H-NMR (PMR) analyses. X-ray analysis of GA_4 was also reported.[205]

Structural determinations of C_{20}-GAs such as GA_{12-15}[13,15,16,48,205-207] which were isolated in the early stages of gibberellin research, have been conducted by combination of chemical degradation and extensive PMR analyses. The structure of GA_{12} was further confirmed by conversion of the structure known as ent-7β-hydroxykaurenolide to GA_{12}. Thereafter, GA_{13} was converted into GA_4[207,208] whose structure had been definitely established; later X-ray analysis of GA_{13}-Me[205] was reported.

MS was used to clarify the structure of GA_{17}[48] and later GAs. Thus, PMR and mass spectra data have been accumulated, and in several cases structures of new GAs have been determined without degradation studies. Such an example has been seen in the characterization of GA_{19}.[40] The spectrometric properties of GAs are summarized and discussed in Section II.B.1.b. The outline of the structural elucidation of GA_{19} is herein described.

The molecular formula of GA_{19} was determined as $C_{20}H_{26}O_6$ from a high resolution mass spectrum of its methyl ester. The PMR spectrum of GA_{19}-Me in deuterochloroform showed the presence of a tertiary methyl (δ 1.15), two methoxycarbonyls (δ3.69, 3.77), a tertiary aldehyde (δ9.73), an exocyclic methylene (δ4.98, 5.20), and two vicinal methines (δ2.43, 3.88) characteristic of the methine protons on C-5 and C-6 of ent-gibberellane. While in the PMR spectrum of GA_{19}-Me in deuteropyridine, the signals due to the exocyclic methylene were observed at δ5.10 and 5.57 and those due to the methine protons on C-5 and C-6, at δ2.46 and 4.16, respectively. Such large deshielding effects on one of the exocyclic methylene protons and the methine proton on C-6 in deuteropyridine indicate the presence of 13-hydroxyl and two carbonyl functions at the C-4 and C-10 positions. The presence of 13-hydroxyl in GA_{19} was further supported by the high resolution mass spectrum, in which a fragment ion peak at m/z 241 ($C_{17}H_{21}O$) due to the largest hydrocarbon containing an oxygen was much more intense than the largest hydrocarbon peak at m/z 225 ($C_{17}H_{21}$). From the above evidence and structural characteristics of known GAs, GA_{19} was identified as ent-13-hydroxylgibberell-16-ene-10-formyl-4,7-dioic acid (Figure 2).

^{13}C-NMR (CMR) is also very useful for structural and biosynthetic studies of natural products. Although the natural abundance of ^{13}C is very low (approximately 1%), application of FT-NMR techniques enables CMR measurement. Yamaguchi et al.[209] measured CMR spectra of various GAs and succeeded in the assignment of the carbons of GAs (see Tables 23 and 24). Evans et al.[210] also reported CMR analyses of GAs.

Yamaguchi et al.[23] further applied CMR to the structural determination of GA_{40} isolated from a culture filtrate of G. fujikuroi. It has the molecular formula $C_{19}H_{26}O_5$, suggesting it is a structural isomer of GA_4. From comparison of a CMR spectrum of GA_{40}-Me with those of GA_4-Me and GA_9-Me, GA_{40} was successfully assigned as 2-hydroxylated GA_9 (2-OH-GA_9). Configuration of the 2-hydroxyl was determined from the PMR spectrum of GA_{40}-Me, i.e., the carbinol proton of GA_{40}-Me was observed at δ4.14 as an ill-defined multiplet ($W^{1/2}$ = 10 Hz) in deuterochloroform, while the signal due to the carbinol proton on C-2 in GA_{29} (which has a β-oriented 2-hydroxyl) appeared at δ4.30 as a broad multiplet ($W^{1/2}$ = 20 Hz).[25] Thus, GA_{40} was characterized as 2β-OH-GA_9 (Figure 2). The structure of GA_{40} was further confirmed by chemical conversion of GA_{40}-Me into 13-deoxy-GA_5-Me. CMR was utilized effectively in the structural determination of $GA_{50,52}$.[68]

c. Identification of New GAs by GC/MS Analysis

GC/MS can be used for identification of minute quantities of chemical compounds based on their mass spectra and their retention times on GC.

Pryce and MacMillan[48] first applied this method to clarify endogenous GAs in the immature

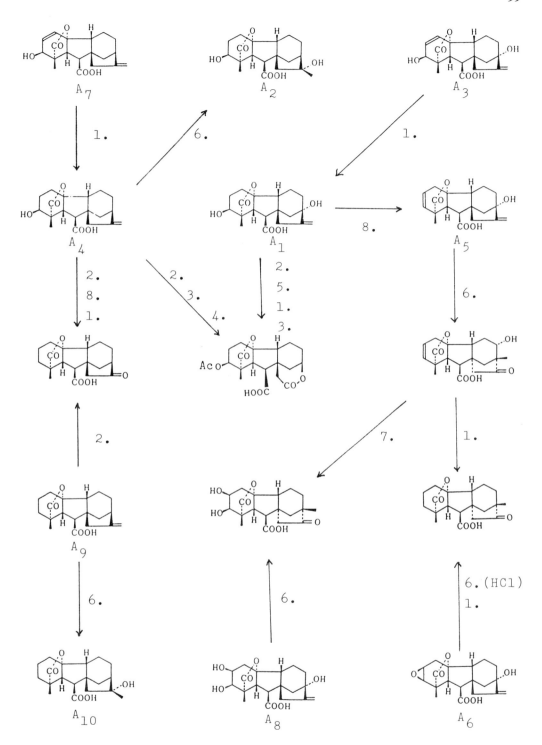

FIGURE 29. Chemical correlation among GA_1 to GA_{10}: (1) catalytic reduction, (2) ozonolysis, (3) acetylation, (4) oxidation with peracid, (5) oxidation with sodium bismutate, (6) acid treatment, (7) oxidation with osmium tetroxide, (8) tosylation and collidine treatment.

seeds of *Phaseolus coccineus*. This resulted in the identification of several known GAs and the discovery of a new GA, GA_{17}, which was an isomer of GA_{13}. This discovery was based

Table 23
ASSIGNMENT OF CMR SPECTRA OF C_{19}-GAs IN DEUTEROPYRIDINE

Carbon	GA_1	GA_3	GA_4	GA_5	GA_7	GA_9	GA_{16}	GA_{30}	GA_{35}	GA_{40}	GA_{50}
1	28.1 (t)	132.3(d)	28.2 (t)	35.6 (t)	132.3 (d)	30.9 (t)	71.8 (d)	132.4	29.4 (t)[a]	39.1	38.9 (t)
2	29.2 (t)	134.3 (d)	29.2 (t)	128.1 (d)[a]	134.1 (d)	19.8 (t)	39.7 (t)	134.3	29.9 (t)[a]	64.6	68.1(d)
3	70.0 (d)	70.0 (d)	70.0 (d)	133.1 (d)[a]	70.0 (d)	34.6 (t)	70.4 (d)	70.0	69.8 (d)	45.3	73.1 (d)
4	55.5 (s)	54.5 (s)	55.5 (s)	48.5 (s)	54.4 (s)	49.1 (s)	54.8 (s)	54.7	55.6 (s)	46.3	54.4 (s)
5	52.5 (d)[a]	51.6 (d)	51.9 (d)	56.1 (d)	52.1 (d)	58.2 (d)	51.1 (d)	52.0	52.2 (d)	58.7	52.1 (d)
6	52.7 (d)[a]	52.1 (d)	52.8 (d)	52.2 (d)	52.4 (d)	52.8 (d)	53.0 (d)	53.4	52.9 (d)	52.0	52.1 (d)
7	175.2 (s)	174.8 (s)	175.2 (s)	N.O.	174.8 (s)	174.8 (s)	174.9 (s)	174.9	175.2 (s)	175.3	174.9 (s)
8	49.8 (s)	50.6 (s)	51.5 (s)	50.7 (s)	52.1 (s)	51.3 (s)	52.1 (s)	52.9	51.9 (s)	51.7	52.7 (s)
9	53.5 (d)	53.5 (d)	54.0 (d)	53.5 (d)	52.9 (d)	54.1 (d)	54.0 (d)	50.8	61.0 (d)	54.2	60.5 (d)
10	93.9 (s)	91.1 (s)	94.1 (s)	91.9 (s)	91.4 (s)	93.0 (s)	96.2 (s)	91.3	94.6 (s)	92.8	94.7 (s)
11	18.0 (t)	17.6 (t)	16.5 (t)	18.0 (t)	16.3 (t)	16.4 (t)	18.7 (t)	27.8	64.0 (d)	16.6	63.9 (d)
12	39.9 (t)	39.9 (t)	31.8 (t)	39.8 (t)	31.8 (t)	31.7 (t)	32.3 (t)	75.4	43.0 (t)	31.8	43.0 (t)
13	77.9 (s)	77.7 (s)	39.4 (d)	77.8 (s)	39.2 (d)	39.3 (d)	39.1 (d)	49.0	39.4 (d)	39.4	39.4 (d)
14	46.3 (t)	45.6 (t)	37.4 (t)	46.6 (t)	36.9 (t)	37.2 (t)	37.0 (t)	34.7	37.4 (t)	37.3	37.5 (t)
15	44.0 (t)	44.0 (t)	45.0 (t)	43.7 (t)	45.0 (t)	44.8 (t)	45.2 (t)	45.6	45.6 (t)	45.0	45.6 (t)
16	159.2 (s)	159.1 (s)	157.7 (s)	158.7 (s)	157.7 (s)	157.6 (s)	158.2 (s)	153.4	158.0 (s)	157.8	157.9 (s)
17	106.5 (t)	106.7 (t)	107.2 (t)	106.4 (t)	107.4 (t)	107.1 (t)	107.4 (t)	109.3	107.4 (t)	107.3	107.4 (t)
18	15.6 (q)	15.5 (q)	15.5 (q)	15.8 (q)	15.5 (q)	17.7 (q)	14.9 (q)	15.7	15.8 (q)	18.4	15.9 (q)
19	179.0 (s)	179.6 (s)	179.0 (s)	N.O.	179.6 (s)	178.8 (s)	179.3 (s)	N.O.	179.5 (s)	179.8	178.7 (s)

Note: Chemical shifts are expressed in ppm downfield from internal tetramethylsilane.[23,63,209] N.O. Not observed because of the short pulse interval.

[a] Chemical shifts thus indicated may be reversed in each vertical column.

Table 24
ASSIGNMENT OF CMR SPECTRA OF C_{20}-GAs IN DEUTEROPYRIDINE

Carbon	GA_{12}	GA_{14}	GA_{17}	GA_{18}	GA_{23}	GA_{28}	GA_{36}	GA_{37}	GA_{38}	GA_{39}	GA_{52}
1	30.8 (t)	28.4 (t)	37.7	28.5	27.6	31.0	27.5 (t)	29.7 (t)	29.7	31.3	44.5 (t)
2	32.2 (t)	34.6 (t)	22.6	34.6	30.2	31.9	30.1 (t)	33.6 (t)	33.7	32.2[a]	68.3 (d)
3	71.2 (d)	71.0 (d)	39.0	70.9	72.9	71.1	73.0 (d)	73.5 (d)	73.5	71.6	77.0 (d)
4	50.6 (s)	49.5 (s)[a]	46.4	49.6	49.7	51.3	50.2 (s)	49.0 (s)	49.1	51.2[b]	48.3 (s)[a]
5	50.4 (d)	50.2 (d)	56.5	51.0	48.2	50.7	47.7 (d)	46.8 (d)	45.8	50.6[c]	45.3 (d)
6	52.1 (d)	51.9 (d)	52.2	51.9	51.8	52.1	51.7 (d)	52.7 (d)	52.7	52.2	51.8 (d)
7	177.1 (s)	177.9 (s)[b]	177.0[a]	177.9[a]	176.4[a]	177.5[a]	176.4(s)[a]	174.9 (s)[a]	174.9[a]	177.2[d]	174.7 (s)[b]
8	50.6 (s)	50.0 (s)[a]	48.7	48.6	52.8	48.9	50.2 (s)	49.9 (s)	47.8	51.0[c]	51.4 (s)[a]
9	57.0 (d)	57.6 (d)	56.9	57.2	56.7	56.6	56.7 (d)	56.1 (d)	55.9	54.1	62.8 (d)[c]
10	57.6 (s)	44.6 (s)	57.7	44.7	48.4	57.3	49.5 (s)	41.9 (s)	41.6	58.4	42.8 (s)
11	19.4 (t)	17.3 (t)	20.4	18.9	19.7	20.6	18.6 (t)	16.1 (t)	17.2	31.8[a]	64.3 (d)[c]
12	32.2 (t)	32.6 (t)	40.1	40.8	40.2	40.3	32.2 (t)	31.9 (t)	40.0	75.7	43.1 (t)
13	40.2 (d)	40.6 (d)	78.3	78.3	78.2	78.4	39.9 (d)	39.9 (d)	78.3	51.9[b]	39.8 (d)
14	36.9 (t)	39.8 (t)	46.0	49.1	45.9	46.4	37.2 (t)	36.7 (t)	45.8	33.4	37.1 (t)
15	47.3 (t)	47.1 (t)	45.7	46.2	45.9	46.4	46.9 (t)	45.4 (t)	45.1	48.4	47.3 (t)
16	157.9 (s)	157.3 (s)	158.9	158.7	159.1	159.3	157.6 (s)	157.9 (s)	159.6	154.7	157.8 (s)
17	105.9 (t)	105.7 (t)	105.1	105.3	105.4	105.0	106.2 (t)	106.4 (t)	105.4	107.7	106.5 (t)
18	24.8 (q)	25.4 (q)	29.9	25.1	22.6	24.9	22.6 (q)	21.1 (q)	21.2	25.3	21.2 (q)
19	178.0 (s)[a]	180.6 (s)[b]	177.9[a]	180.7[a]	177.0[a]	177.9	N.O.[a]	175.8 (s)[a]	175.6[a]	178.0[d]	175.7 (s)[b]
20	178.5 (s)[a]	15.5 (q)	178.4[a]	15.3	N.O.	178.6[a]	N.O.	74.4 (t)	74.4	178.7	74.4 (t)

Note: Chemical shifts are expressed in ppm downfield from internal tetramethylsilane.[63,209] N.O.: Not observed because of the short pulse interval.

[a—d] Chemical shifts thus indicated may be reversed in each vertical column.

on the fact that the new GA-Me showed a different mass spectrum from that of GA_{13}–Me despite the same retention time on GC.

Thereafter, GC/MS and a modified technique, GC/SIM, have been utilized widely in the identification and/or quantification of GAs (see Section II.A). Of new GAs identified recently, $GA_{43-47,51,53,55-58,60-68}$ [25-27,64,67,70,72,74,75,211,212] were identified by GC/MS using fully characterized compounds prepared from known GAs or *ent*-kaurenoic acid derivatives by chemical and microbiological techniques (see Section III).

2. Conjugated GAs

Yokota et al.[55] isolated seven GAs-GEt (tentatively named F-I to F-VII) from the immature seeds of *Pharbitis nil* and succeeded in their structural elucidation, indicating that they were composed of five natural GAs-GEt and two artefacts derived from F-I. Herein, structural elucidation of GA_3-GEt (F-I) is described.

F-I was hydrolyzed with 6 *N* HCl to gibberic acid and glucose, while enzymatic hydrolysis of F-I afforded GA_3. The signals observed in a PRM spectrum of F-I in deuteroacetone-D_2O were shown to correspond to those of GA_3 and β-glucose. The absorption due to the anomeric proton characteristic of β-glucopyranoside was observed δ4.48 as a 1H doublet (J = 8 Hz). Thus F-I was clearly GA_3-β-glucopyranosyl ether.

On the other hand, in the PRM spectrum of F-I methyl ester tetraacetate in deuterochloroform, the signal due to the proton on C-3 of GA_3-Me moiety was observed as a 1H doublet (J = 4 Hz) at δ3.98 while that of the proton on C-3 of F-I was observed as a 1H doublet (J = 4 Hz) at δ4.05. This indicates that the signal due to the proton on C-3 of F-I methyl ester was not shifted by acetylation. Thus, the structure of F-I was shown to be 3-*O*-β-D-glucopyranosyl-GA_3 (Figure 3).

The other GAs-GEts isolated from natural sources have been characterized[59,87,90] by similar methods to GA_3-GEt.

It should be noted that CMR was very useful for structural determination of GAs-GEt. GA_{35}-GEt was isolated from immature seeds of yellow bloom, *Cytisus scoparius*, and its structure was determined based on spectrometric and chemical evidence. Later, the CMR spectra of GA_{35}-GEt and its aglycon were well studied. Chemical shifts are shown in Table 25.[213] The chemical shift of the proton on C-11 of the aglycone moiety exhibited a clear downfield shift compared with that of GA_{35}, showing that the glucosyl group attaches to the C-11 oxygen. The chemical shifts of C-1′ to C-6′ were characteristic of those of β-glucose. It means that the glucosyl ether could be assigned as 11-*O*-β-D-glucosyl-GA_{35} simply by CMR (Table 25), and suggests the possibility that the structures of new GAs-GEt may be elucidated by CMR without degradation studies.

Hiraga et al.[61] obtained several GA-active substances from an NE and an NB fraction of mature seeds of *Phaseolus vulgaris*. One of these GA-active substances, tentatively named NE-I, was indicated to be composed of two components by GC trace using its trimethylsilyl derivative. On acid hydrolysis, NE-I yielded glucose, GA_4, GA_{37}, GA_2 (a hydrate at C-16 exomethylene of GA_4), and GA_{37} hydrate, while on alkaline hydrolysis, NE-I yielded glucose, GA_4, GA_{37}, and C-3 epimers of GA_4 and GA_{37}. In a PMR spectrum of NE-I in deuteroacetone, the signal characteristic of the anomeric proton of β-glucose was observed at δ5.53 as a 1H doublet (J = 7.5 Hz) and the other signals were assigned to those of glucose, GA_4, and GA_{37}. Considering the above chemical and spectrometric evidence together with a mass spectrum of the NE-I trimethylsilyl derivative. NE-I was shown to be composed of GA_4-GEs and GA_{37}-GEs (see Figure 3). The other GAs-GEs were also characterized by similar methods.[91,92]

Gibberethione (see Figure 3), another type of GA conjugate isolated from the immature seeds of *Pharbitis nil*, was fully characterized[93] by X-ray analyses and CD spectroscopy involving absolute configuration.

Table 25
ASSIGNMENT OF CMR SPECTRA OF GA_{35} AND GA_{35}-GEt[213]

Carbon	GA_{35}	GA_{35} glucoside
1	29.43	29.49
2	29.85	29.49
3	69.77	69.83
4	55.64	55.76
5	52.18	52.18
6	52.91	52.97
7	175.16	175.35
8	51.87	51.80
9	61.04	59.46
10	94.59	94.77
11	64.01	62.86
12	42.96	42.41
13	39.38	39.13
14	37.43	37.80
15	45.62	45.57
16	157.99	157.63
17	107.39	107.57
18	15.77	15.78
19	179.53	179.59
1'	—	107.57
2'	—	75.60
3'	—	78.03
4'	—	74.57
5'	—	78.57
6'	—	71.78

3. Physicochemical Properties of GAs
a. UV Spectra

In general, GAs show no strong UV absorption because they do not have a conjugated system of unsaturated bonds in their molecules. However, GAs with C-2~C-3 double bonds such as $GA_{5,22,31}$ exhibit medium strength of absorption around 225 nm (ϵ = approximately 1000) due to the transannular interaction between the C-2~C-3 double bond and the C-4 → C-10 lactone carbonyl, such absorption not being observed in GAs with C-1~C-2 double bond such as $GA_{3,7,30,32}$.

Authentic samples of GA_3 often exhibit an absorption at 253 nm. This is due to the presence of gibberellenic acid (see Figure 5), an artefact of GA_3, which shows an intense absorption at 253 nm (ϵ = 22,400) due to heteroannular diene. Allogibberic and gibberic acids, which are acid degradation products derived from GA_3, exhibit absorption around 260 nm characteristic of an aromatic ring.

GAs with a ketone in the molecule, such as GA_{26} and GA_{33}, are known. Methyl esters of the former show an absorption at 288 nm (ϵ = 280) due to β,γ-unsaturated ketone, and the latter shows a weak absorption at 290 nm (ϵ = 60).

b. IR Spectra

IR spectroscopy is very useful for the analyses of structural units as well as for the identification of unknown compounds.

In the spectra of C_{19}-GAs-Me, absorptions due to the γ-lactone appeared in the range of 1770 to 1780/cm, although those of free acids were observed around 1750/cm. Signals of 1650 to 1670 and 850 to 900/cm region are ascribed to the exocyclic methylene. The hydroxyl frequency occurs between 3200 and 3600/cm. GAs sometimes form dimorphic or hydrated crystals and give different IR spectra from those of authentic specimens.

c. NMR Spectra

NMR spectroscopy gives the most informative evidence for structural analysis of GAs. Hanson[214] reported the first assignment of chemical shifts of protons in GA_{1-9}. Hanson's work and PMR studies on GAs conducted thereafter indicated that the structural characteristics of GAs are well reflected in the signals due to a methyl at C-4, two methine protons at C-5 and C-6, and an exocyclic methylene at C-16. In Table 26, chemical shifts of protons in GAs-Me are presented.

The absorptions due to the methyl at C-4 are observed in the range δ1.03 to 1.24 in deuterochloroform, those of GAs-Me with 3-hydroxyl or C-2 to C-3 double bonds appearing at comparatively lower field. In the spectra of GAs-Me determined in deuteropyridine, the deshielding effect on the methyl at C-4 by 3-hydroxyl further increased as compared with the spectra in deuterochloroform, while that by C-2~C-3 double bond was not observed. A pair of doublets (J = approximately 10 Hz) in the range δ1.8 to 4.0 due to the methine protons on C-5 and C-6 is a characteristic feature of the PMR of many GAs, while the two methine protons of GA_{27}-Me are observed as a 2H singlet. The large coupling constant of the doublets indicates that the dihedral angle between two protons on C-5 and C-6 is almost 180°. The chemical shift of the proton on C-5 is affected by 3β-hydroxyl, because the proton on C-5 is in a β-axial conformation and therefore the 1,3-diaxial transannular effect from 3β-hydroxyl causes the deshielding of the proton on C-5. The deshielding of the proton on C-6 is caused by a carbonyl (C-20) at C-10.

The signals due to the exocyclic methylene protons on C-17 are observed in the range δ4.9 to 5.3 as two 1H broad singlets while in a few GAs they are observed as a 2 H broad singlet. The adjacent 13-hydroxyl causes a significant downfield shift in the position of one of the exocyclic methylene proton resonances. The deshielding effect was amplified by the use of deuteropyridine as an NMR solvent. The absorptions due to exocyclic methylene protons of 13-OH-GAs appeared as a pair of 1H broad singlets at intervals of approximately 0.3 and 0.5 ppm in the spectra determined in deuterochloroform and deuteropyridine, respectively, while in those of GAs without hydroxyls at the C-12, C-13, and C-15 positions the interval is around 0.1 ppm.

As already described in Chapter 2, Section II.B.2, Yamguchi et al.[209] and Evans et al.[210] analyzed the CMR of a series of GAs to indicate its applications to structural elucidation. The assignment of the chemical shifts of carbons in GAs was made by analyzing the proton noise and off-resonance-decoupled spectra. Tables 25 and 26 show the results by Yamaguchi and co-workers and by Fukui et al.[63,68]

The carbonyl carbon signals are observed at 174.8 to 180.7 ppm (downfield shift from tetramethylsilane). In the spectra of C_{19}-GAs, the C-7 signal appears at 174.8 to 175.2 ppm and in the C-19 signal, at 178.8 to 179.6 ppm. Tricarboxy-GAs exhibit three carbonyl carbons (C-7, C-19, and C-20) at 177.0 to 178.7 ppm. Two carbonyl carbons of δ-lactone GAs such as $GA_{37,38}$ resonance at 174.9 (C-7) and 175.6 to 175.8 ppm (C-19). In the spectra of $GA_{23,36}$, the signals due to aldehydic carbonyls (C-20) are not observed because of their long relaxation time and the aldehyde-lactol interconversion.

The resonance of exocyclic double bond carbons appears at 153.4 to 159.6 (C-16) and 105.1 to 109.3 ppm (C-17). It should be noted that the 12-hydroxyl causes an upfield shift of the C-16 resonance and a downfield shift of the C-17 resonance. The 13-hydroxyl exhibits a similar effect to that of the 12-hydroxyl on the C-16 and C-17 resonances, the effect being comparatively small. The other olefinic carbon signals are observed at 128.1 to 134.3 ppm.

The position of the carbinol carbon resonance is in the range of 64.0 to 78.4 ppm. The C-13 signals of 13-OH-GAs show almost the same chemical shift at about 78 ppm. The C-3 signals in the spectra of 3-hydroxylated C_{19}-GAs (3-OH-C_{19}-GAs) and 3-OH-C_{20}-GAs are observed at 69.8 to 70.4 and 70.9 to 73.5 ppm, respectively.

The C-10 signal also reflects the structural feature of GAs. Though the C-10 in C_{19}-GAs resonances at 91.1 to 96.2 ppm, the C-10 chemical shift of C_{20}-GAs changes according to

Table 26
ASSIGNMENT OF PMR OF GAs

GAs	H$_3$-18	Hs-20	H$_2$-17	H-1	H-2	H-3	H-5	H-6	H-11	H-12	H-15
A$_1$	1) 1.13		5.23, 4.92			3.75	3.17	2.65			
	2) 1.44		5.59, 5.04			4.08	3.72	2.97			
A$_2$	2) 1.45		1.45 (CH$_3$)			4.08	3.79	3.02			
	3) 1.03		1.28			3.85	3.22	2.70			
A$_3$	1) 1.23		5.25, 4.94	6.30	5.87	4.08	3.17	2.75			
	2) 1.54		5.45, 5.02	6.41	6.11	4.49	3.69	3.05			
A$_4$	1) 1.13		4.95, 4.83			3.80	3.15	2.65			
	2) 1.43		4.96, 4.86			4.06	3.68	2.93			
A$_5$	1) 1.22		5.21, 4.92		5.80	5.65	2.60	2.77			
	2) 1.33		5.60, 5.06		5.69	5.69	2.88	2.99			
A$_7$	1) 1.23		4.95, 4.83	6.32	5.87	4.12	3.25	2.73			
	2) 1.54		4.97, 4.85	6.37	6.04	4.42	3.62	2.97			
A$_8$	2) 1.71		5.61, 5.03		4.38	4.22	4.01	3.17			
A$_9$	1) 1.07		4.91, 4.79				2.47	2.68			
	2) 1.12		4.97, 4.87				2.61	2.86			
A$_{10}$	1) 1.10		1.38 (CH$_3$)				2.47	2.77			
A$_{11}$	1) 1.23		5.18, 5.18	5.86	4.80		3.15	2.52			
A$_{12}$	5) 1.03	0.64 (CH$_3$)	4.80, 4.80				1.82	3.22			
A$_{13}$	1) 1.21		4.80, 4.73			3.89	2.51	3.79			
A$_{14}$	2)^a 1.17	1.93 (CH$_3$)	4.78, 4.78			4.62	4.05	3.05			
	5) 1.30	0.83 (CH$_3$)	4.96, 4.96				2.49	3.39			
A$_{15}$	1) 1.15	4.03, 4.42 (CH$_2$OCO)	4.90, 4.75				2.21	2.79			
A$_{16}$	1) 1.12		4.95, 4.95	4.20		3.95	3.20	2.72			
A$_{17}$	1) 1.09		5.06, 4.83					3.72			
	2) 1.23		5.42, 4.95					4.05			
	3)^a 1.16		5.03, 4.76					3.81			
A$_{18}$	5)^b 0.97	0.70 (CH$_3$)	5.10, 4.87			4.71	2.26	3.26			
A$_{19}$	1) 1.15	9.73 (CHO)	5.02, 4.80				2.43	3.87			
	2) 1.28		5.57, 5.10				2.46	4.16			
A$_{20}$	1) 1.07		5.21, 4.90				2.50	2.67			
	2) 1.15		5.58, 5.05				2.70	2.91			
A$_{21}$	1)		5.20, 4.90				3.13	2.76			
	2)		5.58, 5.06				3.42	3.05			
A$_{22}$	1) 3.75	⎫ CH$_2$OH	5.21, 4.90				2.83	2.83			
	2) 4.33, 4.10	⎭	5.56, 5.03				3.44	3.27			
A$_{23}$	1) 1.20	9.72 (CHO)	5.17, 4.93				2.43	3.87			
	2) 1.60		5.53, 5.03				2.46	4.16			
A$_{24}$	1) 1.11	9.62 (CHO)	4.84, 4.76				2.19	3.81			
A$_{25}$	1) 1.10		4.80, 4.73				2.07	3.78			
A$_{26}$	1) 1.21		5.20, 5.05		3.87	3.75	3.37	2.73			
	2) 1.50		4.95, 4.83		3.86	3.72	2.76	2.76			
A$_{27}$	1) 1.24	4.46, 4.42 (CH$_2$OCO)	4.95, 4.83		3.86	3.72	2.76	2.76			
	2) 1.69	4.58, 4.19	4.93, 4.79		4.35	4.18	3.27	3.04			
A$_{28}$	1) 1.22		5.14, 4.91			3.99	2.58	3.78			
	2) 1.67		5.53, 5.05			4.40	3.22	4.28			
A$_{29}$	2) 1.26		5.55, 5.06		4.40		2.95	3.10			
	3) 1.06		5.18, 4.84		3.87		2.56	2.73			
A$_{30}$	2)^a 1.44		5.09, 4.99	6.43	6.13	4.48	3.73	3.13		4.01	
	3)^a 1.22		5.03, 4.94	6.35	5.86	4.03	3.22	2.69		3.70	
A$_{31}$	2)^a 1.44		5.07, 4.94		5.73	5.73	3.00	3.00		4.03	
	3)^a 1.22		5.05, 4.93		5.85	5.67	2.61	2.82		3.70	
A$_{32}$	2) 1.68		5.77, 5.72	6.48	5.95	4.50	4.05	3.37	4.10		5.05
	3)^a 1.25		5.33, 5.23	6.40	5.84	3.97	3.16	2.65	3.58		4.36

Table 26 (continued)
ASSIGNMENT OF PMR OF GAs

GAs	H$_3$-18	H$_S$-20	H$_2$-17	H-1	H-2	H-3	H-5	H-6	H-11	H-12	H-15
A$_{33}$	3)^a 1.11		5.08, 4.98	4.42			3.37	2.87		3.74	
A$_{34}$	2)^a 1.71		5.00, 4.87		4.40	4.23	3.98	3.16			
	3)^a 1.15		4.96, 4.84		3.75	3.62	3.25	2.55			
A$_{35}$	1) 1.15		5.04, 4.91			3.86	3.26	2.74	4.22		
A$_{36}$	1) 1.22	9.68 (CHO)	4.92, 4.84			4.11	2.75	3.91			
	2) 1.59	9.96	4.94, 4.86			4.48	3.34	4.29			
A$_{37}$	1) 1.19	4.42, 4.07 (CH$_2$OCO)	4.90, 4.78			3.74	2.74	2.74			
	2)^a 1.84	4.61, 4.13	4.92, 4.78			4.13	3.28	3.28			
A$_{38}$	2)^a 1.86	4.83, 4.20 (CH$_2$OCO)	5.55, 4.95			4.20	3.21	3.44			
A$_{39}$	1) 1.23		5.02, 4.96			^{c}	2.59	3.95		^{c}	
	2) 1.66		5.01, 4.96			^{c}	3.23	4.42		^{c}	
	3)^a 1.35		4.97, 4.89			3.88	2.68	4.10		3.63	
A$_{40}$	2) 1.26		4.98, 4.90		4.50		2.83	2.95			
A$_{42}$	1) 1.17	0.67 (CH$_3$)	1.35 (CH$_3$)			4.12	2.26	3.30			
A$_{43}$	1) 1.28		4.90, 4.81		4.40	3.84	2.64	3.86			
A$_{45}$	1) 1.08		5.08, 5.08				2.45	2.75			3.93
A$_{46}$	1) 1.21		4.92, 4.84		4.26		2.23	3.88			
A$_{47}$	2) 1.59		5.00, 4.90		4.62	4.31	3.83	3.07			
A$_{48}$	1) 1.18		5.10, 5.10	^{c}		^{c}	3.23	2.60		^{c}	
	2) 1.52		5.30, 5.10	^{c}		^{c}	3.80	2.94		^{c}	
	3)^a 1.15		4.95, 4.95	^{c}		3.58	3.20	2.48		3.90	
A$_{49}$	1) 1.16		5.00, 4.90	^{c}		^{c}	3.24	2.61		^{c}	
	2) 1.53		5.00, 5.00	^{c}		^{c}	3.88	2.90		^{c}	
	3) 1.10		4.97, 4.88		3.60—3.90	3.59	3.21	2.53		3.60—3.90	
A$_{50}$	1) 1.14		5.02, 4.89	^{c}		^{c}	3.26	2.65	^{c}		
	2) 1.56		5.04, 4.86	^{c}		^{c}	3.95	3.08	^{c}		
	3)^a 1.13		5.08, 4.96		3.79	3.63	3.26	2.58	4.14		
A$_{51}$	2)^a 1.38		4.90, 4.80		4.28		3.04	3.04			
A$_{52}$	1) 1.19	^{c}	4.93, 4.82	^{c}		^{c}	2.76	2.76	^{c}		
	2) 1.70	^{c}	4.94, 4.76	^{c}		^{c}	3.40	3.15	^{c}		
	3)^a 1.22	4.52, 4.28 (CH$_2$OCO)	4.90, 4.77	^{c}		3.50	2.77	2.77	^{c}		
A$_{53}$	1) 1.14	1.62 (CH$_3$)	5.56, 5.08					4.19			
A$_{54}$	1) 1.22		4.99, 4.78	4.02		3.84	3.56	2.70			
	2) 1.52		4.92, 4.79	4.30		4.13	4.32	3.06			
A$_{55}$	2) 1.51		5.54, 4.97	4.35		4.13	4.37	3.09			
A$_{56}$	2) 1.57		5.58, 5.02		4.60	4.28	3.84	3.08			
A$_{57}$	2) 1.52		5.60, 5.04	4.60		4.20	3.88	3.16			
A$_{59}$	1))		5.26, 4.95		5.91	6.34	3.06	3.19			
	2)		5.63, 5.06		5.85	6.50	3.34	3.50			
A$_{60}$	3) 1.03		5.22, 4.91	3.99			3.12	2.63			
A$_{61}$	3) 1.09		5.00, 4.87	4.05			3.15	2.73			
A$_{62}$	3) 1.20		4.99, 4.87	4.16	5.89	5.89	2.88	2.67			
A$_{63}$	1) 1.15		5.13, 5.10			3.84	3.09	2.74			3.96
A$_{64}$	1) 1.11	4.41, 4.14 (CH$_2$OCO)	5.12, 5.05				2.11	2.81			3.88
A$_{65}$	1) 1.10	9.63 (CHO)	5.14, 5.04				2.13	3.96			3.97
A$_{66}$	1) 1.10		5.10, 5.00				2.05	3.95			3.89

Note: Chemical shifts are expressed in δ-value (ppm downfield shift from internal tetramethylsilane). Solvents: (1) CDCl$_3$, (2) C$_5$D$_5$N, (3) CD$_3$OCD$_3$, (4) CCl$_4$.

[a] PMR spectra of free acids.
[b] PMR spectrum of the acetate methyl ester.
[c] Chemical shifts were not reported in References 62, 63, and 68.

the variation of the C-10 substituents; the position of C-10 resonances of GAs with a carboxyl, an aldehyde, a methyl, and an hydroxylmethyl (δ-lactone) is in the ranges 57.3 to 58.4, 48.4 to 49.5, 44.6 to 44.7, and 41.6 to 41.9 ppm, respectively.

d. Mass Spectra

In the early stage of mass spectrometric studies of GAs, Wulfson et al,[215] Takahashi et al.[216] and Brinks et al.[217] reported fragmentation patterns of GAs and their derivatives. Now, mass spectra have became an essential tool for the structural determination and identification of GAs.

Mass spectra of GAs have been determined as methyl esters,[215-217] methyl ester trimethylsilyl ethers (GAs-Me-TMSi)[217] and trimethylsilyl ester trimethylsilyl ethers (GAs-TMSi-TMSi).[170] The fragmentation patterns of those GA derivatives are presented in Tables 27, 28, and 29.

Mass spectra of GAs and their derivatives show fragmentation patterns characteristic of the common structural features of GAs. These types of fragments can be used to decide whether unknown compounds are GAs or not. Furthermore the mass spectra of GAs afford useful information not only on structural modifications in the GA molecules but on the position of the functional groups.

In the spectra of C_{19}-GAs-Me, the loss of the following mass units from the molecular ion is generally observed: 18 (H_2O), 31/32 (CH_3O, CH_3OH), 44 (CO_2), 46 (HCOOH), 50 ($CH_3OH + H_2O$), 59/60 ($COOCH_3$/$HCOOCH_3$), 78 ($HCOOCH_3 + H_2O$), 104 ($HCOOCH_3 + CO_2$), 106 ($HCOOCH_3$ + HCOOH), and 122 ($HCOOCH_3 + CO_2 + H_2O$). The prominent fragment ion peaks at M-31/32, M-50, and M-59/60 are related to the fragmentation of the methoxycarbonyl group, while the peaks at M-44 and M-46, due to loss of γ-lactone. The spectra of GAs-Me with C-2, C-3 double bond show a prominent M-44 peak; the M-46 peak is weak or is not observed. GAs-Me containing 3-hydroxyl show moderate M-46 and intense M-62 peaks, while the M-44 peak is rarely observed. Methyl esters of $GA_{9,20}$ lacking 3-hydroxyl as well as $GA_{3,7}$-Me containing both C-1 to C-2 double bonds and 3-hydroxyl exhibit both M-44 and M-46 peaks. This suggests that C-1~C-2 and C-2~C-3 double bonds induce the elimination of the γ-lactone in the form of CO_2 and a 3-hydroxyl, in the form of HCOOH or $CO_2 + H_2O$. The intense peaks at M-63, which are characteristic of methyl esters of $GA_{3,7,30}$, are considered to arise from the loss of H_2O, CO_2, and H from A ring and this causes aromatization of A ring.

Methyl esters of C_{20}-GAs with di- or tricarboxylic acid functionality show intense peaks related to the elimination of methylcarbonyl groups at M-91/92 ($COOCH_3 + CH_3$/$HCOOCH_3 + CH_3OH$) and M-119/120 ($COOCH_3 + HCOOCH_3$/$HCOOCH_3 \times 2$) in addition to M-31/32 and M-59/60. Tricarboxy-GAs-Me further show prominent peaks at M-179/180 ($COOCH_3 + HCOOCH_3 \times 2$/$HCOOCH_3 \times 3$). GA_{23}-Me exhibits a prominent peak at M-64 ($CH_3OH \times 2$) due to the elimination of methanol from two methoxycarbonyls. Dimethyl esters of GAs with a methyl at C-10 such as $GA_{14,18}$ show intense peaks at M-93 ($HCOOCH_3 + H_2O + CH_3$) in addition to M-60 and M-78.

Methyl esters of aldehydic GAs such as $GA_{19,23,24,36}$ show characteristic peaks which arise from the loss of 28 (CO) mass units from the molecular ion and the prominent fragment ions. Thus, in the spectra of these aldehydic GAs-Me, M-28, M-60 (32 + 28), M-88 (60 + 28), and M-148 (120 + 28) peaks are observed. The M-46 peak observed in the spectra of methyl esters of $GA_{19,23,36}$ are due to the combined loss of 28 (CO) and 18 (H_2O) from the molecular ions.

Hydrocarbon ion peaks with the largest number of carbon atoms represent the basic skeleton of GAs. Most of the C_{19}-GAs-Me except GA_{21}-Me show ions with composition $C_{17}H_{16-24}$ arising from the loss of all oxygen-containing functionalities. On the other hand, in the case of C_{20}-GAs-Me the composition of the hydrocarbon ions with the largest number of carbon atoms varies with the nature of C-10 substituents. GAs-Me with δ-lactone or a methyl at

Table 27
RELATIVE INTENSITY OF IONS FORMED IN THE MS FRAGMENTATION OF GAs-Me

GA_1 362(M^+,40) 344(20) 330(100) 316(20) 312(30) 303(45) 302(30) 300(20) 298(12) 284(30)
GA_2 364(M^+,2) 346(5) 332(18) 318(5) 314(10) 307(1) 307(20) 304(12) 285(30)
GA_3 360(M^+,10) 342(10) 328(30) 314(10) 310(15) 300(25) 297(95)
GA_4 346(M^+,8) 328(20) 314(100) 300(20) 296(8) 286(30) 284(60) 268(30)
GA_5 344(M^+,15) 329(2) 312(30) 300(50) 298(5) 285(20) 284(12)
GA_6 360(M^+,30) 342(2) 328(100) 314(5) 301(30) 300(20)
GA_7 344(M^+,5) 326(5) 312(25) 300(5) 298(10) 284(25) 281(100)
GA_8 376(M^+,12) 360(3) 358(3) 346(18) 342(2) 332(6) 328(8) 319(10)
GA_9 330(M^+,8) 298(100) 286(15) 284(5) 270(80) 243(45)
GA_{10} 348(M^+,1) 330(5) 316(15) 302(3) 291(20) 288(15) 286(15) 259(40)
GA_{11} 344(M^+,100) 326(2) 312(95) 300(5) 299(18) 298(10) 284(40)
GA_{12} 360(M^+,2)328(20) 300(100) 285(20) 276(8) 240(30)
GA_{13} 420(M^+,2) 402(1) 388(20) 360(10) 356(25) 328(100) 311(20) 300(35)
GA_{14} 376(M^+,2) 344(60) 326(8) 316(15) 298(55) 283(80) 256(20)
GA_{15} 344(M^+,23) 326(2) 312(23) 298(15) 284(50) 239(100)
GA_{16} 362(M^+,30) 344(8) 330(95) 316(5) 312(15) 302(100) 300(60) 298(60)
GA_{17} 420(M^+,1) 402(1) 388(45) 360(5) 356(1) 328(90) 310(15) 300(100)
GA_{18} 392(M^+,3) 360(100) 332(22) 314(62) 299(55) 273(18)
GA_{19} 390(M^+,3) 372(2) 362(8) 358(32) 344(25) 330(50) 312(40) 302(50) 298(30)
GA_{20} 346(M^+,50) 328(8) 314(100) 303(50) 300(25) 286(45)
GA_{21} 390(M^+,5) 372(3) 358(32) 340(25) 330(30) 326(20) 312(25) 298(12) 286(60)
GA_{22} 360(M^+,3) 342(5) 328(100) 314(2) 310(8) 300(20) 298(25)
GA_{23} 406(M^+,3) 388(3) 378(2) 374(90) 360(5) 356(30) 346(35) 342(65) 328(40) 318(20) 314(45)
GA_{24} 374(M^+,3) 346(12) 342(45) 314(100) 310(15) 286(90) 284(35)
GA_{25} 404(M^+,1) 372(15) 344(2) 312(85) 284(100) 269(8) 253(10) 225(50)
GA_{26} 376(M^+,100) 356(15) 344(25) 330(30) 316(40)
GA_{27} 376(M^+,60) 356(15) 344(25) 330(10) 326(35) 316(100) 298(80)
GA_{28} 436(M^+,2) 418(2) 404(29) 386(3) 376(14) 358(8) 344(100) 326(39) 316(46) 298(52)
GA_{29} 362(M^+,50) 344(6) 330(100) 316(16) 303(50) 300(20) 284(28)
GA_{30} 360(M^+,17) 342(15) 329(25) 328(22) 316(20) 310(22) 300(20) 297(42)
GA_{31} 344(M^+,15) 326(20) 312(25) 300(20) 294(32) 284(44)
GA_{32} 392(M^+,10) 374(25) 361(12) 360(16) 356(20) 348(20) 342(35)
GA_{33} 376(M^+,17) 358(100) 344(25) 340(22) 326(43) 317(34)
GA_{34} 362(M^+,10) 344(7) 330(100) 316(3) 392(19) 300(15) 284(45)
GA_{35} 362(M^+,10) 344(50) 330(50) 312(20) 303(10) 300(15) 282(100)
GA_{36} 390(M^+,8) 372(7) 362(13) 358(39) 344(14) 340(37) 330(100) 326(52) 312(32) 302(40)
GA_{37} 360(M^+,60) 342(20) 328(50) 314(20) 310(50) 300(50)
GA_{38} 376(M^+,25) 356(20) 348(25) 344(15) 330(15) 326(25)
GA_{39} 436(M^+) 418 404 386 372 358 344 326 298[a]
GA_{40} 346(M^+,10) 328(5) 314(71) 302(55) 300(9) 286(70)
GA_{41} 438(M^+,1) 420(2) 406(1) 388(13) 360(11) 328(100) 310(21)
GA_{43} 436(M^+,13) 418(2) 404(56) 386(3) 358(21) 344(100) 326(20) 316(39)
GA_{46} 420(M^+,1) 388(31) 360(2) 356(10) 342(11) 328(100) 300(43) 282(51)
GA_{47} 362(M^+,19) 344(+) 330(100) 312(14) 302(42) 300(54) 284(34)
GA_{48} 378(M^+,16)) 360(100) 346(35) 342(34) 328(61) 318(58) 300(72)
GA_{49} 378(M^+,23) 360(72) 346(38) 342(29) 328(57) 318(100) 300(63)
GA_{50} 378(M^+) 360 346 342 328 314 310 298[a]
GA_{51} 346(M^+,10) 328(5) 314(100) 302(3) 286(35) 285(50) 268(100)
GA_{52} 392(M^+) 374 361 346 342 332 328 324 314[a]
GA_{53} 376(M^+,1) 344(11) 316(100) 301(14) 284(12) 257(25) 241(11)
GA_{54} 362(M^+,12) 344(22) 330(37) 326(9) 312(20) 302(17) 300(100) 284(16)
GA_{55} 378(M^+,54) 360(12) 346(100) 328(37) 319(43) 316(30) 304(42)
GA_{56} 378(M^+,72) 360(26) 346(80) 342(12) 332(17) 328(100) 319(56) 304(49) 300(38)
GA_{57} 378(M^+,22) 360(5) 349(6) 346(20) 328(80) 317(12) 304(100) 300(11) 275(30)
GA_{59} 388(M^+,10) 356(89) 338(11) 328(17) 312(21) 310(19) 284(100) 225(80)
GA_{60} 362(M^+,48) 344(10) 330(100) 312(27) 303(85)
GA_{61} 346(M^+,24) 315(27) 314(83) 296(23) 287(40) 286(100) 225(25)

Table 27 (continued)
RELATIVE INTENSITY OF IONS FORMED IN THE MS FRAGMENTATION OF GAs-Me

GA_{62} 344(M^+,1) 313(2) 300(4) 282(12) 223(100) 222(84) 195(16)
GA_{63} 362(M^+,6) 344(5) 331(25) 330(100) 316(5) 312(7) 302(7) 284(20)
GA_{64} 360(M^+,3) 342(2) 329(27) 328(100) 314(3) 310(12) 300(27) 282(24)
GA_{65} 390(M^+,6) 362(4) 358(17) 344(4) 340(82) 330(61) 312(51) 302(60) 284(97) 270(100)
GA_{66} 420(M^+,3) 388(16) 370(3) 360(20) 328(100) 300(75) 268(49) 240(42)

[a] Relative intensity was not reported in References 62, 63, and 68.

Table 28
RELATIVE INTENSITIES OF IONS FORMED IN THE MS FRAGMENTATION OF GAs-Me-TMSi

GA_1 506(M^+,100)491(15) 447(20) 416(5) 390(5)
GA_2 508(M^+,30) 493(15) 476(2) 451(10) 449(5) 418(20) 386(10)
GA_3 504(M^+,100) 489(10) 473(2) 445(5) 431(5) 414(3)
GA_4 418(M^+,25) 403(5) 400(10) 390(10) 386(20)
GA_5 416(M^+,100) 401(20) 385(3) 372(10) 357(15) 343(15)
GA_6 432(M^+,100) 417(20) 403(5) 391(2) 373(20)
GA_7 416(M^+,20) 401(5) 384(50) 370(10) 356(70) 343(10)
GA_8 594(M^+,100) 579(10) 565(5) 553(5) 547(5) 535(10) 521(50) 519(5) 504(5)
GA_{10} 420(M^+,40) 405(25) 389(10) 376(10) 363(15) 361(15) 345(10) 331(30)
GA_{13} 492(M^+,5) 477(10) 460(10) 436(10) 432(5) 400(20) 372(10)
GA_{14} 448(M^+,5) 433(15) 416(40) 388(20) 358(5)
GA_{16} 506(M^+,15) 491(5) 475(5) 459(10) 416(10) 390(100)
GA_{17} 492(M^+,80) 477(5) 460(40) 432(40) 401(25) 372(30)
GA_{18} 536(M^+,50) 521(15) 504(10) 477(15) 461(8) 446(10) 417(8)
GA_{19} 462(M^+,15) 447(10) 434(100) 431(10) 402(40) 375(60) 374(60)
GA_{20} 418(M^+,100) 403(15) 387(5) 375(50)
GA_{21} 462(M^+,100) 447(15) 430(20) 403(40) 389(10)
GA_{22} 504(M^+,100) 489(20) 472(10) 444(8) 429(10) 414(10) 401(55)
GA_{23} 550(M^+,20) 535(20) 522(100) 519(20) 491(20) 463(20)
GA_{26} 520(M^+,100) 505(2) 489(3) 473(2) 461(3) 445(3)
GA_{27} 520(M^+,70) 505(8) 489(5) 461(2) 445(2) 430(15)
GA_{28} 580(M^+,30) 565(15) 548(15) 533(2) 521(10) 520(10) 505(3) 488(5) 461(8)
GA_{29} 506(M^+,100) 491(10) 477(5) 447(5) 389(10)
GA_{30} 504(M^+,10) 489(3) 473(3) 460(3) 444(5) 429(3) 414(10)
GA_{31} 416(M^+,10) 401(2) 385(8) 384(5) 369(10) 361(5) 357(3)
GA_{32} 680(M^+,20) 665(5) 636(2) 621(2) 607(2) 590(20) 577(8)
GA_{33} 520(M^+,10) 505(10) 488(2) 473(2) 465(15) 461(5) 445(1) 430(50)
GA_{34} 506(M^+,100) 475(2) 459(5) 431(5) 416(5)
GA_{35} 506(M^+,25) 491(8) 475(5) 446(5) 416(35) 390(8)
GA_{36} 462(M^+,5) 447(15) 430(50) 402(25) 384(3) 374(15)
GA_{37} 432(M^+,10) 417(5) 401(3) 386(1) 372(2) 343(15)
GA_{38} 520(M^+,70) 505(10) 489(2) 477(1) 461(10) 447(2) 430(10)
GA_{39} 580(M^+,1) 565(25) 548(10) 521(5) 520(10) 488(20) 461(7) 458(15)
GA_{40} 418(M^+,0) 403(5) 387(5) 371(100) 343(85)
GA_{41} 582(M^+,3) 567(10) 550(5) 523(5) 492(5) 477(3) 460(8) 452(3)
GA_{42} 538(M^+,5) 523(30) 506(2) 491(3) 481(15) 448(15) 433(5) 416(25) 408(5) 376(100)
GA_{43} 580(M^+,10) 565(5) 548(2) 520(2) 505(7) 490(7) 458(3)
GA_{44} 432(M^+,65) 417(15) 403(5) 401(3) 389(2) 373(20) 359(5)
GA_{45} 418(M^+,100) 403(20) 389(3) 375(2) 358(25) 343(2)
GA_{46} 492(M^+,10) 477(5) 460(40) 445(15) 435(15) 432(10) 417(5) 400(80) 372(25)
GA_{47} 506(M^+,100) 474(4) 459(10) 431(8) 416(4)
GA_{48} 594(M^+,60) 579(8) 562(3) 547(3) 535(3) 504(20)
GA_{49} 594(M^+,75) 579(5) 562(3) 547(2) 535(2) 504(20)
GA_{50} 594(M^+,50) 579(5) 563(2) 547(2) 535(2) 519(4) 504(35)

Table 28 (continued)
RELATIVE INTENSITIES OF IONS FORMED IN THE MS FRAGMENTATION OF GAs-Me-TMSi

GA_{51} 418(M^+,3) 403(8) 386(30) 371(8) 358(3) 343(10)
GA_{52} 608(M^+,70) 593(5) 577(2) 565(1) 549(1) 536(1) 518(15) 503(2)
GA_{53} 448(M^+,34) 419(8) 416(9) 389(22) 251(30) 241(16) 235(25)
GA_{54} 506(M^+,15) 491(1) 416(20) 390(64) 375(20) 300(29) 223(43) 217(100)
GA_{55} 594(M^+,100) 579(8) 553(9) 535(9) 504(4) 489(3)
GA_{56} 594(M^+,100) 579(6) 547(5) 535(5) 519(5) 504(5)
GA_{57} 594(M^+,51) 579(7) 535(6) 504(6) 478(6) 448(100) 376(31) 375(29)
GA_{58} 506(M^+,27) 491(8) 416(51) 384(44) 356(45) 326(22) 317(20)
GA_{59} 460(M^+,38) 445(4) 428(7) 401(13) 356(8) 343(12) 238(6)
GA_{60} 506(M^+,66) 491(12) 447(14) 375(75) 321(8) 309(10)
GA_{61} 418(M^+,7) 403(14) 359(41) 347(26) 296(96) 284(30) 274(26)
GA_{62} 416(M^+,0) 401(4) 282(23) 223(100) 222(93)
GA_{63} 506(M^+,6) 491(32) 474(4) 446(23) 416(15) 384(6) 377(8)
GA_{64} 432(M^+,29) 417(24) 400(18) 386(4) 385(5) 372(16) 310(27)
GA_{65} 462(M^+,16) 447(36) 434(17) 430(26) 402(79) 312(59) 284(100)
GA_{66} 492(M^+,10) 473(20) 460(15) 432(100) 400(82) 372(85) 342(41)

C-10 exhibit $C_{18}H_{19-26}$ peaks and those with an aldehyde or a carboxyl at C-10, $C_{17}H_{16-25}$ peaks. It should be noted that 13-OH-GAs-Me show intense peaks composed of one oxygen-containing hydrocarbon with the largest carbon number, the corresponding hydrocarbon peaks being rather weak. In the mass spectra of 13-H-GAs-Me, such relation between the two species of ions is not observed.

As described in Section II.A.1 and B.2, GC/MS has been recognized as a powerful tool in identifying GAs in minute quantities. Since GAs-Me-TMSi not only exhibit excellent separation in GC, but afford useful information on GA structures, utility of GAs-Me-TMSi in GC/MS is highly valuable. The molecular ion peaks are observed in all the spectra of GAs-Me-TMSi and hence the number of hydroxyls of GAs can be determined from the comparison with the corresponding GAs-Me. Methyl ester trimethylsilyl ethers of 13-OH-GAs show intense molecular ion peaks except 10-aldehydic-13-OH-GAs-Me-TMSi, whose spectra exhibit intense peaks at M-28 and M-87/88 (28 + 59/60).

Fragment ions observed in common with GAs-Me-TMSi are M-15, M-31/32, M-59/60, and M-89/90. The first and last ions are due to the fragmentation of trimethylsilyl ether groups, and the second and third due to that of methyoxycarbonyl groups.

The prominent peaks observed at m/z 207/208 in the spectra of 13-OH-GAs-Me-TMSi can be interpreted to arise from the cleavage of ring D as shown in **(17)**. The other characteristic fragment ions, **(18)**, **(19)**, and **(20)** are observed in the spectra of methyl ester trimethylsilyl ethers of 3-OH-GAs, 16-OH-GAs, and GAs with glycol systems, respectively. Methyl ester trimethylsilyl ethers of GAs with a 1,3-dihydroxy system give rise to intense ions by elimination of fragment **(21)** from the molecular ions (Figure 30).

GAs-TMSi-TMSi[170] are also excellent derivatives for GC/MS because of good separation in GC and facile derivatization. Particularly, in GC/SIM using GAs-TMSi-TMSi, fragment ions with a large mass number can be selected to get clear results free from the effects of impurities and background noise. It should be noted that in trimethylsilylation of 10-aldehydic-GAs two isomeric derivatives derived from a lactol and an aldehyde-acid form, due to tautomerization between a carboxyl at C-4 and an aldehyde group at C-10.

Yokota et al.[218] analyzed mass spectra of six GAs-GEt-Me-TMSi and five GAs-GEs-TMSi. THe fragmentation patterns of GAs-GEt-Me-TMSi are presented in Table 30. The fragment ions due to collapse of the trimethylsilyl groups are observed at M-15 (CH_3), M-58 (CH_3Si-CH_3), M-72 (($CH_3)_2Si=CH_2$), M-87 (15 + 72) and M-105 (($CH_3)_3-SiOH$ + 15). THese ions and the molecular ions are more abundant in 13-OH-GAs-GEt-Me-TMSi than in 13-H-GAs-GEt-Me-TMSi. The fragment ions at M-335/336, M-349, M-378, M-

Table 29
RELATIVE INTENSITY OF IONS FORMED IN THE MS FRAGMENTATION OF GAs-TMSi-TMSi

GA_1	564(M^+,100) 549(21) 447(47)
GA_3	562(M^+,100) 547(21) 472(8) 445(34) 428(19) 427(11) 297(14) 296(13)
GA_4	476(M^+,29) 461(49) 386(48) 358(14) 356(11) 347(31) 342(33) 339(35) 224(100)
GA_5	474(M^+,100) 459(53) 343(11) 312(15) 299(29)
GA_7	474(M^+,7) 459(23) 384(24) 356(49) 339(95) 311(33) 295(15) 221(100)
GA_8	652(M^+,100) 637(11) 535(39) 506(7)
GA_9	388(M^+,10) 373(33) 298(86) 270(100) 242(11) 226(31)
GA_{12}	476(M^+,3) 461(2) 386(26) 358(100) 343(9) 268(12) 241(51) 240(45)
GA_{13}	666(M^+,6) 651(78) 576(81) 548(39) 532(11) 458(100) 447(27) 430(64)
GA_{14}	564(M^+,5) 549(54) 507(4) 474(78) 446(63) 384(21) 356(100) 345(99)
GA_{15}	402(M^+,21) 387(19) 312(100) 284(40) 239(73) 226(11)
GA_{16}	564(M^+,15) 549(46) 474(59) 448(100) 418(25) 358(91) 340(48) 240(80)
GA_{17}	666(M^+,61) 651(79) 576(89) 549(45) 548(36) 458(99) 430(65) 368(48)
GA_{18}	652(M^+,51) 647(46) 652(54) 535(100) 444(49) 433(28) 405(22) 355(22) 327(55)
GA_{19}	578(M^+,20) 563(47) 550(85) 488(18) 460(92) 443(56) 432(100) 370(39)
GA_{20}	476(M^+,100) 461(20) 433(22) 359(46)
GA_{23}	666(M^+,8) 651(20) 638(28) 576(11) 548(25) 520(13) 458(16) 431(17)
GA_{24}	490(M^+,0) 475(41) 462(22) 400(24) 385(12) 372(100) 310(28) 281(52) 226(95)
GA_{26}	578(M^+,100) 563(23) 488(4) 473(3) 460(6) 445(3) 401(3) 371(5) 343(4)
GA_{27}	578(M^+,100) 563(23) 488(19) 432(12) 370(27) 343(33) 316(20)
GA_{29}	564(M^+,100) 549(21) 447(21) 389(3) 299(4)
GA_{30}	562(M^+,16) 547(36) 472(36) 444(10) 427(68) 382(20) 337(52) 310(42) 221(100)
GA_{34}	564(M^+,100) 549(28) 474(16) 430(8) 389(25) 356(9) 229(14)
GA_{36}	578(M^+,20) 563(47) 550(85) 488(69) 460(48) 432(25) 398(44) 370(55) 342(100)
GA_{37}	490(M^+,39) 475(38) 400(100) 372(10) 342(12) 327(14) 316(27) 310(62) 282(51)
GA_{40}	476(M^+,0) 461(33) 386(36) 371(48) 343(100) 314(25) 299(79) 225(35)
GA_{44}	490(M^+,61) 475(14) 400(4) 384(68) 296(14)
GA_{47}	564(M^+,100) 549(24) 474(38) 459(17) 446(7) 431(28)
GA_{51}	476(M^+,7) 461(36) 386(63) 342(15) 296(6) 268(100) 227(27)
GA_{53}	564(M^+,51) 549(35) 474(30) 447(100) 356(15) 328(34) 267(9) 239(35)
GA_{54}	564(M^+,19) 549(21) 474(100) 448(81) 384(15) 340(14) 330(22)
GA_{55}	652(M^+,100) 637(35) 611(13) 562(16) 547(11) 535(79) 515(11)
GA_{56}	652(M^+,100) 637(16) 562(26) 547(11) 535(46) 519(11) 472(7) 445(19)
GA_{57}	652(M^+,80) 637(23) 562(8) 547(6) 536(11) 535(11) 506(100) 434(40)
GA_{60}	564(M^+,24) 549(14) 444(47) 433(23) 309(6)
GA_{61}	476(M^+,1) 461(15) 386(5) 358(41) 296(19) 225(26)
GA_{62}	474(M^+,0) 459(22) 340(12) 223(64) 222(82)

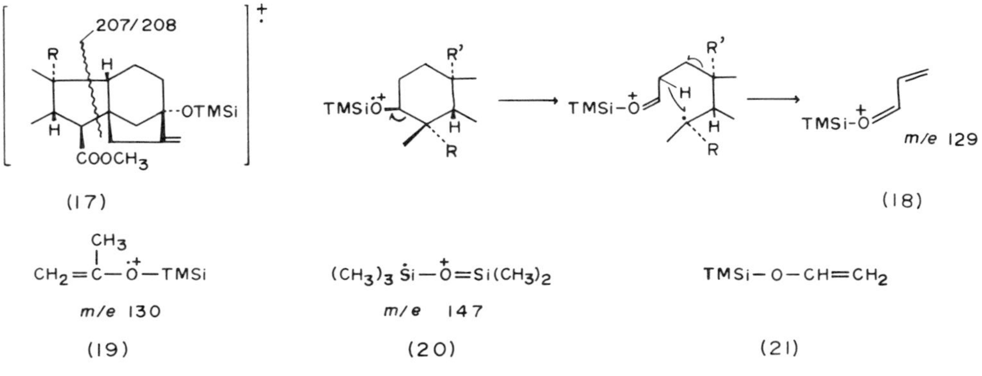

FIGURE 30. Characteristic fragment ions of GAs-Me-TMSi.

Table 30
RELATIVE INTENSITY OF IONS FORMED IN THE MS FRAGMENTATION OF GAs-GEt-Me-TMSi[218]

Mass lost	GAs-GEt-Me-TMSi					
	GA_3-GEt (882)	GA_8-GEt (972)	GA_{26}-GEt (898)	GA_{27}-GEt (898)	GA_{29}-GEt (884)	GA_{35}-GEt (884)
0(M^+)	28	4	t	1	18	t
15	6	2	t	1	9	1
31		t			1	1
32	3		t	1		
41	1	t			3	
44					3	
58	6	2			3	
59					2	
72	25	23			6	
87	5	7	1	3	3	1
103	4	1	4	3		1
105	2	1	1	1	7	1
335		1	12	8		5
336	8				8	5
349	11	57	29	52	93	100
378	18	t	t	10	t	43
381	6	22			38	3
393	12	18	18	10		9
394	12	16	7	6		10
395	7	13	7	2	55	19
408		27	18			
409	15	18	4	2	11	
421	7	91	100	100	13	6
422	8	28			30	
425	8	17	12			4
427					16	7
437				3	14	7
439	34	35	19		26	21
450		16	11	16		2
451	100		6	3	13	3
453	10	25				5
455					17	6
467	36	50	26	27	100	74
468	18	9		5	13	16
481	8	17				
483	16		7	7	15	
485	4	23	10	3	12	35
494		41				
495	24			5	11	
497	10			5		
499	10	19	7	4	35	13
511	50	100	48	69	68	8
512	40	10	6	2	16	12
513	28	15	8	2	25	25
523	10	25			25[a]	90[a]
524				17		
525	12	14	8	7	10	
527	24	15	10	6	29	10
529			11		7	5
531	12					
539	8	42	12	54	20	7
543	8	15	6	5		7

Table 30 (continued)
RELATIVE INTENSITY OF IONS FORMED IN THE MS FRAGMENTATION OF GAs-GEt-Me-TMSi[218]

GAs-GEt-Me-TMSi

Mass lost	GA_3-GEt (882)	GA_8-GEt (972)	GA_{26}-GEt (898)	GA_{27}-GEt (898)	GA_{29}-GEt (884)	GA_{35}-GEt (884)
545				3		12
555	12	25	6	5	10	
557	6	26	12	10	41	27
571	38	34	23	19	30	8
583		20			6	
585	16	18	19	14	16	7
599		30	17	27	16	13
601	28	59	32		27	65
615		24		57	16	

^a These ions contain the ion (m/e 361) arising from glucosyl group.

t = trace.

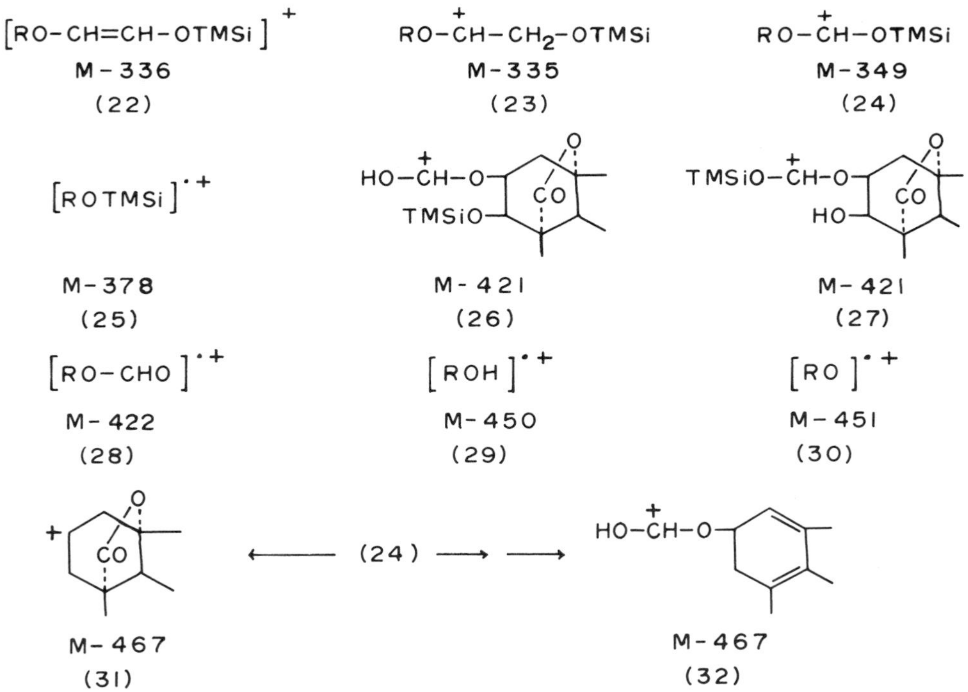

[RO-CH=CH-OTMSi]⁺
M-336
(22)

RO-ĊH-CH$_2$-OTMSi
M-335
(23)

RO-ĊH-OTMSi
M-349
(24)

[ROTMSi]^{•+}
M-378
(25)

HO-ĊH-O-[ring with CO, TMSiO]
M-421
(26)

TMSiO-ĊH-O-[ring with CO, HO]
M-421
(27)

[RO-CHO]^{•+}
M-422
(28)

[ROH]^{•+}
M-450
(29)

[RO]^{•+}
M-451
(30)

[ring with CO]⁺
M-467
(31)

⟵ (24) ⟶

HO-ĊH-O-[ring]
M-467
(32)

FIGURE 31. Characteristic fragment ions of GAs-GEt-Me-TMSi.

421/422, M-450/451, and M-467/468 retain the GA skeleton with or without a part from glucopyranosyl moiety as shown in Figure 31; they further decompose with the elimination of CH_3, CH_3O/CH_3OH, CO_2, HCOOH, $COOCH_3/HCOOCH_3$, $CH_3Si=CH_2$, $(CH_3)_3Si$, $(CH_3)_3SiOH$, and their combination to give rise to the following ions at M-393, M-395, M-408/409, M-439, M-483, M-485, M-495, M-499, M-511, M-512/513, M-523, M-525, M-527, M-539, M-557, M-571, M-585, M-599, and M-601. The ions at m/z 361, 332/331,

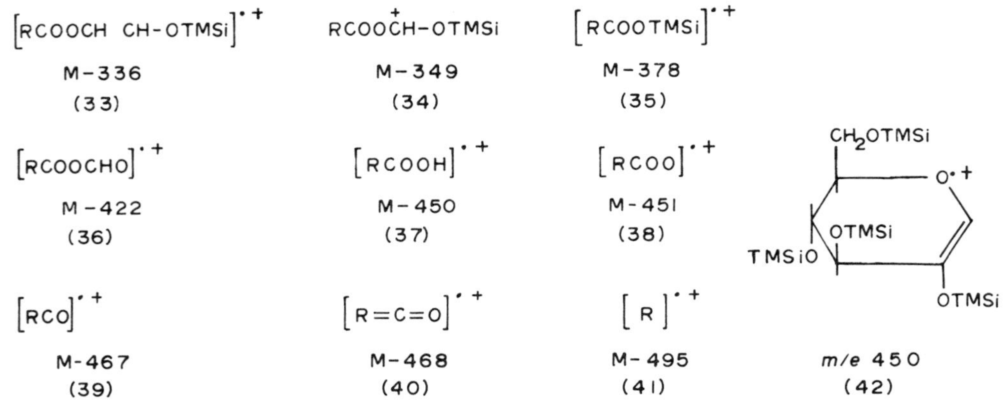

114 Chemistry of Plant Hormones

$[RCOOCH\ CH\text{-}OTMSi]^{\cdot+}$ $RCOO\overset{+}{C}H\text{-}OTMSi$ $[RCOOTMSi]^{\cdot+}$
M-336 M-349 M-378
(33) (34) (35)

$[RCOOCHO]^{\cdot+}$ $[RCOOH]^{\cdot+}$ $[RCOO]^{\cdot+}$
M-422 M-450 M-451
(36) (37) (38)

$[RCO]^{\cdot+}$ $[R=C=O]^{\cdot+}$ $[R]^{\cdot+}$ m/e 450
M-467 M-468 M-495 (42)
(39) (40) (41)

FIGURE 32. Characteristic fragment ions of GAs-GEs-TMSi.

319, 305, 289, 271, 217, 204, and 191, which are not listed in Table 30, are due to fragmentation from the glucopyranosyl moiety.

In the spectra of GAs-GEs-TMSi, the molecular ions and the ions at M-15, M-72, M-88, M-90, M-103/104, and M-144/145 due to the decomposition of the trimethylsilyl groups are observed in the high mass region, the intensities of these ions being rather low in all the compounds. However, the fragment ions at M-336 (**33**), M-349 (**34**), M-378 (**35**), M-442 (**36**), M-450/451 (**37,38**), M-467/468 (**39,40**), and M-495 (**41**) as shown in Figure 32 are more abundant and informative on the structures of GAs-GEs. The ions at M-378 corresponding to the trimethylsilyl ester of aglycone is a base peak in all the compounds, while the M-349 peak, which is prominent in the spectra of GAs-GEt-Me-TMSi, is weak or not observed. The above abundant ions give rise to the ions at M-393, M-395, M-423, M-465/466, M-513, M-539/540, M-557, M-567, and M-585 by further breakdown of them (Table 31).

Other prominent peaks are observed at m/z 450 of the fragment ion (**42**), 435, 378, 361, 332/331, 319, 305, and so on. These ions are due to the decomposition of the sugar moiety, the first three not being observed in the spectra of GAs-GEt-Me-TMSi.

C. Synthesis

In the process of the structural determination of GAs, syntheses of degradation products of GAs were also carried out. The object of such syntheses was to confirm structures assigned to the degradation products. However, after achieving the structural elucidation of GAs it is not too much to say that total synthesis of GAs is a human challenge for natural creativity. Since GAs contain not only many asymmetric centers but also labile partial structures under acidic and/or basic conditions, GAs have been one of the most impregnable synthetic targets.

On the other hand, partial syntheses of GAs or interconversions among GAs have also been carried out by chemical, biochemical, and microbiological methods (see Section III). GAs or GA analogues thus obtained have been used for the characterization of new naturally occurring GAs, or for the elucidation of structure-activity relationships in GA bioassay systems. Information obtained from these partial syntheses has been utilized in total synthesis of GAs as well as in the preparation of isotope-labeled GAs or GA precursors which are available in research of biosynthesis, metabolism, and mode of action of GAs. Preparation of rarer GAs and isotopically labeled GAs by chemical conversion of readily available GAs is described elsewhere.[305]

1. Degradation Products of GAs

Degradation products of GA_3, gibberic acid (**10**), epigibberic acid (**14**), and gibberone

Table 31
RELATIVE INTENSITY OF IONS FORMED IN THE MS FRAGMENTATION OF GAs-GEs-TMSi[218]

	GAs-GEs-TMSi (MW)				
Mass lost	GA_1-GEs (942)	GA_3-GEs (940)	GA_4-GEs (854)	GA_{37}-GEs (868)	GA_{38}-GEs (956)
0(M$^+$)	11	9	3	1	2
15	3	4	2	6	2
72	9	13	2		2
87	5	8	4	2	1
90	3	5	4	1	
103	1			2	
105	1	2	7	4	2
144	1	2		8	6
145		3	1	5	5
336	10	25	8	1	4
349			3	2	3
378	100	100	100	100	100
393	8	14	22	7	14
395	10	10	42	7	11
422	7	8	15	19	14
423	6	9	24		
450	50	90	8	10	33
451	13	21	5	3	8
465	19	27	12	5	34
466	28	27	26	9	62
467	20	25	72	28	34
468	11	15	45	6	
479		7			6
483	4	6		1	
485	4		14		
495	33	60	30	8	32
513	7	17	26	2	3
527	9	8	3		
539	5	27	14	5	20
540	3	8	18	8	18
557	5	18	8		3
567	3	6	10		8
585	8	25	44	17	35

(15) were key compounds in the structural elucidation of GA_3 and now all have been synthesized. In 1963 Lowenthal et al.[219] synthesized ($\pm$)-gibberone using 4-methyl-1-oxo-indanone as a starting material via 4-methyl-1-oxoindanon-2-ylacetic acid methyl ester. They further achieved a total synthesis of ($\pm$)-gibberic acid via a similar synthetic route to that of ($\pm$)-gibberone employing *o*-tolylacetic acid as a starting material[220] (Figure 33).

Mori et al.[221] synthesized ($\pm$)-epigibberic acid from an intermediate in the synthesis of ($\pm$)-gibberic acid by seven steps (Figure 33).

2. Partial Synthesis of GAs

Cross et al.[222] succeeded in the chemical conversion of the methyl ester of gibberellin C (16) into GA_4-Me as shown in Figure 34. Gibberellin C-Me was reduced with sodium borohydride to a diol. When treated with phosphorous pentoxide the diol afforded GA_4-Me in 5% yield. 3-Deoxygibberellin C-Me was also converted to GA_9-Me in approximately 12% yield by the same procedure as above.

The first GA biosynthetic intermediate possessing the *ent*-gibberellane skeleton, GA_{12}-

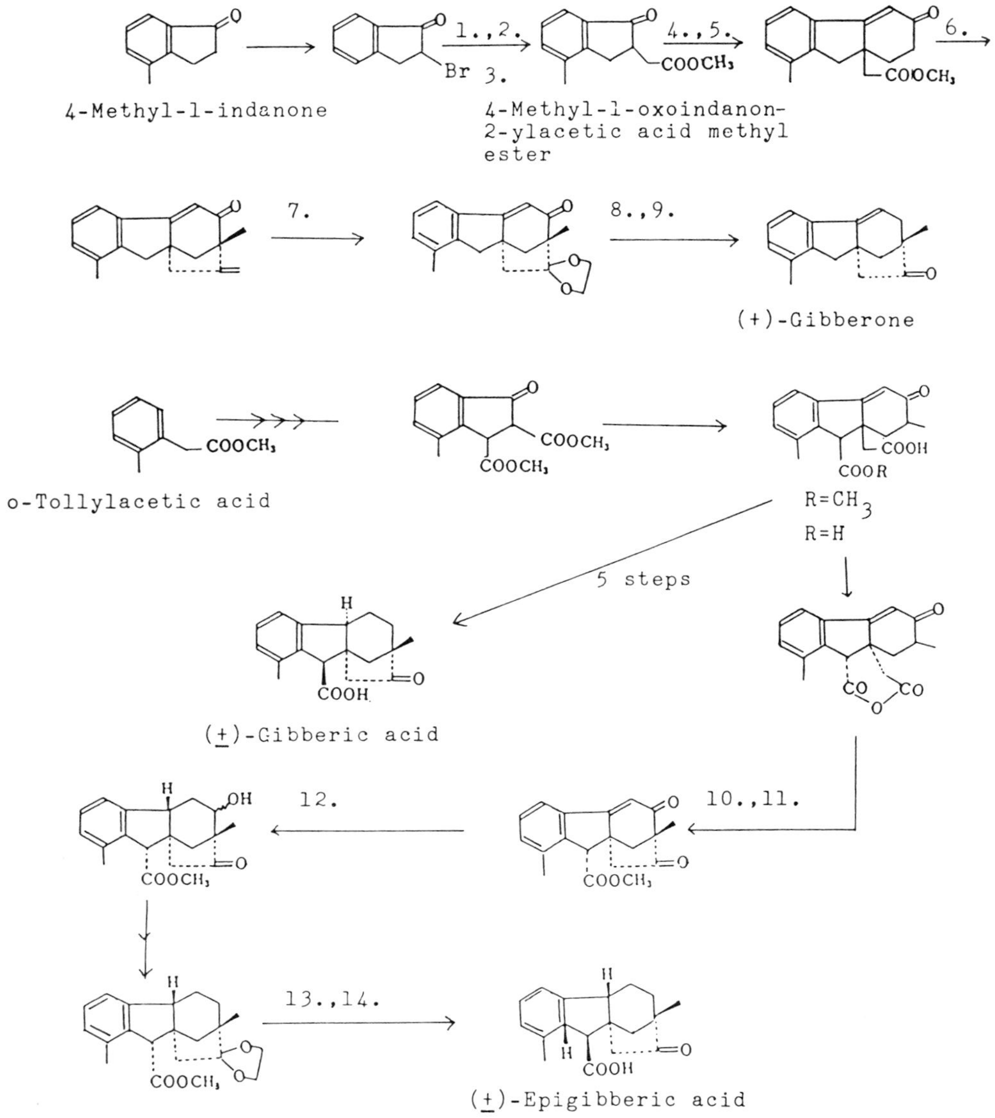

FIGURE 33. Synthetic route of (±)-gibberone, (+)-gibberic acid, and (±)-epigibberic acid: (1) NaH, CH$_2$(COO-*t*-Bu)$_2$, (2) H$^+$, (3) CH$_2$N$_2$, (4) CH$_3$ONa, CH$_3$COC(CH$_3$)=CH$_2$, (5) CH$_2$N$_2$, (6) BF$_3$, (CH$_3$CO)$_2$O, (7) ethylene glycol, H$^+$, (8) Wolf-Kishner reduction, (9) H$^+$, (10) BF$_3$, (11) CH$_2$N$_2$, (12) H$_2$, (13) Hough-Minlon reduction, (14) H$^+$.

7-aldehyde, was prepared[223] from *ent*-7β-hydroxykaurenolide (**18**) by the synthetic route shown in Figure 35. GA$_{12}$-7-aldehyde can be easily converted into GA$_{12}$ by Jones oxidation. GA$_{14}$ aldehyde was also synthesized[224] from *ent*-3α,7β-dihydroxykaurenolide by the route similar to that for GA$_{12}$-7-aldehyde.

Mori et al.[225,226] successfully converted a degradation product (**43**) derived from GA$_3$-Me into gibberellin C-Me as shown in Figure 36. The ketal of (**44**) was treated with trityl sodium and carbon dioxide to introduce a carboxyl at the C-4 position. The compound thus obtained was hydrogenated over palladium charcoal, esterified, reduced by sodium borohydride, and treated with acid to yield the C-3 epimer of gibberellin C-Me, which was further treated with 0.01 *N* sodium hydroxide to afford a mixture of gibberellin C-Me (**16**) and its C-3 epimer at the ratio of 1:9.

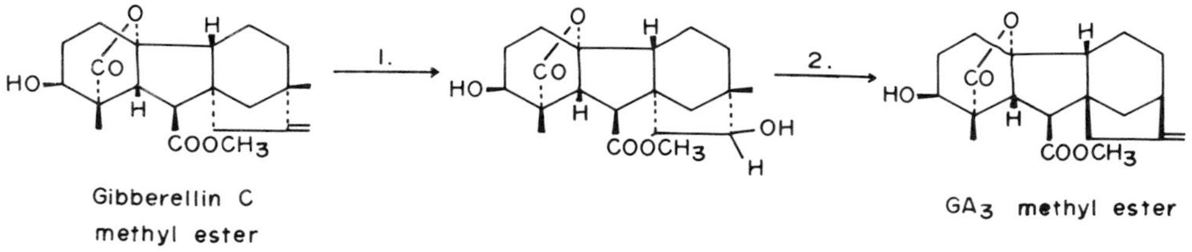

FIGURE 34. Synthetic route of GA_4 methyl ester from gibberellin C methyl ester: (1) $NaBH_4$, (2) PCl_5.

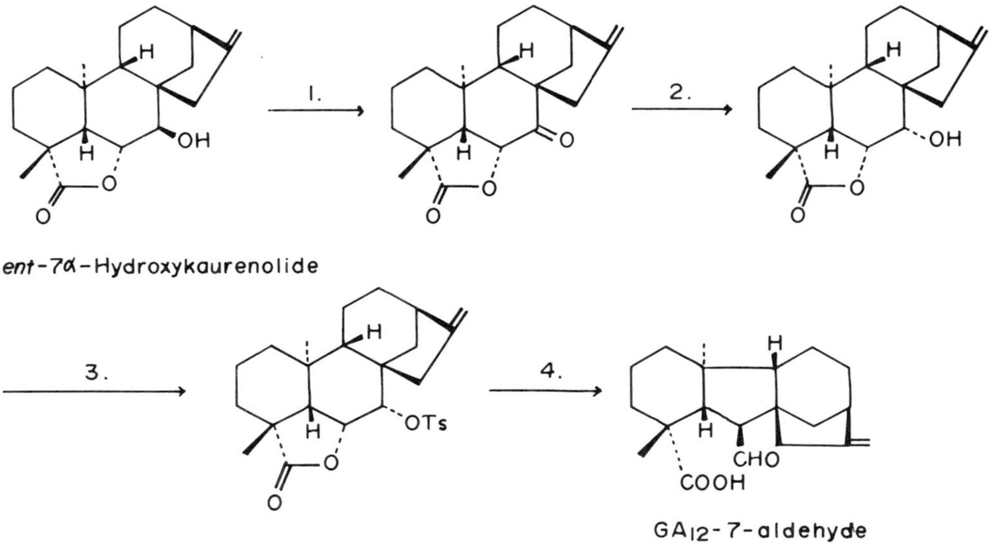

FIGURE 35. Synthetic route of GA_{12}-7-aldehyde from *ent*-7α-hydroxykaurenolide by Galt and Hanson: (1) Jones oxidation, (2) $NaBH_4$, (3) tosylchloride and pyridine, (4) OH^-.

3. Total Synthesis of GAs

In 1968, Mori et al.[227] successfully synthesized the compound **(43)** (Figure 36) from epigibberic acid, in which total synthesis had been accomplished. Since chemical conversion of **(43)** into gibberellin C-Me and gibberellin C into $GA_{4,9}$ was already described (see Section II.C.2) Figure 34), this constitutes a formal total synthesis of the above 13-H-GAs.

Somei and Okamoto[228] succeeded in conversion of enmein **(45)**, a diterpene isolated from *Isoden trichocarpus*, into GA_{15}. Node et al.[229] also synthesized $GA_{15,37}$ from enmein by a different synthetic route. Nakata and Tahara[230] reported chemical conversion of 1-abietic acid **(46)**, a major diterpene of pine resin, into GA_{12}. Since total syntheses of enmein and 1-abietic acid had been achieved,[231,232] these constitute formal total syntheses of $GA_{12,15,37}$.

Mori et al.[233] accomplished a formal synthesis of GA_{12} using a racemic tricyclic enone **(47)**, which is an intermediate in the syntheses of ($\pm$)-kaur-16-ene-19-oic acid and ($\pm$)-steviol, as a starting material. A stereocontrolled total synthesis of ($\pm$)-GA_{15} was completed by Nagata et al.[234] In this synthesis a tetracyclic α,β-unsaturated ketone **(48)**, a key intermediate in their total synthesis of diterpene alkaloids, was employed as a starting material (Figure 37).

In 1978, Corey et al.[235] achieved a stereospecific total synthesis of GA_3, which had been considered one of the most brilliant achievements in the synthesis of complicated natural products. The synthetic route is shown in Figure 38.

Mander and his co-workers[308-310] have also been engaged in intensive studies of syntheses

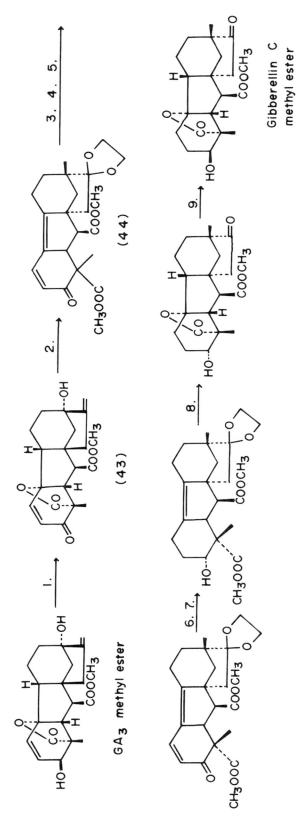

FIGURE 36. Synthetic route of gibberellin C methyl ester from compound (43) by Mori et al.[227]

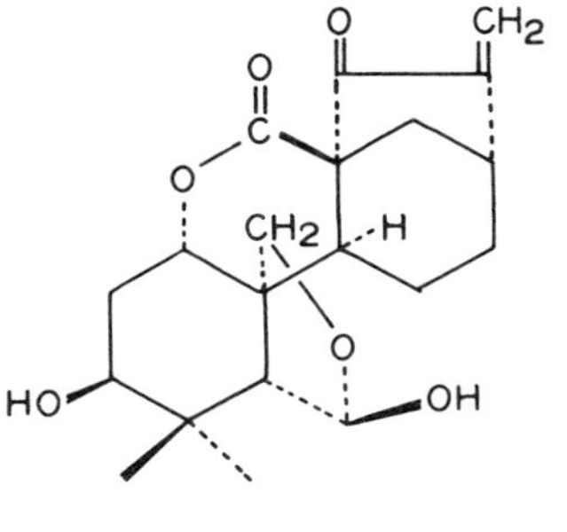

(45) Emmein

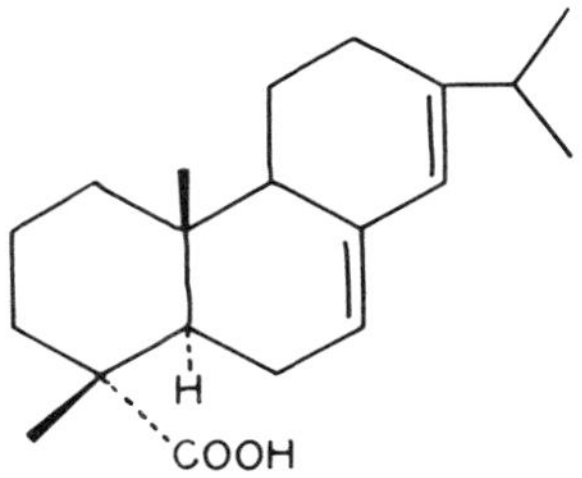

(46) l-Abietic acid

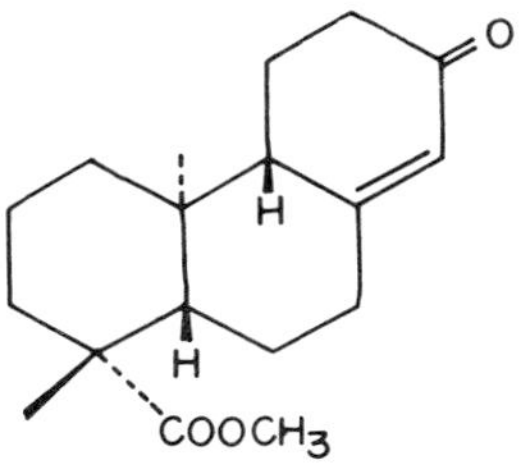

(47)

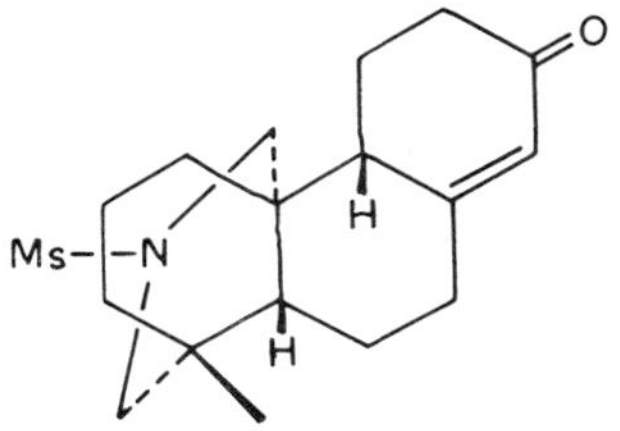

(48)

FIGURE 37. Structures 45 to 48.

of gibberellins. They have succeeded in total synthesis of GA_1, GA_3,[308] and GA_8-Me.[309] Recently, chemical preparation of GA_{19} from GA_3 was also successfully demonstrated by his group.[310]

4. GA Conjugates

GA_1-GEt was synthesized[236] by reaction of GA_1-Me with α-bromoacetoglucose followed by deacetylation with sodium methoxide and hydrolysis according to the method of Bartlett and Johnson.[237] GAs-GEs were also prepared by the coupling of free GAs and α-bromoacetoglucose followed by deacetylation under the critical condition. Preparation of GA_1-GEt, GA_3-GEt and GA_1-GEs, GA_4GEs, GA_{37}-GEs, and GA_{38}-GEs has been reported.[159,238]

Gibberethione isolated from immature seeds of *Pharbitis nil* was synthesized[239] by Michael addition of mercaptopyruvic acid to 3-keto-GA_3 followed by a concerted aldol condensation and a lactol formation.

III. BIOSYNTHESIS AND METABOLISM

The study of biosynthesis and metabolism of GAs in the fungus, *Gibberella fujikuroi*, and in higher plants was initiated rather late in the history of chemical research on GAs. The progress of the study in this field, however, seems to be rapid and glorious.

The diterpenoid nature of GA was first suggested by Cross et al.[240] Biosynthetic study was begun by Birch et al.[241] with the feeding of isotopically labeled precursors in a culture medium of *G. fujikuroi* and observing their distribution in GA_3. They showed that 2-^{14}C-acetic acid and 2-^{14}-C-mevalonate fed in culture media of *G. fujikoroi* were incorporated into GA_3 and determined the location of ^{14}C in the labeled GA_3. Similar procedures have been applied to the study using cell-free systems from some higher plants.[242-248,304] The process of the biosynthesis and metabolism of GAs in the fungus is believed to consist of the following three stages: (A) conversion of mevalonic acid to *ent*-kaurene, (B) conversion

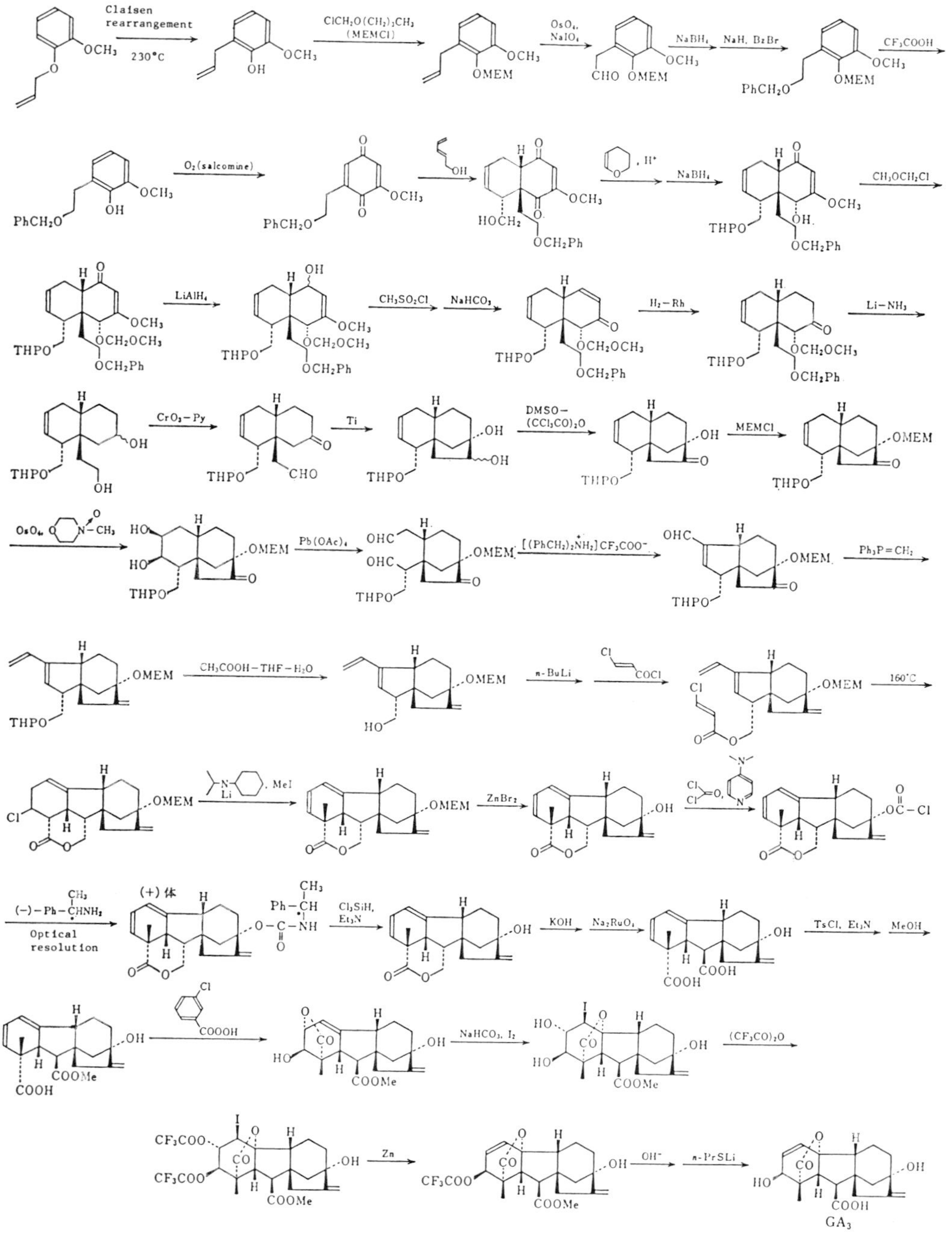

FIGURE 38. Total synthesis of GA_3 by Corey et al.[235]

of *ent*-kaurene to GA_{12}-7-aldehyde, (C) conversion of GA_{12}-7-aldehyde to C_{19}-GAs and interconversion of GAs (introduction of hydroxyl groups into *ent*-gibberellane skeleton).

In the biosynthetic study of GAs, both mutant strains of *G. fujikuroi* and plant growth retardants have been effectively used. The oxidation process from *ent*-kaurenal to *ent*-kaurenoic acid is blocked in the mutant, B1-41a,[249] thus facilitating feeding experiments which have clarified the pathway from *ent*-kaurenoic acid to the GAs.

Plant growth retardants, which were shown to block GA biosynthesis, have been applied to the biosynthetic study of GAs. The action of plant growth retardants has been reviewed by Cathey.[250] The plant growth retardants shown in Figure 39 block the cyclization step(s) from geranylgeranyl pyrophosphate to *ent*-kaurene and/or oxidation of *ent*-kaurene to *ent*-kaurenoic acid.[251,252] These plant growth retardants have been used to study details of the metabolic pathway after *ent*-kaurene in both *G. fujikuroi* and cell-free systems from higher plants.

There are several reviews on the biosynthesis and metabolism of GAs. Each of them generally describes aspects of the biosynthetic studies of gibberellins and contains topics of the day.[72,253-260]

A. Biosynthesis of GAs in *Gibberella fujikuroi*
1. Stage A: Conversion of Mevalonic Acid to ent-Kaurene

ent-Kaurene is the first tetracyclic diterpene produced from mevalonate. By feeding isotopically labeled compounds Hanson and White[261] showed that pyrophosphates of geraniol, geranylgeraniol, and labdadienol were direct precursors of *ent*-kaurene. Evans and Hanson[262] also showed that neither geometric isomer of pimaradiene was incorporated into *ent*-kaurene by a cell-free system from *G. fujikuroi*.

Schechter and West[263] prepared a soluble enzyme system from a cell-free extract of mycelia of *G. fujikuroi*, and incubated with 2-^{14}C-mevalonate, ATP, and $MgCl_2$. They successfully identified ^{14}C-labeled *trans*-(−)-labda-8(17),13-diene-15-ol (copalol), geranylgeraniol, and *ent*-kaurene in the incubation mixture. Both *trans*-geranylgeranyl pyrophosphate and copalyl pyrophosphate were converted by this enzyme system into *ent*-kaurene, but the free geranylgeraniol was not. The pathway of stage A is summarized in Figure 40.

2. Stage B: Conversion of ent-Kaurene to GA$_{12}$-7-Aldehyde

This stage consists of two processes. The first is the oxidation of *ent*-kaurene to *ent*-kaurenoic acid and the introduction of hydroxyl group(s) into the *ent*-kaurene skeleton. The second is the ring contraction step where the six-member B ring of *ent*-kaurene is converted into a five-member ring to form the *ent*-gibberellane skeleton.

The conversion of *ent*-kaurene, *ent*-kaurenol, *ent*-kaurenoic acid, and *ent*-7α-hydroxykaurenoic acid into GA$_3$ was shown by feeding those compounds labeled to the culture media of *G. fujikuroi*.[264-269] Cross et al.[270] showed that GA$_{12}$-7-aldehyde was converted by *G. fujikuroi* into GA$_3$ with a high incorporation ratio; the aldehyde was later shown to be the first intermediate with an *ent*-gibberellane skeleton.[261] The oxidation of the B ring determines the fate of *ent*-kaurenoic acid, i.e., *ent*-gibberellane or to other metabolites without the *ent*-gibberellane skeleton such as *ent*-7α-hydroxykaurenolide and fujenal, *ent*-7α-Hydroxykaurenoic acid was shown to be converted into *ent*–gibberellan-7-al-19-oic acid (GA$_{12}$-7-aldehdye), and also into *ent*-7α-hydroxykaurenolide and *ent*-6α,7α-dihydroxykaurenoic acid, by microsomes from *G. fujikuroi*.[256] *ent*-6α,7α- Dihydroxykaurenoic acid does not serve as a precursor of GAs but it is converted into fujenal.[271] These observations indicate that *ent*-7α-hydroxykaurenoic acid is a branching point leading either to the GAs or to other kaurenoids as shown in Figure 41. Intermediates between *ent*-7α-hydroxykaurenoic acid and GA$_{12}$-7-aldehyde have not been found. The ring contraction step has been well studied by Hanson and White[264] and their co-workers[272-274] using ^{14}C,^{3}H-doubly labeled mevalonic acid (MVA). The labeling patterns of GA$_3$ and intermediates from 2R-(2-^{3}H;2-^{14}C)-MVA and 5R-(5-^{3}H;2-^{14}C)-MVA are shown in Figures 41 and 42. From the labeling patterns it was shown that the σ-bond between C-7 and C-8 attacked the cationic center at C-6 to expel C-7 from the ring giving GA$_{12}$-7-aldehyde.

Hanson et al.[274] fed (1-^{3}H$_2$, 1-^{14}C)-geranylgeranyl pyrophosphate to the culture of *G. fujikuroi* to study the fate of 6-hydrogen in the ring contraction process. The phosphate was converted to GA$_{12}$-7-aldehyde with a low but specific incorporation (0.005%). Based on the

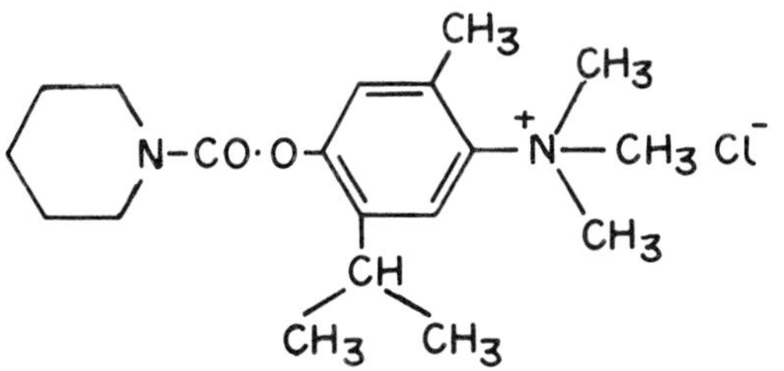

(49) AMO-1618

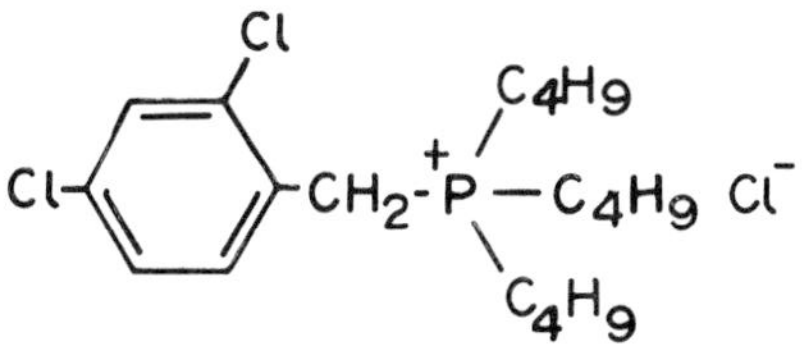

(50) Phospon D

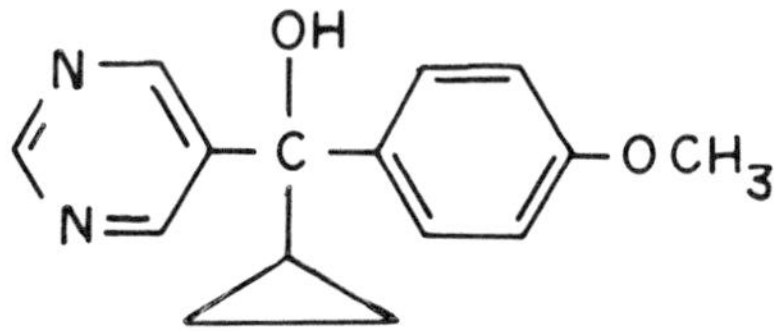

(51) Ancymidol

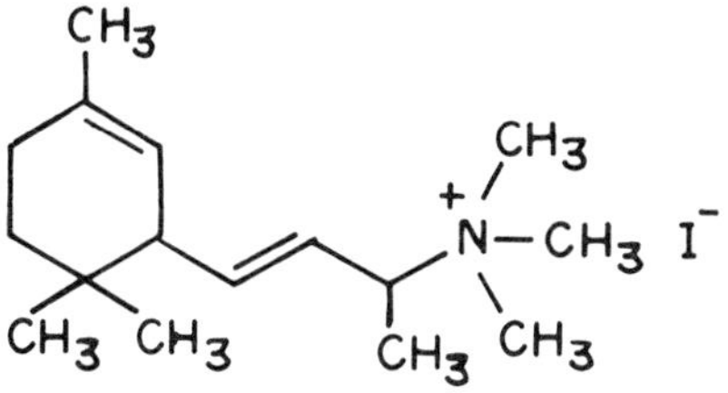

(52)

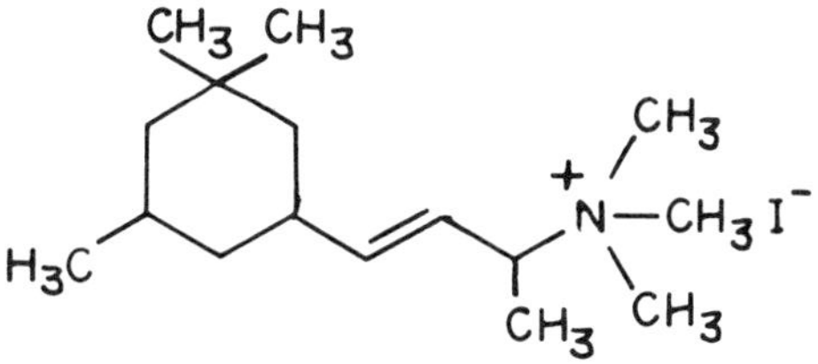

(53)

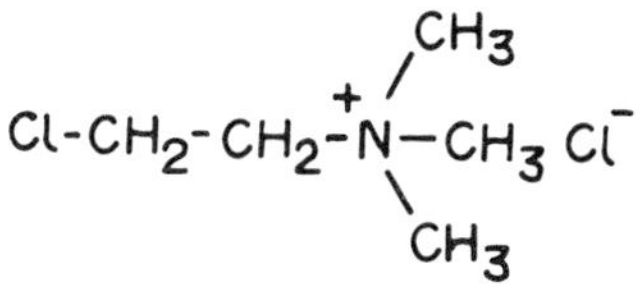

(54) CCC

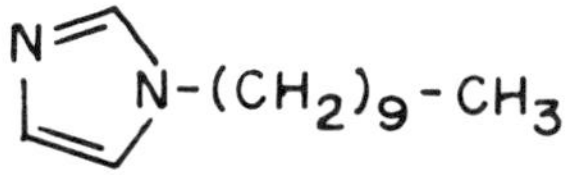

(55)

FIGURE 39. Typical plant growth retardants inhibiting GA biosynthesis. **(49)** AMO-1618; 2′-isopropyl-4′-(trimethylammonium chloride)-5′-methylphenylpiperidine-1-carboxylate. **(50)** Phosphon D; trimethyl-2,4-dichlorobenzylphosphonium chloride. **(51)** Ancymidol; α-cyclopropyl-4-methoxy-α-(pyrimidin-5-yl)benzyl alcohol. **(52)** *N,N,N*-trimethyl-1-methyl(3′,3′,5′-trimethyl-cyclohex-2′-en-1′-yl)prop-2-enylammonium iodide. **(53)** *N,N,N*-trimethyl-1-methyl-(3′,3′,5′-trimethylcyclohexan-1′-yl)prop-2-enylammonium iodide. **(54)** C C C; 2-chloroethyltrimethyl ammonium chloride. **(55)** 1-decylimidazole.

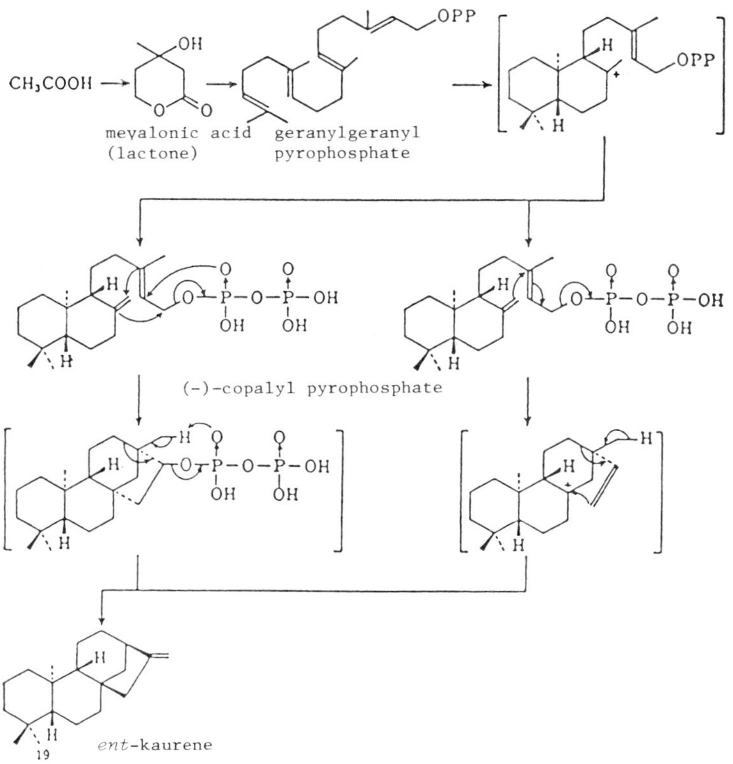

FIGURE 40. Biosynthetic pathway from MVA to *ent*-kaurene (stage A).

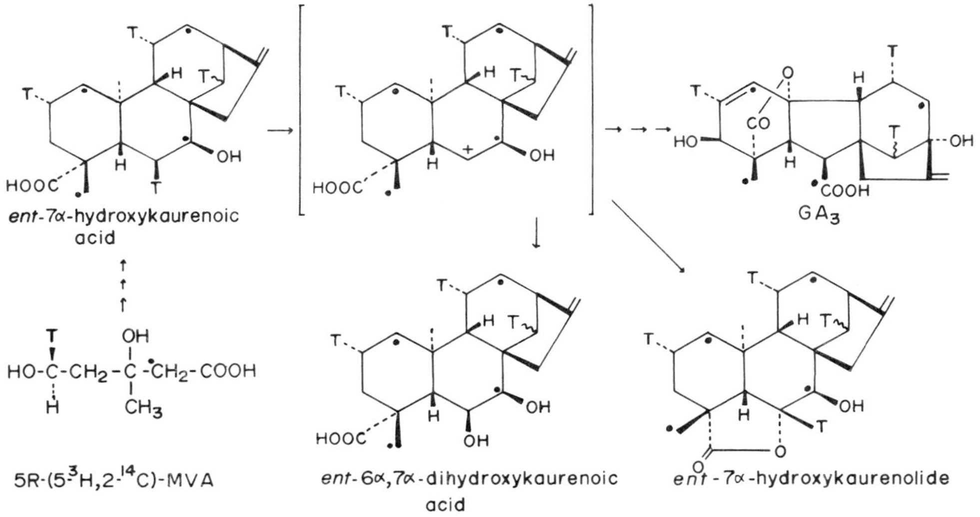

FIGURE 41. Incorporation of 5R-(5-^{3}H, 2-^{14}C)-MVA into GA$_3$ and kaurenoides.

^{3}H:^{14}C ratio in the aldehyde and its degradation study, they reported that the ring contraction proceeded accompanying a shift of *ent*-6α-hydrogen to C-7, thus becoming the formyl hydrogen of GA$_{12}$-7-aldehyde. However, this migration mechanism could be incorrect and revised because it is based on the GA$_{12}$-7-aldehyde with so little incorporation from (^{3}H$_2$, ^{14}C)-geranylgeranyl pyrophosphate and is different from the one established in higher plants.

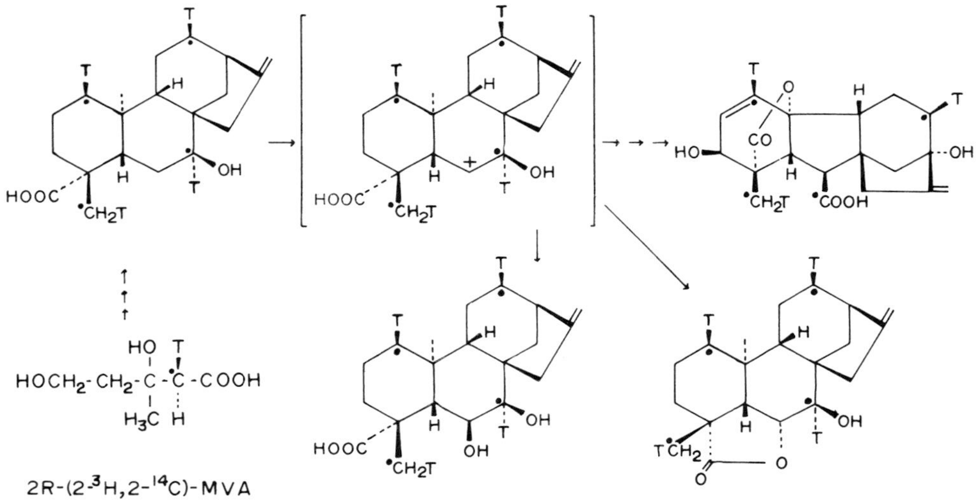

FIGURE 42. Incorporation of 2R-(2-^{3}H, 2-^{14}C)-MVA to GA$_3$ and kaurenoids.

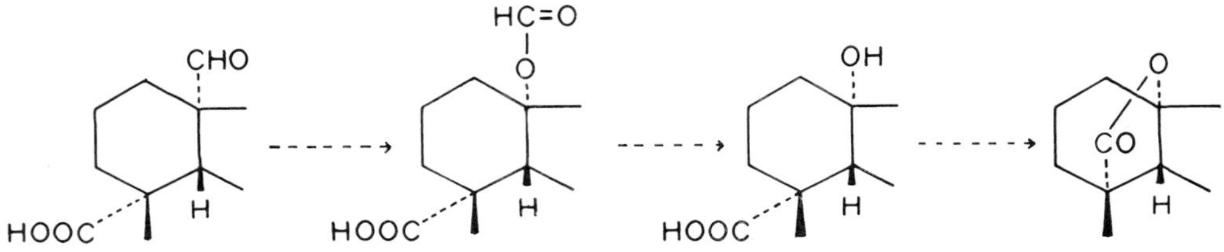

FIGURE 43. Hypothetical mechanism of the removal of C-20 as formate.[275] One of the oxygens on C-19 is released.

The details of this process have been studied using cell-free systems of higher plants (see Section III.C).

3. Stage C-1: Formation of C_{19}-GAs from GA_{12}-7-Aldehyde via C_{20}-GAs

The mechanism for the loss of C-20 in the formation of the C_{19}-GAs from the C_{20}-GAs is not known. Studies are being focused to resolve this process.

The substituents on C-10 in the C_{20}-GAs are the methyl, hydroxyl methyl (forming δ-lactone), aldehyde, and carboxyl groups. This variation suggests a mechanism for the loss of C-20. Thus, the methyl group on C-10 is considered to be stepwise oxidized to a carboxyl group, and the C_1-unit may be released during the oxidation process.

Since GA$_{13}$, carrying a carboxyl group on C-10, was not converted into GA$_3$ by *G. fujikuroi*,[270,272] the process of the removal of a C_1-unit by decarboxylation after the introduction of a carboxyl group at C-10 was ruled out. An alternative process was proposed by Hanson and White[275] (Figure 43). In this mechanism the C_1-unit can be removed by hydrolysis of the formate produced by a Baeyer-Villiger type oxidation of the aldehyde group on C-10 and following γ-lactonization with the carboxyl group on C-4. Recently, this formate hypothesis was also ruled out. A process in which the C_1-unit is removed after oxidation into a carboxyl group is again suggested based on the following observations: (a) the ^{18}O-labeled GA$_{12}$-7-alcohol, which carries ^{18}O in the carboxyl group at C-4 was converted into C_{19}-GAs, retaining the ^{18}O atom on C-19.[276] However, in a process via a formate, one of the two oxygens on C-19 would be removed. (b) the *ent*-kaurene labeled with ^{14}C at the

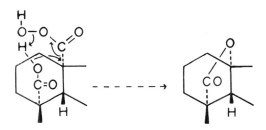

FIGURE 44. Hypothetical mechanism of the removal of C-20 via peracid.[278] Two oxygens on C-19 are retained.

C_1-unit on C-10 released $^{14}CO_2$ instead of $H^{14}COOH$ or $H^{14}CHO$ when it was incorporated into C_{19}-GAs.[277,278]

The above results suggest that the C_1-unit on C-10 is removed by decarboxylation after formation of a carboxyl group. Based on these results and their previous work that GA_{13}-7-aldehdye was incorporated into GA_3, Dockerill et al.[277] proposed another mechanism whereby the C_1-unit on C-10 is removed after the formation of the peracid as shown in Figure 44. However, the possibility of GA_{13}-7-aldehyde as a precursor to 3-OH-C_{19}-GAs has been revoked by Bearder et al.[306] with feeding experiments of GA_{13}-7-aldehyde and its related compounds to the culture media of two different strains of G. fujikuroi.

Further intensive studies are required to clarify the conversion mechanism from C_{20}-GAs to C_{19}-GAs because the possibility remains that the C_1-unit is released as HCOOH from the molecule and then oxidized in the media to be trapped as CO_2, as Hedden[307] pointed out.

4. Stage C-2: Biosynthetic Conversion of GAs

This stage consists of the introduction of hydroxyl group(s) and an olefinic double bond into the ent-gibberellane skeleton. The hydroxylation of GAs produced by G. fujikuroi mainly occurs by three different procedures.

One procedure is 3β-hydroxylation. The presence or absence of 3β-hydroxylation of GA_{12}-7-aldehdye seems to be a branching point to two pathways leading either to 3β-OH-GAs such as GA_4, GA_{14}, and GA_{36} or to GAs not carrying 3-hydroxyl such as GA_9, GA_{12}, and GA_{24}.

3β-hydroxylation of GA_{12}-7-aldehyde gives GA_{14}-7-aldehyde. This was shown by Hedden et al.[279] to be a direct precursor of 3β-OH-GAs in a feed of 6-^{3}H-GA_{12}-7-aldehyde to the mutant B1-41a of G. fujikuroi.

A second procedure is the hydroxylation on C-13. In the genetic mutant R-9 of G. fujikuroi, production of GA_1 and GA_3 is not observed, but GA_4, GA_7, and other GAs are produced. This shows that the hydroxylation on C-13 occurs after the formation of C_{19}-GAs such as GA_4 and GA_7, and in the mutant R-9 the 13-hydroxylation pathway is blocked.[280] This is also supported by the fact that none of the 13-OH-C_{20}-GAs have been found to be metabolites of G. fujikuroi. Most 13-OH-GAs found as metabolites of G. fujikuroi are C_{19}-GAs with a 3β-hydroxyl group; one exception is GA_{20} which was recently found as a metabolite of the GA-producing strain ACC 917 of G. fujikuroi.[24]

Hydroxylation at C-1 or C-2 seems to occur late in the pathway. The hydroxyl group at C-1 of fungal GAs is found in both orientations as in GA_{16} and GA_{55}. In contrast the hydroxyl group at C-2 for fungal GAs is only in the α-configuration, a characteristic contrast to the β-only configuration of the 2-hydroxyl group in plant GAs.

The introduction of an olefinic double bond between C-1 and C-2 as in $GA_{3,7}$ was studied by feeding isotopically labeled $GA_{1,4,16}$ to G. fujikuroi (ACC 917). GA_1 was not converted to GA_3, or converted in very small amounts depending on the culture media.[281] The low or nonconversion of GA_1 into GA_3 was also observed in cultures using the G. fujikuroi mutant

R-9. The mutant B1-41a efficiently converted GA_4 and GA_7 to GA_3, but not GA_1 and GA_{16}.[282] These observations suggest that the C-1~C-2 double bond is introduced into GA_4 to give GA_7 and not GA_1 to GA_3. Evans et al.[272] showed that the dehydrogenation process from GA_4 to GA_7 is a *cis*-elimination of 1 and 2 hydrogens by analysis of stereospecifically labeled GA_4 and GA_7 from 2*R*-, 2*S*, and 5*R*-^{3}H-MVAs. The biosynthetic pathway of GAs in *G. fujikuroi* is summarized in Figure 45 based mainly on the experiments by Bearder et al.[282] and Evans and Hanson.[283]

B. Metabolism of Nonfungal GAs by *G. fujikuroi*

The enzyme systems for GA biosynthesis in *G. fujikuroi* show a broad substrate specificity. These properties have been used in the preparation of GAs and GA-like compounds which are not native to the fungus. Mutant strain B1-41a of *G. fujikuroi* blocked at the step *ent*-kaurenol to *ent*-kaurenal has been used. Plant growth retardants that block the early step in GA biosyntheses have also been used. Bearder et al.[284] fed steviol to the culture medium of a mutant B1-41a and obtained *ent*-7α,13-dihydroxykaurenoic acid, *ent*-6α,7α,13-trihydroxykaurenoic acid, *ent*-7α,13-dihydroxykaurenolide, GA_{53} and GA_1 as major metabolites, and GA_{17-20} as minor metabolites, while steviol acetate was converted mainly into the acetates of GA_{17} and GA_{20} and *ent*-6α,7α-dihydroxy-13-acetoxykaurenoic acid by this mutant.[285] GA_{45} and some other 15β-hydroxy GA analogues were successfully prepared from *ent*-15α-hydroxykaurenoic acid using this mutant. Bearder et al.[286] fed trachylobanic acid to the culture medium of the mutant B1-41a and prepared the 12,16-cyclo-analogues of $GA_{4,9,12-15,24,25,37,47}$ (Figures 46 and 47).

The addition of plant growth retardants to the culture medium of GA producing strains of *G. fujikuroi* has also been used to prepare rare GAs and their analogues. Murofushi et al.[287] fed steviol to the culture medium of GA-producing strain (G-2) containing the plant growth retardant *N,N,N*-trimethyl-3-(3′,3′,5′-trimethylcyclohexyl)-2-propenylammonium iodide (**53**).[252] They obtained *ent*-7α,13-dihydroxykaurenoic acid, *ent*-6α,7α,13-trihydroxykaurenoic acid, and $GA_{1,18,19,53}$. Wada et al.[288] incubated *ent*-15α-hydroxykaurenoic acid in the medium of *G. fujikuroi* (G-2) containing 1-decylimidazole (**55**) and obtained the metabolites, GA_{45}, 15β-hydroxy GA_{12}, 15β-hydroxy GA_7, and GA_{66} (Figure 48). Hanson et al.[289] prepared atisagibberellin A_{12} and atisagibberellin A_{14} by feeding *ent*-7α-hydroxyatis-16-en-19-oic acid to the culture medium of *G. fujikuroi* (ACC 917) in the presence of AMO 1618 (**49**) (Figure 49).

While the above experiments demonstrate the low substrate specificity of the enzyme systems of *G. fujikuroi*, atisagibberellins were not susceptible to oxidation at C-20, a major site of oxidation in natural GAs; the metabolism of 12,16-cyclogibberellin A_{12} to 12,16-cyclogibberellin A_9 shows the requirement of stereochemistry for the enzyme systems controlling the biosynthesis of GAs in *G. fujikuroi*.

C. Biosynthesis of GAs in Higher Plants

The biosynthetic pathway of GAs in higher plants is basically similar to that in *G. fujikuroi*; however, more variations in the structures of plant GAs are noticed than in fungal GAs. The variety of positions and stages in biosynthesis, where hydroxyl groups and an olefinic double bond are introduced, provides a wide variety of plant GAs. In fact, GAs isolated from Leguminosae contain many 13-OH-C_{20}-GAs such as $GA_{18,19,23,44,53}$ suggesting that 13-hydroxylation occurs at the C_{20}-GAs stage in Leguminosae in contrast to the 13-hydroxylation at the last stage of biosynthetic pathway in *G. fujikuroi*. Many 13-H-GAs have been isolated from Cucurbitaceae, showing that the biosynthetic pathway lacking 13-hydroxylation to C_{20}-GAs is also present in higher plants.

The biosynthetic pathway from MVA to GA_{12}-7-aldehyde is probably the same in higher plants and the fungus *G. fujikuroi*. However, the hydroxylation process after the formation of GA_{12}-7-aldehyde is different. The biosynthetic pathway from MVA to GA_{12}-7-aldehyde

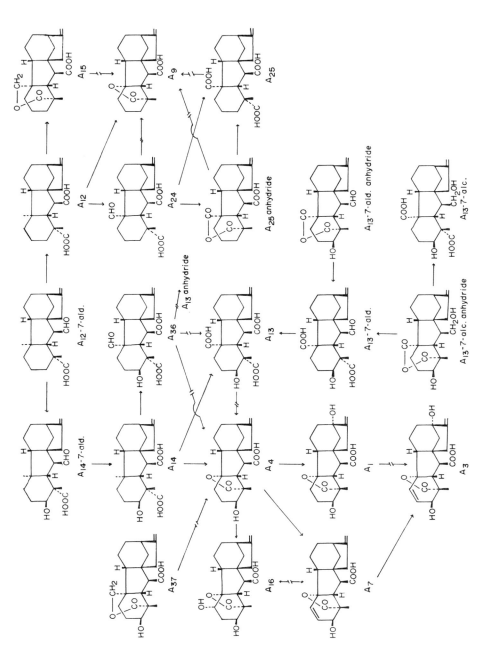

FIGURE 45. Metabolism of GAs and homologs fed to culture media of *G. fujikuroi*.[282,283,306]

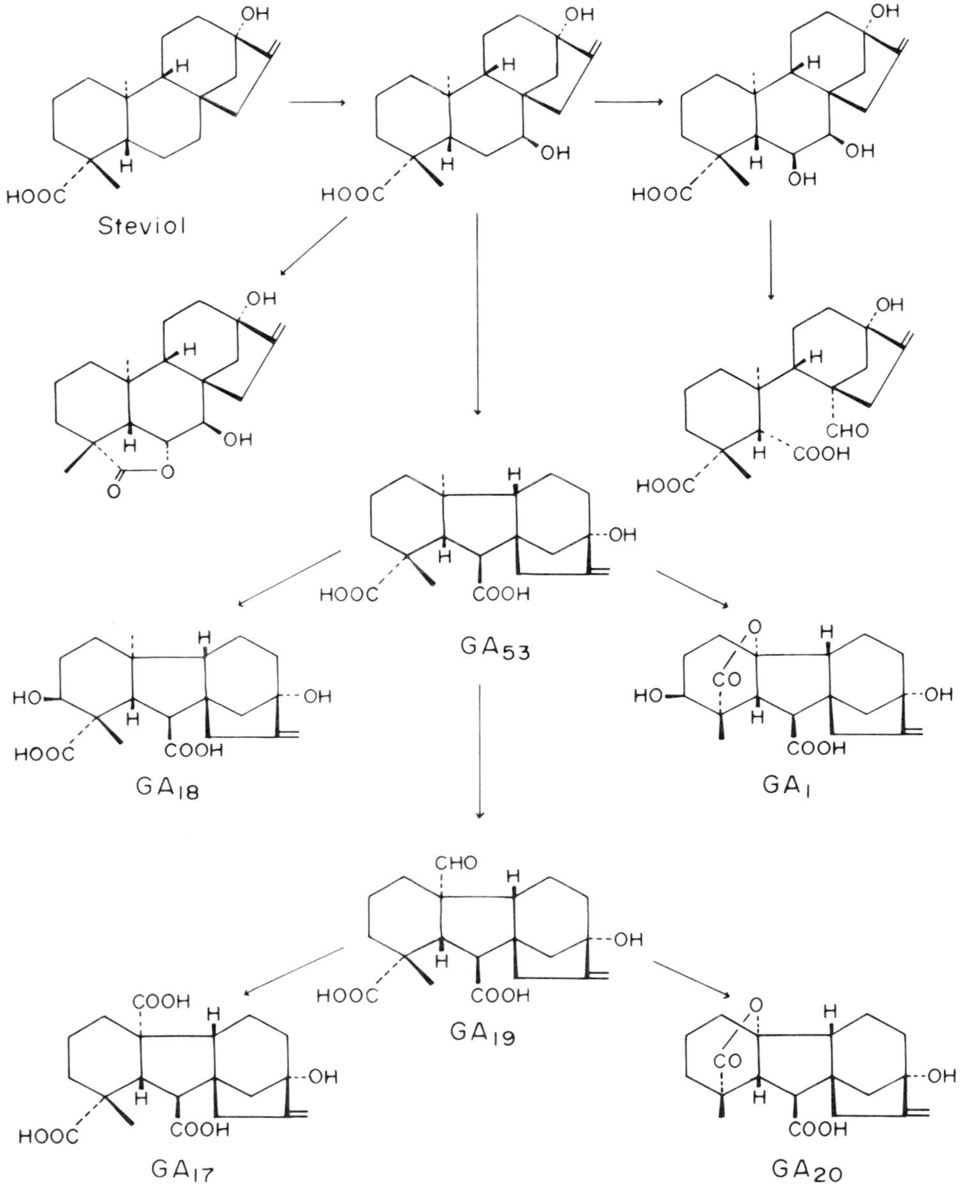

FIGURE 46. Metabolites from steviol G. fujikuroi B1-41a.[284]

has been studied using cell-free systems prepared from several species of higher plants. The steps subsequent to GA_{12}-7-aldehyde have been studied using both cell-free systems and intact plants.

1. Stages A and B: Formation of GA_{12}-7-Aldehyde from MVA

West and co-workers[242,243,290-294] confirmed that MVA was metabolized into *ent*-kaurene via intermediates such as geranylgeranyl pyrophosphate and copalyl pyrophosphate using cell-free systems prepared from the endosperm of wild cucumber (*Marah macrocarpus*, former *Echinocystis macrocarpa*) and from seedlings of caster bean (*Ricinus communis* L.). The cell-free system prepared from *M. macrocarpus* converted *ent*-kaurene into *ent*-7α-

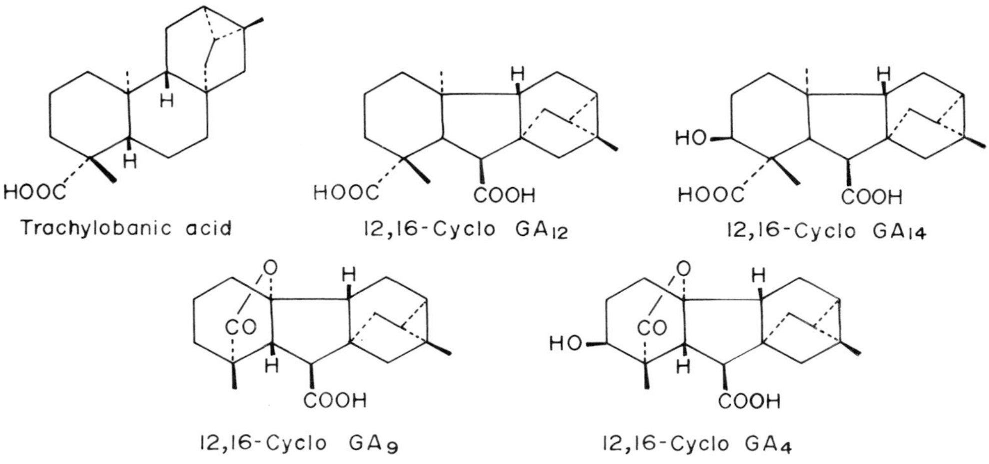

FIGURE 47. Trachylobanic acid and its metabolites by *G. fujikuroi* B1-41a.[286]

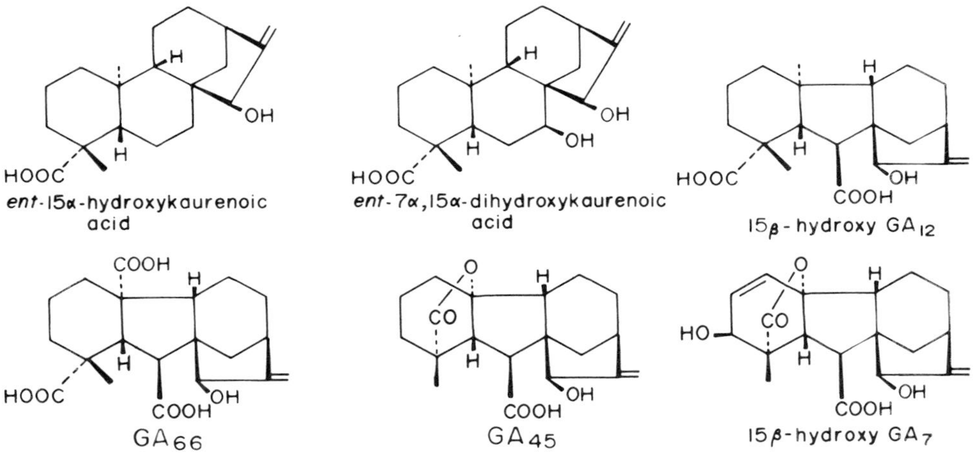

FIGURE 48. *ent*-15α-Hydroxykaurenoic acid and its metabolites by *G. fujikuroi* treated with 1-decylimidazole.[288]

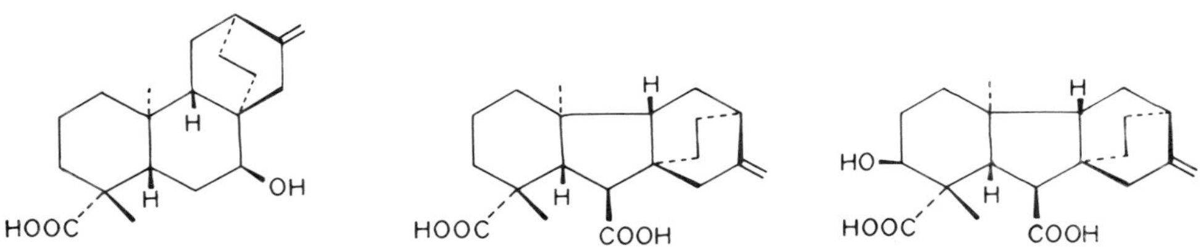

FIGURE 49. *ent*-7α-Hydroxyatis-16-en-19-oic acid and its metabolites by AMO-1618 treated *G. fujikuroi*.[289]

hydroxykaurenoic acid via intermediates of *ent*-kaurenol, *ent*-kaurenal, *ent*-kaurenoic acid, but not further into GA_{12}-7-aldehyde.[290,294]

Graebe et al.[295] first reported the biosynthesis of GA_{12}-7-aldehyde from MVA using cell-free systems from *Curcurbita maxima*. Graebe and co-workers[72,110,244,295,296] showed that the cell-free system required ATP and Mn^{2+} or Mg^{2+} to be active in the conversion of mevalonate to *ent*-kaurene,[72] and that the enzymes are located in the soluble fraction of cell-free extracts. The microsomal enzymes that convert *ent*-kaurene to GA_{12}-7-aldehyde, require NADPH and

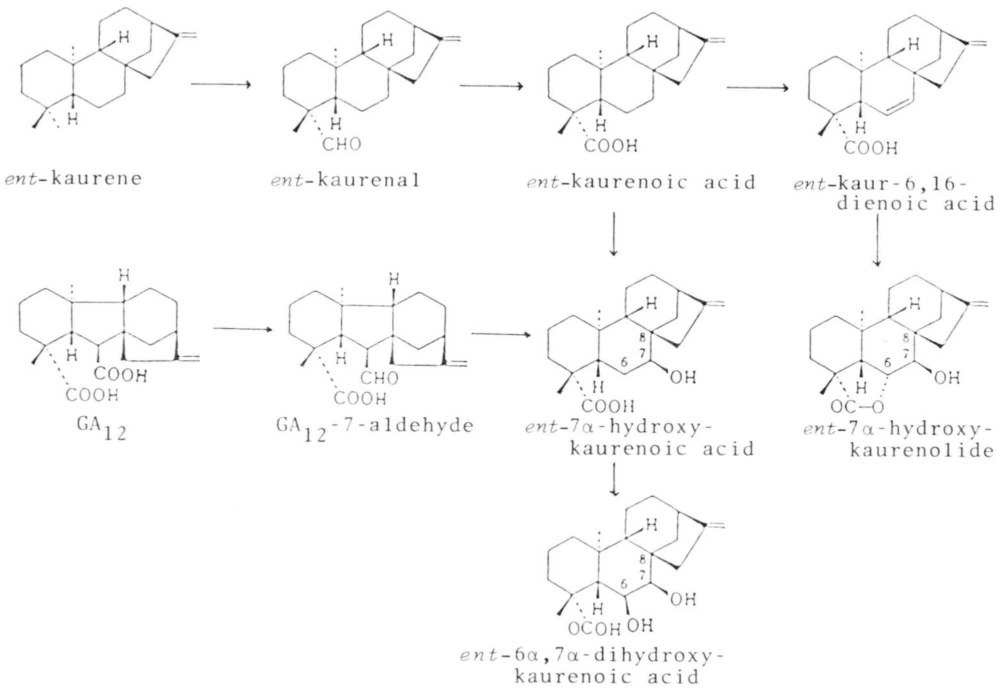

FIGURE 50. Metabolism of *ent*-kaurene by a cell-free system of *C. maxima*.[72]

O_2.[72] They recently demonstrated steps in the pathways from *ent*-kaurene to GA_{12}-7-aldehyde, *ent*-6α,7α-dihydroxykaurenoic acid, and *ent*-7α-hydroxy-kaurenolide using a *C. maxima* system.[72] A new kaurenoid, *ent*-kaur-6,16-dienoic acid, was identified as an intermediate to *ent*-7α-hydroxylkaurenolide but not to GA_{12}-7-aldehyde or to *ent*-6α,7α-dihydroxykaurenoic acid. In a cell-free system of *C. maxima*, *ent*-7α-hydroxykaurenoic acid was not converted into *ent*-7α-hydroxykaurenolide.[72] This is different from the pathway reported for *G. fujikuroi* (Figure 50).

For the interpretation of the ring contraction step, Hedden[307] proposed a free radical mechanism as shown in Figure 51.

The pathway from *ent*-kaurene to *ent*-7α-hydroxykaurenoic acid has been demonstrated in cell-free systems from germinating grains of *Hordum distichon*,[245] immature seed of *Pisum sativum*,[309] and the suspensors of *Phaseolus coccineus* seeds.[246,298]

2. Stages C: Conversion of GA_{12}-7-Aldehyde to C_{19}-GAs and Interrelationship of GAs

The conversion of GA_{12}-7-aldehyde into C_{20}- and C_{19}-GAs in higher plants has been studied using cell-free systems from only a few higher plants. The interrelationship of GAs has been studied using some intact plants.

Graebe et al.[72] did intensive work on the metabolism of GAs using a cell-free system of *C. maxima*. GA_{12}-7-aldehyde was converted into GAs using a supernatant with added Fe^{2+} and pyridine nucleotide in the absence of Mn^{2+} and Mg^{2+}. The results are shown in Figure 52 where the dotted lines represent the metabolism in an excess of GA_{12}-7-aldehyde. The "natural" pathway is indicated by solid lines.

Ceccarelli et al.[298] have also shown that a cell-free system from suspensors of *P. coccineus* seeds converts *ent*-7α-hydroxykaurenoic acid into $GA_{1,5,8}$.

Kamiya and Graebe[247] prepared a cell-free system from immature pea (*Pisum sativum*) embryos and showed that this system converted GA_{12} sequentially to GA_{15} (open lactone), GA_{24}, GA_9, and GA_{51}. They also showed that GA_{12} was converted to GA_{53} by the microsomal

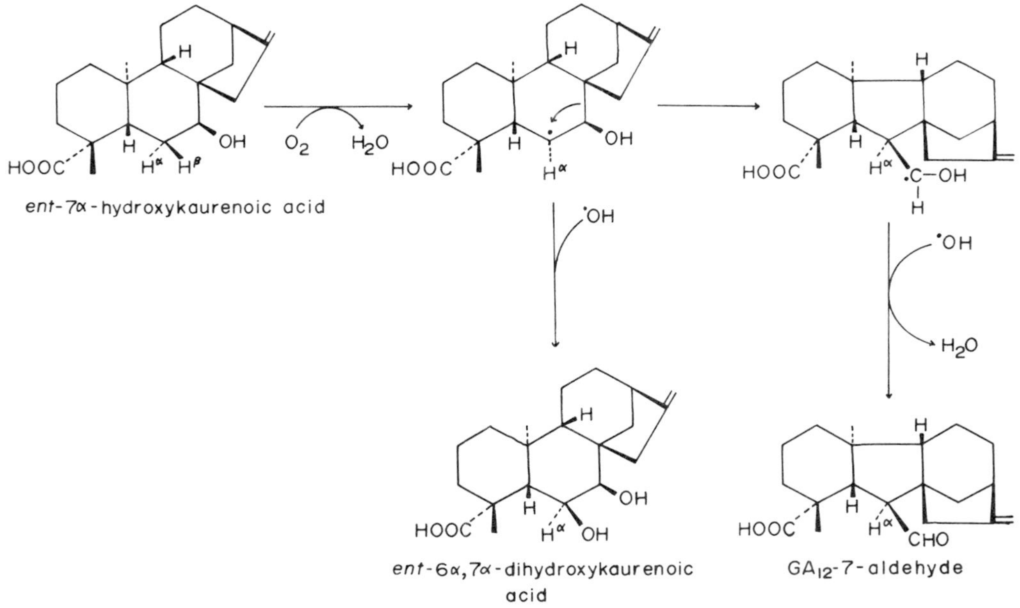

FIGURE 51. Interpretation of ring contraction *via* hypothetical radical intermediates.[307]

fraction, and the substrate was sequentially metabolized into GA_{44} (open lactone), GA_{19}, GA_{20}, and GA_{29} by the soluble fraction. They showed that GA_{24} was a precursor of GA_9, whereas GA_{25} and GA_{25}-anhydride were not. They concluded that the conversion of C_{20}-GAs to C_{19}-GAs took place with C-20 at the stage of an aldehyde, at least for nonhydroxylated GAs in *P. sativum*. They also mentioned that the conversion of GA_{13}-7-aldehyde to C_{19}-GAs in *G. fujikuroi*, which had been reported by Dockerill et al.,[277,278] was less likely based on their result and the studies by Graebe et al.[72] and by Metzger and Zeevaart.[164] The pathway observed in a cell-free system from *P. sativum* is shown in Figure 53.

By using a cell-free system prepared from immature seeds of *Phaseolus vulgaris*, Kamiya et al.[248] showed that GA_{20} labeled with ^{14}C or ^{18}O was converted into GA_1 and GA_5, which required oxygen, Fe^{2+}, and α-ketoglutarate. As GA_1 and GA_{29} were not converted into GA_5, they concluded that GA_{20} was directly converted to GA_5.

It is quite important to clarify the biosynthesis and metabolism of GAs for an understanding of their physiological roles in the control of plant growth. It is difficult to carry out metabolic studies with intact plants because of low incorporation. Nevertheless, some progress is being made on the metabolism of GAs. Yamane et al.[299,300] fed tritiated C_{19}-GAs, such as $GA_{1,4,5,8,20}$ to developing seeds of *Phaseolus vulgaris*, from which $GA_{1,4-6,8,17,20,37,38,44}$, GA_8-GEt, GA_1-GEs, GA_4-GEs, GA_{37}-GEs, and GA_{38}-GEs were identified as endogenous GAs. Analyses were made after various periods of incubation. The pathway they confirmed is summarized in Figure 54. The conversions from GA_4 to GA_1 and GA_8, and from GA_{20} to GA_1 and GA_8 were observed in all incubations. However, the conversion from GA_5 to GA_8 and the metabolism to conjugated GAs were not observed in 3-day incubations with early immature seeds, suggesting that metabolism from free GAs to conjugated GAs occurred at later stages of seed maturation.

Sponsel and MacMillan studied the metabolism of GAs in immature seeds of *P. sativum*, from which $GA_{9,17,20,29,44,51}$ were identified as endogenous GAs.[301-303] GA_9 fed to "summer" grown seeds was metabolized into GA_{51} and dihydro-GA_{31} and its conjugate, the latter two of which are not endogenous in *P. sativum*. GA_9 fed to "winter" grown seeds was metab-

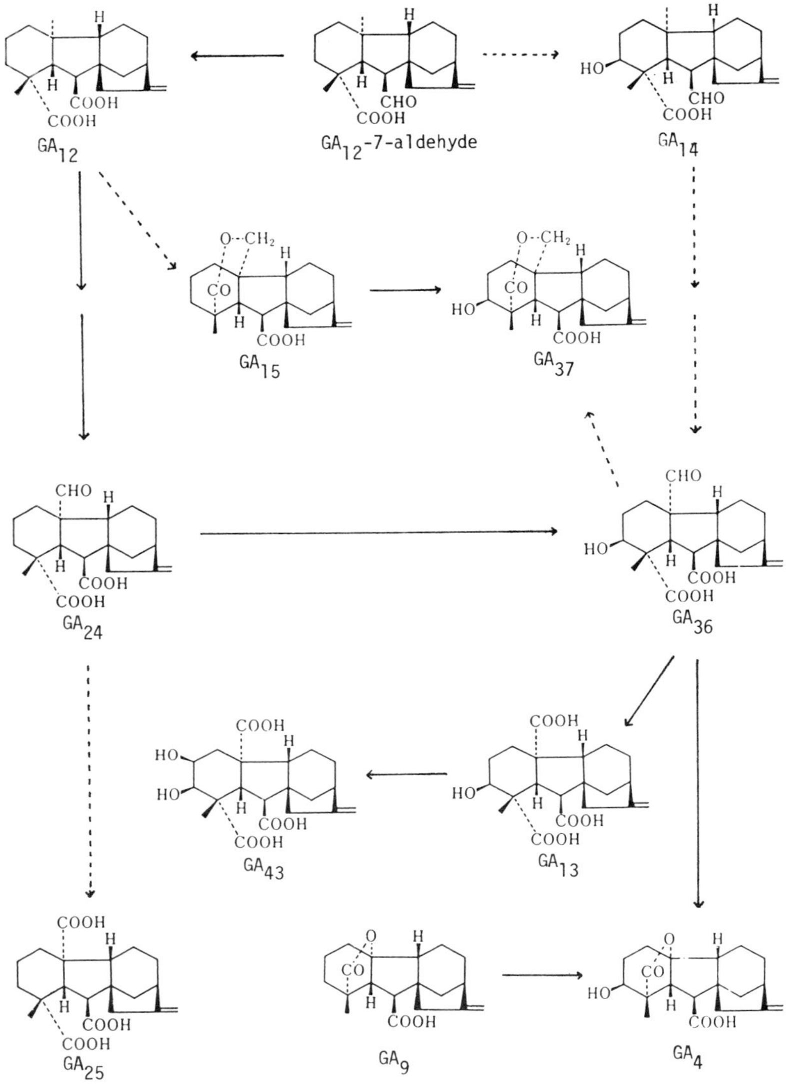

FIGURE 52. Conversion of GA_{12}-7-aldehyde by the 200,000 g supernatant fraction of the *C. maxima* cell-free system. Broken-line arrow pathways are observed when excess GA_{12}-7-aldehyde is added as a substrate. GA_9 is not found in the system, but is readily converted into GA_4.[72]

olized into GA_{20} and GA_{51}. The metabolism of GA_{20} and GA_{29} in the seeds was also studied in detail, and they showed the pathway summarized in Figure 55.

Another unique metabolite is the gibberellethione isolated from immature seeds of *Pharbits nil*.[93,239] Yokota et al.[93] suggested that this compound was formed from GA_3 and mercaptopyruvic acid in the process shown in Figure 56.

IV. PHYSIOLOGY

A. Plant Growth Promoting Effect

The most prominent physiological effect of GAs is shoot elongation in intact plants. This response is clearly observed when GAs are applied to young plants. The growth promoting

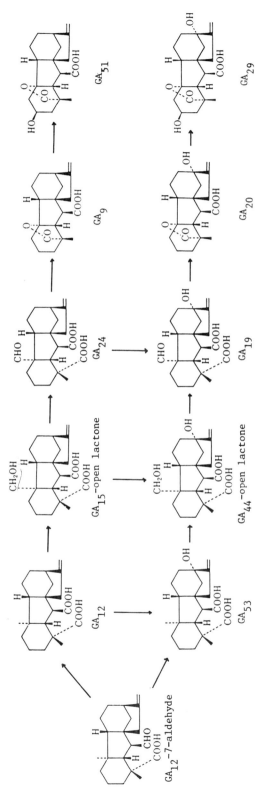

FIGURE 53. Metabolism of GA_{12}-7-aldehyde by the cell-free system of *P. sativum*.[247]

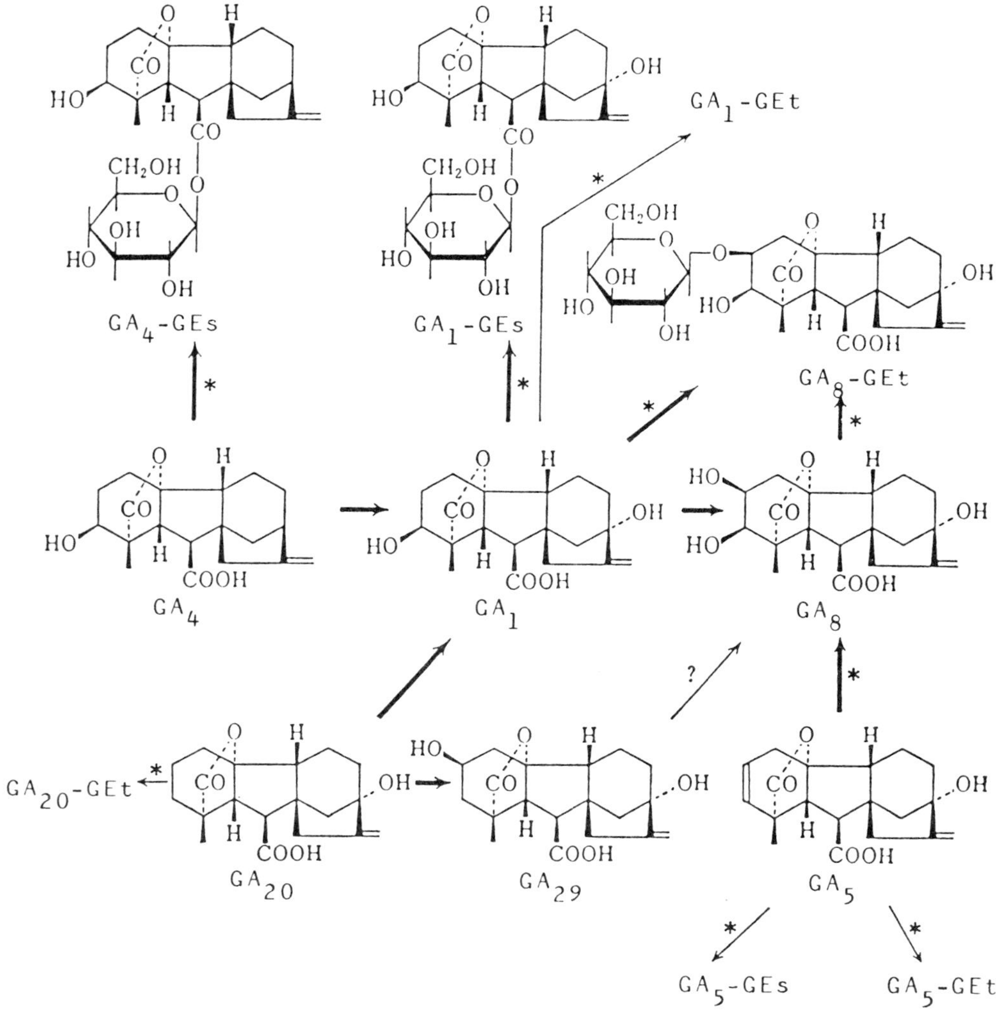

134 *Chemistry of Plant Hormones*

FIGURE 54. Metabolism of GAs in *P. vulgaris* seeds.[230,299] Thick arrows show the metabolic pathway of endogenous GAs confirmed by feeding experiments. Thin arrows show the pathway noticed only in feeding experiments. The pathways with asterisk are not observed in the early immature seeds.

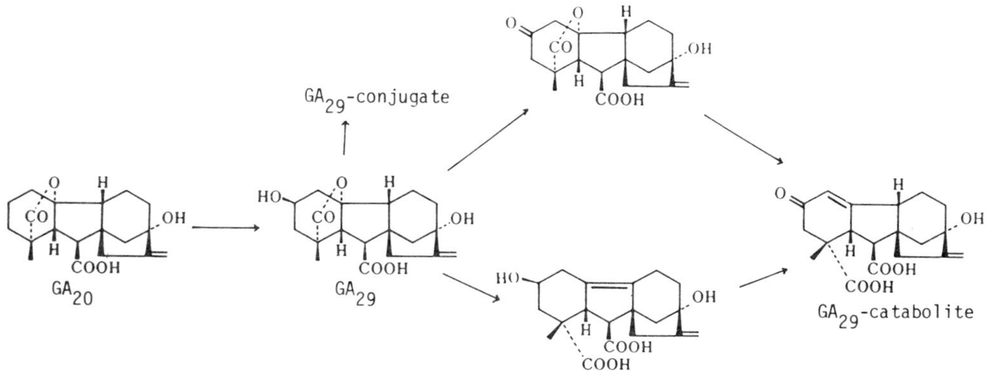

FIGURE 55. Metabolism of GAs in *P. sativum* seeds.[301—303]

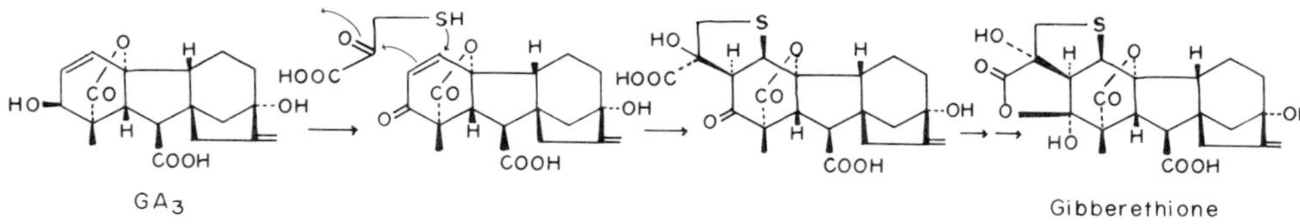

FIGURE 56. Speculated biosynthetic pathway of gibberethione.[93]

effect of GAs is caused mainly by cell elongation and partially by cell division. Of the 68 known free GAs only a limited number show high biological activity.

Most bioassay systems for GAs have been developed by taking advantage of their growth-promoting effect on intact plants.[311] Dwarf mutants are often used as assay plants to estimate the physiological activity of GAs. The dwarf mutants strain d_1, d_2, d_3, and d_5 of maize (*Zea mays*),[312] the dwarf mutants Tan-ginbozu, Waito-C, and Kotaketamanishiki of rice (*Oryza sativa*),[313] and dwarf pea (*Pisum sativum* L.) cv. Progress No. 9 [314] are generally used. In addition to these mutants, both lettuce (*Lactuca sativa* L.) hypocotyl assay[315] and the cucumber (*Cucumis sativas* L.) hypocotyl assay[316] are used. The barley (*Hordeum vulgare*) endosperm[317] assay is also employed in which GA-induced α-amylase is measured. In the dwarf maize bioassay, the lengths of the first and second leaf sheaths are measured and their sum is used as an indication of growth-promoting effect. In the rice bioassay, both the immersion method and microdrop method[318] are used. Free GAs show higher activity in the microdrop method, conjugated GAs higher activity in the immersion method.[319] In the rice bioassay the second leaf sheath is measured as an indication of activity. In the barley endosperm assay, the activity of α-amylase induced by GA is measured.

The biological activities of GAs in the three assay systems are given in Table 32.[326] In general, the following relationships are observed between structure and biological activity:

1. GAs possessing the partial structure (11) in Figure 57, such as $GA_{3,7,30,32}$ show very high activity.
2. GAs with a hydroxyl group at C-2 such as $_{8,29,34,40,}$ etc., show very low activity.
3. C_{20}-GAs show lower activity than C_{19}-GAs. This is because C_{20}-GAs are generally biosynthetic precursors of C_{19}-GAs, and show activity after their conversion into C_{19}-GAs in the assay plants.
4. GAs carrying carboxyl groups at C-4, C-6, and C-10 (so-called tricarboxylic GAs) such as $GA_{13,17,25,28,39,}$ etc., show extremely low activity. These tricarboxylic GAs are end products and are not converted into active C_{19}-GAs.
5. Methyl esters of GAs are mostly inactive in most of the bioassay systems. The only exception is the antheridium-inducing activity of GA methyl esters in *Lygodium*. GA methyl esters show higher activity than respective free GAs in antheridium induction as shown in Table 33.[82]

Gibberellin glucosyl derivatives show very weak activity compared to their aglycons. Some of the GAs-GEt and GAs-GEs occasionally show high activity in the dwarf rice assay (immersion method), (Tables 34 and 35).[159] It has been shown that the aglycons produced from the conjugates by the action of glucosidase in the assay plant or by microbial degradation show the activity.

The biological activity of GAs varies depending on the assay plants (see Table 32). For example, GAs such as $GA_{4,7,9}$ which do not carry a hydroxyl group at C-13, show higher activity in the cucumber assay than corresponding 13-OH–GAs such as $GA_{1,3,20}$.

As mentioned before, dwarf mutants are more sensitive to exogenously applied GAs than their counterparts. In some dwarf mutants, the GA biosynthetic pathway is genetically

Table 32
RELATIVE ACTIVITIES OF GAs IN FIVE BIOASSAY SYSTEMS[326]

GAs	Barley aleurone	Dwarf pea	Lettuce hypocotyl	Dwarf rice	Cucumber hypocotyl
GA_1	+ + + +	+ + +	+ + +	+ + +	+ +
GA_2	+ + + +	+ +	+ +	+ + +	+ +
GA_3	+ + + +	+ + + +	+ + +	+ + + +	+ +
GA_4	+ + +	+ + +	+ +	+ +	+ + +
GA_5	+ +	+ + +	+ +	+ + +	+
GA_6	+ +	+ +	+ +	+ + +	+
GA_7	+ + +	+ + +	+ + + +	+ + +	+ + + +
GA_8	+	+	+	+	0
GA_9	+	+ +	+ + +	+ +	+ + +
GA_{10}	+	0	0	+ + +	+ +
GA_{11}	+	0	0	+	+
GA_{12}	0	0	0	+	+
GA_{13}	+	0	+	+	0
GA_{14}	0	0	0	+	0
GA_{15}	0	+	+ +	+ +	+ +
GA_{16}	+	+ +	+	+	+
GA_{17}	0	0	0	+	0
GA_{18}	0	+ +	+	+ + +	0
GA_{19}	0	0	+	+ + +	0
GA_{20}	+	+	+ + +	+ + +	0
GA_{21}	0	0	+	0	0
GA_{22}	+ + +	+ + +	+ +	+ + +	0
GA_{23}	+ +	+ +	+	+ + +	0
GA_{24}	0	+	0	+ + +	+ + +
GA_{25}	0	0	0	0	+
GA_{26}	0	0	0	0	0
GA_{27}	0	+	0	+	0
GA_{28}	0	0	0	+	0
GA_{29}	+	0	0	+	0
GA_{30}	+ + +	+ + +	+	+ +	+
GA_{31}	+	+ +	0	+ +	0
GA_{32}	+ + +	+ + +	+ +	+ + + +	+ + +
GA_{33}	+	+	+	+	0
GA_{34}	0	+	0	+	0
GA_{35}	+ + +	+ +	+	+ +	+ + +
GA_{36}	+ +	+ +	+	+ + +	+ + +
GA_{37}	+ +	+ +	+ +	+ + +	+ + +
GA_{38}	+	+ + +	0	+	+
GA_{40}		0	+	+	+ +
GA_{43}		0	0	0	0
GA_{46}		0	0	0	0
GA_{47}		+ +	+	+	+ +
GA_{51}		0	0	0	0
GA_{54}	+ + +			+ +	
GA_{55}	+ + +			+ +	
GA_{60}	+			+	
GA_{61}	+			+ +	
GA_{62}	+ +			+ +	

Note: + + + +, Very high; + + +, high; + +, moderate; +, low; 0, very low/inactive.

blocked at certain points, and GAs are not biosynthesized in these mutants. Therefore, when GAs located after the block in the biosynthetic pathway are applied exogenously, they can be converted into active GAs to induce elongation of the dwarf mutants.

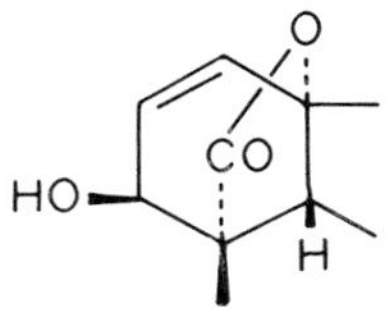

FIGURE 57. Partial structure showing high biological activity.

Table 33
EFFECT OF GA_3, GA_4, GA_7, AND GA_9 AND THEIR METHYL ESTERS ON ANTHERIDIUM FORMATION (%) IN *LYGODIUM JAPONICUM* PROTONEMATA[82]

Sample	Concentration						
	0	10^{-10}	10^{-9}	10^{-8}	10^{-7}	10^{-6}	10^{-5}
GA_3	0	0	0	0	5 ± 2	89 ± 7	100 ± 0
GA_3-Me	0	0	0	11 ± 3	83 ± 2	96 ± 5	99 ± 1
GA_4	0	0	0	69 ± 2	96 ± 5	100 ± 0	—
GA_4-Me	0	2 ± 1	66 ± 5	98 ± 1	99 ± 0	100 ± 0	—
GA_7	0	0	3 ± 1	91 ± 3	99 ± 0	100 ± 0	—
GA_7-Me	0	2 ± 1	52 ± 1	87 ± 7	96 ± 0	76 ± 22	—
GA_9	0	0	0	17 ± 16	80 ± 7	98 ± 1	—
GA_9-Me	0	5 ± 3	69 ± 12	99 ± 0	99 ± 0	100 ± 0	—

Note: The values mean the average % protonema with antheridia of three replicates with S.E. —: Not tested.

Table 34
RELATIVE ACTIVITIES OF GAs-GEt IN FOUR ASSAY SYSTEMS[159]

Samples	Dwarf maize d_5	Dwarf rice (Kotaketa-manishiki) immersion method		Dwarf pea	Cucumber hypocotyl
		Aseptic	Nonaseptic		
GA_3-GEt	+ +	+	+ + +	0	0
GA_8-GEt	+	+	+	0	0
GA_{26}-GEt	0	0	0	0	0
GA_{27}-GEt	0	0	0	0	0
GA_{29}-GEt	0	0	0	0	0
GA_{35}-GEt	0	—	—	+	+
GA_3	+ + + +	+ + + +	+ + + +	+ + + +	+ +
GA_8	+	+	+	+	0

Note: + + + +, Very high; + + +, high; + +, moderate; +, 0, very low/inactive. —: Not tested.

B. Effect on Flowering[320,321]

It is well known that differentiation of flower buds in higher plants is affected by daylength (photoperiodism) and temperature. For example, some plants differentiate flower buds after receiving single or repetitive dark periods longer than a certain length called the critical dark

Table 35
RELATIVE ACTIVITIES OF GAs-GEs IN TWO ASSAY SYSTEMS[159]

Samples	Dwarf maize d_5	Dwarf rice (Tanginbozu)	
		Nonaseptic immersion method	Microdrop
GA_1-GEs	+ +	+ + +	+ +
GA_3-GEs	+ + +	+ + + +	+ + + +
GA_4-GEs	+ +	+ +	+ +
GA_{37}-GEs	+	+ +	+
GA_{38}-GEs	+	+ + +	+ +
GA_3	+ + + +	+ + + +	+ + + +
GA_4	+ +	+ +	+ +

Note: + + +, Very high; + + +, high; + +, moderate; +, low; 0, very low/inactive.

period (which differs depending upon the plant species). These plants are called "short day" plants. *Pharbitis*, *Xanthium*, and *Perilla* are typical short day plants. Some plants differentiate flower buds after receiving repetitive dark periods shorter than the critical dark period. These plants are called "long day" plants. *Spinacia*, *Silene*, *Lactuca*, and *Raphanus* are typical long day plants. Some plants differentiate flower buds regardless of the photoperiodism and these are called "day neutral" plants. Some plants are required to pass low temperature period for flower induction. This low temperature treatment is called "vernalization", and many plants such as *Hyascyamus nigeer*, *Paucus carpa*, *Brasica napus*, *Digitalis purpurea*, and *Viola tricolor* (among others) are known as cold-requiring plants.

GAs are capable of inducing flower formation in some species of long day plants under noninductive conditions, as well as in long day plants that require low temperature for flower induction. This is especially true for plants that grow as rosettes during the vegetative stage or those plants that require low temperature for floral initiation.

Generally, applied GAs cannot induce flower formation in short day plants. There are some exceptions. GA treatment initiates flower formation in short day plants under noninductive conditions in *Impatiens balsamina* and *Zinnia elegans*.

GA has a prominent effect on the promotion of flowering where floral buds have already been differentiated. GA_3 is of practical use in the culture of flowering plants such as *Cyclamen*, *Primula*, *Matricaria*, *Gymnaster*, and *Tulipa*.

C. Effect on Dormancy Break and Germination[322]

When dormant potato tubers are treated with 2 to 3 ppm solution of GA_3, they start to germinate. GAs can also replace light and low temperature in the germination of seeds of *Lactuca*, *Nicotiana*, *Perilla*, *Prunus*, and *Malus*.

D. Effect on Parthenocarpy[323]

In nature, fruit set generally occurs as the result of pollination and fertilization. After fertilization, an ovary develops to a fruit containing seed(s). However, in some plants, fruit set occurs without fertilization, and in this case an ovary develops to a seedless fruit (parthenocarpy). Plant hormones such as GA, auxin, and cytokinin are known to cause parthenocarpy, e.g., tomato, cucumber, peach, pear, apple, and grapes. GA_3 is used commercially to increase the size of produced seedless grapes such as the varieties, Delaware, and mascut berry A in Japan and Thompson seedless grapes in the U.S. and Europe. A corrola of Delaware treated twice with 1000 ppm water solution of GA_3 10 days before and after flowering develops completely seedless fruits and the ripening period is considerably shortened. Though the mechanism of seedless fruit set in Delaware is ambiguous, it is suspected that blocking the development of male or female organs by treatment of GA accompanies parthenocarpy and produces seedless fruits.

E. Effect on Sex Expression[321-322]

Most flowering plants produce bisexual flowers containing both male and female organs; however, some plants produce unisexual flowers containing either male or female organs, and in some species male and female flowers are developed in different plants.

The sex expression of these plants is genetically controlled, but can be changed by treatment with GA, auxin, cytokinin, and ethylene. GA has an effect contrary to that of auxin and ethylene. Cucumbers treated with GA increase the number of male flowers, and unisexual female cucumbers develop both male and female flowers when treated with GA. The unisexual female hemp plant also develops both male and female flowers with treatment by GA. In the unisexual pumpkin and melon, a female top grafted to a male plant develops both male and female flowers. These observations suggest that GA induce male sex expression in flower development.

F. Activation of Hydrogenase Activity[324]

GA activates some hydrolysases in the seeds of grains. In the aleurone layers of barley, several hydrolysases are produced. In 1960, Yomo and Paleg independently found that the activity of α-amylase in the barley endosperm was greatly enhanced when it was treated with GA. Later it was proved that GA activates the *de novo* synthesis of α-amylase in barley aleurone layers in the following experiments. In barley seeds without embryo, ^{14}C-labeled phenylalanine incubated with GA_3 was incorporated into a protein fraction showing α-amylase activity, and α-amylase snythesis was inhibited when barley aleurone layers were treated with GA_3 together with actinomycin D which is an inhibitor of RNA synthetase.

The hypothesis that regulation of α-amylase synthesis by GA_3 involves the regulation of gene expression and the production of α-amylase mRNA is supported by several experiments. In aleurone incubated with GA_3, the level of translatable mRNA for α-amylase increases. It has also been shown that accumulation of translatable mRNA in aleurone layers incubated with GA_3 occurs by the increase in mRNA abundance and not by activation of pre-existing inactive mRNA. Studies are being continued to clarify the action of GA in the protein synthesis. The relationship between the promoting effects of GA in plant growth and protein synthesis is a quite important subject to be clarified in the future.

ACKNOWLEDGMENTS

The authors wish to express their thanks to Professor Phinney and Dr. Spray of the University of California, Los Angeles, for their review of this chapter prior to publication.

REFERENCES

1. **Kurosawa, E.,** Experimental studies on the secretion of *Fusarium heterosporum* on rice plants, *Trans. Natl. Hist. Soc. Formosa*, 16, 213, 1926.
2. **Yabuta, T. and Sumiki, Y.,** Biochemical studies on "Bakanae" fungus. Crystals with plant growth promoting activity (in Japanese), *J. Agric. Chem. Soc. Jpn.*, 14, 1526, 1938.
3. **Curtis, P. J. and Cross, B. E.,** Gibberellic acid. A new metabolite from the culture filtrates of *Gibberella fujikuroi*, *Chem. Ind.*, 1066, 1954.

4. **Stodola, F. H., Raper, K. B., Fennell, D. I., Conway, H. F., Sohns, V. E., Langford, C. T., and Jackson, R. W.**, The microbiological production of gibberellin A and X, *Arch. Biochem. Biophys.*, 54, 240, 1955.
5. **Takahashi, N., Kitamura, H., Kawarada, A., Seta, Y., Takai, M., Tamura, S., and Sumiki, Y.**, Biochemical studies on "Bakanae" fungus. XXXIV. Isolation of gibberellins and their properties, *Bull. Agric. Chem. Soc. Jpn.*, 19, 267, 1955.
6. **Takahashi, N., Seta, Y., Kitamura, H., and Sumiki, Y.**, Biochemical studies on "Bakanae" fungus. XLII., *Bull. Agric. Chem. Soc. Jpn.*, 21, 396, 1957.
7. **Cross, B. E., Galt, R. H. B., and Hanson, J. R.**, Gibberellin A_7. A new fungal gibberellin, *Tetrahedron Lett.*, 18(15), 1960.
8. **Cross, B. E., Galt, R. H. B., and Hanson, J. R.**, Gibberellin A_9, *Tetrahedron Lett.*, 22(23), 1960.
9. **Cross, B. E., Galt, R. H. B., and Hanson, J. R.**, New metabolites of *Gibberella fujikuroi-1*, gibberellin A_7 and gibberellin A_9, *Tetrahedron*, 18, 451, 1962.
10. **Cross, B. E., Galt, R. H. B., and Hanson, J. R.**, *Regulateurs Natureles de la Croissance Vegetale*, Centre Nionae de la Recherche Scientifique, Paris, 1964, 265.
11. **Hanson, J. R.**, New metabolites of *Gibberella fujikuroi* -X. Gibberellin A_{10}, *Tetrahedron*, 22, 701, 1966.
12. **Brown, J. C. and Cross, B. E.**, New metabolites of *Gibberella fujikuroi*-XIII. Two gibbane 1,3-lactones, *Tetrahedron*, 23, 4095, 1967.
13. **Cross, B. E. and Norton, K.**, New metabolites of *Gibberella fujikuroi*. VIII. Gibberellin A_{12}, *J. Chem. Soc.*, 1570, 1965.
14. **Galt, R. H. B.**, New metabolites of *Gibberella fujikuroi*, *J. Chem. Soc.*, 3143, 1965.
15. **Cross, B. E.**, New metabolites of *Gibberella fujikuroi*. XI. Gibberellin A_{14}, *J. Chem. Soc. C:*, 502, 1966.
16. **Hanson, J. R.**, New metabolites of *Gibberella fujikuroi*-XII. Gibberellin A_{15}, *Tetrahedron*, 23, 733, 1967.
17. **Galt, R. H. B.**, New metabolites of *Gibberella fujikuroi*-XIV. Gibberellin A_{16} methyl ester, *Tetrahedron*, 24, 1377, 1968.
18. **Harrison, D. M., MacMillan, J., and Galt, R. H. B.**, Gibberellin A_{24}, an aldehydic gibberellin from *Gibberella fujikuroi*, *Tetrahedron Lett.*, 3137, 1968.
19. **Harrison, D. M. and MacMillan, J.**, Two new gibberellins, A_{24} and A_{25}, from *Gibberella fujikuroi* : their isolation, structure and correlation with gibberellin A_{13} and A_{15}, *J. Chem. Soc. C:*, 631, 1971.
20. **Bearder, J. R. and MacMillan, J.**, Gibberellin A_{36}. Isolation from *Gibberella fujikuroi*, structure and conversion to gibberellin A_{37}, *Agric. Biol. Chem.*, 36, 342, 1972.
21. **Bearder, J. R. and MacMillan, J.**, Fungal products IX. Gibberellins A_{16}, A_{36} A_{37} A_{41} and A_{42} from *Gibberella fujikuroi*, *J. Chem. Soc. Perkin Trans.*, 1: 2824, 1973.
22. **Yamaguchi, I., Miyamoto, M., Yamane, H., Takahashi, N., Fujita, K., and Imanari, M.**, Structure of gibberellin A_{40}, *Agric. Biol. Chem.*, 37, 2453, 1973.
23. **Yamaguchi, I., Miyamoto, M., Yamane, H., Murofushi, N., Takahashi, N., and Fujita, K.**, Elucidation of the structure of gibberellin A_{40} from *Gibberella fujikuroi*, *J. Chem. Soc. Perkin Trans. 1:*, 996, 1975.
24. **Beeley, L. J. and MacMillan, J.**, Partial synthesis of 2-hydroxylgibberellins; characterization of two new gibberellins, A_{46} and A_{47}, *J. Chem. Soc. Perkin Trans. 1:*, 1022, 1976.
25. **McInnes, A. G., Smith, D. G., Durley, R. C., Pharis, R. P., Arsenault, G. P., MacMillan, J., Gaskin, P., and Vining, L. C.**, Biosynthesis of gibberellins in *Gibberella, fujikuoi* Gibberellin A_{47}, *Can. J. Biochem.*, 55, 728, 1977.
26. **Murofushi, N., Sugimoto, M., Itoh, K., and Takahashi, N.**, Three novel gibberellins produced by *Gibberella fujikuroi*, *Agric. Biol. Chem.*, 43, 2179, 1979.
27. **Murofushi, N., Sugimoto, M., Itoh, K., and Takahashi, N.**, A novel gifbberellin, A_{57}, produced by *Gibberella fujikuroi*, *Agric. Biol. Chem.*, 44, 1538, 1980.
28. **Schreiber, K., Weiland, J., and Sembdner, G.**, Isolierung von O-(2)-Acetyl-Gibberellinsaure als Stoffwechselproduckt von *Fusarium Moniliforme* Sheld., *Phytochemistry*, 5, 1221, 1966.
29. **Vancura, V.**, Detection of gibberellic acid in *Azotobacter cultures, Nature (London)*, 192, 88, 1961.
30. **Katznelson, H., Sirois, J. C., and Cole, S. E.**, Production of a gibberellin-like substance by *Arthrobacter golbiformis, Nature,(London)*, 196, 1012, 1962.
31. **Katznelson, H. and Cole, S. E.**, Production of gibberellin-like substances by Bacteria and Actinomycetes, *Can. J. Microbiol.*, 11, 733, 1965.
32. **Aube, C. and Sackston, W. E.**, Distribution and prevalence of *Verticillium* species producing substances with gibberellin-like biological properties, *Can. J. Bot.*, 43, 1335, 1965.
33. **Rademacher, W. and Graebe, J. E.**, Gibberellin A_4 produced by *Sphaceloma manthoticola*, the cause of the superelongation disease of Casava *(Manthot esculenta)*, *Biochem. Biophys. Res. Commun.*, 91, 35, 1979.
34. **Kawanabe, Y., Yamane, H., Murayama, T., Takahashi, N., and Nakamura, T.**, Identifiication of gibberellin A_3 in mycelia of *Neurospora crassa, Agric. Biol. Chem.*, in press.
35. **Mitchell, J. W., Skaggs, D. P., and Anderson, W. P.**, Plant growth stimulating hormones in immature seeds, *Science*, 114, 159, 1951.

36. **Radley, M.,** Occurrence of substances similar to gibberellin acid in higher plants, *Nature (London),* 178, 1070, 1956.
37. **West, C. R. and Phinney, B. O.,** Properties of gibberellin-like factors from extracts of higher plants, *Plant physiol.,* 31(Suppl.), 20, 1956.
38. **MacMillan, J. and Suter, P. J.,** The occurrence of gibberellin A_1 in higher plants: isolation from seed of runner bean *(Phaseolus multiflorus), Naturwissenschaften,* 45, 46, 1958.
39. **MacMillan, J., Seaton, J. C., and Suter, P. J.,** A plant-growth promoting acid - gibberellin A_5 from the seeds of *Phaseolus multiflorus, Proc. Chem. Soc. London,* 325, 1959.
40. **MacMillan, J., Seaton, J. C., and Suter, P. J.,** Plant hormones. I. Isolation of gibberellin A_1, and gibberellin A_5 from *Phaseolus multiflorus, Tetrahedron,* 11, 60, 1960.
41. **MacMillan, J., Seaton, J. C., and Suter, P. J.,** Plant hormones. II. Isolation and structures of gibberellin A_6 and A_8, *Tetrahedron,* 18, 349, 1962.
42. **West, C. A. and Phinney, B. O.,** Gibberellins from flowering plants. I. Isolation and properties of a gibberellin from *Phaseolus vulgaris* L., *J. Am. Chem. Soc.,* 81, 2424, 1959.
43. **Kawarada, A. and Sumiki, Y.,** The occurrence of gibberellin A_1 in water sprauts of *Citrus, Bull. Agric. Chem. Soc. Jpn.,* 23, 343, 1959.
44. **Tamura, S., Takahashi, N., Murofushi, N., Iriuchijima, S., Kato, J., Wada, Y., Watanabe, E., and Aoyama, T.,** Isolation and structure of a novel gibberellin in bamboo shoots *(Phyllostochys edulis), Tetrahedron Lett.,* 2465, 1966.
45. **Murofushi, N., Iriuchijima, S., Takahashi, N., Tamura, S. Kato, J., Wada, Y., Watanabe, E., and Aoyama, T.,** Isolation and structure of a novel C_{20} gibberellin in bamboo shoots, *Agric. Biol. Chem.,* 30, 917, 1966.
46. **Koshimizu, K., Fukui, H., Kusaki, T., Mitsui, T., and Ogawa, T.,** A new C_{20} gibberellin in immature seeds of *Lupinus luteus, Tetrahedron Lett.,* 2459, 1966.
47. **Koshimizu, K., Fukui, H., Kusaki, T., Ogawa, Y., and Mitsui, T.,** Isolation and structure of gibberellin A_{18} from immature seeds of *Lupinus luteus, Agric. Biol. Chem.,* 32, 1135, 1968.
48. **Pryce, R. J. and MacMillan, J.,** A new gibberellin in the seed of *Phaseolus multiflorus, Tetrahedron Lett.,* 4173, 1967.
49. **Murofushi, N., Takahashi, N., Yokota, T., and Tamura, S.,** Gibberellins in immature seeds of *Pharbitis nil,* I. Isolation and structure of a novel gibberellin, gibberellin A_{20}, *Agric. Biol. Chem.,* 32, 1239, 1968.
50. **Murofushi, N., Takahashi, N., Yokota, T., Kato, J., Shiotani, Y., and Tamura, S.,** Gibberellin in immature seeds of *Canavalia.* I. Isolation and biological activity of gibberellin A_{21} and A_{22}, *Agric. Biol. Chem.,* 33, 592, 1969.
51. **Murofushi, N., Takahashi, N., Yokota, T., and Tamura, S.,** Gibberellin in immature seeds in *Canavalia.* II. Structures of gibberellin A_{21} and A_{22}, *Agric. Biol. Chem.,* 33, 598, 1969.
52. **Fukui, H., Ishii, H., Koshimizu, K., Katsumi, M., Ogawa, Y., and Mitsui, T.,** The structure of gibberellin A_{23} and the biological properties of 3, 13-dihydroxy C_{20}-gibberellins, *Agric. Biol. Chem.,* 36, 1003, 1972.
53. **Yokota, T., Murofushi, N., Takahashi, N., and Tamura, S.,** Gibberellins in immature seeds of *Pharbitis nil.* II. *Isolation and structure of novel gibberellins, gibberellins A_{26} and A_{27}, Agric. Biol. Chem.,* 35, 573, 1971.
54. **Fukui, H., Koshimizu, K., and Mitsui, T.,** Gibberellin A_{28} in the fruits of *Lupinus luteus, Phytochemistry,* 10, 671, 1971.
55. **Yokota, T., Murofushi, N., Takahashi, N., and Tamura, S.,** Gibberellins in immature seeds of *Pharbitis nil.* III. Isolation and structures of gibberellin glucosides, *Agric. Biol. Chem.,* 35, 583, 1971.
56. **Murofushi, N., Yokota, T., Watanabe, A., and Takahashi, N.,** Isolation and characterization of gibberellins in *Calonyction aculeatum* and structures of gibberellin A_{30}, A_{31}, A_{33} and A_{34}, *Agric. Biol. Chem.,* 37, 1101, 1973.
57. **Yamaguchi, I., Yokota, T., Murofushi, N., Takahashi, N., and Ogawa, Y.,** Isolation of gibberellin A_5, A_{32}, A_{32} acetonide and (+)-abscisic acid from *Prunus persica, Agric. Biol. Chem.,* 39, 2399, 1975.
58. **Yamaguchi, I., Yokota, T., Murofushi, N., and Takahashi, N.,** Structure elucidation of gibberellin A_{32} and its acetonide, *Agric. Biol. Chem.,* 39, 2405, 1975.
59. **Yamane, H., Yamaguchi, I., Murofushi, N., and Takahashi, N.,** Isolation and structures of gibberellin A_{35} and its glucoside from immature seed of *Cytisus scoparius, Agric. Biol. Chem.,* 38, 649, 1974.
60. **Hiraga, K., Yokota, T., Murofushi, N., and Takahashi, N.,** Isolation and characterization of a free gibberellin and glucosyl esters of gibberellins in mature seeds of *Phaseolus vulgaris, Agric. Biol. Chem.,* 36, 345, 1972.
61. **Hiraga, K., Yokota, T., Murofushi, N., and Takashi, N.,** Isolation and characterization of gibberellins in mature seeds of *Phaseolus vulgaris, Agric. Biol. Chem.,* 38, 2511, 1974.
62. **Fukui, H., Koshimuzu, K., Usuda, S., and Yamazaki, T.,** Isolation of plant growth regulators from seeds of *Cucurbita pepo* L., *Agric. Biol. Chem.,* 41, 175, 1977.

63. **Fukui, H., Nemori, R., Koshimizu, K., and Yamazaki, Y.,** Structures of gibberellins A_{39}, A_{48}, A_{49} and a new kaurenolide in *Cucurbita pepo* L., *Agric. Biol. Chem.*, 41, 181, 1977.
64. **Beeley, L., Gaskin, P., and MacMillan, J.,** Gibberellin A_{43} and other terpenes in endosperm of *Echinocystis macrocarpa*, *Phytochemistry*, 14, 779, 1975.
65. **Frydman, U. M., Gaskin, P., and MacMillan, J.,** Qualitative and quantitiative analysis of gibberellins throughout seed maturation in *Pisum sativum*. cv. Progress No. 9., *Planta*, 118, 123, 1974.
66. **Martin, G. C., Dennis, F. G., Jr., Gaskin, P., and MacMillan, J.,** Identification of gibberellins A_{17}, A_{25}, A_{45}, abscisic acid, phaseic acid and dihydrophaseic acid in seeds of *Pyrus communis*, *Phytochemistry*, 16, 607, 1977.
67. **Beeley, L. J. and MacMillan, J.,** Partial syntheses of 2-hydroxygibberellins; characterization of two new gibberellins, A_{46} and A_{47}, *J. Chem. Soc. Perkin Trans. 1:*, 1022, 1976.
68. **Fukui, H., Koshimizu, K., and Nemori, R.,** Two new gibberellins A_{50} and A_{52} in seeds of *Lagenaria leucantha*, *Agric. Biol. Chem.*, 42, 1571, 1978.
69. **Sponsel, V. M. and MacMillan, J.,** Further studies on the metabolism of gibberellins (GAs) A_9, A_{20} and A_{29} in immature seeds of *Pisum sativum* cv. Progress No. 9, *Planta* 135, 129, 1977.
70. **Sponsel, V. M., Gaskin, P., and MacMillan, J.,** The identification of gibberellins in immature seeds of *Vicia faba*, and some chemotaxonomic considerations, *Planta*, 146, 101, 1979.
71. **Gaskin, P., Kirkwood, P. S., Lenton, J. R., MacMillan, J., And Radley, M. E.,** Identification of gibberellins in developing wheat grain, *Agric. Biol. Chem.*, 44, 1589, 1980.
72. **Graebe, J. E., Hedden, P., and Rademacher, W.,** Gibberellins biosynthesis, *Br. Plant Growth Group Monogr.*, 5, 31, 1980.
73. **Yokota, T. and Takahashi, N.,** Gibberellin A_{59}: a new gibberellin from *Canavalia gladiata*, *Agric. Biol. Chem.*, 45, 1251, 1981.
74. **Kirkwood, P. S. and MacMillan, J.,** Gibberellins A_{60}, A_{61} and A_{62}: partial syntheses and natural occurrence, *J. Chem. Soc. Perkin Trans 1:* 689, 1982.
75. **MacMillan, J.,** personal communication.
76. **Jennings, R. C.,** Gibberellins as endogenous growth regulators in green and brown algae, *Plant*, 80, 34, 1968.
77. **Abe, H., Uchiyama, M., Sato, R., and Muto, S.,** Plant growth regulators in marine algae, in *Plant Growth Substances 1973: Proc. 8th Int. Conf. Plant Growth Substances*, Hirokawa Publishing, Tokyo, 1973, 201.
78. **Taylor, I. E. P. and Wilkinson, A. J.,** The occurrence of gibberellins and gibberellin-like substances in algae *Phycologia*, 16, 37, 1977.
79. **Takahashi, N., Yamane, H., Seto, Y., Takahashi, N., and Iwatsuki, K.,** Identification of GA_{36} in *Psilotum rudum*, *Phytochemistry*, in press.
80. **Endo, M., Nakanishi, K., Naf, U., Me Keon, W., and Walker, R.,** Isolation of the antheridiogen of *Anemia phyllitidis*, *Physiol. Plant.*, 26, 183, 1972.
81. **Nakanishi, K., Endo, M., Naf, U., and Johnson, L. F.,** Structure of the antheridium-inducing factor of the fern *Anemia phyllitidis*, *J. Am. Chem. Soc.*, 93, 5579, 1971.
82. **Yamane, H., Takahashi, N., Takeno, K., and Furuya, N.,** Identification of gibberellin A_9 methyl ester as a natural substance regulating formation of reproductive organs in *Lygodium japonicum*, *Planta*, 147, 251, 1979.
83. **McComb, A. J.,** Bound gibberellin in mature runner bean seeds, *Nature (London)*, 192, 1548, 1961.
84. **Murakami, Y.,** Occurrence of "water-soluble" gibberellin in higher plants, *Bot. Mag. Tokyo*, 75, 102, 1962.
85. **Ogawa, Y.,** Ethyl acetate-soluble and water-soluble gibberellin-like substances in the seeds of *Pharbitis nil*, *Lupinus luteus* and *Prunus persica*, *Bot. Mag. Tokyo*, 79, 69, 1976.
86. **Zeevaart, J. A. D.,** Reduction of the gibberellin content of *Phabitis* seeds by CCC, *Plant Physiol.*, 41, 856, 1966.
87. **Schreiber, K., Weiland, J., and Sembdner, G.,** Isolierung und struktur eines Gibberellinglucosids, *Tetrahedron Lett.*, 4285, 1967.
88. **Schreiber, K., Weiland, J., and Sembdner, G.,** Isolierung von Gibberellin-A_8-O-(3)-β-D-glucopyranosid aus Fruchten von *Phaseolus coccineus*, *Phytochemistry*, 9, 189, 1970.
89. **Tamura, S., Takahashi, N., Yokota, T., Murofushi, N., and Ogawa, Y.,** Isolation of water-soluble gibberellins from immature seeds of *Pharbitis nil*, *Planta*, 78, 208, 1968.
90. **Yokota, T., Kobayashi, S., Yamane, H., and Takahashi, N.,** Isolation of a novel gibberellin glucoside, 3-O-β-D-glucopyranosyl-gibberellin A_1 from *Dolichos lablab* seed, *Agric. Biol. Chem.*, 42, 1811, 1978.
91. **Yamaguchi, I., Kobayashi, M., and Takahahsi, N.,** Isolation and characterization of glucosyl esters of gibberellin A_5 and A_{44} from immature seeds of *Pharbitis purpurea*, *Agric. Biol. Chem.*, 44, 1975, 1980.
92. **Lorenzi, R., Horgan, R., and Heald, J. K.,** Gibberellins A_9 glucosyl ester in needles of *Picea sitchensis*, *Phytochemistry*, 15, 789, 1976.

93. **Yokota, T., Yamazaki, S., Takahashi, N., and Iitaka, Y.,** Structure of pharbitic acid, a gibberellin-related diterpenoid, *Tetrahedron Lett.*, 2957, 1974.
94. **Hiraga, K., Kawabe, S., Yokota, T., Murofushi, N., and Takahashi, N.,** Isolation and characterization of plant growth substances in immature seeds and etiolated seedling of *Phaseolus vulgaris, Agric. Biol. Chem.*, 38, 2521, 1974.
95. **Yamane, H., Murofushi, N., Osada, H., and Takahashi, N.,** Metabolism of gibberellins in early immature bean seeds, *Phytochemistry*, 16, 831, 1977.
96. **Durley, R. C., MacMillan, J., and Pryce, R. J.,** Investigation of gibberellins and other growth substances in the seed of *Phaseolus multiflorus* and of *Phaseolus vulgaris* by gas chromatography and by gas chromatography-mass spectrometry, *Phytochemistry*, 10, 1891, 1971.
97. **Crozier, A., Bower, D. H., MacMillan, J., Reid, D. M., and Most, B. H.,** Characterization of gibberellins from dark-grown *Phaseolus coccineus* seedlings by gas-liquid chromatography and combined gas chromatography-mass spectrometry, *Planta*, 97, 142, 1971.
98. **Bowen, D. H., Crozier, A., MacMillan, J., and Reid, D. M.,** Characterization of gibberellin from light-grown *Phaseolus coccineus* seedings by combined GC-MS, *Phytochemistry*, 12, 2935, 1973.
99. Unpublished data, referred from *Encyclopedia of Plant Physiology, New Ser. Vol. 9, Hormonal Regulation of Development*, MacMillan, J., Ed., Springer-Verlag, Berlin, 1980, 30.
100. **Koshimizu, K., Ishii, H., Fukui, H., and Mitsui, T.,** Gibberelin A_{18} and A_{23} from immature seeds of *Wistaria floribonda, Phytochemistry*, 11, 2355, 1972.
101. **Yokota, T. and Yamane, H.,** Unpublished data.
102. **Frydman, V. M. and MacMillan, J.,** Identification of gibberellin A_{20} and A_{29} in seed of *Pisum sativum* cv. Progress No. 9 by combined gas chromatography-mass spectrometry, *Planta*, 115, 11, 1973.
103. **Komada, Y., Isogai, Y., and Okamoto, T.,** Isolation of gibberellin A_{20} from pea pods, *Sci. Pap. Coll. Gen. Educ. Univ. Tokyo*, 18, 221, 1968.
104. **Adesmoju, A. A.,** An Investigation of Some Hormonal Bases for Abscission in Couspea (*Vigna unguiculata* L. Walp), Ph.D. thesis, University of Ibadan, Nigeria, 1977.
105. **Sircar, P. K., Dey, B., Sanyal, T., Ganguly, S. N., and Sircar, S. M.,** Gibberellic acid in the floral parts of *Cassia fistula, Phytochemistry*, 9, 735, 1970.
106. **Arigayo, S., Sakata, K., Fujisawa, S., Sakurai, A., Adisewojo, S. S., and Takahashi, N.,** Characterization of gibberellins in immature seeds of *Leucaena leucocephala* (Lmk) De Wit, *Agric. Biol. Chem.*, 47, 2939, 1983.
107. **Yamaguchi, I., Fujisawa, S., and Takahashi, N.,** Qualitative and semi-quantitative analysis of gibberellins, *Phytochemistry*, 821, 2049, 1982.
108. **Matsuo, T., Itoo, S., and Murofushi, N.,** Identification of gibberellins in the seeds of sweet potato (*Ipomea batatas* Lam.) and other several Convolvulaceae plants, *Agric. Biol. Chem.*, in press.
109. **Yamaguchi, I., Yokota, T., Yoshida, S., and Takahashi, N.,** High pressure liquid chromatography of conjugated gibberellins, *Phytochemistry*, 18, 1699, 1979.
110. **Hemphill, D. D., Jr., Baker, L. R., and Sell, H. M.,** Isolation and identification of the gibberellins of *Cucumis sativus* and *Cucumis melo, Plant*, 103, 241, 1972.
111. **Graebe, J. E., Hedden, P., Gaskin, P., and MacMillan, J.,** The biosynthesis of a C_{19}-gibberellin from mevalonic acid in a cell-free system from higher plants, *Planta*, 120, 307, 1974.
112. **Yuda, E., Yamaguchi, I., Murofushi, N., and Takahashi, N.,** Fruit set and development of three pear species induced by gibberellins, *Acta Horticult.*, 137, 277, 1983.
113. **Kobayashi, M. and Yamaguchi, I.,** unpublished data.
114. **Sinska, I., Lewak, St., Gaskin, P., and MacMillan, J.,** Reinvestigation of apple-seeds gibberellins, *Planta* 114, 359, 1973.
115. **Coomb, B. G. and Tate, M. E.,** A polar gibberellin from apricot seed, in *Plant Growth Substances 1970*, Carr, D. J., Ed., Springer-Verlag, Berlin, 1972, 158.
116. **Bukovac, M. J., Yuda, E., Murofushi, N., and Takahashi, N.,** Endogenous plant growth substances in developing fruit of *Prunus cerasus* L., *Plant Physiol.*, 63, 129, 1979.
117. **Kurogochi, S., Murofushi, N., Ota, Y., and Takahashi, N.,** Gibberellins and inhibitors in rice plant, *Agric. Biol. Chem.*, 42, 207, 1978.
118. **Kurogochi, S., Murofushi, N., Ota, Y., and Takahashi, N.,** Identification of gibberellins in rice plant and quantitative changes of gibberellin A_{19} throughout its life cycle, *Planta*, 146, 185, 1979.
119. **Kobayashi, M., Yamaguchi, I., Murofushi, N., Takahashi, N., and Ota, Y.,** Endogenous gibberellins in immature seeds and flowering ears of rice, *Agric. Biol. Chem.*, submitted.
120. **Eckert, H., Schilling, G., Podesak, W., and Franke, P.,** Extraction and identification of gibberellins (GA_1 and GA_3) from *Triticum aestivum* L. and *Seale cereale* L. and changes in contents during autogenesis, *Biochem. Physiol. Pflanz.*, 172, 475, 1978.
121. **Browning, G. and Saunders, P. F.,** Membrane localised gibberellins A_9 and A_4 in wheat chloroplasts, *Nature (London)*, 265, 375, 1977.

122. **Sato, Y.**, Studies on Biologically Active Substances Regulating Life-Cycle of Fern, *Ph.D. thesis, University of Tokyo*, 1975, 58.
123. **Gaskin, P., Kirkwood, P. S., Lenton, J. R., MacMillan, J., and Radley, M. E.**, Identification of gibberellins in developing wheat grain, *Agric. Biol. Chem.*, 44, 1589, 1980.
124. **Dathe, W., Schneider, G., and Sembdner, G.**, Endogenous gibberellins and inhibitors in caryopses of rye, *Phytochemistry*, 17, 963, 1978.
125. **Yamada, K.**, Determination of endogenous gibberellins in germinating barley by combined gas chromatography-mass spectrometry, *Am. Soc. Brew. Chem. J.*, 40, 18, 1982.
126. **Murphy, G. J. P. and Briggs, D. E.**, Gibberellin estimation and Biosynthesis in germinating *Hordeum distichon*, *Phytochemistry*, 12, 1299, 1973.
127. **Kaufman, P. B., Ghosheh, N. S., Nakosteen, L., Pharis, R. P., Durley, R. C., and Morf, W.**, Analysis of native gibberellins in the internode, leaves and inflorescence of developing Avena plants. *Plant Physiol.*, 58, 131, 1976.
128. **Hedden, P., Phinney, B. O., Heupel, R., Fujii, D., Cohen, H., Gaskin, P., MacMillan, J., and Graebe, J. E.**, Hormones of young tassels of *Zea Mays*, *Phytochemistry*, 21, 391, 1982.
129. **Harada, H. and Nitsch, J. P.**, Isolation of gibberellins A_1, A_3, A_9 and of a fourth growth substance from *Althaea rosea* cav., *Phytochemistry*, 6, 1695, 1967.
130. **Harada, H. and Yokota, T.**, Isolation of gibberellin A_8-glucoside from shoot apices of *Althaea rosa*, *Planta*, 92, 100, 1970.
131. **Shindy, W. W. and Smith, O. E.**, Identification of plant hormones from cotton ovules, *Plant Physiol.*, 55, 550, 1975.
132. **Kamienska, A., Durley, R. C., and Pharis, R. P.**, Isolation of gibberellin A_3, A_4, A_7 from *Pinus attenuata* pollen, Phytochemistry, 15, 421, 1976.
133. **Lorenzi, R., Horgan, R. and Heald, J. K.**, Gibberellins in *Picea sitchensis* Carriere: seasonal variation and partial characterization, *Planta*, 126, 75, 1975.
134. **Khalifah, Lewis, L. N., and Coggins, C. W., Jr.**, Isolation and properties of gibberellin-like substances from Citrus fruits, *Plant Physiol.*, 40, 441, 1965.
135. **Noma, M., Huber, J., and Pharis, R. P.**, Occurrence of $\Delta^{1(10)}$ gibberellin A_1 counterpart, GA_1, GA_4, and GA_7 in somatic cell embryo cultures of carrot and anise, *Agric. Biol. Chem.*, 43, 1793, 1979.
136. **Noma, M., Huber, J., Ernst, D., and Pharis, R. P.**, Quantification of gibberellins and the metabolism of (^{3}H) gibberellin A_1 during somatic embryogenesis in carrot and anis cell cultures, *Planta*, 155, 369, 1982.
137. **Ganguly, S. N., Ganguly, T., and Sircar, S. M.**, Gibberellins of *Echydra fluctuans*, *Phytochemistry*, 11, 3433, 1972.
138. **Metzger, J. D. and Zeevaart, J. A. D.**, Identification of six endogenous gibberellins in spinach shoots, *Plant Physiol.*, 65, 623, 1980.
139. **Gaskin, P., MacMillan, J., and Zeevaart, J. A. D.**, Identification of gibberellin A_{20}, abscisic acid and phaseic acid from flowering *Bryophyllum daigremonitanum* by combined gas chromatography-mass spectrometry, *Planta*, 111, 347, 1973.
140. **Williams, P. M., Bradbeeer, J. W., Gaskin, P., and MacMillan, J.**, Studies in seed dormancy. VIII. The identification and determination of gibberellins A_1 and A_9 in seeds of *Caryus avellana* L., *Planta*, 117, 101, 1974.
141. **Watanabe, N., Yokota, T., and Takahashi, N.**, Identification of N^6-(3-methyl-but-2-enyl)adenosine, zeatin, zeatin riboside, gibberellin A_{19} and abscisic acid in shoots of hop plant, *Plant Cell Physiol.*, 19, 1263, 1978.
142. **Park, K.-H., Fujisawa, S., Sakurai, F. A., Yamaguchi, I., and Takahashi, N.**, Gibberellin production in cultured cells of *Nicotiana tabacum*, *Plant Cell Physiol.*, 24, 1241, 1983.
143. **Ganguly, S. N. and Sircar, S. M.**, Gibberellins from mangrove plants, *Phytochemistry*, 13, 1911, 1974.
144. **Gaskin, P., MacMillan, J., Gauguly, S. N., Saryal, T., Sircar, P. K., and Sircar, S. M.**, Identification of the gibberellin from *Sonneratia apelata* Ham. as tetrahydrogibberellin A_3, *Chem. Ind.*, 424, 1972.
145. **Yokota, T., Murofushi, N., and Takahashi, N.**, Extraction, purification and identification, *Hormonal Regulation of Development I. Encyclopedia of Plant Physiology*, New Ser., Vol. 9, MacMillan, J., Eds., Springer-Verlag, Berlin, 1980 chap 2.
146. **Reeve, D. R. and Crozier, A.**, Quantitative analysis of plant hormones, in *Hormonal Regulation of Development I. Encyclopedia of Plant Physiology* New Ser., Vol. 9, MacMillan, J., Ed., Springer-Verlag, Berlin, 1980, chap 3.
147. **Moffatt, J. S.**, Gibberellic acid. XVI. The chromophore of gibberellenic acid, *J. Chem. Soc.*, 3045, 1960.
148. **Cross, B. E., Grove, J. F., and Morrison, A.**, Gibberellic acid. XVIII. Some rearrangements of ring A, *J. Chem. Soc.*, 2498, 1961.
149. **Takahashi, N., Kiamura, N., Kawarada, A., Seta, Y., Takai, M., Tamura, S., and Sumiki, Y.**, Biochemical studies on "Bakanae" fungus. XXXIV. Isolation of gibberellins and their properties. *Bull. Agric. Chem. Soc. Jpn.*, 19, 267, 1955.

150. **Durley, R. C. and Pharis, R. P.**, Partition coefficient of 27 gibberellins, *Phytochemistry*, 11, 317, 1972.
151. **Pitel, D. W., Vinig, L. C., and Arsenault, G. P.**, Improved method for preparing pure gibberellins from cultures of *Gibberella fujikuroi*. Isolation by adsorption or partition chromatography on Sephadex columns, *Can. J. Biochem.*, 49, 185, 1971.
152. **Kagawa, T., Fukibara, T., and Sumiki, Y.**, Thin layer chromatography of gibberellins, *Agric. Biol. Chem.*, 27, 598, 1963.
153. **MacMillan, J. and Suter, P. J.**, Thin layer chromatography of gibberellins, *Nature (London)*, 197, 790, 1963.
154. **MacMillan, J. and Wels, C. M.**, Partition chromatography of gibberellins and related diterpenes on columns of Sephadex LH-20, *J. Chromatogr.*, 87, 271, 1973.
155. **Powell, L. E. and Tautvydas, K. J.**, Chromatography of gibberellins on silica gel partition columns, *Nature (London)*, 213, 292, 1967.
156. **Durley, R. C., Crozier, A., Pharis, R. P., and MacLaughlin, G. E.**, Chromatography of 33 gibberellins on a gradient eluted silical gel partition column, *Phytochemistry*, 11, 3029, 1972.
157. **Schliemann, W.**, Hydrolysis of gibberellin-*O*-glucosides by glucosidase of *Pharbitis nil. Biochem. Physiol. Pflanz.*, 178, 359, 1983.
158. **Knoefel, H.-D., Mueller, P., and Sembdner, G.**, Studies on the enzymatic hydrolysis of gibberellin -*O*-glucosides, in *Biochemistry and Chemistry of Plant Growth Regulators*, Schrieber, K., Schuuette, H. R., and Sembdner, G., Eds., Institute of Biochemistry Academy of Science, Halle/S., West Germany, 1974, 121.
159. **Hiraga, K., Yamane, H., and Takahashi, N.**, Biological activity of some synthetic gibberellin glucosyl esters, *Phytochemistry*, 13, 2371, 1974.
160. **Reeve, D. R., Yokota, T., Nask, L. J., and Crozier, A.**, The development of a high performance liquid chromatography with a sensitive on-stream radioactivity monitor for the analysis of ^{3}H- and ^{14}C-labeled gibberellins, *J. Exp. Bot.*, 27, 1243, 1976.
161. **Crozier, A. and Reeve, D. R.**, The application of high performance liquid chromatography to the analysis of plant hormones. in *Plant Growth Regulation, Proc. 9th Int. Conf. Plant Growth Substances*, Pilet, P.-E., Ed., Springer-Verlag, Berlin, 1977, 67.
162. **Jones, M. G., Metzger, J. D., and Zeevaart, J. A. D.**, Fractionation of gibberellins in plant extracts by reversed phase high performance liquid chromatography, *Plant Physiol.*, 65, 218, 1980.
163. **Metzger, J. D. and Zeevaart, J. A. D.**, Effect of photoperiod on the levels of endogenous gibberellins in spinach as measured by combined gas chromatography-selected ion current monitoring, *Plant Physiol.*, 66, 844, 1980.
164. **Metzger, J. D. and Zeevaart, J. A. D.**, Photoperiodic control of gibberellin metabolism in spinach, *Plant Physiol.*, 69, 287, 1983.
165. **Barendse, G. W. M. and Van de Werken, P. H.**, High-performance liquid chromatography of gibberellins, *J. Chromatogr.*, 198, 449, 1980.
166. **Yamaguchi, I., Fujisawa, S., and Takahashi, N.**, Qualitative and semi-quantitative analysis of gibberellins, *Phytochemistry*, 21, 2049, 1982.
167. **Yamaguchi, I., Fujisawa, S., and Takahashi, N.**, Systematic ultra-micro analysis of plant growth regulators, in *IUPAC Pesticide Chemistry, Human Welfare and the Environment*, Miyamot, J., Ed., Pergamon Press, Oxford, 1983, 145.
168. **Koshioka, M., Harada, J., Takeno, K., Noma, M., Sassa, T., Ogiyama, K., Taylor, J. S., Road, S. B., Legge, R. L., and Pharis, R. P.**, Reversed phase C_{18} high pressure/performance liquid chromatography of acidic and conjugated gibberellins, *J. Chomatogr.*, 256, 101, 1983.
169. **Crozier, A., Zaerr, J. B., and Morris, R. O.**, Reversed- and normal-phase high-performance liquid chromatography of gibberellin methoxycoumaryl esters, *J. Chromatogr.*, 238, 157, 1982.
170. **Fujisawa, S., Yamaguchi, I., Park, K.-H., Kobayashi, M., and Takahashi, N.**, Qualitative and semi-quantitative analysis of gibberellins in immature seeds of *Pharbitis purpurea, Agric. Biol. Chem.*, submitted.
171. **Dathe, W., Sembdner, G., Yamaguchi, I., and Takahashi, N.**, Gibberellins and growth inhibitors in spring bleeding sap, roots and branches of *Juglans regia* L., *Plant Cell Physiol.*, 23, 115, 1982.
172. **Laurent, R., Gaskin, P., Albone, K. Y., and MacMillan, J.**, GC-MS identification of endogenous gibberellins and gibberellin conjugates as their permethylated derivatives, *Phytochemistry*, 20, 687, 1981.
173. **Fuchs, S., Haimovich, J., and Fuchs, Y.**, Immunological studies of plant hormones, *Eur. J. Biochem.*, 18, 384, 1971.
174. **Weiler, E. W. and Wieczorek, U.**, Determination of femtomol quantities of gibberellic acid by radioimmunoassay, *Planta*, 152, 159, 1981.
175. **Atzorn, R. and Weiler, E. W.**, The immunoassay of gibberllins. II. Quantitation of GA_3, GA_4, and GA_7 by ultra-sensitive solid-phase enzyme immunoassays, *Planta*, 159, 7, 1983.
176. **Atzorn, R. and Weiler, E. W.**, The immunoassays of gibberellins A_1, A_3, A_4, A_7, A_9, and A_{20}, *Planta*, 159, 1, 1983.

177. **Nakagawa, R., Yamaguchi, I., Kurogochi, S., Murofushi, N., Weiler, E. W., and Takahashi, N.**, in preparation.
178. **Rowe, J. W.**, The common and systematic nomenclature of cyclic diterpenes, *Proposal IUPAC Comm. Org. Nomenclature,* 3rd Rev., Forest Products Laboratory, U.S.Department of Agriculture, Madison, Wisc., 1968, 57.
179. **MacMillan, J. and Takahashi, N.**, Proposed procedure for the allocation of trivial names to the gibberellins, *Nature (London),* 217, 170, 1968.
180. **Cross, B. E., Grove, J. F., MacMillan, J., Moffatt, J. S., Mulholland, T. P. C., Seaton, J. C., and Sheppard, N.**, A revised structure for gibberellic acid, *Proc. Chem. Soc. London,* 302, 1959.
181. **Cross, B. E.**, Gibberellic acid. I., *J. Chem. Soc.*, 4670, 1954.
182. **Mulholland, T. P. C. and Ward, G.**, Gibberellic acid. II. The structure and synthesis of gibberene, *J. Chem. Soc.*, 4676, 1954.
183. **Cross, B. E., Grove, J. F., MacMillan, J., and Mulholland, T. P. C.**, Gibberellic acid. VII. The structure of gibberic acid, *J. Chem. Soc.*, 2520, 1958.
184. **Morrison, A. and Mulholland, T. P. C.**, Gibberellic acid. VIII. Synthesis of methyl ($\pm$)-α- and -β-3:5-dimethoxycarbonyl-6-(2-methoxylcarbonyl-6-methylphenyl)-3-methylhexanoate, *J. Chem. Soc.*, 2536, 1958.
185. **Mulholland, T. P. C.**, Gibberellic acid. IX. The structure of allogibberic acid, *J. Chem. Soc.*, 2639, 1958.
186. **Grove, J. F. and Mulholland, T. P. C.**, Gibberellic acid. XII. the stereochemistry of allogibberic acid, *J. Chem. Soc.*, 3007, 1960.
187. **Cross, B. E.**, Gibberellic acid. XIII. The structure of ring A. *J. Chem. Soc.*, 3022, 1960.
188. **Grove, J. F., MacMillan, J., Mulholland, T. P. C., and Turner, W. B.**, Gibberellic acid. XVII. The stereochemistry of gibberic and epigibberic acid, *J. Chem. Soc.*, 3049, 1960.
189. **Cross, B. E., Grove, J. F., and Morrison, A.**, Gibberellic acid. XVIII. Some rearrangements of ring A. *J. Chem. Soc.*, 2498, 1960.
190. **Speake, R. N.**, Gibberellic acid. XIX. The degradation of gibberellic acid in sulphuric acid, *J. Chem. Soc.*, 7, 1963.
191. **Aldridge, D. C., Grove, J. F., Speake, R. N., Tidd, B. K., and Klyne, W.**, Gibberellic acid. XX. The stereochemistry of ring A, *J. Chem. Soc.*, 143, 1963.
192. **Bourn, P. M., Grove, J. F., Mulholland, T. P. C., Tidd, B. K., and Klyne, W.**, Gibberellic acid. XXI. The stereochemistry of rings B, C and D, *J. Chem. Soc.*, 154, 1963.
193. **Kitamura, H., Seta, Y., Takahashi, N., Kawarada, A., and Sumiki, Y.**, Biochemical studies on Bakanae fungus, Part 49. Chemical structure of gibberellins. XIV., *Bull. Agric. Chem. Soc. Jpn.*, 23, 408, 1959.
194. **Seta, Y., Takahashi, N., Kawarada, A., Kitamura, H., and Sumiki, Y.**, Biochemical studies on Bakanae fungus. Part 50. Chemical structure of gibberellins. XV. *Bull. Agric. Chem. Soc. Jpn.*, 23, 412, 1959.
195. **Takahashi, N., Seta, Y., Kitamura, H., Kawarada, A., and Sumiki, Y.**, Biochemical studies on Bakanae, fungus. Part 51. Chemical structure of gibberellins. XVII., *Bull. Agric. Chem. Soc. Jpn.*, 23, 493, 1959.
196. **Seta. Y., Takahashi, N., Kitamura, H., Takai, M., Tamura, S., and Sumiki, Y.**, Biochemical studies on bakanae fungus. Part 52. Chemical structure of gibberellins. XVIII. *Bull. Agric. Chem. Soc. Jpn.*, 23, 499, 1959.
197. **Takahashi, N., Seta, Y., Kitamura, H., and Sumiki, Y.**, Biochemical studies on Bakanae fungus. Part 53. Chemical structure of gibberellins. XIX., *Bull. Agric. Chem. Soc. Jpn.*, 23, 509, 1959.
198. **Takahashi, N., Hsu, Y., Kitamura, H., Miyano, K., Kawarada, A., Tamura, S., and Sumiki, Y.**, Biochemical studies on Bakanae fungus. Part 56, Chemical structure of gibberellins. XXI., *Agric. Biol. Chem.*, 25, 860, 1961.
199. **Hsu, Y., Takahashi, N., Miyano, K., Kawarada, A., Kitamura, H., Tamura, S., and Sumiki, Y.**, Biochemical studies on Bakanae fungus. Part 57. Chemical structure of gibberellins. XXII., *Agric. Biol. Chem.*, 25, 865, 1961.
200. **McCapra, F., Scott, A. I., Sim, G. A., and Young, D. W.**, The structure and stereochemistry of gibberellic acid, *Proc. Chem. Soc.*, 185, 1962.
201. **McCapra, F., McPhail, A. T., Scott, A. I., Sim. G. A., and Young, D. W.**, Fungal metabolites. VII. Stereochemistry of gibberellic acid: X-ray analysis of methyl bromogibberellate, *J. Chem. Soc. C:.* 1577, 1966.
202. **Hartsuck, J. A. and Lipscomb, W. N.**, Molecular and crystal structure of the di-*p*-bromobenzoate of the methyl ester of gibberellic acid, *J. Am. Chem. Soc.*, 85, 3414, 1963.
203. **Grove, J. F.**, Gibberellin A_2, *J. Chem. Soc.*, 3545, 1961.
204. **MacMillan, J., Seaton, J. C., and Suter, P. J.**, Isolation and structures of gibberellin A_6 and gibberellin A_8, *Tetrahedron,* 18, 349, 1962.
205. **Ellames, G., Hanson, J. R., Hitchcock, P. B., and Thomas, S. A.**, The molecular and crystal structures of gibberellin A_4 and A_{13} methyl esters, *J. Chem. Soc. Perkins Trans. 1:.* 1922, 1979.
206. **Galt, R. H. B.**, New metabolites of *Gibberella fujikuroi*. IX. Gibberellin A_{13}, *J. Chem. Soc.*, 3143, 1965.

207. **Murofushi, N., Yamaguchi, I., Ishigooka, H., and Takahashi, N.**, Chemical conversion of gibberellin A_{13} to gibberellin A_4, *Agric. Biol. Chem.*, 40, 2471, 1976.
208. **Bearder, J. R., MacMillan, J., Cartenn-Lichterfelde, C., and Hanson, R.**, The removal of C(20) in gibberellins, *J. Chem. Soc. Perkin Trans 1:*, 1918, 1979.
209. **Yamaguchi, I., Takahashi, N., and Fujita, K.**, Application of ^{13}C nuclear magnetic resonance to the study of gibberellins, *J. Chem. Soc. Perkin Trans 1:*, 992, 1975.
210. **Evans, R., Hanson, J. R., and Siverns, M.**, The ^{13}C nuclear magnetic resonance spectra of some gibberellins. *J. Chem. Soc.. Perkin. Trans. 1:*, 1514, 1975.
211. **Bearder, J. R., Dennis, F. G., MacMillan, J., Martin G. C., and Phinney, B. O.**, A new gibberellin A_{45} from seed of *Pyrus communis* L., *Tetrahedron Lett.*, 669, 1975.
212. **Sponsel, V. M., Gaskin, P., and MacMillan, J.**, The identification of gibberellins in immature seeds of *Vicia faba* and some chemotaxonomic considerations, *Planta*, 146, 101, 1979.
213. **Yamane, H.**, Studies on Gibberellins in Higher Plants, Ph.D. thesis, University of Tokyo, 1975, 29.
214. **Hanson, J. R.**, Gibberellic acid. XXXI. The nuclear magnetic resonance spectra of gibberellin derivatives, *J. Chem. Soc.*, 5036, 1965.
215. **Wulfson, N. S., Zaretskii, V. I., and Papernaja, I. B.**, Mass spectrometry of gibberellins, *Tetrahedron Lett.*, 4209, 1965.
216. **Takahashi, N., Murofushi, N., Tamura, S., Wasada, N., Hoshino, H., Tsuchiya, T., Sasaki, S., Aoyama, T., and Watanabe, E.**, Mass spectrometric studies on gibberellins, *Org. Mass Spectrom.*, 2, 711, 1969.
217. **Brinks, R., MacMillan, J., and Pryce, R. J.**, Plant hormones-VIII. Combined gas chromatography-mass spectrometry of the methyl esters of gibberellin A_1 to A_{24} and their trimethylsilyl ethers, *Phytochemistry*, 8, 271, 1969.
218. **Yokota, T., Hiraga, K., Yamane, H., and Takahashi, N.**, Mass spectrometry of trimethylsilyl derivatives of gibberellin glucosides and glucosyl esters, *Phytochemistry*, 14, 1569, 1975.
219. **Loewenthal, H. J. E.**, The synthesis of gibberone, *Proc. Chem. Soc. London*, 355, 1960.
220. **Loewenthal, H. J. E. and Mulhotra, S. K.**, Synthesis of compounds related to gibberellic acid. II. ($\pm$)-gibberic acid, *J. Chem. Soc.*, 990, 1965.
221. **Mori, K., Matsui, M., and Sumiki, Y.**, Biochemical studies on Bakanae fungus. X. Total synthesis of racemic epigibberic acid, *Agric. Biol. Chem.*, 27, 537, 1963.
222. **Cross. B. E., Hanson, J. R., and Speake, R. N.**, Gibberellic acid. XXX. The preparation of 7-deoxy-gibberellins from gibberellic acid, *J. Chem. Soc.*, 3555, 1965.
223. **Galt, R. H. B. and Hanson, J. R.**, New metabolites of Gibberella fujikuroi, VII. The preparation of some ring B nor-derivatives, *J. Chem. Soc.*, 1565, 1965.
224. **Hedden, P., MacMillan, J., and Grinsted, M. J.**, Fungal products. VIII. New kaurenolides from *Gibberella fujikuroi*, *J. Chem. Soc. Perkin Trans 1:*, 2773, 1973.
225. **Mori, K., Matsui, M., and Sumiki, Y.**, Biochemical studies on Bakanae fungus. XII. Reactions of compounds derived from gibberellin C, *Agric. Biol. Chem.*, 28, 179, 1964.
226. **Mori, K., Matsui, M., and Sumiki, Y.**, Synthesis of substances related to gibberellins. XV. A partial synthesis of gibberellin C, *Tetrahedron Lett.*, 1803, 1964.
227. **Mori, K., Shiozaki, M., Itaya, N., Ogawa, T., Matsui, M., and Sumiki, Y.**, Synthesis of substances related to gibberellins. XVIII. Total synthesis of ($\pm$)-gibberellin A_2, A_4, A_9 and A_{10}, *Tetrahedron Lett.*, 2183, 1968.
228. **Somei, M. and Okamoto, T.**, A novel method for attacking non-activated C-H bond and its application to the synthesis of gibberellin A_{15} from enmein, *Chem. Pharm. Bull.*, 18, 2135, 1970.
229. **Node, M., Hori, H., and Fujita, E.**, Syntheses of methyl esters of gibberellin A_{15} and gibberellin A_{37}, *J. Chem. Soc. Chem. Commun.*, 898, 1975.
230. **Nakata, T. and Tahara, A.**, Synthesis of gibberellin A_{12} from *l*-abietic acid,, *Tetrahedron Lett.*, 1515, 1976.
231. **Wenkert, E., Afonso, A., Brendenberg, J. B., Kaneko, C., and Tahara, A.**, Synthesis of some resin acids, *J. Am. Chem. Soc.*, 86, 2038, 1964.
232. **Fujita, E., Shibuya, M., Nakamura, S., Okada, Y., and Fujita, T.**, Terpenoids. XXVIII. Total synthesis of enmein, *J. Chem. Soc. Perkin Trans 1:*, 165, 1974.
233. **Mori, K., Takemoto, I., and Matsui, M.**, A total synthesis of ($\pm$)-7,16-dioxo-17-norkauran-19-oate. *Tetrahedron*, 32, 1497, 1976.
234. **Nagata, W., Wakabayashi, T., Narisada, M., Hayase, Y., and Kamata, S.**, The stereocontrolled total synthesis of *dl*-gibberellin A_{15}, *J. Am. Chem. Soc.*, 93, 5740, 1971.
235. **Corey, E. J., Danheiser, R. L., Chandrasekaran, S., Keck, G. E., Gopalan, B., Larsen, S. D., Siret, P., and Gras, J.-L.**, Stereospecific total synthesis of gibberellic acid, *J. Am. Chem. Soc.*, 100, 8034, 1978.
236. **Schneider, G.**, Partialsynthese von Gibberellin-A_1-$O(3)$-β-D-Glucopyranoside, *Tetrahedron Lett.*, 4053, 1972.

237. **Bartlett, P. A. and Johnson, W. S.,** An improved reagent for O-alkyl cleavage of methyl esters by nucleophilic displacement, *Tetrahedron Lett.,* 4459, 1970.
238. **Schreiber, K., Wieland, J., and Sembdner, G.,** Gibberellin-XV Synthese von O(3)-β-D-Glucopyranosyl-Gibberellin-A$_3$-Methylester, *Tetrahedron,* 25, 5541, 1969.
239. **Yokota, T., Yamane, H., and Takahashi, N.,** The synthesis of gibberethiones, gibberellin-related diterpenoids, *Agric. Biol. Chem.,* 40, 2507, 1976.
240. **Cross, B. E., Grove, J. F., MacMillan, J., Mulholland, T. P. C., and Sheppard, N.,** The structure of gibberellic acid, *Proc. Chem. Soc. London,* 221, 1958.
241. **Birch, A. J., Richards, R. W., Smith, A., Harris, A., and Whalley, W. B.,** Studies in relation to biosynthesis. XXI. Rosenolactone and gibberellic acid, *Tetrahedron,* 7, 241, 1959.
242. **Graebe, J. E., Dennis, D. J., Upper, C. D., and West, C. A.,** Biosynthesis of gibberellins. I. The isolation of (−)-kaurene, (−)-kauren-19-ol and trans-geranylgeraniol in endosperm nucleus of *Echinocystis macrocarpa* Greene, *J. Biol. Chem.,* 240, 1847, 1965.
243. **Robinson, D. R. and West, C. A.,** Biosynthesis of cyclic diterpenes in extracts from seedlings of *Ricinus communis.* L. I. Identification of diterpene hydrocarbon formed from mevalonate, *Biochemistry,* 9, 70, 1970.
244. **Graebe, J. E., Hedden, P., Gaskin, P., and MacMillan, J.,** Biosynthesis of gibberellins A$_{12}$, A$_{15}$, A$_{24}$, A$_{36}$ and A$_{37}$ by a cell-free system from *Cucurbita maxima, Phytochemistry,* 13, 1433, 1974.
245. **Murphy, G. J. P. and Briggs, D. E.,** Metbolism of *ent*-kaurenol-(17-^{14}C), ent-kaurenal-(17-^{14}C) and ent-kaurenoic acid-(17-^{14}C), by germinating *Hordeum distichon* grains, *Phytochemistry,* 14, 429, 1975.
246. **Ceccarelli, N., Lorenzi, R., and Alpi, A.,** Kaurene and kaurenol biosynthesis in cell-free system of *Phaseolus coccineus* suspensor, *Phytochemistry,* 18, 1657, 1979.
247. **Kamiya, Y. and Graebe, J. E.,** The biosynthesis of all major pea gibberellins in a cell-free system from *Pisum sativum, Phytochemistry,* 22, 681, 1983.
248. **Kamiya, Y., Takahashi, M., Takahashi, N., and Graebe, J. E.,** Conversion of gibberellin A$_{20}$ to gibberellin A$_1$ and A$_5$ in a cell-free system from *Phaseolus vulgaris, Planta,* submitted.
249. **Bearder, J. R., MacMillan, J., Weis, C. N., Chaffey, M. B., and Phinney, B. O.,** Position of the metabolic block for gibberellin biosynthesis in mutant B1-41a of *Gibberella fujikuroi, Phytochemistry,* 13, 911, 1974.
250. **Cathey, H. M.,** Physiology of growth retarding chemicals, *Ann. Rev. Plant Physiol.,* 15, 271, 1964.
251. **Hedden, P., Phinney, B. O., MacMillan, J., and Sponsel, V. M.,** Metabolism of kaurenoids by *Gibberella fujikuroi* in the presence of the plant growth retardant N,N,N-trimethyl-1-methy-(2′,6′,6′-trimethylcyclohex-2′-en-1′-yl)prop-2-enyl-ammonium iodide, *Phytochemistry,* 16, 1913, 1977.
252. **Cho, K. Y., Sakurai, A., Kamiya, Y., Takahashi, N., and Tamura, S.,** Effects of the new plant growth retardants of quarternary ammonium iodide on gibberellin biosynthesis in *Gibberella fujikuroi, Plant Cell Physiol.,* 20, 75, 1979.
253. **Lang, A.,** Gibberellins: structure and metabolism, *Ann. Rev. Plant Physiol.,* 21, 537, 1970.
254. **MacMillan, J.,** Diterpenes - the gibberellins, in *Aspects of Terpenoid Chemistry and Biochemistry,* Goodwin, T. W., Ed., Academic Press, New York, 1971, 153.
255. **Hanson, J. R.,** The biosynthesis of the diterpenes, *Fortschr. Chem. Org. Naturst.,* 29, 395, 1971.
256. **West, C. A.,** Biosynthesis of gibberellins, in *Biosynthesis and Its Control in Plants,* Milborrow, B. V., Ed., Academic Press, London, 1973, 143.
257. **MacMillan, J.,** Metabolic process related to gibberellin biosynthesis in mutants of *Gibberella fujikuroi, Planta,* 120, 33, 1974.
258. **Barendse, G. W. H.,** Biosynthesis, metabolism, transport and distribution of gibberellins, in *Gibberellin and Plant Growth,* Krishnamoorthy, H. N., Ed., Wiley Eastern, New Delhi, 1975, 65.
259. **Bearder, J. R. and Sponsel, V. M.,** Selected topics in gibberellin metabolism, *Biochem. Rev.,* 5, 569, 1977.
260. **Hedden, P., MacMillan, J., and Phinney, B. O.,** The metabolism of the gibberellins, *Ann. Rev. Plant Physiol.,* 29, 149, 1978.
261. **Hanson, J. R. and White, A. F.,** Studies in terpenoid biosynthesis, IV. Biosynthesis of the kaurenolides and gibberellic acid, *J. Chem. Soc. C:,* 981, 1969.
262. **Evans, R. and Hanson, J. R.,** The formation of (−)-kaurene in a cell-free system of *Gibberella fujikuroi, J. Chem. Soc. Perkin Trans 1:,* 2382, 1972.
263. **Schecter, I. and West, C. A.,** Biosynthesis of gibberellins. IV. Biosynthesis of cyclic diterpenes from *trans*-geranylgeranyl pyrophosphate, *J. Biol. Chem.,* 244, 3200, 1969.
264. **Hanson, J. R. and White, A. F.,** The oxidation modification of the kaurenoid ring B during gibberellin biosynthesis, *Chem. Commun.,* 410, 1969.
265. **Geissman, T. A., Verbiscar, A. J., Phinney, B. O., and Cragg, G.,** Studies on the biosynthesis of gibberellins from (−)-kaurenoic acid in cultures of *Gibberella fujikuroi, Phytochemistry,* 5, 933, 1966.
266. **Galt, R. H. B.,** New metabolites of *Gibberella fujikuroi,* IX, Gibberellin A$_{13}$, *J. Chem. Soc.,* 3143, 1965.

267. **Graebe, J. E., Dennis, D. T., Upper, C. D., and West, A. A.**, Biosynthesis of gibberellins. I. The biosynthesis of (−)-kaurene, (−)-kaurenol, and trans-geranylgeraniol in endosperm nucellus of *Echinocytsis macrocarpa*, Greene, *J. Biol. Chem.* 240, 1847, 1965.
268. **Dennis, D. T. and West, C. A.**, Biosynthesis of gibberellins, III. The conversion of (−)-kaurene to (−)-kauren-19-oic acid in endosperm of *Echinocystis macrocarpa* Greene, *J. Biol. Chem.*, 242, 3293, 1967.
269. **West, C. A., Oster, M., Robinson, D., Lew, F., and Murphy, P.**, Biosynthesis of gibberellin precursors and related diterpenes, in *Biochemistry and Physiology of Plant Growth Substances*, Wightman, F. and Setterfield, G., Eds., Runge Press, Ottawa, 1968, 313.
270. **Cross, B. E., Norton, K., and Stewart, J. C.**, The biosynthesis of gibberellins. III., *J. Chem. Soc. C:*, 1054, 1968.
271. **Cross, B. E., Stewart, J. C., and Stoddart, J. L.**, 6β,7β-Dihydroxykaurenoic acid: its biological activity and possible role in the biosynthesis of gibberellic acid. *Phytochemistry*, 9, 1065, 1970.
272. **Evans, R., Hanson, J. R., and White, A. F.**, Studies in terpenoid biosynthesis. VI. The stereochemistry of some stages in tetracyclic diterpene biosynthesis, *J. Chem. Soc. C:*, 2601, 1970.
273. **Hanson, J. R. and Hawker, J.**, The ring contraction stage in gibberellin biosynthesis, *Chem. Commun.*, 208, 1971.
274. **Hanson, J. R., Hawker, J., and White, A. F.**, Studies in terpenoid biosynthesis, IX. The sequence of oxidation on ring B in kaurene-gibberellin biosynthesis, *J. Chem. Soc. Perkin Trans. 1:*, 1892, 1972.
275. **Hanson, J. R. and White, A. F.**, Studies in terpenoid biosynthesis, IV. Biosynthesis of the kaurenolides and gibberellic acid, *J. Chem. Soc. C:*, 981, 1969.
276. **Bearder, J. R. and MacMillan, J.**, Origin of the oxygen atoms in the lactone bridge of C_{19}-gibberellins, *Chem. Commun.*, 834, 1976.
277. **Dockerill, B., Evans, R., and Hanson, J. R.**, Removal of C-20 in gibberellin biosynthesis, *Chem. Commun.*, 919, 1977.
278. **Dockerill, B. and Hanson, J. R.**, The fate of C-20 in C_{19}-gibberellin biosynthesis, *Phytochemistry*, 17, 701, 1978.
279. **Hedden, P., MacMillan, J., and Phinney, B. O.**, Fungal products. XII. Gibberellin A_{14}-aldehyde, an intermediate in gibberellin biosynthesis in *Gibberella fujikuroi*, *J. Chem. Soc. Perkin Trans. 1:*, 587, 1974.
280. **Bearder, J. R., MacMillan, J., and Phinney, B. O.**, Conversion of gibberellin A_1 into gibberellin A_3 by the mutant R-9 of *Gibberella fujikuroi*, *Phytochemistry*, 12, 2655, 1973.
281. **Pitel, D. W., Vining, L. C., and Arsenault, G. P.**, Biosynthesis of gibberellins in *Gibberella fujikuroi*. The sequence after gibberellin A_4, *Can. J. Biochem.*, 49, 194, 1971.
282. **Bearder, J. R., MacMillan, J., and Phinney, B. O.**, Fungul products. XIV. Metabolic pathway from *ent*-kaurenoic acid to the fungal gibberellins in mutant B1-41a of *Gibberella fujikuroi*, *J. Chem. Soc. Perkin Trans 1:*, 721, 1975.
283. **Evans, R. and Hanson, J. R.**, Studies in terpenoid biosynthesis. XIII. The biosynthetic relationship of gibberellin in *Gibberella fujikuroi*, *J. Chem. Soc. Perkin Trans. 1:*, 663, 1975.
284. **Bearder, J. R., MacMillan, J., Wells, C. M., and Phinney, B. O.**, Fungal products. XV. The metabolism of steviol to 13-hydroxylated *ent*-gibberellanes and *ent*-kaurenes, *Phytochemistry*, 14, 1741, 1975.
285. **Bearder, J. R., Frydman, V. M., Gaskin, P., MacMillan, J., and Phinney, B. O.**, Fungal products. XVI. Conversion of isosteviol and steviol acetate into gibberellin analogs by mutant B1-41a of *Gibberella fujikuroi* and the preparation of (^{3}H)-gibberellin A_{20}, *J. Chem. Soc. Perkin Trans. 1:*, 173, 1976.
286. **Bearder, J. R., MacMillan, A., and Matsuo, A.**, Conversion of trachyrobanic acid into novel pentacyclic analogues of gibberellins by *Gibberella fujikuroi*, Mutant B1-41a, *Chem. Commun.*, 649, 1979.
287. **Murofushi, N., Nagura, S., and Takahashi, N.**, Metabolism of steviol by *Gibberella fujikuroi*, in the presence of plant growth retardant, *Agric. Biol. Chem.*, 43, 1159, 1979.
288. **Wada, K., Imai, T., and Shibata, K.**, Microbial production of unnatural gibberellins from (−)-kaurene derivatives in *Gibberella fujikuroi*, *Agric. Biol. Chem.*, 43, 1157, 1979.
289. **Hanson, J. R., Sarah, F. Y., Fraga, B. M., and Hernandez, M. G.**, The microbial preparation of two atisagibberellins, *Phytochemistry*, 18, 1875, 1979.
290. **Upper, U. D. and West, C. A.**, Biosynthesis. II. Enzymic cyclization of geranylgeranyl pyrophosphate to kaurene. *J. Biol. Chem.*, 242, 3285, 1967.
291. **Dennis, D. J. and West, C. A.**, Biosynthesis. III. The conversion of (−)-kaurene, to (−)-kauren-19-oic acid in endosperm of Echinosystis macrocarpa Greene, *J. Biol. Chem.*, 242, 3293, 1967.
292. **Shechter, I. and West, C. A.**, Biosynthesis of gibberellins. IV. Biosynthesis of cyclic diterpenes from *trans*-geranylgeranyl pyrophosphate. *J. Biol. Chem.*, 244, 3200, 1969.
293. **Robinson, D. R. and West, C. A.**, Biosynthesis of cyclic diterpenes in extract from seedlings of *Ricinus communis* L. II. Conversion of geranylgeranyl pyrophosphate into diterpene hydrocarbons and partial purification of the cyclization enzyme, *Biochemistry*, 9, 80, 1970.
294. **Lew, F. T. and West, C. A.**, (−)-Kaur-16-en-7β-ol-19-oic acid, an intermediate in gibberellin biosynthesis, *Phytochemistry*, 10, 2065, 1971.

295. **Graebe, J. E., Hedden, P., and MacMillan, J.,** The ring contraction step in gibberellin biosynthesis, *Chem. Commun.*, 161, 1975.
296. **Graebe, J. E., Hedden, P., and MacMillan, J.,** Gibberellin biosynthesis: new intermediates in the *Cucurbita* system, in *Plant Growth Sbustances 1973: Proc. 8th Int. Conf. Plant Growth Substances*, Hirokawa Publishing, Tokyo, 1974, 260.
297. **Graebe, J. E.,** GA-Biosynthesis: the development and application fo cell-free system for biosynthetic studies, in *Plant Growth Substances 1979: Proc. 10th Int. Conf. Plant Growth Substances*, Springer-Verlag, Berlin, 1980, 180.
298. **Ceccarelli, N., Lorenzi, R., and Alpi, A.,** Gibberellin biosynthesis in *Phaseolus coccineus* suspensor, *Z. Pflanzenphysiol.*, 102, 37, 1981.
299. **Yamane, H., Murofushi, N., and Takahashi, N.,** Metabolism of gibberellins in maturing and germinating bean seeds, *Phytochemistry*, 14, 1195, 1975.
300. **Yamane, H., Murofushi, N., Osada, H., and Takahashi, N.,** Metabolism of gibberellins in early immature bean seeds, *Phytochemistry*, 16, 831, 1977.
301. **Frydman, V. M. and MacMillan, J.,** The metabolism gibberellins A_9, A_{20} and A_{29} in immature seeds of *Pisum sativum* cv. Progress No. 9, *Planta*, 125, 181, 1975.
302. **Sponsel, V. M. and MacMillan, J.,** Metabolism of gibberellin A_{29} in seeds of *Pisum sativum* cv. Progress No. 9: use of (^{2}H) and (^{3}H) GAs and the identification of a new GA catabolite, *Planta*, 144, 69, 1978.
303. **Sponsel, V. M.,** Metabolism of gibberellins in immature seeds of *Pisum sativum*, in *Plant Growth Substances 1979:, Proc. 10th Int. Cong. Plant Growth Substances*, Springer-Verlag, Berlin, 1980, 170.
304. **Ceccarelli, N. and Lorenzi, R.,** Gibberellin biosynthesis in endosperm and cotyledons of *Sechium edule* seeds, *Phytochemistry*, 22, 2203, 1983.
305. **Takahashi, N. and Yamaguchi, I.,** Preparation and isotropic labeling of gibberellins, in *The Biochemistry and Physiology of Gibberellins*, Vol. 1, Crozier, A. Ed., Praeger, New York, 1983, 457.
306. **Bearder, J. R., MacMillan, J., Phinney, B. O., Hanson, J. R., Rivett, D. E. A., and Willis, C. L.,** Gibberellin A_{13}-7-aldehyde: a proposed intermediate in the fungal biosynthesis of gibberellin A_3, *Phytochemistry*, 21, 2225, 1982.
307. **Hedden, P.,** *In vitro* metabolism of gibberellins, in *The Biochemistry and Physiology of Gibberellins*, Vol. 1, Crozier, A., Ed., Praeger, New York, 1983, 99.
308. **Lombardo, L., Mander, L. N., and Turner, J. V.,** General strategy for gibberellin synthesis: total synthesis of (+)-gibberellin A_1 and gibberellic acid, *J. Am. Chem. Soc.*, 102, 6626, 1980.
309. **Lombardo, L. and Mander, L. N.,** A new strategy for C_{20} gibberellin synthesis: Total synthesis of (+)-gibberellin A_{38} methyl ester, *J. Org. Chem.*, 48, 2298, 1983.
310. **Mander, L. N.,** Personal communication.
311. **Bailiss, K.. W. and Hill, T. A.,** Biological assays for gibberellins, *Bot. Rev.*, 37, 437, 1971.
312. **Phinney, B. O.,** Biochemical mutants in maize: dwarfism and its reversal with gibberellins, *Plant Physiol.*, 31(Suppl.), 20, 1956.
313. **Ogawa, Y.,** Studies on the conditions for gibberellin assay using rice seedling, *Plant Cell Physiol.*, 4, 227, 1963.
314. **Hayashi, F., Blumenthal-Goldsmith, S., and Rappaport, L.,** Acid and neutral gibberellin-like substances in potato tubers, *Plant Physiol.*, 37, 774, 1962.
315. **Frankland, B. and Wareing, P. F.,** Effect of gibberellic acid on hypocotyl growth of lettuce seedling, *Nature (London)*, 194, 255, 1960.
316. **Brian, P. W., Hemming, H. G., and Lowe, D.,** Comparative potency of nine gibberellins, *Ann. Bot.*, 28, 369, 1964.
317. **Nicholls, P. B. and Paley, L. G.,** A barley endosperm bioassay for gibberellins, *Nature (London)*, 199, 823, 1963.
318. **Murakami, Y.,** The microdrop method, a new rice seedling test for gibberellins and its use for testing extracts of rice and morning glory, *Bot. Mag. Tokyo*, 81, 33, 1968.
319. **Yokota, T., Murofushi, N., and Takahashi, N.,** Biological activities of gibberellins and their glucosides in *Pharbitis nil*, *Phytochemistry*, 10, 2943, 1971.
320. **Zeevaart, J. A. D.,** Physiology of flower formation, *Ann. Rev. Plant Physiol.*, 27, 321, 1976.
321. **Zeevaart, J. A. D.,** Physiology of flower formation in *The Biochemistry and Physiology of Gibberellins*, Vol. 1, Crozier, A., Ed., Praeger, New York, 1983, 99.
322. **Saunders, P.,** Phytohormones and bud dormancy, in *Phytohormones and Related Compounds: A Comprehensive Treatise*, Vol. 2, Letham, D. S., Goodwin, P. B., and Higgins, T. J. V., Eds., Elsevier/North-Holland, Amsterdam, 1978, 423.
323. **Witter, S. H.,** Phytohormones and chemical regulators in agriculture, in *Phytohormones and Related Compounds: A Comprehensive Treatise*, Vol. 2, Letham, D. S., Goodwin, P. B., and Higgins, T. J. V., Eds., Elsevier/North-holland, Amsterdam, 1978, 599.

324. **Jacobsen, J. V., Chandler, P. M., Higgins, T. J. V., and Zwar, j. A.,** Control of protein synthesis in barley aleurone layers by gibberellin, in *Plant Growth Substances 1982: Proc. 11th Int. Conf. Plant Growth Substances,* Wareing, P. F., Ed., Academic Press, London, 1982, 111.
325. **Hoad, G. V.,** The role of seed derived hormones in the control of flowering in apple, *Acta Horticult.,* 80, 93, 1978.
326. **Crozier, A. and Durley, R. C.,** Modern methods of analysis of gibberellins, in *The Biochemistry and Physiology of Gibberellins,* Vol. 1, Crozier, A., Ed., Praeger, New York, 1983, 485.

Chapter 4

CYTOKININS

Koichi Koshimizu and Hajime Iwamura

TABLE OF CONTENTS

I.	Occurrence (History)		154
II.	Chemistry		161
	A.	Isolation and Characterization	161
		1. Extraction	161
		2. Purification and Separation	161
		3. Identification	163
	B.	Structural Determination	164
		1. UV Spectroscopy	164
		2. Mass Spectroscopy (MS)	164
		3. Other Spectroscopy	165
	C.	Synthesis	166
III.	Biosynthesis and Metabolism		173
	A.	Biosynthesis of Free Cytokinins	173
	B.	Metabolism of Natural Cytokinins	175
IV.	Biological Activity		178
References			189

I. OCCURRENCE (HISTORY)

Cytokinins are a group of phytohormones that regulates cell division and differentiation in certain plant tissue cultures. They have recently been shown to participate in the control of development and senescence of plants.

The existence of cell division factors was suggested by Wiesner[1] as long ago as 1892. Experimental evidence for this concept was obtained by Haberlandt (1913),[2] who found that mature parenchymatous cells of a potato tuber divided in the presence of phloem diffusates. He also demonstrated that cell division induced on cut surfaces of certain plants was prevented by rinsing these surfaces and restored by spreading juice from other leaves on the wound surface tissue.[3] Bonner and English[4] isolated a fatty acid from bean pods capable of evoking cell division in the parenchymatous cells of the bean pod mesocarp, and identified it as 1-decene-1,10-dicarboxylic acid. Since this compound (termed traumatin) was found on wounded tissues, it has been referred to as a wound hormone.

Later, van Overbeek et al.[5] demonstrated (1941) the presence of a factor in coconut milk which was necessary for the growth of excised *Datura* embryos. Subsequent work by Caplin and Steward[6] showed that coconut milk markedly stimulated the growth by cell division of explants from the secondary phloem tissue of carrot root. After extensive work they identified the constituents which considerably influenced growth in the carrot assay system. No single compound, however, was found that could account for the growth activity of coconut milk.

The first discovery of a cytokinin, kinetin, was made by Skoog and a colleague using the tobacco tissue culture technique. Jablonski and Skoog[7] found that tobacco pith tissue responded by an enormous cell enlargement entirely unaccompanied by cell division in the presence of suitable concentration of auxin in a synthetic basal medium. Cell division did occur, however, in pith tissue with attached vascular strands, and in severed pith tissue placed in contact with vascular tissue. The material active in inducing cell division was found by these workers to be contained in extracts of the vascular tissue, coconut milk or malt extract. A potent cell division promoting activity was later found in aged DNA or autoclaved DNA under acidic conditions. In 1955, Miller et al.[8] isolated a highly active compound from a rich source of autoclaved herring sperm DNA and identified it as 6-furfurylaminopurine.[9] It was given the trivial name kinetin because of its specific ability to bring about cytokinesis in cells of tobacco pith at concentrations as low as 1 $\mu g/\ell$. Kinetin artificially formed could be generated either by direct reaction between the NH_2-group of adenine and deoxyribose[10] or by the sequence of reactions involving cyclonucleoside formation and rearrangement of deoxyadenosine.[11]

The term kinin was suggested for all synthetic and naturally occurring substances with kinetin-like biological activity.[8] This term, however, was first used for a group of materials of animal origin. To avoid confusion kinin was replaced by cytokinin.[12] A cytokinin is defined[13] as a compound which, in the presence of optimal auxin, induces cell division in tobacco pith or similar tissue cultures, and in its other activities also resembles kinetin.

In 1963, the first natural cytokinin, termed zeatin (Z), was isolated from *Zea mays* kernels by Letham[14] and identified as 6-(4-hydroxy-3-methyl-2-*trans*-butenylamino)purine.*[15,16] Subsequently Letham[17] isolated 9-β-D-ribofuranosyl zeatin ([9R]Z) and its 5′-monophosphate ([9R-5′P]Z) along with minor derivatives of hydroxylated Z. Extracts from a number of different plants were also assumed to contain Z and [9R]Z from the results of bioassay and chromatographic analysis. Roots, xylem sap, developing fruits, germinating seeds, and tumor tissue are rich sources of cytokinins.[13,18]

---

* According to the Z and E designation for configuration about a double bond, *trans*-zeatin should be 6-((E)-4-hydroxy-3-methyl-2-bytenyamino)purine. In this chapter, however, the *cis* and *trans* notation is used because it is used in most of the literature cited and may be more convenient for readers who are concerned with physiological and biochemical, rather than chemical aspects of cytokinins.

Recently a number of free, natural cytokinins have been efficiently purified by high performance liquid chromatography (HPLC) and identified unequivocally by mass spectrometry (MS). The principal cytokinins that have been isolated from higher plants and/or explicitly identified by MS coupled with other physicochemical evidences are listed in Table 1. The abbreviations used for cytokinins are based on a system proposed by Letham,[13] and their structures, names and abbreviations are summarized in Figure 1.

All these cytokinins possess an N^6-substituted adenine base and can be grouped according to the carbon skeleton of N^6-substituents on the adenine ring into two classes: N^6-isoprenoid and N^6-benzyl adenine analogues.

N^6-Isoprenoid adenine analogues, the major class of cytokinins in plants (Table 1), are further divided into three groups. The first group is made up of Z and its derivatives whose N^6-isoprenoid side chain is either 4-hydroxy-3-methyl-2-*trans*-butenylamino or the *cis* isomer. Z and its riboside have now been satisfactorily identified in a wide variety of higher plants, as shown in Table 1, and are the dominant cytokinins in extracts of flowering plants. Z also occurs in mosses.[82,83] The riboside is an important translocation form of cytokinin in the xylem.[13] The major cytokinin in crown gall tissue of *Vinca rosea* is [9R]Z.[62] The ribotide ([9R-5′P]Z) has been identified only in several plants (Table 1). The nucleotide of Z, however, has been widely found in higher plants by chromatographic and GC/MS analysis of nucleosides derived by phosphatase hydrolysis of plant extracts.[48,73,75,84,85] The comparatively high content of the ribotide in the root exudate of the rice plant suggests that this is the form of cytokinin transported from roots to other parts of the plant.[43]

Z and [9R]Z have also been found in the culture media of microorganisms which form symbiotic or parasitic associations with higher plants, e.g., *Rhizopogen roseolus*,[86,87] *Suilus punctipes*,[88,89] *Corynebacterium fascians*,[89,90] and *Agrobacterium tumefaciens*,.[91–93] Production of Z by *A. tumefaciens* is a function of the nopaline Ti plasmid.[94] High pathogenicity of *C. fascians* strains and a high cytokinin activity level in the culture medium are associated with the presence of this plasmid.[95] In the culture media of the pathogenic bacteria, the following *cis* isomer and 2-methylthio derivatives have been identified (*cis*)Z, (*cis*)[9R]Z, (*cis*)[2MeS]Z and [2MeS 9R]Z.[92,93,95] Fully referenced lists of cytokinin production by microorganisms are given by Greene.[96] (*cis*)[9R]Z is a common cytokinin nucleoside present in plant tRNA hydrolysates. Cytokinin bases in tRNA have been found to contain a 2-methylthio group, i.e., [2MeS 9R]Z and (*cis*)[2MeS 9R]Z.[97–99] Although (*cis*)[9R]Z has recently been identified as a minor cytokinin in plant extracts (Table 1) along with [2MeS 9R]Z, the possibility that they were derived from tRNAs by enzyme hydrolysis during extraction procedure has not been excluded. The presence of Z and [9R]Z in mushroom (*Agricus bisporus*) is supported by chromatographic evidence.[100]

N-Glucosides of Z, in which the glucose moiety is attached to the purine ring nitrogen at position 7 or 9, have been found in some plants. [7G]Z, termed raphanatin, has been identified as a metabolite of Z in radish cotyledons[101,102] and subsequently shown to be an endogenous cytokinin in radish seed (Table 1). The 9-glucoside, [9G]Z, occurs in *Zea mays* kernels and crown gall tissue (Table 1). Both glucosides have been shown to be β-glucopyranoside.[103—106] 7- and 9-Glucosides are metabolites with enhanced metabolic stability and much lower activity than the parent cytokinin.[13]

O-Glucosides, in which a glucosyl moiety is conjugated to the oxygen on the isoprenoid side chain of Z, have recently been shown to be endogenous plant cytokinins. (OG)Z and (OG)[9R]Z have been identified as O-glucopyranosides of Z (Table 1) and synthesized.[107—109] The cytokinins in crown gall tissue of *Vinca rosea* are almost entirely glycosides.[62] (OG)Z is susceptible to enzymatic cleavage by β-glycosidases while N-glucosides are not. Cytokinin O-glucosides appear to be storage forms of cytokinins which release free, active cytokinin bases and/or ribosides when these are required.[13]

An unusual metabolite of Z was isolated from *Lupinus* spp. and shown to be the alanine conjugate of Z, L-2-[6-(4-hydroxy-3-methyl-2-*trans*-butenylamino)-purin-9-yl]alanine

Table 1
FREE NATURALLY OCCURRING CYTOKININS IN HIGHER PLANTS

Compound	Source
Z	*Castanea* spp. (chestnut shoots;[19] *Humulus luplus* (hop) cones,[20,21] and shoots;[22] *Actinidia chinensis* fruits;[23] *Raphanus sativus* root;[24] *Pyrus communis* (pear) receptacles;[25] *Prunus cerasus* fruits;[23,26] *Lupinus luteus* seeds and pod walls;[27] *Glycine max* (soybean) root exudate;[28] *Phaseolus vulgaris* (pinto bean) fruits;[29] *Dolichos lablab* immature seeds;[30] *Mercurialis ambigua* shoot apices;[31] *M. annua* buds;[32] *Mangifera indica* (mango fruit) immature seeds;[33] *Acer pseudoplatanus* spring sap;[34] *Gossypium hirsutum* (cotton) ovules;[35] *Vinca rosea* crown gall tissue;[36,37] *Lycopersicon esculentum* (tomato) xylem sap;[38] *Helianthus annus* leaves;[39] *Hordeum rulgare* (malt) extracts;[40] *Zea mays* kernels;[16,17,41] *Cocos nucifera* coconut milk)[42]
(*cis*) Z	*Humulus lupulus* (hop) cones;[21] *Mangifera indica* (mango fruit) immature seeds;[33] *Oryza sativa* root exudate[43]
[9R]Z	*Picea sitchensis* needles;[44] *Castanea* spp. (chestnut) shoots;[19] *Humulus lupulus* (hop) cones,[20,21] shoots;[22] *Beta vulgaris* crown gall tissue;[45] c *Actinidia chinensis* fruits;[23] *Brassica oleracea* hearts;[46] *Pyrus communis* (pear) receptacles;[25] *Prunus cerasus* fruits;[23] *Lupinus luteus* seeds and pod walls;[27] *Glycine max* (soybean) root exudate;[28] *Dolichos lablab* immature seeds;[30] *Mercurialis annua* buds;[32] *Mangifera indica* (mango fruit) immature seeds;[33] *Acer pseudoplatanus* spring sap;[34,47] *Gossypium hirsutum* (cotton) ovules;[35] *Daucus carota* (carrot) root callus tissue;[48] *Vinca rosea* crown gall tissue;[36,37,49] *Lycopersicon esculentum* (tomato) xylen sap;[38] *Nicotiana tabacum* roots and plant tops;[50] *Cichorium intybus* roots;[51] *Oryza sativa* root exudate;[43] *Zea mays* kernels;[17,52] *Cocos nucifera* (coconut milk)[42,53]
(*cis*)[9RZ]	*Humulus lupulus* (hop) cones[21,54] and various tissues;[50] *Dolichos lablab* immature seeds;[30] *Mercurialis ambigua* shoot apices;[31] *Mangifera indica* (mango fruit) immature seeds;[33] *Ipomoea batatas* (sweet potato) tubers;[56] *Solanum tuberosum* (potato) aerial and subterranean parts;[57] *Nicotiana tabacum* roots[50] and plant tops;[50,58] *Oryza sativa* root exudate[43]
[9R-5′P]Z	*Vinca rosea* crown gall tissue;[59] *Datura innoxia* crown gall tissue;[60] *Oryza sativa* root exudate;[43] b*Zea mays* kernels[17,52]
[9G]Z	*Vinca rosea* crown gall tissue;[61-63] *Datura innoxia* crown gall tissue;[64] *Zea mays* kernels[65,66]
[7G]Z	*Raphanus sativus* seeds;[67] *Datura innoxia* crown gall tissue[64]
(OG)Z	*Castanea* spp. (chestnut) shoots;[19] a*Lupinus luteus* seeds and pod walls[27] and pods;[68] *Dolichos lablab* immature seeds;[30] *Vinca rosea* crown gall tissue;[69,70] a*Datura innoxia* crown gall tissue;[64] *Zea mays* kernels[66]
(OG)[9R]Z	*Humulus lupulus* (hop) cones;[21] *Lupinus luteus* seeds and pod walls;[27,68] *Dolichos lablab* immature seeds;[30] *Vinca rosea* crown gall tissue;[62,63,69,70] a*Datura innoxia* crown gall tissue;[64] *Zea mays* kernels[66]
[2MeS]Z	*Nicotiana tabacum* roots[50]
(*cis*)[2MeS]Z	*Nicotiana tabacum* roots[50]
[2MeS 9R]Z	*Humulus lupulus* (hop) various tissues[55]
[2OH]Z	*Zea mays* kernels[17]
[9Ala]Z	*Lupinus luteus* seeds and pod walls[27]
(diH OG)Z	*Lupinus luteus* seeds and pod walls[27] and pods;[68] *Phaseolus vulgaris* leaves;[74] a*Datura innoxia* crown gall tissue;[64] *Zea mays* kernels[66]
(diH OG)[9R]Z	*Lupinus luteus* seeds and pod walls[27] and pods;[68] *Vinca rosea* crown gall tumor tissue;[63] a*Datura innoxia* crown gall tumor tissue;[64] *Zea mays* kernels[66]
(diH)[9Ala]Z	*Lupinus luteus* seeds and pod walls[27]
(3,4-diOH)Z	*Zea mays* kernels[17]
(2,3,4-TriOH)Z	*Zea mays* kernels[17]
iP	*Castanea* spp. (chestnut) shoots;[19] *Gossypium hirsutum* (cotton) ovules[35]
[9R]iP	*Castanea* spp. (chestnut) shoots;[19] *Humulus lupulus* (hop) shoots;[22] *Beta vulgaris* crown gall tissue;[45] *Mercurialis ambigua* shoot apices;[31] *M. annua*[75] buds;[32] *Mangifera indica* (mango fruit) immature seeds;[33] *Gossypium hirsutum* (cotton) ovules;[35] *Nicotiana tabacum* (tobacco) autonomous callus tissue;[76] *Eleocharis tuberosa* (water chestnuts) corms[77]
[2MeS 9R]iP	*Brassica oleracea* hearts[46]
[9G]iP	*Ipomoea batatas* (sweet potato) tubers[78]
(*o*OH)[9R]BAP	*Populus robusta* leaves;[79] *Zantedeschia aethiopica* fruits[80]
(*o*OH)[2MeS 9G]BAP	*Zantedeschia aethiopica* fruits[81]

Table 1 (continued)
FREE NATURALLY OCCURRING CYTOKININS IN HIGHER PLANTS

Note: The abbreviations used in this table are summarized in Figure 1.

[a] The purine aglycone, derived by hydrolysis of glucosides with β-glucosidase, was characterized by MS.
[b] The riboside produced by phosphatase hydrolysis was characterized by MS.
[c] Determined by RIA.

([9Ala]Z), termed lupinic acid.[27] This alanine conjugate is a stable metabolite and resistant to enzymatic degradation.

The second group of N^6-isoprenoid analogues consists of dihydrozeatin (diH)Z, the reduction metabolite of the side chain of Z, and its derivatives. (diH)Z isolated from immature lupin seed was optically active, and the configuration of the asymmetric carbon has been established as S.[110] Recently, the riboside [(diH)[9R]Z], N-glucoside [(diH)[9G]Z], and O-glucosides [(diH OG)Z] and (diH OG)[9R], have been identified conclusively in a number of plant tissues (Table 1). In leaves of *Phaseolus vulgaris*[72,73] and immature seeds of *P. coccineus*,[111] the side chain-saturated cytokinins are estimated to be predominant. The ribotide appears to occur in some plant tissues.[64,73] The alanine conjugate (diH)[9Ala]Z is also shown to occur along with [9Ala]Z in *Lupinus* spp.[27]

The third group of N^6-isoprenoid cytokinins contains 6-(3-methyl-2-butenylamino)purine, [N^6-($\triangle^2$-isopentenyl)adenine, iP], and its ring substitution products. First obtained by synthesis and shown to possess a high degree of cytokinin activity,[112] iP was later isolated from the culture media of the plant pathogen *Corynebacterium fascians*.[113,114] iP and its riboside ([9R]iP) have since been identified in several species of plants (Table 1), but are not commonly detected in extracts of many higher plants. iP is, however, the major free cytokinin produced in certain bacteria,[95,96] mosses,[82,83] and slime mold.[116] [2MeS 9R]iP has been identified in the culture medium of pathogenic bacteria.[92,95] Bacterial tRNA usually contains predominantly [2MeS 9R]iP and smaller amounts of [9R]iP.[96] In plant tRNA hydrolysates, [2MeS 9R]iP is the minor cytokinin and (*cis*)[9R]Z is commonly the major cytokinin.[97—99] [9R]iP is probably the most widely distributed nucleoside in the tRNA species isolated from microorganisms, animals, and plants.[117] In all tRNAs which have been sequenced, one of the following purines has been found adjacent to the first letter of the anticodon triplet: adenine, N^6-methyladenine, iP, [2MeS]iP, N-(purine-6-yl-carbamoyl)threonine, and modified guanines.[117] TMV RNA does not contain cytokinin nucleosides.[118] The occurrence and function of cytokinin molecules in tRNA have been reviewed by Letham and his colleagues[13,117,119] The novel cytokinin, 3-(3-amino-3-carboxylpropyl)iP, termed discadenine, is produced by the slime mold *Dictyostelium discoideun*.[120] Recently, a cytokinin linked to ecdysone [(22-N^6-(isopentenyl)adenosine monophosphoric ester of ecdysone] has been found in newly laid eggs of an insect, *Locusta migratoria*.[121]

In addition to the predominant cytokinins described above, smaller amounts of hydroxylated or oxygenated metabolites have been identified in *Zea mays* kernels,[13,17] but are probably not widely distributed in higher plants. Among them, (2,3,4-TriOH)Z, identified in sweet corn kernels, was formed by the oxidation of (*cis*)Z with potassium permanganate. Its activity as a cell division factor, when examined by the soybean callus assay, equalled that of the parent compound.[122]

N^6-Benzyl adenine analogues constitute the second class of cytokinins. 6-Benzylaminopurine (BAP) was first known as a synthetic compound with high cytokinin activity.[123.] A cytokinin with this nonisoprenoid side chain was first isolated from poplar leaves and identified as (*o*OH)[9R]BAP.[79] This cytokinin has been found in high concentrations (estimated; 0.5 μg/g fresh weight) in fruits of *Zantedeschia aethiopica*.[80] Synthetic (*o*OH)[9R]BAP has also been found to be an active cytokinin in vitro.[124] The fact that it possesses a

Figure 1
STRUCTURES OF NATURALLY OCCURRING CYTOKININS

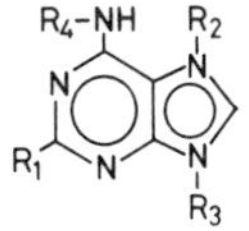

	Substituents			Common name	Systematic name	Abbreviation
R_1	R_2	R_3	R_4			
H	—	H	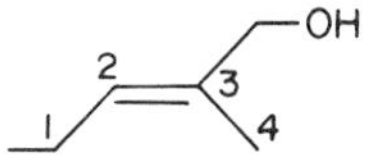 -CH₂- C(CH₃)=CH-CH₂-OH (trans)	Zeatin	6-(4-Hydroxy-3-methyl-*trans*-2-butenylamino)purine	Z
H	—	H	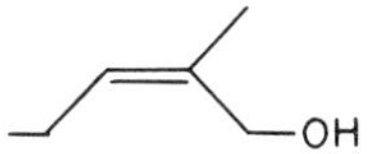 (cis isomer, -OH)	*cis*-Zeatin	6-(4-Hydroxy-3-methyl-*cis*-2-butenylamino)purine	(*cis*)Z
H	—	Ribosyl	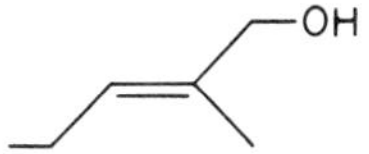 (trans, -OH)	Zeatin riboside, ribosylzeatin	6-(4-Hydroxy-3-methyl-*trans*-2-butenylamino)-9-β-D-ribofuranosylpurine	[9R]Z
H	—	Ribosyl	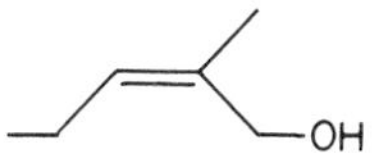 (cis, -OH)	*cis*-Zeatin riboside, ribosyl-*cis*-zeatin	6-(4-Hydroxy-3-methyl-*cis*-2-butenylamino)-9-β-D-ribofuranosylpurine	(*cis*)[9R]Z
H	—	5'-Phosphoribosyl	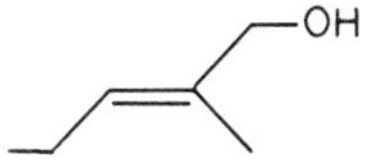 (trans, -OH)	Zeatin ribotide	6-(4-Hydroxy-3-methyl-*trans*-2-butenylamino)-9-β-D-ribofuranosylpurine-5'-monophosphate	[9R-5'P]Z
H	—	Glucosyl	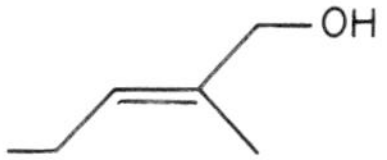 (trans, -OH)	Zeatin-9-glucoside, 9-glucosylzeatin	6-(4-Hydroxy-3-methyl-*trans*-2-butenylamino)-9-β-D-glucopyranosylzeatin	[9G]Z
H	Glucosyl	—	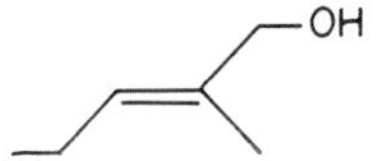 (trans, -OH)	Zeatin-7-glucoside, 7-glucosylzeatin	6-(4-Hydroxy-3-methyl-*trans*-2-butenylamino)-7-β-D-glucopyranosylzeatin	[7G]Z

			Trivial name	Systematic name	Abbreviation
H	—	H	—O-Glucosyl (trans-butenyl chain)	Zeatin O-glucoside, O-glucosylzeatin	6-(4-O-β-D-glucopyranosyl-3-methyl-*trans*-2-butenylamino)purine	(OG)Z
H	—	Ribosyl	—O-Glucosyl (trans-butenyl chain)	O-Glucosylzeatin riboside	6-(4-O-β-D-glucopyranosyl-3-methyl-*trans*-2-butenylamino)-9-β-D-ribofuranosylpurine	(OG)[9R]Z
CH$_3$S—	—	H	—OH (trans-butenyl chain)	2-Methylthiozeatin	2-Methylthio-6-(4-hydroxy-3-methyl-*trans*-2-butenylamino)purine	[2MeS]Z
CH$_3$S—	—	H	—OH (cis-butenyl chain)	2-Methylthio-*cis*-zeatin	2-Methylthio-6-(4-hydroxy-3-methyl-*cis*-2-butenylamino)purine	(*cis*)[2MeS]Z
CH$_3$S—	—	Ribosyl	—OH (trans-butenyl chain)	2-Methylthiozeatin riboside	2-Methylthio-6-(4-hydroxy-3-methyl-*trans*-2-butenylamino)-9-β-D-ribofuranosylpurine	[2MeS 9R]Z
OH	—	H	—OH (trans-butenyl chain)	2-Hydroxyzeatin	2-Hydroxy-6-(4-hydroxy-3-methyl-*trans*-2-butenylamino)purine	[2OH]Z
H	—	—CH$_2$CH(NH$_2$)COOH	—OH (trans-butenyl chain)	Lupinic acid	L-β-[6-(4-Hydroxy-3-methyl-*trans*-2-butenylamino)purin-9-yl]-alanine	[9Ala]Z
H	—	H	—OH (saturated chain)	Dihydrozeatin	6-(4-Hydroxy-3-methylbutylamino)purine	(diH)Z
H	—	Ribosyl	—OH (saturated chain)	Dihydrozeatin riboside, ribosyl dihydrozeatin	6-(4-Hydroxy-3-methylbutylamino)-9-β-D-ribofuranosylpurine	(diH)[9R]Z
H	—	H	—O-Glucosyl (saturated chain)	Dihydrozeatin O-glucoside, O-glucosyldihydrozeatin	6-(4-O-β-D-glucopyranosyl-3-methyl-butylamino)purine	(diH OG)Z

Figure 1. Continued.

Substituents				Common name	Systematic name	Abbreviation
R_1	R_2	R_3	R_4			
H	—	Ribosyl	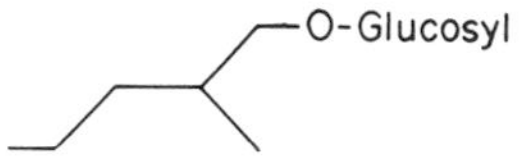 O-Glucosyl	O-Glucosyldihydrozeatin riboside	6-(4-O-β-D-glucopyranosyl-3-methyl-butylamino)-9-β-D-ribofuranosylpurine	(diH OG)[9R]Z
H	—	—CH$_2$CH(NH$_2$)COOH	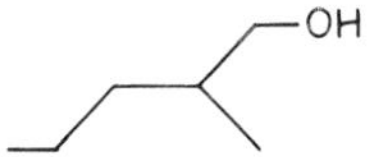 —OH	Dihydrolupinic acid	L-β-[6-(4-Hydroxy-3-methyl-butylamino)purin-9-yl]-alanine	(diH)[9Ala]Z
H	—	H	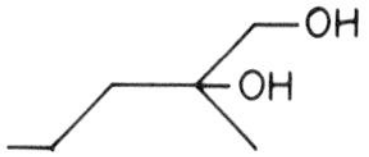 —OH, —OH		6-(3,4-Dihydroxy-3-methylbutyl-amino)purine	(3,4-diOH)Z
H	—	H	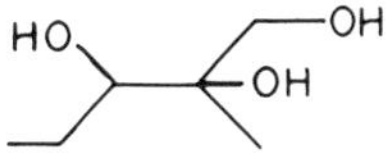 —OH, —OH, —OH		6-(2,3,4-Trihydroxy-3-methyl-butylamino)purine	(2,3,4-triOH)Z
H	—	H	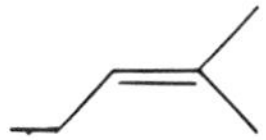 (isopentenyl)	N^6-(Δ^2-Isopentenyl)-adenine, isopentenyl-adenine	6-(3-Methyl-2-butenylamino)-purine	iP
H	—	Ribosyl	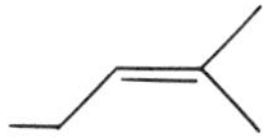 (isopentenyl)	N^6-(Δ^2-Isopentenyl)-adenosine, isopentenyl-adenosine	6-(3-Methyl-2-butenylamino)-9-β-D-ribofuranosylpurine	[9R]iP
CH$_3$S—	—	Ribosyl	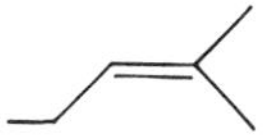 (isopentenyl)		2-Methylthio-6-(3-methyl-2-butenylamino)-9-β-D-ribofuranosylpurine	[2MeS 9R]iP
H	—	Glucosyl	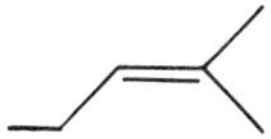 (isopentenyl)		6-(3-Methyl-2-butenylamino)-9-β-D-glucopyranosylpurine	[9G]iP
H	—	Ribosyl	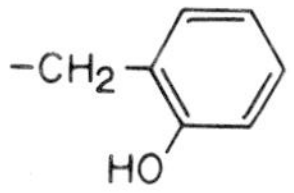 —CH$_2$—C$_6$H$_4$—OH		6-(o-Hydroxybenzylamino)-9-β-D-ribofuranosylpurine	(oOH)[9R]BAP
CH$_3$S—	—	Glucosyl	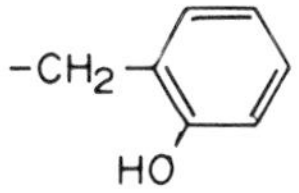 —CH$_2$—C$_6$H$_4$—OH		2-Methylthio-6-(o-hydroxy-benzylamino)-9-β-D-glucofuranosylpurine	(oOH)[2MeS 9G]BAP

Note: Abbreviations are based on a system proposed previously by Letham.[13,119] Substitution on the purine ring of these compounds is indicated within square brackets, while substitution or modification on the isoprenoid side chain is presented within parentheses. 6-Benzylaminopurine (BAP) derivatives are denoted in analogous manner.

hydroxybenzylic side chain raises the question of the biochemical sources for the side chain synthesis of natural cytokinins. From the same source (oOH)[2MeS 9G]BAP was isolated as a minor cytokinin.[81] The presence of (oOH)BAP in grapevine infloresences has been suggested by paper and column chromatographic analysis.[125]

Reviews of naturally occurring cytokinins are given by Letham,[13] Letham and Palni,[119] Horgan,[126] Skoog and Schmitz,[127] and Bearder.[128]

II. CHEMISTRY

A. Isolation and Characterization

1. Extraction

Plant tissues are known to contain certain nonspecific phosphatases which hydrolyze cytokinin nucleotides to their corresponding nucleosides. These enzymes are not completely inactivated during the usual extraction method of plant growth substances by the homogenation of plant tissues in ice-cold 80% methanol or ethanol. Horgan[129] recommends the simple extraction method of adding plant tissues to ethanol at 0°C so as to bring the volume ratio of ethanol to tissue water to 4:1, and allowing the solution to stand at 0°C for 12 hr. When the tissue is not homogenized, overall recovery of cytokinin nucleotides is more than 80% by this method. Parker and Letham[130] have minimized nucleotide breakdown by extraction with methanol-water (4:1, v/v) at 65°C. Bieleski[131] has developed a solvent (chloroform:methanol:formic acid:water = 5:12:1:2, v/v, −20°C) suitable for plant soft tissues to minimize the effects of phosphatase activity during the tissue extraction. Perchloric acid has also been used for cytokinin nucleotide extraction.[132,133] Although the extraction procedures of Bieleski[131] and Brown[132] employed subzero temperatures to inactivate phosphatases, this did not completely prevent cytokinin nucleotide breakdown; even with the Bieleski extraction method, 13.7% [^{14}C]AMP was degraded to adenosine and adenine.[59]

2. Purification and Separation

Purification is monitored by cytokinin bioassay. For preliminary clean-up of crude extracts, solvent partitioning is useful. n-Butanol extraction is effective for separation of free-base cytokinins and nucleoside cytokinins;[59,62,64] however, the recovery of each cytokinin species from the butanol partition step must be considered independently. A significant portion of the O-glucosides in an aqueous extract fails to partition into butanol due to the unfavorable partition coefficients [(OG)Z, 0.64; (OG)[9R]Z, 0.27[62]]. Partition data of some cytokinins are given by Letham[134] and summarized by Horgan.[129]

Ion exchange resins are widely used for the purification of cytokinins, and cation exchange resins (e.g., Dowex 50[135] and Zeokarb 225,[17] equilibrated to pH 3.0 to 3.4 in NH_4^+ form) are useful in the initial purification stage to obtain a cytokinin-rich fraction from crude extracts. Cellulose phosphate (pI floc type, equilibrated to pH 3.0 to 3.2 in the NH_4^+ form) is recommended for later stages of purification or small-scale studies.[130] Cytokinin nucleotides pass through the cellulose phosphate column in weak acid (pH 3), and cytokinin bases, ribosides, and glucosides retained on the column can be subsequently eluted with dilute ammonia.[65,73]

Anion exchange resins (e.g. De-Aciditie FF in acetate form, Dowex 1 in formate form) have been used in the purification of cytokinin nucleotides and nucleosides carrying carboxyl group[17] as well as zeatin glucoside.[136] DEAE-cellulose is used for the purification of cytokinin glucosides[101] and the fractionation of mononucleotides.[59]

The chromatographic material most commonly used is Sephadex® LH-20 (for a recent application, see Reference 59). A wide range of cytokinins can be separated almost quantitatively on a Sephadex® LH-20 column by elution with 35% ethanol.[137] More polar cytokinins are eluted earlier from the column by use of reversed phase partitioning. The elution values of some cytokinins and adenine metabolites on Sephadex® LH-20 are summarized by Horgan.[129]

Polyvinylpyrrolidone (PVP) has been used for the chromatographic separation of cytokinins with phosphate buffer[138] or water[70] as the developing solvent. A major class of UV interferants, phenols, and pigments, can be partially excluded from crude plant extracts by chromatography on PVP.[138,139] PVP chromatography is of great value in the initial purification step,[138] since the extract is generally aqueous. Cytokinins can be concentrated on a column of C_{18} Porasil B, on which such impurities as sugars are not retained.[140]

As cytokinins are present in most plant tissues and some microorganisms in very small amounts (nanogram to microgram range), the isolation of a cytokinin of sufficient purity for characterization has required extensive use of chromatographic methods such as paper (PC), thin layer (TLC), gas (GC), and HPLC.

Most naturally occurring cytokinins can be separated efficiently by virtue of their polar nature by PC[141,142] and cellulose TLC.[66,143] Silica gel layers are effective for the separation of cytokinins of very similar structure.[109] Anion-exchange layers (DEAE-cellulose containing green fluorescent indicator on cellulose layers impregnated with polyethyleneimine) have been used for separation of cytokinin nucleotides.[59]

HPLC is used not only to separate but also to identify cytokinins, by comparing their elution volumes with those of authentic samples. The combination of rapid analysis, high resolution, and the elimination of derivatization steps gives HPLC a marked advantage over other methods of separation PC, TLC, and GC. Fractions purified by HPLC are often sufficiently pure for direct introduction into the mass spectrometer as trimethylsilylated or permthylated derivatives, a GC step being unnecessary.[66]

The packing materials of HPLC for adsorption, partition (including reversed phase), ion-exchange, and gel filtration chromatography are commercially available. Ion-exchange columns were used in some of the early attempts to separate cytokinins by HPLC, but the resins do not provide high resolution of cytokinins.[144] Improved resolution and increased sample capacity have been achieved by using reversed phase packing materials which consist of organic stationary phases bonded to silica supports. Octadecyl silane (ODS or C_{18}) type is widely used as a packing for separation and identification of cytokinins.[62,66,140,145-148] Preparative C_{18} columns have been used in initial purification steps for the concentration of cytokinins from crude extracts.[65,66,139] Other bonded phase packings used are C_2,[147] C_8,[149] and phenyl.[110] Cytokinin nucleotides have been shown to separate on a column of an aminopropyl-bonded silica with weak anion-exchange properties (Hypersil APS) using volatile buffers.[150]

Improved separation of geometric isomers of cytokinins in the reversed phase mode has been obtained by ion-pair chromatography with an acidic solvent system and a counter ion (heptanesulfonate).[149,150] Cytokinin nucleotides, which are commonly determined as their corresponding nucleosides or bases derived by enzymatic or chemical hydrolysis, can be separated on a column of ODS (Hypersil ODS) in an ion-pair reversed phase system without hydrolysis.[150]

Normal phase partition HPLC using Partisil PAC has been used to separate a mixture of cytokinin glucosides.[147] Adsorption chromatography by HPLC grade silica (Partisil 10, Hypersil) has been used for the analysis of cytokinin standards.[147] Steric exclusion chromatography on a gel (μSpherogel support) with a molecular exclusion limit (2000) has been developed,[151] and the technique has been shown to be suitable for routine use at an early stage in the purification of trace quantities of endogenous plant growth substances in crude plant extracts.

The polar nature of most cytokinins places limits on the types of HPLC systems usable. Polarity, however, can be reduced by permethylation, and permethylated cytokinins have been separated with good resolution on HPLC columns of adsorptive Spherisorb and polar Partisil 10 PAC.[152]

GC has been used extensively for the separation and identification of cytokinins. Analysis requires the formation of volatile derivatives of which the most widely used is trimethylsilylate. After rigorous drying, cytokinins can be trimethylsilylated with bis-trimethylsilylacetamide (BSA)[72,73,153] or bis-trimethylsilyltrifluoroacetamide (BSTFA)[106] in aprotic and polar solvents such as acetonitrile, dimethyl formamide, and pyridine. The N-9 of the purine bases is trimethylsilylated if the position is free of a substituent. Hydroxyl groups in the N^6-substituent and in the sugar moiety are simultaneously trimethylsilylated. The N-6 position is not usually trimethylsilylated under these reaction conditions, but it can be trimethylsilylated by the treatment of cytokinins with BSA-trimethylchlorosilane-acetonitrile, yielding fully trimethylsilylated (TMSi)derivatives.[55] TMSi cytokinins purified by preparative GC can easily be converted back to the parent compounds by mild acid hydrolysis, and the collected products may be used for bioassay.

Permethylation by the methylsulfinyl carbanion-methyl iodide procedure has also been used to prepare volatile derivatives.[23,154] The advantages of permethyl over TMSi derivatives include their chemical stability and lower molecular weight. Permethylation of cytokinin bases and cytokinin glucosides gives good yields of homogeneous products[23,30,61,69,106] with attachment of methyl groups to all hydrolysis of the side chain and the sugar moiety as well as to the exocyclic nitrogen and the N-9 of the adenine residue if the position is not substituted.[23,154] However, the 7-glucosides of BAP and Z form a mixture of the products of hydrolysis and methylation in poor yields.[106] Both the TMSi and permethylated derivatives of [3G]BAP have been found to undergo thermal rearrangement during GC.[106] Trifluoroacetylated (TFA) derivatives of cytokinins have been analyzed by GC coupled with the electron capture detector.[155]

Several silicone-type stationary phases of the nonpolar (OV-1,[62,73,129] OV-101,[64] SE-54[156—158]) and intermediate polar type (OV-17[66,95,159]) have normally been used for the separation of the TMSi derivatives. Improved GC separations have been achieved by use of fused-silica capillary columns.[156—158]

In GC of the TMSi derivatives, packed columns and flame ionization detectors (FID) have generally been employed, yielding a detection limit in the nanogram range.[34,153] An alkali FID (nitrogen-phosphorus detector, NPD) and flame-photometric detector (FPD) (selective for nitrogen and sulfur-containing compounds, respectively) have been found to give greatly enhanced sensitivity and specificity over the FID for cytokinins which are rich in nitrogen and some cytokinins containing sulfur.[159] In combination with a fused-silica capillary column, the NPD have been shown to permit the detection of permethylated cytokinins at low picogram levels with good resolution.[160] By electron capture (EC) capillary GC, quantities as low as 1 pg of the TFA derivatives have been detected.[155]

The extraction and purification of plant hormones including cytokinins are discussed in detail by Yokota et al.[161]

3. Identification

The identification and quantitation of cytokinins in plant tissue are necessary for the physiological study on their mode of action and metabolism. Significant progress has recently been made in the identification of cytokinins by the combined GC/MS technique. This system has both the resolving power of the GC and the selective detection characteristic of the MS. GC/MS technqiues have been reviewed by Yokota et al.,[161] Horgan,[162] and Brenner.[163]

For the identification of cytokinins at low levels in relatively impure extracts, GC/MS is the most suitable technqiue. This type of analysis has been reported by Palni et al,[64] who have unambiguously identified 16 cytokinins, including 7- and 9-glucosides, *O*-glucosides, bases, ribosides, and nucleotides, in the *Datura innoxia* crown gall tissues. The identifications are based on comparisons with synthetic compounds by GC/MS as well as TLC, Sephadex® LH-20 chromatography, and HPLC. The quantification of these cytokinins was achieved by using relevant ^{2}H-labeled internal standards.

In GC/MS analysis, the GC can be used to monitor, as a function of time the intensity of particular fragment ions in the mass spectrum of a compound being studied. This mode is called GC/SIM and the GC trace obtained is extremely simplified. GC/SIM has been used to identify and quantify known cytokinins. Quantitative analysis by GC/SIM was first carried out by Thompson et al.,[164] who measured the levels of (oOH)[9R]BAP in attached leaves of poplar using the unnatural *para* isomer as an internal standard. However, losses of the internal standard added may occur during the purification of the endogeous cytokinin by high resolution chromatographic techniques prior to analysis. ^{2}H- and ^{15}N-labeled internal standards have been developed recently for the analysis of endogenous cytokinins in plant tissues.[46,62,65,67]

Scott et al.[62] have measured the levels of cytokinins in the *Vinca rosea* crown gall by a mass spectrometric isotope dilution technique using analogues labeled with four ^{15}N atoms as internal standards. The labeled and natural compounds can be distinguished in the analysis because the m/z values of their mass spectral fragments are displaced by four units. A known quantity of the ^{15}N-labeled analogue is added to the initial tissue extract, and after purification the molar ratio of natural cytokinin to labeled internal standard in the extract is determined by GC/SIM.

Besides physicochemical methods, radioimmunoassay (RIA) has also been used for cytokinin analysis. Weiler[165] has developed RIA for Z-type cytokinins using high affinity antisera produced against BSA conjugates of [9R]Z. The high specificity and sensitivity (measuring ranges from 0.02 to 10 ng of [9R]Z) of this assay have allowed the quantitation of the Z levels present in crude plant extracts. Weiler and Spanier[166] and Vold and Leonard[167] have detected the low levels of cytokinins by use of RIA for direct analysis of unfractionated extracts.

A technique of combined HPLC-RIA has been developed for the rapid and sensitive (to the picogram range) estimation and identification of the multiple cytokinins in natural sources.[145,168]

B. Structural Determination

The structural determination of unknown cytokinins from natural sources has relied on spectroscopic techniques, particularly UV spectroscopy, MS, and/or GC/MS.

1. UV Spectroscopy

Naturally occurring cytokinins show strong UV absorption in the region from 250 to 280 nm due to the adenine ring. The spectra in acidic, neutral, or basic solutions exhibit hyper- or bathochromic shifts in the relative positions of the maxima and minima. These spectral features reflect the substitution pattern of the adenine ring. The structural assignment of unknown cytokinins can be made by comparision of their features with those of variously substituted purines reported in the literature.[169,170] Examples of the application of UV spectroscopy for the structural studies have been published.[30,61,66,79,80]

In early experiments[15] the assignment of the substitution pattern of the adenine ring made use of the acid dissociation constant (pK_a), which are determined spectrophotometrically.[171] The pK_a values due to the imidazole-NH group and the exocyclic nitrogen vary with the position of the substituents.[172]

2. Mass Spectrometry (MS)

MS is the most reliable technique for structural determination of naturally occurring cytokinins belonging to the class of N^6-substituted adenines. Most of the cytokinins summarized in Table 1 have been identified on the basis of the MS studies.

Highly purified samples for identification are introduced into the MS source by direct insertion probe and usually ionized by high-energy electron impact (electron impact ionization, EI). The mass spectra of the underivatized cytokinins thus obtained exhibit structurally significant fragments at m/z 135 ($C_5H_5N_5^+$, adenine ion), (119 purine ion), and 108 (adenine-HCN) which characterize them as an adenine derivative.[173] The difference between the molecular ion (M^+) and m/z 135 peaks indicates the combined atomic weights of the substituents on the adenine ring.

Shannon and Letham[173] have shown that the appearance of ions at m/z 149, 148, 121, 120, and 119 is indicative of the presence of 6-alkyladenines. The ion at m/z 148 is characteristic of cytokinins with a methylene group adjacent to the N^6 atom of the adenine ring. This ion has been observed in the mass spectra of a new cytokinin, (oOH)[9R]BAP, in the leaves of *Populus robusta*[79] and the fruits of *Zantedeschia aethiopica*,[80] providing proof of the presence of a methylene group between the N^6 atom of the adenine ring and an aromatic residue of the N^6 side chain.

The presence of a ribosyladenine residue in a cytokinin molecule has been deduced from the appearance of ions attributable to $(M-CHOH)^+$, $(M-C_3H_5O_3)^+$, and $(M-C_4H_7O_3)^+$, together with those at m/z 178 (adenosine-$C_3H_5O_3)^+$ and 164 (adenosine-$C_4H_7O_3)^+$,[79,80] which have been proposed to arise from initial transfer of sugar hydroxyl hydrogens to the charge-localized purine base.[174]

Direct probe insertion of underivatized cytokinin glucosides has occasionally given irreproducible mass spectra due to thermal instability. MacLeod et al.[106] have analyzed a series of glucosides of Z and BAP as volatile TMSi and permethylated derivatives by GC/MS. The mass spectra of the TMSi and permethylated glucosides show a molecular ion varying in intensity between 0.1 and 20% of the base peak. In the spectra of the TMSi glucosides, ions are observed at m/z 73, 103, 129, 147, 169, 204, 217, 305, and 319, which are common to TMSi carbohydrates and related compounds. The sugar ring size in those glucosides can be assigned by the presence or absence of certain critical ions in the spectra. The TMSi derivatives give spectra with well-defined fragment ions for the identification of known compounds by comparison with the mass spectra of authentic compounds, but much of the information for the structural determination of unknown compounds is lost due to the effect of the TMSi group on the fragmentation.[129]

The permethylated derivatives are more stable than TMSi derivatives and are also more readily resolved on GC. Permethylated glucosides also exhibit a similar series of characteristic ions at m/z 71, 75, 88, 101, and 111, which are common to permethylated sugar derivatives. Permethylated derivatives have proven useful in the identification of (OG)Z and (OG)[9R]Z in *Vinca rosea* crown gall tissue.[69]

Chemical ionization mass spectrometry (CIMS) is of particular value for the identification of cytokinin ribosides. The CIMS spectra of the TMSi derivatives exhibit intense protonated-molecular ion (MH^+) peaks and also prominent structure-related fragmentation that is either not evident or very minor in the EI-MS spectra. The CI- and EI-MS techniques have been used successfully to determine the structure of cytokinin metabolites in sweet-corn kernels.[66] A cytokinin ribotide, [9R-5'P]Z, in *Datura innoxia* crown-gall tissue has been detected and quantified as an intact molecule by CIMS.[60]

3. Other Spectroscopy

The very small amount of pure cytokinins available has limited the investigation to the measurement and interpretation of nuclear magnetic resonance (NMR) and infrared (IR) spectra.

^{1}H-NMR spectroscopy has been used to determine the structures of Z^{15} and (diH)Z.[71] The spectra exhibit characteristic signals due to the 2 and 8 ring protons on the adenine ring if their positions are not substituted. An unusual cytokinin, discadenine, has been isolated from the culture media of the slime mold *Dictyostelium discoideum* and its structure has been fully confirmed by both the ^{1}H- and ^{13}C-NMR spectra.[120] The NMR spectra of (oOH)[9R]BAP obtained from fruits of *Zantedeschia aethiopica* have been also measured for the structural determination.[80]

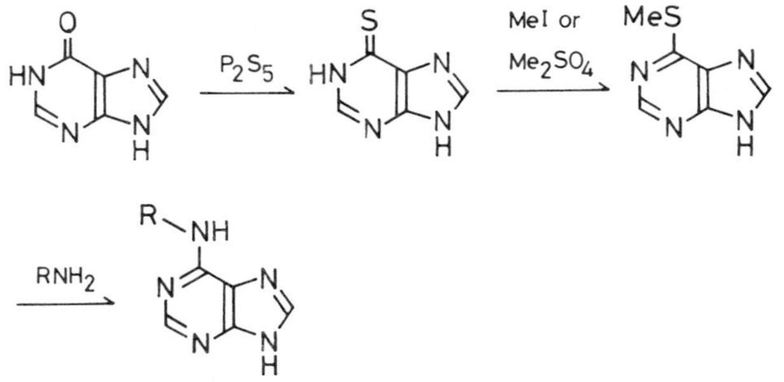

SCHEME 1.

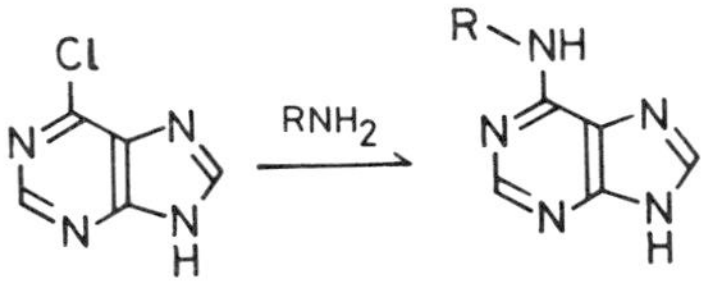

SCHEME 2.

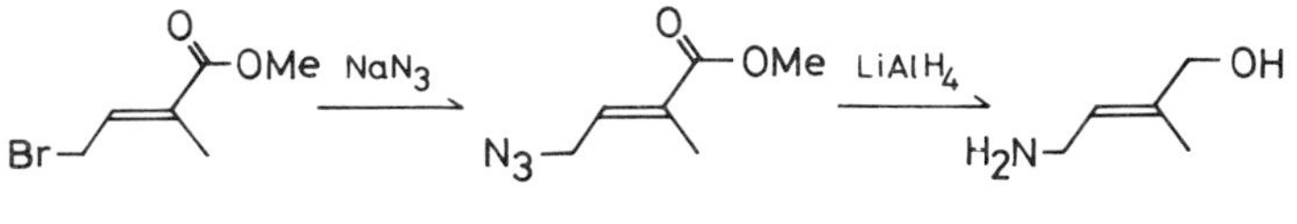

SCHEME 3.

C. Synthesis

Kinetin, the first member of the class, is prepared with a fair yield by reacting furfurylamine and 6-methylthiopurine at elevated temperatures.[9] The method is based on that developed by Elion et al.[175] and is applicable to the synthesis of other N^6-substituted adenine derivatives (Scheme 1). The starting 6-methylthiopurine is prepared by treating hypoxanthine with phosphorus pentasulfide, followed by methylation by methyl iodide or dimethyl sulfide.[175] The reaction of 2,6-bis-methylthiopurine with an amine such as 4-hydroxy-3-methylbutylamine or isopentylamine affords the corresponding cytokinin-active N^6-substituted-2-methylthioadenine.[176] 6-Methylthio function is more susceptible to nucleophilic attack than a 2-methylthio group. However, this method fails to produce much 6-(4-hydroxy-3-methyl-cis-2-butenylamino)-2-methylthiopurine,[177] [2MeS]-cis-Z, and even less reactivity is documented at the riboside level, i.e., 2,6-bis-methylthio-9-β-D-ribofuranosylpurine.[178] Thus this method is less useful than that in which a 6-chloropurine derivative and an appropriate amine are reacted.

6-Chloropurine is commercially available and a variety of 6-substituted adenine compounds can be prepared without recourse to sealed tube reactions at extremely high temperatures (Scheme 2). Kinetin was the first cytokinin prepared by this method.[170] Z is readily prepared by refluxing 6-chloropurine with the side chain amine, 4-hydroxy-3-methyl-trans-2-butenylamine, in n-butanol, as is [9R]Z (6-(4-hydroxy-3-methyl-trans-2-butenylamino)-9-β-D-ribofuranosylpurine) from 6-chloro-9-β-D-ribofuranosylpurine.[180] The side chain amine was prepared from methyl γ-bromotiglate[181] via a γ-azido derivative (Scheme 3). The amine for the N^6-side chain of the cis-isomers cis-Z and ribosyl cis-Z was prepared by coupling 1-chloro-1-nitrosocyclohexane[182] with isoprene to give an intermediate 1,2-oxazine which was reductively cleaved to the cis-aminobutenol (Scheme 4).[177] Coupling of the cis-amine with

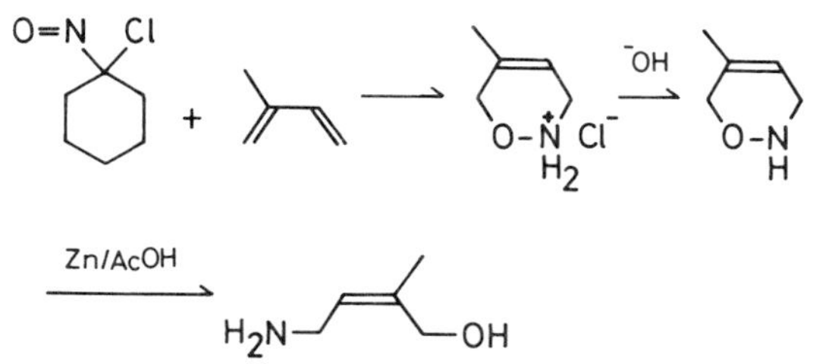

SCHEME 4.

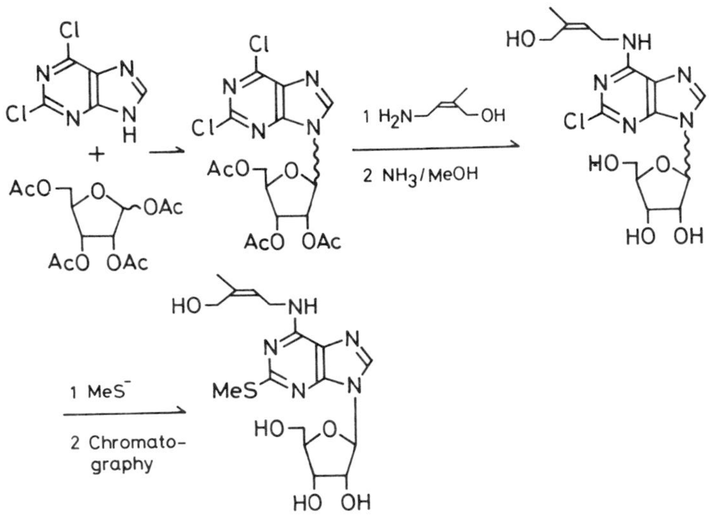

SCHEME 5.

6-chloropurine or its 9-β-D-ribofuranoside afforded cis-Z[183] or its riboside,[11] respectively, in reasonable yields; coupling with 6-chloro-2-methylthiopurine produces [2MeS]-cis-Z.[177] The production of its riboside, [2MeS 9R]Z, was inadequate by the method described above using 6-chloro-2-methylthio-9-β-D-ribofuranosylpurine. There is an alternative route via the 2,6-dichloro derivative in which the cis-amine is reacted with 2,6–dichloro-9-(2,3,5-tri-O-acetyl-β-D-ribofuranosyl)purine[185] to give, after deblocking, 2-chloro-[9R]-cis-Z, which is then treated with sodium methylmercaptide (Scheme 5).[177] The 2,6-dichloro intermediate is the product of a fusion reaction between 2,6-dichloropurine and fully acetylated ribofuranose, and is mixture of α- and β-anomers which must be separated chromatographically at the final step. 6-(3-Methyl-2-butenylamino)purine, iP, was synthesized piror to its isolation from nature.[186] Its 9-β-D-ribofuranoside, [9R]iP, and 2-methylthio analogues can be prepared by either of the methods mentioned above. 3-Methyl-2-butenylamine (Δ^2-isopentenylamine or γ,γ-dimethylalylamine) is prepared either by the reaction of γ,γ-dimethylacrylonitrile with LiALH4,[187] or from γ,γ-dimethylallyl bromide via N-(γ,γ-dimethylallyl)phthalimide.[188]

[9R-5′P]iP and [9r-5′P]BAP have been synthesized by three different methods,[189,190] the most convenient of which is the route via the direct alkylation of the unprotected adenosine 5′-phosphate Na-salt and successive rearrangement of the resultant 1-substituted compound to the product (Scheme 6).[191] 3′,5′-Cyclic phosphates of iP and BAP were also prepared by a similar method,[191] and these nucleotide analogues of cytokinins may be of use in biological and biochemical studies of cytokinin action. The rearrangement reaction of 1-alkyladenine derivatives to N^6-alkyladenines is worthy of note because the cytokinin activity of 1-(3-

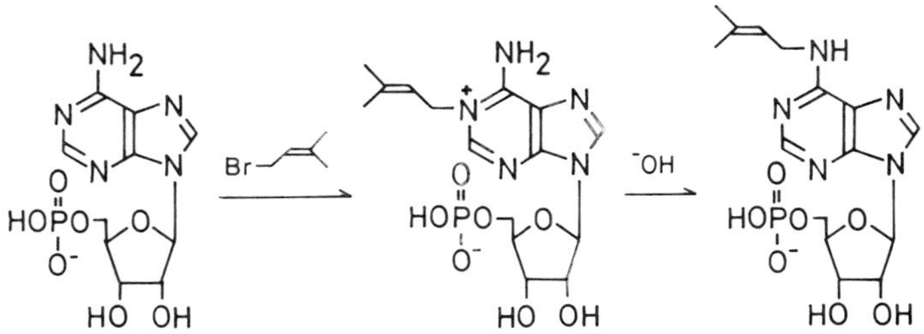

168 Chemistry of Plant Hormones

SCHEME 6.

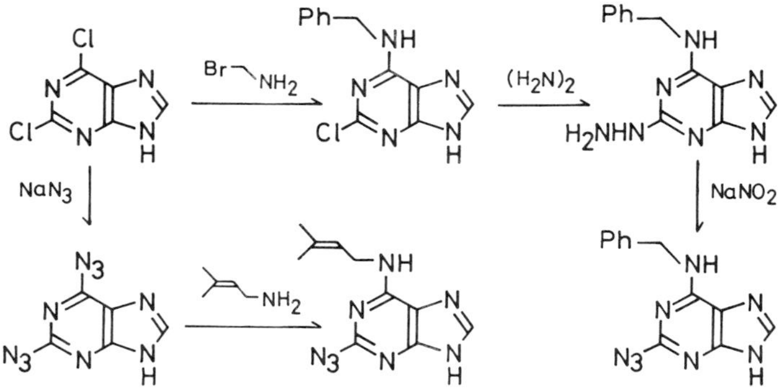

SCHEME 7.

methyl-2-butenyl)- and 1-benzyladenines can be attributed to this novel reaction occurring under conditions for bioassay.[191]

A variety of N^6-substituted adenine analogues which possess one or two substituents at the 2-, 8-, or 2,8-positions were prepared and examined for biological activity.[192,193] Most of these were prepared by refluxing a suitably substituted purine precursor with an appropriate amine, utilizing the difference in reactivity of the 2-, 6-, and 8-positions toward nucleophilic displacement. Good leaving groups such as chloro- or methylsulfonyl at the 6- position are more reactive than the same or a poorer leaving group at either the 2- or 8-position.[193,194] A photolabile azido function was introduced into the 2-postition of BAP and iP to obtain photoaffinity-labeled cytokinins via a 2-chloro intermediate,[195] or a 2,6-diazido derivative[196] (Scheme 7). The latter route appears useful for the preparation of 2-azide compounds. The application of the similar method to the preparation of 8-azido cytokinins from 6,8-diazido precursors was unsuccessful.[196] 8-Azido-BAP, another type of photolabile cytokinin and reportedly more active in a tobacco callus bioassay than the 8-unsubstituted parent compound,[196] was synthesized in a displacement reaction of 8-bromoadenosine by N_3^- followed by benzylation at N-1, rearrangement, and hydrolytic cleavage of the N^9-glycoside bond (Scheme 8).[197] An alternative method reported below utilizes 6,8-dichloropurine,[198] which is first reacted with benzylamine or 3-methyl-2-butenylamine. The N^6-substituted adenine derivative obtained is the n protected by ethoxyethyl at the 9-position and the displacement of the 8-chloro group is effected by treatment with triethylammonium azide.[196] The ethoxyethyl group is readily removed in dilute acid and thus can be utilized as a protecting group in cases where the reactions are hampered by the presence of the acidic =NH proton.[199] When 2-azido-iP was photolysed in the presence of a cytokinin-binding protein isolated from wheat germ,[200] its attachment effectively blocked the binding of (^{14}C)-kinetin.[196] The azido compound is thought to bind to the protein covalently via a nitrene intermediate, $R-N_3 \xrightarrow{h} R-\overline{\underline{N}}$ (or $R-\overline{\underline{\dot{N}}}\cdot$) + N_2.

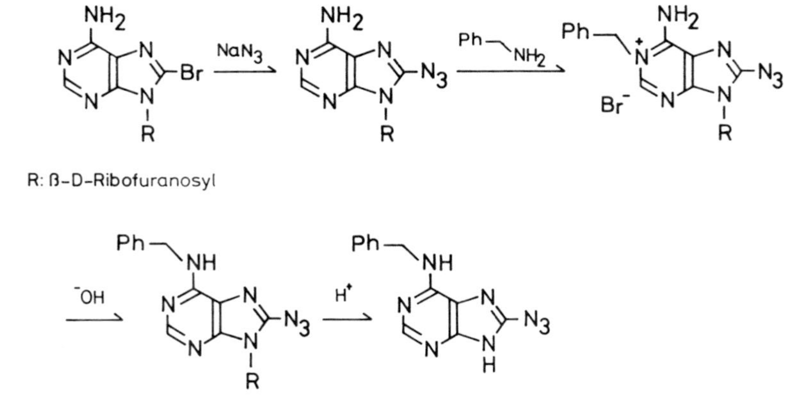

SCHEME 8.

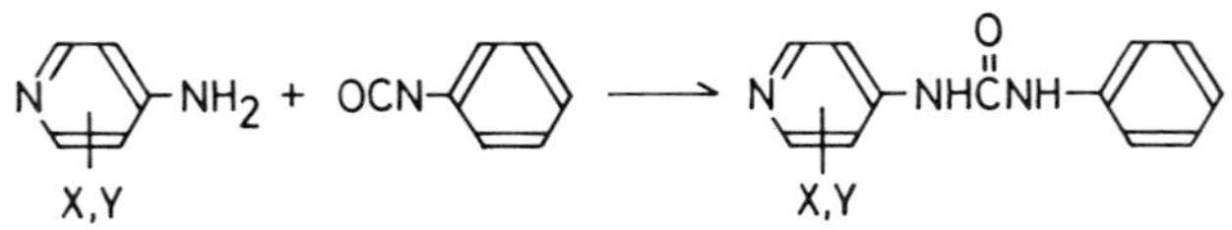

SCHEME 9.

Radioisotopically labeled cytokinins possess wide utility as a tool for physiological and biochemical studies. A few representatives of the N^6-substituted adenylate cytokinins labeled at the 8-position by ^{14}C are commercially available. Those having a label in the side chain moiety, which are not commercially available, can be prepared by either of the methods described above (Schemes 1 and 2) using a labeled precursor. Tritium labeling at the 8-position of BAP derivatives and Z has been performed by a catalytic exchange reaction in ^3H_2O solution to give (8-^3H) compounds.[201—203] BAP labeled with ^3H in the methylene moiety of the side chain has been prepared by lithium aluminun tritiide reduction of benzoylaminopurine,[204] and by the coupling of 6-chloropurine with labeled benzylamine prepared from benzoic acid.[205] Catalytic dehalogenation of 6-(p-bromobenzyl)adenine[206] and 6-(m-iodobenzyl)adenine[207] with tritium gas gave p-(^3H)- and m-(^3H)BAPs, respectively. The latter method appears best for obtaining compounds having high specific activity.

A structurally very different class of compounds, N,N'-diphenylureas, exerts cytokinin activity. They are generally obtained by reaction of an amine with an isocyanate.[108] The compounds in which one of the N-substituents is replaced by pyridyl were prepared similarly (Scheme 9)[208,209] or, in the case of unsubstituted 4-pyridyl, by the reaction of isonicotinoyl azide with an aniline.[209] Although the activity of N,N'-diphenylureas is generally low in the tobacco pith bioassay,[208] the compounds having 2,6-dichloro- and 2,6-dibromo-4-pyridyl groups in place of one of the phenyl groups have very high activity in the tobacco callus assay.[210] Phenyl- and alkylureidopurines, intermediate in structure between N^6-substituted adenines and N,N'-diphenylureas, were also prepared from an appropriate isocyanate and a 9-protected or unprotected adenine derivative.[199,211]

Compounds in which the heterocyclic moiety of N^6-substituted adenines is altered have been targets of syntheses to obtain both more active compounds and more information on the structural requirements for the appearance of activity.[212—214] 8-Aza-kinetin, the oldest one of this kind, has a weak cytokinin activity and is prepared from the pyrimidine precursor via cyclization and alkylation (Scheme 10).[215] The 1-deaza and 3-deaza analogues[213] have been similarly synthesized from the pyrimidine precursor (Scheme 11).[216]

Modification of the imidazole moiety of cytokinins has led to the development of novel

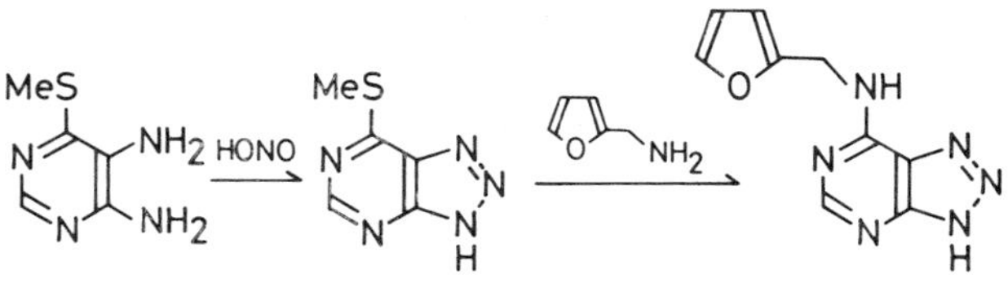

SCHEME 10.

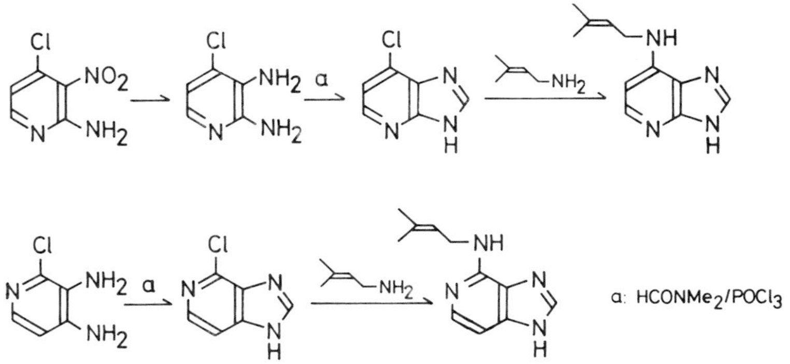

SCHEME 11.

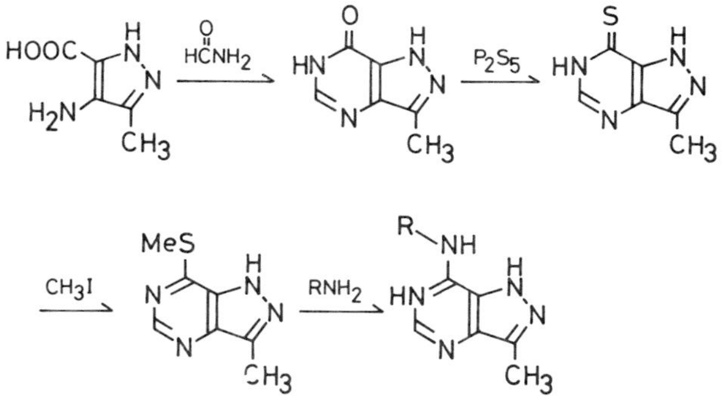

SCHEME 12.

anticytokinin compounds. A class of anticytokinins having 7-substituted 3-methylpryrazolo [4,3-d]pyrimidine structure have been synthesized, similarly to the preparation of N^6-substituted adenines from a 6-methylthio precursor, by the nucleophilic attack of amines on 3-methyl-7-methylthiopyrazolo[4,3-d]pyrimidine,[217] the immediate precursor (Scheme 12).[218] 4-Substituted 2-methylthio- and 4-substituted 2-methylpyrrolo [2,3-d] pyrimidines are strong anticytokinins and may be useful as a tool for investigating the physiological and biochemical roles of cytokinins in plant tissue. They are prepared rather easily from commercially available, simple precursors (Scheme 13).[219,220] 4-Substituted 2-methylthiopyrido[2,3-d]pyrimidines are another class of anticytokinins having an unique, fused 6-6 membered ring system synthesized from a pyridine precursor (Scheme 13).[221]

Cytokinin glucosides have been identified as the dominant metabolites of cytokinins when fed to a number of plant tissues. The structures of the major metabolites of Z and BAP were confirmed unequivocally by synthesis to be 7- and 9-β D-glucopyranosides. The latter compound was prepared by conventional procedures for nucleoside synthesis, i.e., by the con-

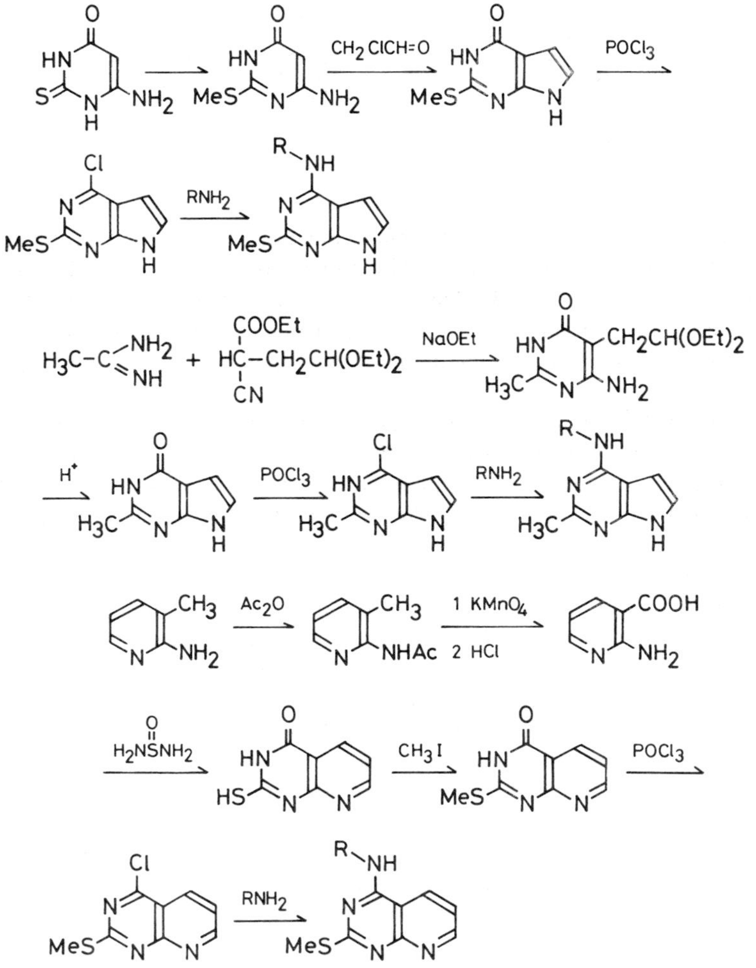

SCHEME 13.

densation of chloromercuripurine with acetobromoglucose followed by deacetylation of the product and treatment with amines (Scheme 14).[103] Of the few possible routes available for the preparation of the rather unusual 7-glucosides, the most successful was the method which utilizes the ring closure of an imidazole derivative to produce directly and unequivocally the 7-substituted glucosides (Scheme 15).[103] The β-configuration of the anomeric carbon atom of the major product in the reaction was substantiated by the H1'-H2' *trans* diaxial coupling constant (J ≃ 9Hz) in the PMR spectra. *O*-β-D-Glucopyranosylzeatin, (OG)Z, is another glucose conjugate isolated from lupin seedlings fed with zeatin. Attempts to couple acetobromoglucose directly with Z under a variety of conditions were reported to be unsuccessful, probably because of the low solubility of zeatin in solvents and the ease of formation of *N*-glucosyl products. The synthesis was achieved by introducing an *O*-glucosyl linkage to the side chain moiety prior to its attachment to the purine nucleus, i.e., the phthaloyl-protected amine was reacted with acetobromoglucose to give glucosylated phthaloylamine which, after deblocking by methanolic ammonia, was condensed with 6-chloropurine (Scheme 16).[108] L-Lupinic acid, L-2-(6-((*E*)-4-hydroxy-3-methylbut-2-enylamino)purin-9-yl)alanine, is an amino acid conjugate of Z, the structure of which has been confirmed by synthesis. To generate the β-alanyl side chain, 2-trifluoroacetylaminoprop-2-enoate was

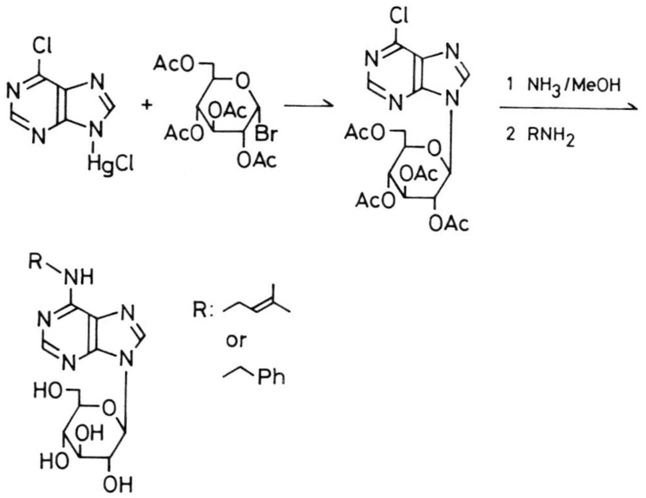

SCHEME 14.

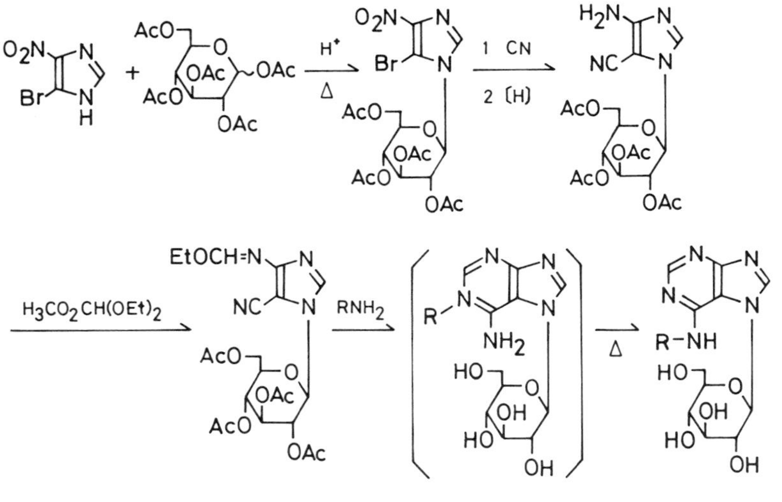

SCHEME 15.

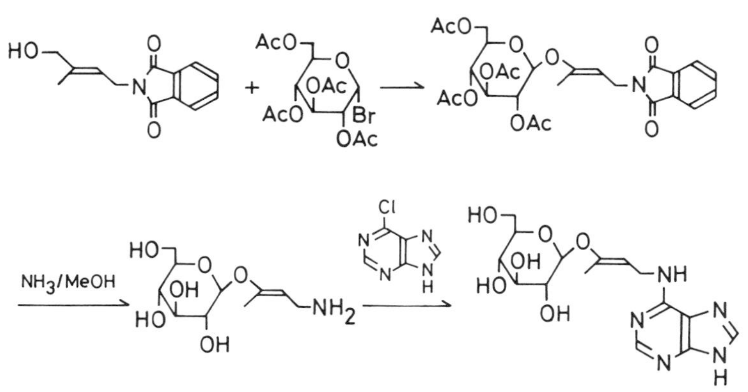

SCHEME 16.

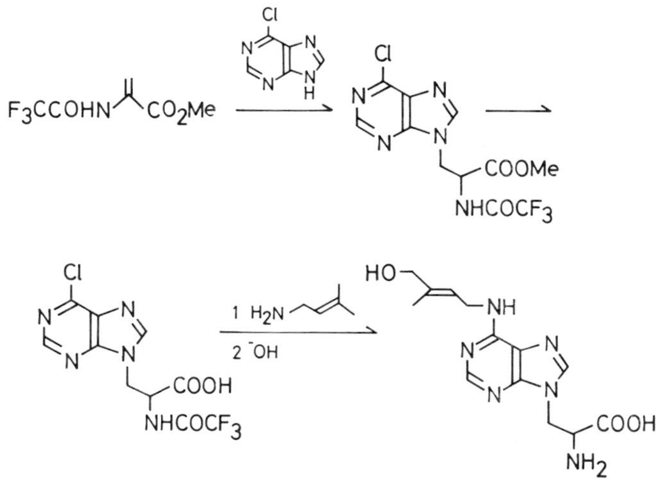

SCHEME 17.

chosen as a precursor which could be prepared from 2-chloroalanine methyl ester by *N*-trifluoroacetylation, followed by dehydrochlorination. Michael condensation of this compound with 6-chloropurine afforded the 6-chloropurin-9-ylalanine derivative. After selective hydrolysis of the methyl ester, it was reacted with the side chain amine, and treatment of the product with a base to remove the trifluoroacetyl-protecting group gave (±)-lupinic acid (Scheme 17).[108] Direct Michael addition to *O*-protected zeatin resulted in a poor yield. Discadenine (3-(3-amino-3-carboxypropyl)-6-(3-methyl-2-butenylamino)purine), a self-germination inhibitor isolated from spores of a cellular slime mold, is also an amino acid conjugate of the cytokinin iP. The synthesis of (±)-compound was achieved by two methods,[222] one of which, utilizing the direct alkylation of the cytokinin by α-phthalimido-γ-bromobutylate followed by deblocking, appears superior in both yield and in the number of steps involved (Scheme 18).

III. BIOSYNTHESIS AND METABOLISM

Recent advances in research on the biosynthesis and metabolism of cytokinins are reviewed by Sembdner et al.[223] and fully discussed by Letham and Palni.[119]

A. Biosynthesis of Free Cytokinins

It is generally accepted that cytokinins in higher plants are synthesized mainly in the root system and transported via the transpiration stream in xylem to the shoot where they regulate the development and senescence (for review, see References 13 and 223). The production of free cytokinins in plants during growth and development can be accounted for either by the turnover of cytokinin-containing tRNA, by *de novo* biosynthesis, or by a combination of both mechanisms.

Cytokinins occur in the tRNA of various plant species. The major cytokinin-active base in tRNA, [9R]iP, is formed by the transfer of an isopentenyl group from Δ^2-isopentenyl pyrophosphate (Δ^2-IPP) to the N^6 nitrogen atom of an unmodified adenine in the tRNA by enzyme systems (for a review, see References 13 and 117). The cytokinin-containing tRNA is therefore a potential source of free cytokinins, which could be released by the turnover of the tRNA.

Klemen and Klämbt[224] determined the half-life of the tRNA in roots of *Zea mays*. Having calculated the potential amount of cytokinins produced within this period, they concluded

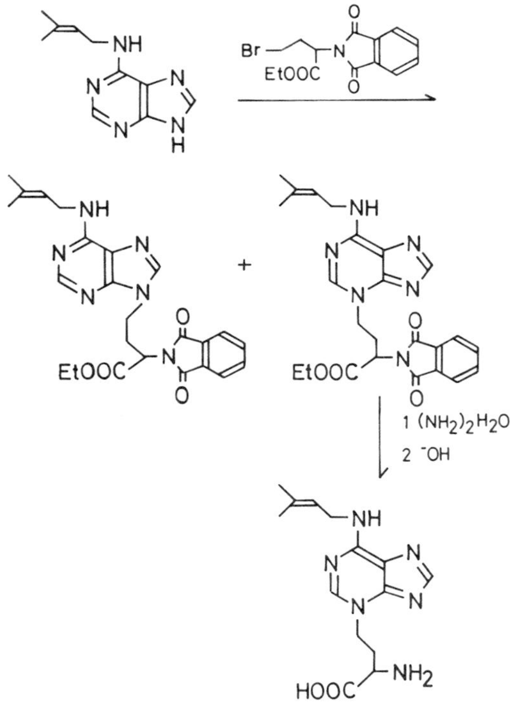

SCHEME 18.

that free cytokinins could be generated through the breakdown of the tRNAs. Subsequent experiments with roots of *Phaseolus vulgaris*[225] also supported the hypothesis that cytokinin-containing tRNA and oligonucleotides could be precursors of free cytokinins. More recently, the tRNA of *Vinca rosea* crown gall tissue has been shown to contain high levels of [9R]Z.[226] This was interpreted as evidence that biosynthesis of free [9R]Z in this tissue could result from an extremely rapid turnover of tRNA-containing [9R]Z. Abnormally rapid turnover of a subpopulation of tRNA has been reported for animal tumors.[227]

While free cytokinins might derive from tRNA, there is much evidence to support the possibility that free cytokinin is formed by *de novo* biosynthesis. When callus cells from moss were supplied with labeled adenine, they produced a labeled cytokinin which has the same chromatographic behavior as iP, but not labeled cytokinin could be isolated from tRNA.[228] In the cytokinin-autotrophic tobacco tissue, adenine and an adenosine analogue modified in the ribose moiety were converted to iP and its derivatives, while no analogue was incorporated into tRNA.[229] In vivo conversions of labeled adenine and adenosine into free cytokinins have also been shown to occur in synchronously dividing tobacco cells[230] and *V. rosea* crown gall tissue.[59,231] These findings suggested a direct biosynthetic pathway to free cytokinins.

Burrows[232] examined whether the cytokinins found in free form are also present in the tRNA of the same plant species. The tRNA isolated form *Populus robusta* leaves contained only (*cis*)[9R]Z, [2MeS 9R]Z, [9R]iP, and [2MeS 9R]iP as the cytokinin-active constituents. It is difficult to visualize a simple mechanism whereby the isopentenyl side chain of these could be converted into the N^6-2-hydroxybenzyl side chain in (*o*OH)[9R]BAP, which has been identified as the predominant free cytokinin in the leaves.[79] Further evidence for *de novo* biosynthesis of free cytokinins has been provided by the small amounts of cytokinins present in tRNA[233] and the low rates of tRNA turnover in many plant tissues.[234]

Direct evidence supporting an alternative pathway to tRNA turnover was demonstrated by the isopentenylation of AMP to a free cytokinin ribonucleotide and subsequent formation of iP in a crude enzyme system from *Dictyostelium discoideum*,[235] which contains a large quantity of free iP. The conversion of AMP to free cytokinins has also been shown to occur in cell-free systems from cytokinin-autotrophic tobacco tissue[236] and cultured tobacco cells.[237] An enzyme called Δ^2-isopentenylpyrophosphate:5′-AMP-Δ^2-isopentenyltransferase (isopentenyltransferase) was partially purified from cytokinin-autotrophic tobacco tissues.[238] This enzyme catalyzed the synthesis of [9R-5′P]iP from Δ^2-IPP and AMP. These experiments with cell-free systems also suggest that cytokinin nucleotides are important intermediates of cytokinin biosynthesis.

In feeding epxeriments using *V. rosea* crown gall tumor tissue,[231] [^{14}C]adenine was incorporated into Z, [9R]Z, [9R-5′P]Z, (OG)Z, and (OG)[9R]Z. More than 90% of the supplied adenine was taken up during the first few hours, and most of this was converted into adenine nucleotides.[231] This occurred probably by the action of adenine phosphoribosyltransferase, as this enzyme catalyzes the first step in the utilization of free purines by most tissues. The maximum radioactivity in cytokinins was found after incubation for 8 hr,[231] and the radioactivity in [9R-5′P]Z was considerably higher than in other Z derivatives.[59] This suggests that cytokinin biosynthesis in vivo also occurs primarily at the mononucleotide level.

Experiments on cytokinin biosynthesis in an in vitro system[235-237] suggested that iP type cytokinins are probably the initial products in the synthesis of Z type cytokinins. However, no incorporation of label from adenine into iP, [9R]iP, or [9R-5′P]iP was detected in *V. rosea* crown gall tissue.[59,213] A very rapid conversion of iP to Z and its derivatives was recently observed in this tissue.[239] Therefore, the steady-state concentration of [9R-5′P]iP or its derivatives may be below the limits of detection. From these results, Palni et al.[59] concluded that the attachment of the isoprenoid side chain occurs at the 5′-AMP level to form [9R-5′P]iP, and Z and [9R]Z are produced subsequently by the action of specific enzymes. This pathway is also supported by the results obtained with cell-free systems;[235,236,238] however, the possibility of side-chain transfer at the base or ribonucleoside level cannot be excluded.

Barnes et al.[240] demonstrated that cytokinins liberated in potato cell culture by the catabolism of tRNA can account for more than 40% of the free cytokinins. Watanabe et al.[21] identified the major free cytokinin in cones of the hop plant as (*cis*)[9R]Z. This cytokinin, known to be a common constituent of plant tRNA,[234] was suspected to be derived from the tRNA turnover. Based on quantitative comparison of free and tRNA-bound cytokinins, it was suggested that (*cis*)[9R]Z and [2MeS 9R]Z might be derived from tRNA turnover, while [9R]Z might be formed independently of the tRNA turnover.[55] These results indicate that endogenous free cytokinins in plants can be synthesized by both pathways. Biosynthesis of (*o*OH)[9R]BAP is unknown.

B. Metabolism of Natural Cytokinins

The metabolism of exogenously applied cytokinins has been studied extensively in intact plants, detached organs, and cultured cells. The schematic pathway of cytokinin biosynthesis and metabolism presented by Letham et al.[241] is shown in Figure 2 with minor modifications. iP- and/or Z-type cytokinins are converted to a diversity of metabolites by one or more of the following basic reactions:[241] (1) *trans*-hydroxylation of the terminal methyl group on the side chain of iP-type cytokinins, (2) side chain reduction of Z-type cytokinins, (3) isoprenoid side chain cleavage, (4) *O*-glucosylation, (5) *N*-glucosylation, (6) ring substitution by alanine moiety, and (7) base-ribonucleoside-ribonucleotide interconversions.

The first reaction, the stereospecific hydroxylation of iP-type cytokinins to Z-type cytokinins, has been found in *Rhizopogon roseolus*,[242,243] *Z. mays* endosperm,[243] tobacco tissue cultures,[238] and *V. rosea* crown gall tissue.[239] The enzyme system involved in this reaction has been found to be present in microsomes. The microsomal enzymes prepared from tobacco tissue cultures and wheat germ have converted iP and [9R]iP to Z and [9R]Z, respectively,

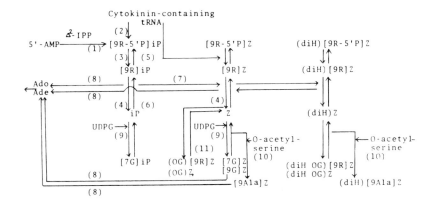

FIGURE 2. Schematic pathway of the biosynthesis and metabolism of natural cytokinins (according to Letham et al.[241] with minor modifications). (1) Δ^2-IPP:5'-AMP-Δ^2-isopentenyltransferase, (2) ribonucleases, (3) 5'-nucleotidase, (4) adenosine nucleosidase, (5) adenosine kinase, (6) adenosine phosphorylase, (6-5) adenine phosphoribosyltransferase, (7) microsomal mixed function oxidases and other enzymes, (8) cytokinin oxidase, (9) cytokinin 7-glucosyltransferase, (10) β-(9-cytokinin)-alanine synthase, (11) β-glucosidase. Abbreviations: Ade, adenine; Ado, adenosine; 5'-AMP, adenosine-5'-monophosphate; Δ^2-IPP, Δ^2-isopentenylpyrophosphate; UDPG, uridine diphosphate glucose. For other abbreviations, see Figure 1.

in the presence of NADPH.[238] This indicates that microsomal mixed function oxidases are possibly involved in these steps.

By the second reaction, the double bond in the side chain of Z-type cytokinins is stereospecifically hydrogenated to yield (diH)Z and its riboside and ribotide. Incorporation of a label from Z into (diH)Z and its derivatives has been demonstrated in lupin seedlings[244] and immature seeds,[245] Z. mays kernels,[245] immature apple seeds,[245] bean (P. vulgaris) plants,[246] poplar (P. alba) leaves,[107] and detached alder (Alnus glutinosa) leaves.[247] The enzyme systems are unknown.

In the soybean callus assay, (diH)Z was found to be more active than Z,[247] while in other assay systems the result is reversed. This difference has been explained in terms of the relative stabilities of the two cytokinins to cytokinin oxidase (see next reaction): (diH)Z is resistant to N^6-side chain cleavage by cytokinin oxidase, while Z is readily oxidized to inactive metabolites.[248] Hence, inactivation by the oxidase may restrict the level of Z in the soybean tissue, but not that of (diH)Z.[119,247] Metabolic studies of (diH)Z in bean plants suggest that a Z—(diH)Z interconversion mechanism in different parts of the plant effectively maintain a decreasing concentration gradient of Z derivatives and an increasing gradient of (diH)Z derivatives from the roots to the laminae.[73]

The third reaction, the N^6-side chain cleavage of cytokinins, leads to the irreversible destruction of cytokinin activity. Enzyme activity catalyzing the side chain cleavage of iP-type cytokinins was demonstrated in tobacco cells,[141,249] and callus.[141] An enzyme (mol wt 88,000) was partially purified from cultured tobacco tissue[250] and Z. mays kernels.[251] Its reaction requires molecular oxygen[251] and liberates 3-methyl-2-butenal from the side-chain of iP,[252] and the enzyme has been thus called cytokinin oxidase.[251] Cytokinin oxidase utilizes not only iP and [9R]iP but also Z and [9R]Z as substrates to give adenine (Ade) and adenosine (Ado).[251,253] However, the saturation of the isopentenyl side chain, the relocation of the double bond from Δ^2 to Δ^3, or the substitution of other functional groups all reduced or completely eliminated reactivity with the enzyme.[251]

Cytokinin oxidase has recently been partially purified from V. rosea crown gall tissue and compared with the enzyme from mature Z. mays kernels.[248] The molecular weights of the two enzymes were very different, 94,400 for Z. mays and 25,100 for V. rosea, but both enzymes showed similar specificity for cytokinin substrates.

Side chain cleavage is the major metabolic fate of exogenously applied cytokinins in *Z. mays* kernels,[66,245] root and de-rooted seedlings of *Z. mays*,[254] detached bean leaves,[255] and *V. rosea* crown gall tissue.[256] This indicates that cytokinin oxidase may play an important role in lowering cytokinin activity levels in plant tissues. Synthetic adenine derivatives with an aromatic substituent at position N^6 of the purine ring, such as kinetin and BAP, are resistant to cytokinin oxidase.[251] However, both cytokinins are degraded by the side chain cleavage in plant tissues,[13,119] indicating the presence of a further enzyme different from cytokinin oxidase.

The most abundant cytokinins in *V. rosea* crown gall tissue[62] and *Z. mays* kernels[66] are Z and [9R]Z, which are readily decomposed by cytokinin oxidase. McGaw and Horgan[248] suggest that compartmentation prevents the cytokinins from coming into contact with the oxidase system.

O-Glucosylation, the fourth reaction, of Z forms *O*-glucosides in which a glucosyl moiety is conjugated in β-configuration to oxygen on the isoprenoid side chain. The enzyme systems are uncharacterized. The *O*-glucosylation process has been substantiated in various plants, e.g., immature apple seeds,[245] de-rooted lupin seedling,[108,244,257] immature lupin seeds,[245] poplar (*P. alba*) leaves,[107,109] and alder (*A. glutinosa*) root nodules.[258] In these plants exogenously applied Z is converted into one or more of the following metabolites: (OG)Z, (OG)[9R]Z, (diH OG)Z, and (diH OG)[9R]Z.

O-Glucosides have also been found to be major endogenous cytokinins in *P. vulgaris*,[246] *V. rosea* crown gall tissue,[62] and *Z. mays* kernels,[66] as shown in Table 1, and appear to occur widely as endogenous cytokinins in plants.

Although (OG)Z and its riboside are resistant to cytokinin oxidase,[248,251] the *O*-glucosides of Z, (diH)Z, and their ribosides are easily hydrolyzed by β-glucosidase with the release of the free cytokinin.[107,108,257] These findings, together with analyses of the accumulation and utilization of the *O*-glucosides at different stages of plant development,[259] provide evidence that the O-glucosides are cytokinin-storage forms which release free cytokinins when required.[119,241,248]

In cytokinin bioassays, (OG)Z and (OG)[9R]Z exhibited activity similar to Z and [9R]Z, respectively.[260] The enhanced activity of the *O*-glucosides has been explained as follows: the *O*-glucosyl moiety renders the side chain resistant to cleavage by cytokinin oxidase during movement of the cytokinin to the subcellular site where hydrolysis of the glucosyl residue releases Z, an active form.

N-Glucosylation of cytokinin base has been demonstrated in experiments on the metabolism of labeled Z in a number of plants, which revealed that cytokinin is converted to *N*-glucosides in which the glucose moiety may be linked to nitrogen atoms 7 or 9 of the adenine ring. The major metabolite of Z in de-rooted radish seedlings[130] and radish roots[261] was [7G]Z, while that in roots of *Z. mays* was [9R]Z.[254,262] The occurrence of endogenous 7- and 9-glucosides has been reported, as shown in Table 1.

Two enzymes capable of forming [7G]Z and traces of [9R]Z from Z have been extracted from radish cotyledons.[245,263] The enzymes, named cytokinin-7-glucosyltransferase, use UDP-glucose as glucose donor. The transferases exhibit a strong preference for adenine derivatives with a N^6 side chain of at least three carbon atoms. One of the transferases converted Z to [7G]Z plus [9G]Z (ratio [7G]Z/[9G]Z, 10.5), and kinetic studies established that UDP-glucose and Z bound to the enzyme to form a ternary complex from which the products, [7G]Z and UDP, were sequentially released.[241,263]

Many metabolic studies have indicated that 7- and 9-glucosides are metabolically stable.[66,130,241,244,245] However, cytokinin oxidase degrades [7G]Z and [9G]Z.[248] Hence the metabolic stabilty of [7G]Z in plant tissues may be due to compartmentation or to the absence or low levels of cytokinin oxidase.[248]

The sixth reaction is ring substitution by an alanine moiety. Exogenously applied Z is metabolized in de-rooted lupin seedlings to a cytokinin-amino acid conjugate termed lupinic acid, [9Ala]Z[244] (see Table 1 and Figure 1). Dihydrolupinic acid (diH)[9Ala]Z, has also been found in seedlings, while [9Ala]Z is formed in immature apple seeds.[245] Isotope dilution methods have revealed [9Ala]Z and (diH)[9Ala]Z to be endogenous metabolites of cytokinins in lupin pod walls and seeds.[27] These two conjugates represent a new type of plant product in which an amino acid moiety is conjugated to a purine ring nitrogen atom. Another natural compound of this type is discadenine[120] produced by the slime mold *D. discoideum.* The formation of [9Ala]Z from Z was demonstrated in a crude enzyme preparation derived from immature *L. luteus* seeds,[265] and the enzyme catalyzing the conversion has since been purified from the same seeds.[266] The enzyme (mol wt 64,500 ± 3000) uses *O*-acetylserine as donor of the alanine moiety and has tentatively been called β-(9-cytokinin)alanine synthase, β-(6-alkylamino-purin-9-yl)alanine synthase, or lupinic acid synthase. It requires an adenine ring for activity and shows optimum activity when the adenine is substituted with approximately five carbon atoms with an olefinic linkage at N^6. This requirement is similar to that of 7-glucosyltransferase.[249,266]

9-Alanyl substitution, like 7- and 9-glucosyl substitution, renders cytokinins resistant to degradation in plant tissues.[66,241,244] Cytokinin alanine conjugates and 7- and 9-glucosides are very weakly active in cytokinin bioassays relative to Z.[260] This low cytokinin activity of conjugates and *N*-glucosides is probably due to their inability to cleave the substituents readily and release the free cytokinin bases.[260] In this way, cytokinin oxidase, 7-glucosyltransferase and β-(9-cytokinin)alanine synthase may provide alternative processes for reducing levels of cytokinin activity.[119]

The final basic reaction comprises cytokinin ribonucleotide—ribonucleoside—base interconversions. Incorporation of labeled cytokinin bases into ribonucleosides (ribosides) and riboside 5′-phosphates (ribotides) has been demonstrated in ash embryos,[267] de-rooted lupin seedlings[244] and stems of the decapitated plants,[73] de-rooted radish seedlings[130] *Z. mays* roots[254] and tobacco cells.[141] Exogenously applied cytokinin ribosides are also converted into bases and ribotides in tobacco cells[133,141] and *Acer* cells.[133] These findings suggest that the free base, riboside, and ribotide forms of cytokinins are interconvertible in plant tissues. The interconversions may be caused by such enzymes as adenosine phosphorylase[268] (in *Escherichia coli*), adenosine kinase[269] (in *Schizosaccharomyces pombe* and *Acer pseudoplatanus*), 5′-nucleotidase[270] (5′-ribonucleotide phosphohydrolase), adenosine nucleosidase[271] (adenosine ribohydrolrase), and adenine phosphoribosyltransferase[272] (AMP:pyrophosphate phosphoribosyltransferase, in *Triticum aestivum* germ) (see Figure 2).

When cytokinin bases are supplied exogenously to tobacco cells, the principal metabolites formed initially are cytokinin ribotides. These metabolites accumulate in the cells which are impermeable to them.[273] The ribotide formation appears to be associated with cytokinin uptake and transport across cell membranes.[119] A number of observations suggest that ribosides are translocation forms of cytokinins in the xylem,[13] and that the free base is one of the active forms of cytokinins.[119] Thus, the enzymic regulation of the interconversion of cytokinin ribotides, ribosides, and bases may play a significant role in maintaining adequate levels of active cytokinin in plant cells.[270—273]

IV. BIOLOGICAL ACTIVITY

The range of biological activity of cytokinins has been extensively examined by a number of plant biologists. Included are cell-division promoting activity in tissue culture systems, the promotion of seed germination, the maintenance of chlorophyll in excised leaves, the promotion of leaf and cotyledon growth, the control of apical dominance, and effects on the morphology of cultured tissues. Many other effects have been reported in a variety of experimental systems. Like other plant hormones, cytokinins are utilized as tools for investigating various aspects of plant growth. The purpose of this section is to describe the chemical aspects rather than the physiological roles of these cytokinin activities. To know

Table 2
EFFECTS OF N^6-n–ALKYLADENINES ON TOBACCO CALLUS GROWTH[123]

	Conc. (μM) required for	
N^6-Substituent	Detectable response	Max. yield
CH_3	100	12.5
C_2H_5	0.5	0.5
C_3H_7	0.02	0.1—0.5
C_4H_9	0.004	0.1
C_5H_{11}	0.0008	0.3
C_6H_{13}	0.2	25
C_7H_{15}	0.1	
$C_{10}H_{21}$	25—100	

the structural factors or requirements which confer such activities is a major target of chemical study. Structure-activity relationship studies have been confined to cultured cell systems because of their high sensitivity and their reliability. Successful results obtained in these systems will be extended to interpretations of the molecular mode of their action in other plant systems, and the approaches explored are considered prototypes.

Since the discovery of kinetin, huge numbers of its analogues, N^6-substituted adenines, have been prepared and tested for activity. Various degrees of effects caused by the modification of the N^6-side chain as well as that caused by the introduction of substituents into other positions on the adenine ring have been summarized in the literature.[123,274] The best-recognized feature is that the activity of structurally simple N^6-n-alkyladenines varies with the length of the side chain in several bioassay systems.[274] Table 2 shows, in the results of the tobacco callus bioassay, that there is an optimum chain length of four or five carbon atoms. Structural factors which govern activity may be steric or hydrophobic; there may be an optimum steric condition for the accomodation of the compounds to the receptor cavity or there may be an optimum hydrophobic-hydrophilic balance in the transport processes to the target site. Further modification, however, makes the situation much more complex, and straightforward interpretations of the structure-activity profile become difficult.

The highest active members of the class, Z, iP, kinetin, and BAP, possess a double bond allylic to the N^6 atom. There is a methyl at the 3-position in the side chains of the first two of these compounds. Among the compounds having a side chain of three to five carbon atoms, introduction of an allylic double bond or a methyl group at the 3-position seems to enhance activity; for example, N^6-allyladenine and iP are more active than N^6-n-propyl and N^6-butyl derivatives, respectively[123] (Tables 2 and 3). Hydroxylation of the *trans*-methyl of the side chain of iP gives the more active Z, and that of *cis*-methyl produces the less active *cis*-Z.[183] In the N^6-(X-hydroxy-3-methyl)adenine series, where X = 2, 3, and 4, the 4-hydroxy compound is higher but the 2- and 3-hydroxy and 2,3- and 3,4-dihydroxy derivatives are weaker in activity than the parent unsubstituted compound.[275] Replacement of methylene groups of the side chain by other atoms or groups like O, S, CO, or NH modifies activity to various degrees. The activity of N^6-acyladenines appears to be a function of the length of the N^6-substituents, as observed in N^6-n-alkyladenines, in both the tobacco and the soybean callus bioassays.[276,277] Introduction of a substituent into the benzene ring of N^6-benzyladenine lowers the activity in the radish leaf expansion test[278] and in the tobacco callus growth assay.[279]

To explain these complex effects of structural modification on cytokinin activity, a quantitative approach has been adopted,[279] where the substituent effects are divided into three classes: hydrophobic, electronic, and steric factors. It has been well-documented that biological response is a combined reflection of these three effects. In the case of cytokinins

Table 3
COMPARISON OF CYTOKININ ACTIVITY OF SOME N^6-SUBSTITUTED ADENINE DERIVATIVES IN THE TOBACCO CALLUS GROWTH TEST

Compound	Conc. (μM) required for max. growth
Kinetin	0.1[123]
N^6-Benzyladenine (BAP)	0.07[123]
N^6-Allyladenine	5[123]
N^6-(3-Methyl-2-butenyl)adenine(iP)	0.02[123]
N^6-Isopentyladenine	0.1[123]
trans-Zeatin(Z)	0.004—0.5[123]
cis-Zeatin(cis-Z)	0.5[183]

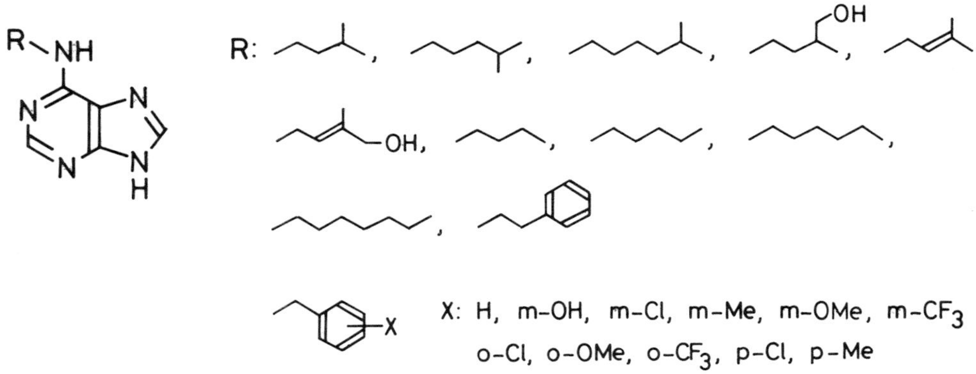

X: H, m-OH, m-Cl, m-Me, m-OMe, m-CF$_3$
o-Cl, o-OMe, o-CF$_3$, p-Cl, p-Me

FIGURE 3. Structure of N^6-alkyl and N^6-benzyladenines.

having N^6-alkyl and N^6-benzyl side chains such as those shown in Figure 3, activity is correlated in a quantitative manner with the combination of steric and electronic factors as shown by Equation 1 below, where log (1/E$_{50}$) is the logarithm of the concentration (× 10^{-7} M) of a compound at which the 50% callus yield of the maximum response is given in the tobacco callus bioassay, n is the number of the compounds analyzed, r is the multiple correlation coefficient, and s is the standard deviation. The independent variable W$_{max}$, in the correlation equation expresses the maximum width in Angstroms of the N^6-substituents from the bond axis (L) that connects the N^6 atom with its α-carbon atom. The W$_{o,m}$ term expresses

$$\log(1/E_{50}) = -0.32W_{max}^2 + 3.35W_{max} - 0.65W_{o,m} + 2.03\sigma^* - 8.05$$

$$n = 22, r = 0.85, s = 0.26 \tag{1}$$

the width of the o- and m-substituents on the benzene ring of the benzyl side chain whose steric bulk is not reflected in the total width parameter, W$_{max}$, as depicted in Figure 4. The significance of the W$_{max}^2$ term with a negative coefficient is that there is an optimum steric condition for activity in terms of the W$_{max}$; its parabolic relation to activity is shown in Figure 5. The optimum value is calculated to be about 5.2 Å and explains the higher activity of Z and iP as well as that of N^6-n-amyladenine in the n-alkyl series of compounds, their W$_{max}$ values being closer to the optimum. The W$_{max}$ value (6.02 Å) of even the unsubstituted

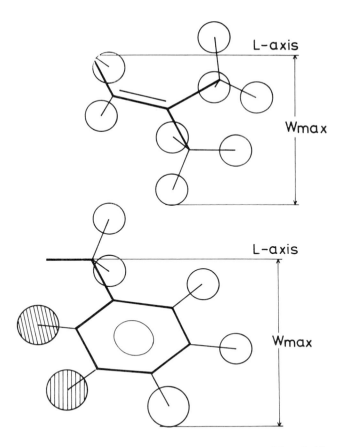

FIGURE 4. Schematic representation along the L-axis of 3-methyl-2-butenyl (upper) and benzyl (lower) substituents showing the W_{max} parameter. Maximum width of the hatched substituents on the benzene ring is expressed by $W_{o,m}$. (From Iwamura, H., Fujita, T., Koyama, S., Koshimizu, K., and Kumazawa, Z., *Phytochemistry*, 19, 1309, 1980. With permission.)

benzyladenine is larger than the optimum and, as indicated by the $W_{o,m}$ term with the negative sign, bulkiness, and thus, the existence of a substituent at *o*- or *m*-position is detrimental to the activity. These facts explain why any compound having a substituent at any position of the benzene ring exhibits lower activity than the unsubstituted compound.[279] The positive sign of the electronic σ^* term indicates that the electron-withdrawing effect of the side chain enhances binding to the receptor, probably by hydrogen-bonding interaction with a basic site on the receptor surface via an N^6H group.

Modification of the purine moiety alters the activity further (Table 4). Replacement of the carbon atom at the 8-position by a nitrogen atom in kinetin, iP, and BAP gives their respective 8-aza analogues with activity drastically lowered to less than 10% of the corresponding adenine derivatives.[123] 1-Deaza analogues of kinetin and iP, where the nitrogen atom at the 1-position is replaced by carbon atom, are 15- and 2-fold less active, and 3-deaza compounds, where the nitrogen atom at the 3-position is replaced by a carbon atom, are 2000- and 1000-fold less active than their parent compounds.[213] The activity of doubly modified 8-aza-1-deaza analogues is as high as or slightly less than that of the corresponding adenine derivatives, whereas that of the 8-aza-3-deaza analogues is significantly lower. These modficiations alter the steric dimension of the compounds very little, so that the change in activity must be attributable to changes in the hydrophobic or electronic properties.

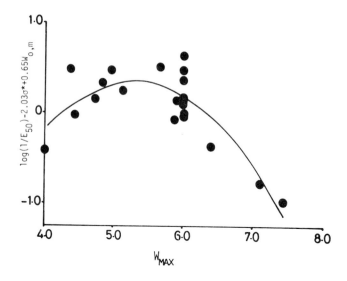

FIGURE 5. Parabolic relation of the cytokinin activity of N^6-substituted adenine to W_{max} as expressed by Equation 1. (From Iwamura, H., Fujita, T., Koyama, S., Koshimizu, K., and Kumazawa, Z., *Phytochemistry*, 19, 1309, 1980. With permission.)

Table 4
ACTIVITY OF AZA AND DEAZA ANALOGUES OF N^6-SUBSTITUTED ADENINES ON TOBACCO CALLUS GROWTH

Compound	Conc.(μM) required for max. growth
8-Aza-kinetin	2^{123}
8-Aza-BAP	0.5^{123}
1-Deaza-kinetin	0.2^{213}
1-Deaza-iP	0.02^{213}
1-Deaza-benzoyladenine	0.01^{280}
3-Deaza-kinetin	Not detectable[213]
3-Deaza-iP	20^{213}
8-Aza-1-deaza-kinetin	1^{213}
8-Aza-1-deaza-iP	0.02^{213}
8-Aza-3-deaza-kinetin	5^{213}
8-Aza-3-deaza-ip	1^{213}

6-Benzoylamino-1-deazapurine, where both the side chain and purine moiety are modified, is reportedly as active as Z, the most active member of the N^6-adenylate cytokinins, in the tobacco callus bioassay.[280]

The aza-deaza interconversion of the imidazole moiety of the purine ring has led to the development of novel, anticytokinin-active compounds, which has brought about a much better understanding of the molecular mechanism of cytokinin action. 7-Isopentylamino-3-methylpyrazolo[4,3-d]pyrimidine (Figure 6) completely inhibits the growth of tobacco callus

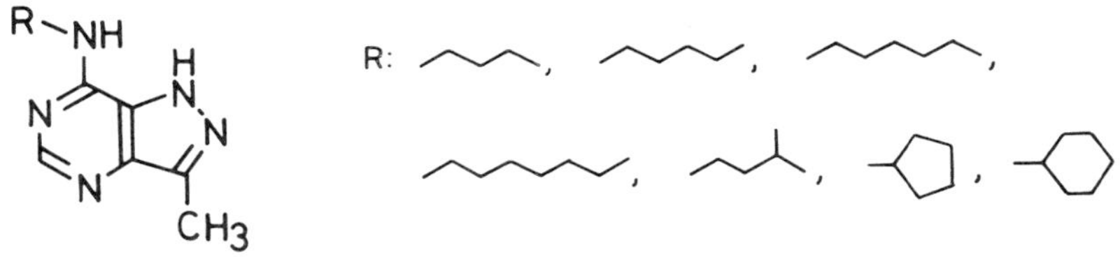

FIGURE 6. Structure of 7-substituted 3-methylpyrazolo[4,3-d]-pyrimidines.

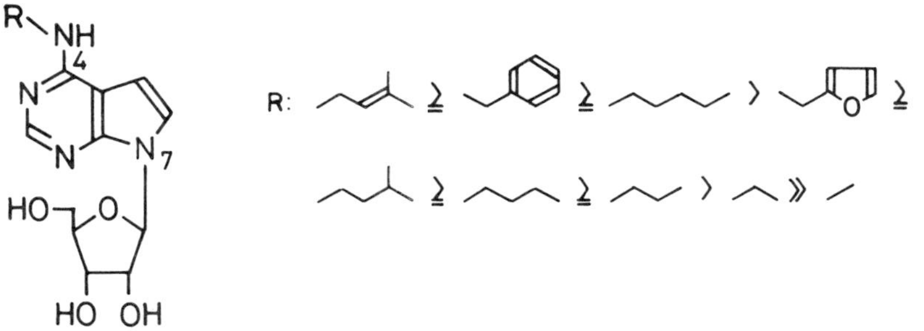

FIGURE 7. Structure of 4-substituted-7-(β-D-ribofuranosyl)pyrrolo[2,3-d]pyrimidine anticytokinins. R is arranged in decreasing order of anticytokinin activity.

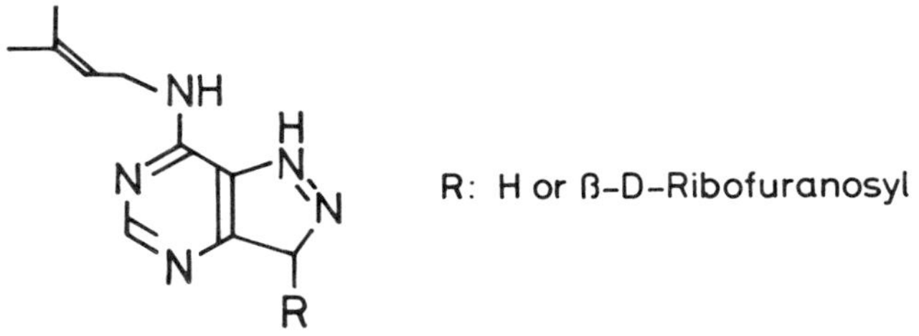

FIGURE 8. Structure of N^6-(3-methyl-2-butenyl)-8-aza-9-deazaadenine.

supplied with an optimal concentration of either iP or BAP when added to the tissue in 100- to 200-fold excess over the cytokinins.[281] Inhibitory activity is highest when the N^7-substituent is n-amyl. Activity of the compounds having a longer or shorter side chain is lower.[282] Some series of compounds having a substituted pyrrolo[2,3-d]pyrimidine(7-deazapurine) structure have also been shown to exhibit anticytokinin activity. N^4-substituted 4-amino-7-(β-D-ribofuranosyl)pyrrolo[2,3-d]pyrimidines (Figure 7) show anticytokinin activity which varies with the systematic transformation of the N^4-side chain in order of increasing cytokinin activity of the corresponding n^6-substituted adenines, i.e., 3-methyl-2-butenyl ≥ benzyl ≥ n-amyl > furfuryl ≥ isoamyl ≥ n-butyl ≥ n-propyl > ethyl >> methyl.[283] This is suggestive of a similar mode of interaction with the cytokinin receptor.

The corresponding deribosylated compounds, N^4-substituted 4-aminopyrrolo[2,3-d]pyrimidines, on the other hand, show cytokinin activity.[284] This indicates that this series of compounds exerts cytokinin activity without being ribosylated in vitro at the 7-position which is analogous to the 9-position in adenine. If glycosylation occurred at the 7-nitrogen atom, the resulting compounds should behave as anticytokinins. To examine the necessity or role of ribosyl moiety at the 9-position of exogenously added cytokinins, N^6-(3-methyl-2-butenyl)-8-aza-9-deazaadenine(7-isopentenylaminopyrazolo[4,3-d]pyrimidine) and its 9-β-D-ribofuranoside (Figure 8) have been prepared and their cytokinin activity has been

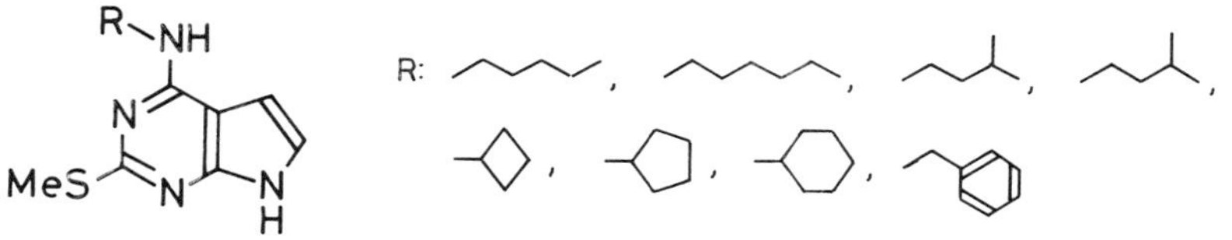

FIGURE 9. Structure of 4-substituted 2-methylthiopyrrolo[2,3-d]pyrimidine anticytokinins.

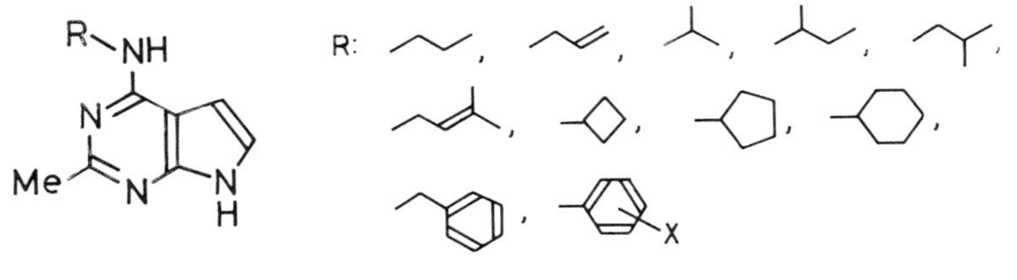

FIGURE 10. Structure of 4-substituted 2-methylpyrrolo[2,3-d]pyrimidine anticytokinins.

identified by tobacco callus assay.[218] Because these derivatives possess a carbon rather than a nitrogen atom at the 9-position, both ribosylation and deribosylation are unlikely. This result and that above together indicate that ribosylation is not a necessary condition for the appearance of cytokinin activity, although plant systems are quite capable of attaching or removing the sugar moiety.[234]

Other classes of anticytokinins having a pyrrolo[2,3-d]pyrimidine structure are the 4-substituted 2-methylthio-[219] and 4-substituted 2-methylpyrrolo[2,3-d]pyrimidines[220] (Figures 9 and 10). The activity of these two classes of compounds is conspicuous, and the highest retarding activity is afforded by N^4-cyclobutyl and N^4-cyclopentyl derivatives in both series.[219,220,285,286] The activity of the latter series of compounds varies continuously from anticytokinin to cytokinin with the transformation of the structure of N^4-side chain (see Figure 12, where the compounds having a side chain indicated by a open circle have cytokinin activity, whereas the others have anticytokinin activity.) The antagonistic nature of the compounds having growth-retarding effects was unequivocally established by the kinetic method of Lineweaver and Burk[286] Results for the N^4-cyclopentyl derivative are shown in Figure 11, where the reciprocal of the growth response is plotted against the reciprocal of the concentration of added kinetin. The fact that the resultant family of straight lines possesses a common intercept fulfills the requisite for competitive inhibition, and indicates that both anticytokinin- and cytokinin-active compounds interact with a common cytokinin receptor.

Structural effects on the extent of both the cytokinin-agonistic and antagonistic activities of the 2-methylpyrrolo[2,3-d]pyrimidine derivatives were analyzed quantitatively in terms of physicochemical properties to give a single correlation equation.[286] Figure 12 shows the results graphically, indicating that the extent of activity of either kind is parabolically related to the W_{max} value, the maximum width of the N^4-substituents, like that observed for the N^6-substituted adenines above (Figure 5). Moreover, these agonists and antagonists are distinguished by the W_{max} values, i.e., the value of the cytokinin-active compounds denoted by open circles in Figure 12 is in the range of 4.0 to 6.5 Å and the compounds having larger or smaller W_{max} values than this range are anticytokinins. That they are at the same time on the common parabolic curve indicates that the maximum width of the substituents is responsible not only for the extent of activity, i.e., for binding to the receptor, but also for the quality of activity, agonistic or antagonistic. These results coincide with and provide evidence for the hypothetical concept for hormonal action that agonist binding causes a conformational change of an otherwise inactive receptor to the active form, and that antag-

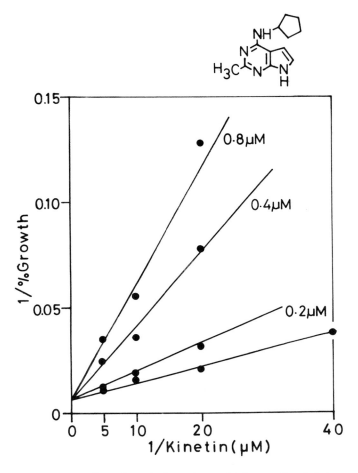

FIGURE 11. The reciprocal of growth rate of tobacco callus plotted as a function of the reciprocal of the concentration of kinetin alone (bottom line) and in the presence of 4-cyclopentylamino-2-methylpyrrolo[2,3-d]pyrimidine. (From Iwamura, H., Masuda, N., Koshimizu, K., and Matsubara, S., *J. Med. Chem.*, 26, 838, 1983. With permission.)

onists are species which bind similarly to the receptor but do not cause effective conformational change. In this case, compounds having W_{max} values out of the range 4.5 to 6.0 bind to the cytokinin receptor but do not make it active.

4-Substituted 2-methylthiopyrido[2,3-d]pyrimidines are anticytokinins having a unique, fused 6-6 membered ring system.[221] Their activity is also parabolically related to the W_{max}, as shown in Figure 13. The optimum value is about 4.6 and coincides well not only with that (5.2 Å) for the N^6-adenylate cytokinins but also with the value of about 4.7 Å calculated for the N^4-substituted 2-methylpyrrolo[2,3-d]pyrimidines above. That the optimum steric condition in terms of the W_{max} is nearly the same in all of these series of compounds, which have different structures but compete for the same receptor, provides us with an insight into the size of the receptor cavity into which compounds have to fit to be active.

N,N'-Diphenylureas, having what seems to be a very different structure from N^6-substituted adenines, exhibit cytokinin activity. Whether or not they can be classified as cytokinins has been a controversy among plant physiologists. To examine the correspondence between diphenylureas and N^6-substituted adenines in terms of chemical structure and biological activity, the data[208] on compounds in which one of the benzene rings is unsubstituted (Figure

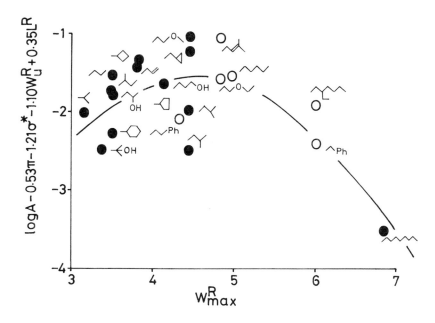

FIGURE 12. Relationship of the cytokinin agonistic and antagonistic activities of 4-alkyl-amino-2-methylpyrrolo[2,3-d]-pyrimidine derivatives to W_{max}, based on the correlation equation:[66] $\log A = 0.53\pi + 1.21\sigma^* + 2.98W_{max} - 0.33(W_{max})^2 + 1.10W_u + 0.53L - 8.27$. The activity term $\log A$ expresses the logarithm of reciprocal of the I_{50} values (μM) when the compounds are anticytokinin active and that of the E_{50} values (μM) when the compounds are cytokinin active. The π and σ^* terms are the hydrophobic and Taft's electronic parameters, respectively, of the N^4-side chains, and W_u and L are steric parameters which express the width upward and the length along the bond axis, respectively, of the N^4-substituents. (From Iwamura, H., Masuda, N., Koshimizu, K., and Matsubara, S., J. Med. Chem., 26, 838, 1983. With permission.)

14) were quantitatively analyzed to obtain Equation 2, where C is the minimum concentration (mM) at which activity is observed in the tobacco pith bioassay. The positive

$$\log(l/C) = 0.95\sigma - 0.85L_o - 0.27L_p + 1.04\pi_m + 2.00$$
$$n = 39, r = 0.91, s = 0.38 \qquad (2)$$

Hammett σ term indicates that electron-withdrawing substituents enhance activity, as with the N^6-substituted adenines. The electronic effect is considered to be directed toward the NH of the ureide bridge in diphenylureas, and it is toward the N^6H in the case of adenylate cytokinins, as described above. These results suggest a hydrogen-bonding interaction with the common basic site of the cytokinin receptor. L_o and L_p are the lengths of the o- and p-substituents, and the negatively larger coefficient of the former term indicates a more strict steric demand at the o-position. The π_m term expresses the hydrophobicity of m-substituents; it enhances the activity. These position-specific effects rationalize and provide a physicochemical basis for the generally recognized trend in activity: meta > para > ortho. Comparison and interpretations of Equation 1 for adenines and Equation 2 for diphenylureas have led to the proposal of the cytokinin-receptor binding model shown in Figure 15, where the stippled lines represent the steric interaction sites or spatial walls of the receptor deduced from the steric parameters incorporated into the correlations, and the smooth line facing the NH groups is the electron-donating site. The striped circle is the hydrophobic region where, as suggested by the π_m term in Equation 2, the m-substituent of the diphenylureas comes when the compound fits with the receptor. Accordingly, the figure displays the structural

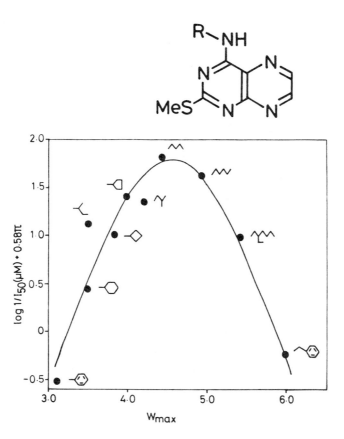

FIGURE 13. Relationship of anticytokinin activity of 4-alkylamino 2-methylthiopyrido[2,3-d]pyrimidines to W_{max}, based on the correlation equation:[48] $\log(1/I_{50}) = -0.96(W_{max})^2 + 8.75W_{max} - 0.58\pi - 18.32$. (From Iwamura, H., Murabami, S., Koga, J., Matsubara, S., and Koshimizu, K., *Phytochemistry*, 18, 217, 1979. With permission.)

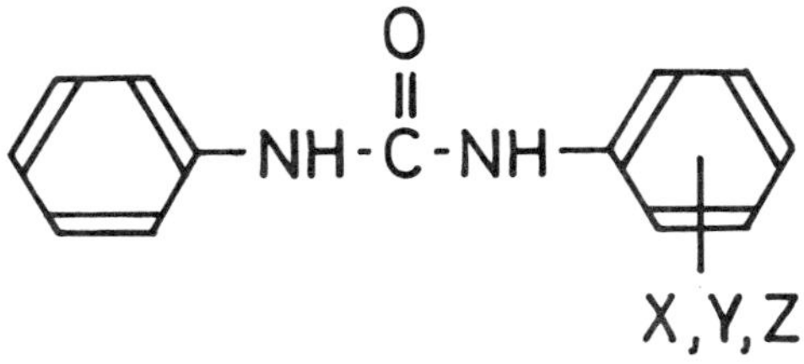

FIGURE 14. Structure of N,N'-diphenylureas.

correspondence or similarity between the two classes of cytokinin-active compounds, as well as the approximate size of the receptor cavity.

Although even the most active of the diphenylureas is still less potent than adenylate cytokinins such as BAP and Z, the replacement of one of the benzene rings by pyridine (Figure 16) considerably enhances activity; activity of N-(2-chloro-4-pyridyl)-N'-phenylurea is about 10 times more active than N^6-benzyladenine, and nearly as active as Z.[209] The structure-activity profile of this series of compounds coincides very well with the results of Equation 2 for diphenylurea derivatives. The fact that the compounds with a hydrophobic or electron-withdrawing substituent at the 2-position (*meta* to the urea bridge) of the pyridine

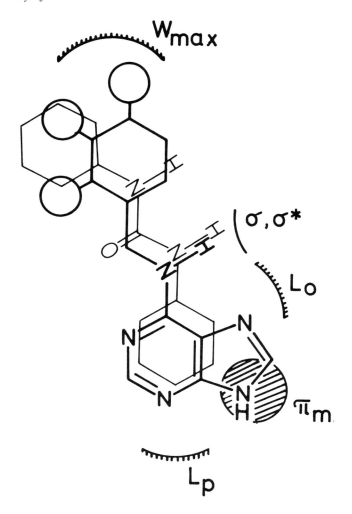

FIGURE 15. Schematic cytokinin-receptor complex where a hexagon represents a benzene ring. Stippled lines show the spatial walls, smooth lines show the electron-donating site, and the dotted square shows the hydrophobic region. (From Iwamura, H., Fujita, T., Koyama, S., Koshimizu, K., and Kumazawa, Z., *Phytochemistry*, 19, 1309, 1980. With permission.)

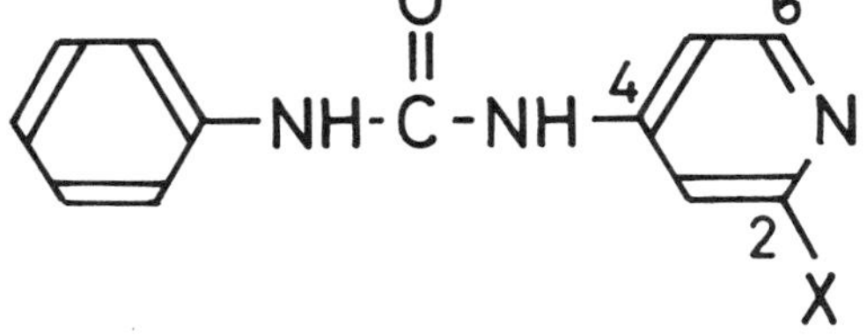

FIGURE 16. Structure of N-(2-substituted-4-pyridyl)-N'-phenylureas.

ring show high activiity, while hydrophilic substituents suppress activity, is explained by the positive σ and π_m terms. The activity of the compounds having a substituent at the 3-position (*ortho* to the urea bridge) is low and this appears due to the steric effect as indicated by the negative L_o term. The correlation in Equation 3 was separately obtained for N-pyridyl-N'-phenylureas having 2-substituents on the pyridine ring,[287] where C is the concentration at which the maximum growth of the tobacco callus is obtained. The π_m term is the hydrophobicity of the 2-substituents, which correspond to the m-substituents in diphenylureas, and σ_m is the Hammett σ value for m-substituents. Coinciding with the results indicated by Equation 3

$$\log(1/C) = 3.81\sigma_m + 0.52\pi_m + 6.50$$
$$n = 11, r = 0.975, s = 0.281 \tag{3}$$

that hydrophobic or electron-withdrawing 2-(or m-) substituents enhance activity, 2,6- (or m,m'-)- dichloro and -dibromo derivatives among the N-(2,6-disubstituted 4-pyridyl)-N'-phenylureas prepared exhibited very high activity.[287]

Our understanding of the molecular mechanism and mode of action of cytokinins has made much progress in the past 10 years through the development of new types of active compounds and interpretations of their structure-activity relationships. The chemical approach in the future will also be continuously directed toward the exploration of action mechanisms, linking more with biochemical studies. In the author's opinion, much effort should and will be directed to finding where the cytokinins act, which should lead to deeper insights. Isolation, and in particular, identification of the cytokinin receptor molecules or sites have not yet succeeded, and it may require new approaches and methodology linked with chemistry to reach this goal.

REFERENCES

1. **Wiesner, J.**, *Die Elementarstruktur und das Wachstum der lebenden Substanz*, Holder, Vienna, 1892.
2. **Haberlandt, G.**, Zur Physiologie der Zellteilung, *Sitzungs. ber. Kl. Preuss. Akad. Wiss.*, 318, 1913.
3. **Haberlandt, G.**, Wundhormone als Errerger von Zellteilungen, *Beitr. Allg. Bot.*, 2, 1, 1921.
4. **Bonner, J. and English, J.**, A chemical and physiolgoical study of traumatin, a plant wound hormone, *Plant Physiol.*, 13, 331, 1938.
5. **van Overbeek, J., Conklin, J. M., and Blakeslee, A. F.**, Factors in coconut milk essential for growth and development of very young *Datura* embryos, *Science*, 94, 350, 1941.
6. **Caplin, S. M. and Steward, F. C.**, Effect of coconut milk on the growth of explants from carrot root, *Science*, 108, 655, 1948.
7. **Jablonski, J. R. and Skoog, F.**, Cell enlargement and cell division in excised tobacco pith tissue. *Physiol. Plant.*, 7,, 16, 1954.
8. **Miller, C. O., Skoog, F., Okumura, F. S., Von Saltza, M. H., and Strong, F. M.**, Structure and synthesis of kinetin, *J. Am. Chem. Soc.*, 77, 2662, 1955.
9. **Miller, C. O., Skoog, F., Okumura, F. S., Von Saltza, M. H., and Strong, F. M.**, Isolation, structure, and synthesis of kinetin, a substance promoting cell division, *J. Am. Chem. Soc.*, 78, 1375, 1956.
10. **Hall, R. H. and de Ropp, R. S.**, Formation 6-furfurylaminopurine from DNA breakdown products, *J. Am. Chem. Soc.*, 77, 6400, 1955.
11. **Scopes, D. I. C., Zarnack, U., Leonard, N. J., Schmitz, R. Y., and Skoog, F.**, Alternative routes for the genesis of kinetin: a synthetic intramolecular route of 2'-deoxyadenosine to kinetin, *Phytochemistry*, 15, 1523, 1976.
12. **Skoog, F., Strong, F. M., and Miller, C. O.**, Cytokinins, *Science*, 148, 532, 1965.
13. **Letham, D. S.**, Cytokinins, in *Phytohormones and Related Compounds: A Comprehensive Treatise*, Vol. 1, Letham, D. S., Goodwin, P. B., and Higgins, T. J. V., Eds., Elsevier/North-Holland, Amsterdam, 1978, 205.
14. **Letham, D. S.**, Zeatin, a factor inducing cell division from *Zea mays*, *Life Sci.*, 2, 569, 1963.
15. **Letham, D. S., Shannon, J. S., and McDonald, I. R. C.**, The structure of zeatin, a factor inducing cell division, *Proc. Chem. Soc. London*, 230, 1964.

16. **Letham, D. S.,** Regulators of cell division in plant tissues. II. A cytokinin in plant extracts: isolation and interaction with other growth regulators, *Phytochemistry,* 5, 269, 1966.
17. **Letham, D. S.,** Cytokinins from *Zea mays, Phytochemistry,* 12, 2445, 1973.
18. **Kende, H.,** The cytokinins, *Int. Rev. Cytol.,* 31, 301, 1971.
19. **Yokota, T. and Takahashi, N.,** Cytokinins in shoots of the chestnut tree, *Phytochemistry,* 19, 2367, 1980.
20. **Watanabe, N., Yokota, T., and Takahashi, N.,** Identification of zeatin and zeatin riboside in cones of the hop plant and their possible role in cone growth, *Plant Cell Physiol.,* 19, 617, 1978.
21. **Watanabe, N., Yokota, T., and Takahashi, N.,** Variations in the levels of *cis*- and *trans*-ribosylzeatins and other minor cytokinins during development and growth of cones of the hop plant, *Plant Cell Physiol.,* 22, 489, 1981.
22. **Watanabe, N., Yokota, T., and Takahashi, N.,** Identification of N^6-(3-methyl-2-butenyl)adenosine, zeatin, zeatin riboside, gibberellin A_{19} and abscisic acid in shoots of the hop plant, *Plant Cell Physiol.,* 19, 1263, 1978.
23. **Young, H.,** Identification of cytokinins from natural sources by gas-liquid chromatography mass spectrometry, *Anal. Biochem.,* 79, 226, 1977.
24. **Koyama, S., Kawai, H., Kumazawa, Z., Ogawa, Y., and Iwamura, H.,** Isolation and identification of *trans*-zeatin from the roots of *Raphanus sativus* L. cv. Sakurajima, *Agric. Biol. Chem.,* 42, 1997, 1978.
25. **Martin, G. C., Scott, I. M., Neill, S. J., and Horgan, R.,** Identification of abscisic acid glucose ester, indole-3-acetic acid, zeatin, and zeatin riboside in receptacles of pear, *Phytochemistry,* 21, 1079, 1982.
26. **Hopping, M. E., Young, H., and Bukovac, M. J.,** Endogenous plant growth substances in developing fruit of *Prunus cerasus* L. VI. Cytokinins in relation to initial fruit development, *J. Am. Hortic. Sci.,* 104, 47, 1979.
27. **Summons, R. E., Letham, D. S., Gollnow, B. I., Parker, C. W., Entsch, B., Johnson, L. P., MacLeod, J. K., and Rolfe, B. G.,** Cytokinin translocation and metabolism in species of the Leguminoseae: studies in relation to shoot and nodule development, in *Metab. Mol. Act. Cytokinins, Proc. Int. Colloq.,* Guern, J. and Peaud-Lenoel, C., Eds., Springer-Verlag, Berlin, 1981, 69.
28. **Heindl, J. C., Carlson, D. R., Brun, W. A., and Brenner, M. L.,** Ontogenetic variation of four cytokinins in soybean root pressure exudate, *Plant Physiol.,* 70, 1619, 1982.
29. **Krasnuk, M., Witham, F. H., and Tegley, J. R.,** Cytokinins extracted from Pinto bean fruit, *Plant Physiol.,* 48, 320, 1971.
30. **Yokota, T., Ueda, J., and Takahashi, N.,** Cytokinins in immature seeds of *Dolichos lablab, Phytochemistry,* 20, 683, 1981.
31. **Dauphin, B., Teller, G., and Durand, B.,** Identification and quantitative analysis of cytokinins from shoot apices of *Mercuralis ambigua* by gas chromatography-mass spectrometry computer system, *Planta,* 144, 113, 1979.
32. **Dauphin, B., Teller, G., and Durand, B.,** Mise au point d'une nouvelle méthode de purification, caracterisation et dosage de cytokinines endogènes, extraites de bourgeons de Mercuriales annuelles, *Physiol. Veg.,* 15, 747, 1977.
33. **Chen, W. S.,** Cytokinins of the developing mango fruit. Isolation, identification, and changes in levels during maturation, *Plant Physiol.,* 71, 356, 1983.
34. **Purse, J. G., Horgan, R., Horgan, J. M., and Wareing, P. F.,** Cytokinins of sycamore spring sap, *Planta,* 132, 1, 1976.
35. **Shindy, W. W. and Smith, O. E.,** Identification of plant hormones from cotton ovules, *Plant Physiol.,* 55, 550, 1975.
36. **Miller, C. O.,** Cell-division factors from *Vinca rosea* L. crown gall tumor tissues, *Proc. Natl. Acad. Sci. U.S.A.,* 72, 1883, 1975.
37. **Peterson, J. B. and Miller, C. O.,** Cytokinins in *Vinca rosea* L. crown gall tumor tissue as influenced by compounds containing reduced nitrogen, *Plant Physiol.,* 57, 393, 1976.
38. **Van Staden, J. and Menary, R. C.,** Identification of cytokinins in the xylem sap of tomato, *Z. Pflanzenphysiol.,* 78, 262, 1976.
39. **Klämbt, D.,** Cytokinins aus *Helianthus annus, Planta,* 82, 170, 1968.
40. **Van Staden, J. and Drewes, S. E.,** Isolation and identification of zeatin from malt extract, *Plant Sci. Lett.,* 4, 391, 1975.
41. **Letham, D. S. and Miller, C. O.,** Identity of kinetin-like factors from *Zea mays, Plant Cell Physiol.,* 6, 355, 1965.
42. **Van Staden, J. and Drewes, S. E.,** Identification of zeatin and zeatin riboside in coconut milk, *Physiol. Plant.,* 34, 106, 1975.
43. **Murofushi, N., Inoue, A., Watanabe, N., Ota, Y., and Takahashi, N.,** Identification of cytokinins in root exudate of the rice plant, *Plant Cell Physiol.,* 24, 87, 1983.
44. **Lorenzi, R., Horgan, R., and Wareing, P. F.,** Cytokinins in *Picea sitchensis* Carriere: identification and relation to growth, *Biochem. Physiol. Pflanz.,* 168, 333, 1975.

45. **Weiler, E. W.,** Dynamics of endogenous growth regulators during the growth cycle of a hormone-autotrophic plant cell culture, *Naturwissenschaften,* 68, 377, 1981.
46. **Hashizume, T., Sugiyama, T., Imura, M., Cory, H. T., Scott, M. F., and McCloskey, J. A.,** Determination of cytokinins by mass spectrometry based on stable isotope dilution, *Anal. Biochem.,* 92, 111, 1979.
47. **Horgan, R., Hewett, E. W., Purse, J. G., Horgan, J. M., and Wareing, P. F.,** Identification of a cytokinin in sycamore sap by gas chromatography-mass spectrometry, *Plant. Sci. Lett.,* 1, 321, 1973.
48. **Mizuno, K. and Komamine, A.,** Isolation and identification of substances inducing formation of tracheary elements in cultured carrot-root slices, *Planta,* 138, 59, 1978.
49. **Miller, C. O.,** Revised methods for purification of ribosyl *trans*-zeatin from *Vinca rosea* L. crown gall tumor tissue, *Plant Physiol.,* 55, 448, 1975.
50. **Basile, B., Kimura, K., and Hashizume, T.,** Presence and levels of *cis*- and *trans*-methylthiozeatin riboside and *cis*- and *trans*-zeatin riboside in tobacco plant determined by mass spectrometry, *Proc. Jpn. Acad. Ser. B:,* 57, 276, 1981.
51. **Bui-Dang-Ha, D. and Nitsch, J. P.,** Isolation of zeatin riboside from chicory root, *Planta,* 95, 119, 1970.
52. **Letham, D. S.,** A new cytokinin bioassay and the naturally occurring cytokinin complex, in *Biochemistry and Physiology of Plant Growth Substances, Proc. 6th Int. Conf. Plant Growth Substances,* Wightman, F. and Setterfield, G., Eds., Runge Press, Ottawa, 1968, 19.
53. **Letham, D. S.,** The cytokinins of coconut milk, *Physiol. Plant.,* 32, 66, 1974.
54. **Watanabe, N., Yokota, T., and Takahashi, N.,** *cis*-Zeatin riboside: its occurrence as a free nucleoside in cones of the hop plant, *Agric. Biol. Chem.,* 42, 2415, 1978.
55. **Watanabe, N., Yokota, T., and Takahashi, N.,** Transfer RNA, a possible supplier of free cytokinins, ribosyl-*cis*-zeatin and ribosyl-2-methylthiozeatin: quantitative comparison between free and transfer cytokinins in various tissues of the hop plant, *Plant Cell Physiol.,* 23, 479, 1982.
56. **Hashizume, T., Suye, S., and Sugiyama, T.,** Isolation and identification of *cis*-zeatin riboside from tubers of sweet potato (*Ipomoea batatas* L.), *Agric. Biol. Chem.,* 46, 663, 1982.
57. **Mauk, C. S. and Langille, A. R.,** Physiology of tuberisation in *Solanum tubersosum* L. *cis*-Zeatin riboside in the potato plant: its identification and changes in endogenous levels as influenced by temperature and photoperiod, *Plant Physiol.,* 62, 438, 1978.
58. **Hashizume, T., Kimura, K., and Sugiyama, T.,** Identification of *cis*-zeatin-D-riboside from the top of tobacco plants, *Heterocycles,* 10, 139, 1978.
59. **Palni, L. M. S., Horgan, R., Darrall, N. M., Stuchbury, T., and Wareing, P. F.,** Cytokinin biosynthesis in crown-gall tissue of *Vinca rosea*. The significance of nucleotides, *Planta,* 159, 50, 1983.
60. **Summons, R. E., Palni, L. M. S., and Letham, D. S.,** Mass spectrometric analysis of cytokinins in plant tissues. IV. Determination of intact zeatin nucleotide by direct chemical ionization mass-spectrometry, *FEBS Lett.,* 151, 122, 1983.
61. **Scott, I. M., Horgan, R., and McGaw, B. A.,** Zeatin-9-glucoside, a major endogenous cytokinin of *Vinca rosea* crown gall tissue, *Planta,* 149, 472, 1980.
62. **Scott, I. M., Martin, G. C., Horgan, R., and Heald, J. K.,** Mass spectrometric measurement of zeatin glycoside levels in *Vinca rosea* L. crown gall tissue, *Planta,* 154, 273, 1982.
63. **Palni, L. M. S. and Horgan, R.,** Cytokinins from the culture medium of *Vinca rosea* crown gall tumor tissue, *Plant Sci. Lett.,* 24, 327, 1982.
64. **Palni, L. M. S., Summons, R. E., and Letham, D. S.,** Mass spectrometric analysis of cytokinins in plant tissues. V. Identification of the cytokinin complex of *Datura innoxia* crown-gall tissue, *Plant Physiol.,* 72, 858, 1983.
65. **Summons, R. E., Duke, C. C., Eichholzer, J. V., Entsch, B., Letham, D. S., MacLeod, J. K., and Parker, C. W.,** Mass spectrometric analysis of cytokinins in plant tissues. II. Quantitation of cytokinins in *Zea mays* kernels using deuterium labelled standards, *Biomed. Mass Spectrom.,* 6, 407, 1979.
66. **Summons, R. E., Entsch, B., Letham, D. S., Gollnow, B. I., and MacLeod, J. K.,** Regulators of cell division in plant tissues. XXVIII. Metabolites of zeatin in sweet-corn kernels: purifications and identifications using HPLC and chemical-ionization mass spectrometry, *Planta,* 147, 422, 1980.
67. **Summons, R. E., MacLeod, J. K., Parker, C. W., and Letham, D. S.,** The occurrence of raphanatin as an endogenous cytokinin in radish seed. Identification and quantitation by gas chromatographic-mass spectrometric analysis using deuterium labelled standards, *FEBS Lett.,* 82, 211, 1977.
68. **Summons, R. E., Entsch, B., Parker, C. W., and Letham, D. S.,** Mass spectrometric analysis of cytokinins in plant tissues. III. Quantitation of the cytokinin glycoside complex of lupin pods by stable isotope dilution, *FEBS Lett.,* 107, 21, 1979.
69. **Morris, R. O.,** Mass spectroscopic identification of cytokinins. Glucosyl zeatin and glucosylribosylzeatin from *Vinca rosea* crown gall, *Plant Physiol.,* 59, 1029, 1977.
70. **Peterson, J. B. and Miller, C. O.,** Glucosylzeatin and glucosylribosylzeatin from *Vinca rosea* L. crown gall tumor tissue, *Plant Physiol.,* 59, 1026, 1977.

71. **Koshimizu, K., Kusaki, T., Mitui, T., and Matsubara, S.,** Isolation of a new cytokinin from immature yellow lupin seeds, *Agric. Biol. Chem.*, 31, 795, 1967.
72. **Wang, T. L. and Horgan, R.,** Dihydrozeatin riboside, a minor cytokinin from the leaves of *Phaseolus vulgaris* L., *Planta*, 140, 151, 1978.
73. **Palmer, M. V., Horgan, R., and Wareing, P. F.,** Cytokinin metabolism in *Phaseolus vulgaris* L. III. Identidication of endogenous cytokinins and metabolism of [8-^{14}C]-dihydrozeatin in stems of decapitated plants, *Planta*, 153, 297, 1981.
74. **Wang, T. L., Thompson, A. G., and Horgan, R.,** A cytokinin glucoside from the leaves of *Phaseolus vulgaris*, *Planta*, 135, 285, 1977.
75. **Dauphin-Guerin, B., Teller, G., and Durand, B.,** Different endogenous cytokinins between male and female *Mercurialis annua*, *Planta*, 148, 124, 1980.
76. **Dyson, W. H. and Hall, R. H.,** N^6-(Δ^2-isopentenyl)adenosine: its occurrence as a free nucleoside in an autonomous strain of tobacco tissue, *Plant Physiol.*, 50, 616, 1972.
77. **Tsui, C., Shao, L. M., Wang, C. M., and Tao, G. Q.,** Identification of a cytokinin in water chestnut (corms of *Eleocharis tuberosa*), *Plant Sci. Lett.*, 32, 225, 1983.
78. **Hashizume, T., Suye, S., Soeda, T., and Sugiyama, T.,** Isolation and characterization of a new glucopyranosyl derivative of 6-(3-methyl-2-butenyl-amino)purine from sweet potato tubers, *FEBS Lett.*, 144, 25, 1982.
79. **Horgan, R., Hewett, E. W., Horgan, J. M., Purse, J., and Wareing, P. E.,** A new cytokinin from *Populus robusta*, *Phytochemistry*, 14, 1005, 1975.
80. **Chaves das Neves, H. J. and Pais, M. S.,** Identification of a spathe regreening factor in *Zantedeschia aethiopica*, *Biochem. Biophys. Res. Commun.*, 95, 1387, 1980.
81. **Chaves das Neves, H. J. and Pais, M. S.,** A new cytokinin from the fruits of *Zantedeschia aethiopica*, *Tetrahedron Lett.*, 21, 4387, 1980.
82. **Beutelmann, P.,** Purification and identification of a cytokinin from moss callus cells, *Planta*, 133, 215, 1977.
83. **Wang, T. L., Horgan, R., and Cove, D.,** Cytokinins from the moss *Physcomitrella patens*, *Plant Physiol.*, 68, 735, 1981.
84. **Rodriguez-Barrueco, C., Miguel, C., and Palni, L. M. S.,** Cytokinins in rootnodules of the nitrogen-fixing nonlegume *Myrica Gale*, *Z. Pflanzenphysiol.*, 95, 275, 1979.
85. **Vonk, C. R., and Davellaar, E.,** 8-^{14}C-zeatin metabolites and their transport from leaf to phloem exudate of Yucca, *Physiol. Plant.*, 52, 101, 1981.
86. **Miller, C. O.,** Zeatin and zeatin riboside from a mycorrhizal fungus, *Science*, 157, 1055, 1967.
87. **Miller, C. O.,** Naturally-occurring cytokinins, in *Biochemistry and Physiology of Plant Growth Substances, Proc. 6th Int. Conf. Plant Growth Substances*, Wightman, F. and Setterfield, F., Eds., Runge Press, Ottawa, 1968, 33.
88. **Croft, C. B. and Miller, C. O.,** Detection and identification of cytokinins produced by mycorrhizal fungi, *Plant Physiol.*, 54, 586, 1974.
89. **Scarbrough, E., Armstrong, D. J., Skoog, F., Frihart, C. R., and Leonard, N. J.,** Isolation of *cis*-zeatin from *Corynebacterium fascians* cultures, *Proc. Natl. Acad. Sci. U.S.A.*, 70, 3825, 1973.
90. **Armstrong, D. J., Scarbrough, E., Skoog, F., Cole, D. L., and Leonard, N. J.,** Cytokinins in *Corynebacterium fascians* cultures. Isolation and identification of 6-(4-hydroxy-3-methyl-*cis*-2-butenylamino)-2-methylthiopurine, *Plant Physiol.*, 58, 749, 1976.
91. **Chapman, R. W., Morris, R. O., and Zaerr, J. B.,** Occurrence of *trans*-ribosylzeatin in *Agrobacterium tumefaciens* tRNA, *Nature (London)*, 262, 153, 1976.
92. **Kaiss-Chapman, R. W. and Morris, R. O.,** *trans*-Zeatin in culture filtrates of *Agrobacterium tumefaciens*, *Biochem. Biophys. Res. Commun.*, 76, 453, 1977.
93. **McCloskey, J. A., Hashizume, T., Basile, B., Ohno, Y., and Sonoki, S.,** Occurrence and levels of *cis*- and *trans*-zeatin ribosides in the culture medium of a virulent strain of *Agrobacterium tumefaciens*, *FEBS Lett.*, 111, 181, 1980.
94. **Regier, D. O. and Morris, R. O.,** Secretion of *trans*-zeatin by *Agrobacterium tumefaciens*: a function determined by the nopaline Ti plasmid, *Biochem. Biophys. Res. Commun.*, 104, 1560, 1982.
95. **Murai, N., Skoog, F., Doyle, M. E., and Hanson, R. S.,** Relationships between cytokinin production, presence of plasmids, and fasciation caused by strains of *Corynebacterium fascians*, *Proc. Natl. Acad. Sci. U.S.A.*, 77, 619, 1980.
96. **Greene, E. M.,** Cytokinin production by microorganisms, *Bot. Rev.*, 46, 25, 1980.
97. **Burrows, W. J., Armstrong, D. J., Kaminek, M., Skoog, F., Bock, R. M., Hecht, S. M., Dammann, L. G., Leonard, N. J., and Occolowitz, J. L.,** Isolation and identification of four cytokinins from wheat germ transfer ribonucleic acid, *Biochemistry*, 9, 1867, 1970.
98. **Burrows, W. J., Skoog, F., and Leonard, N. J.,** Isolation and identification of cytokinins located in the transfer ribonucleic acid of tobacco callus growth in the presence of 6-benzylaminopurine, *Biochemistry*, 10, 2189, 1971.

99. **Edwards, C. A. and Armstrong, D. J.**, Cytokinin-activity ribonucleosides in *Phaseolus* RNA. II. Distribution in tRNA species from etiolated *P. vulgaris* L. seedling, *Plant Physiol.*, 67, 1185, 1981.
100. **Kovacs, E.**, Cytokinin activity of *Agaricus bisporus, Acta Aliment.*, 11, 169, 1982.
101. **Parker, C. W., Letham, D. S., Cowley, D. E., and MacLeod, J. K.**, Raphanatin, an unusual purine derivative and a metabolite of zeatin, *Biochem. Biophys. Res. Commun.*, 49, 460, 1972.
102. **Duke, C. C., Liepa, A. J., MacLeod, J. K., Letham, D. S., and Parker, C. W.**, Synthesis of raphanatin and its 6-benzylaminopurine analogue, *J. Chem. Soc. Chem. Commun.*, 964, 1975.
103. **Cowley, D. E., Duke, C. C., Liepa, A., MacLeod, J. K., and Letham, D. S.**, The structure and synthesis of cytokinin metabolites. I. The 7- amd 9-β-D-glucofuranosides and pyranosides of zeatin and 6-benzylaminopurine, *Aust. J. Chem.*, 31, 1095, 1978.
104. **Cowley, D. E., Jenkins, I. D., MacLeod, J. K., Summons, R. E., Letham, D. S., Wilson, M. M., and Parker, C. W.**, Structure and synthesis of unusual cytokinin metabolites, *Tetrahedron Lett.*, 1015, 1975.
105. **Letham, D. S., Gollnow, B. I., and Parker, C. W.**, The reported occurrence of 7-glucofuranoside metabolites of cytokinins, *Plant Sci. Lett.*, 15, 217, 1979.
106. **MacLeod, J. K., Summons, R. E., and Letham, D. S.**, Mass spectrometry of cytokinin metabolites. Per(trimethylsilyl) and permethyl derivatives of glucosides of zeatin and 6-benzylaminopurine, *J. Org. Chem.*, 41, 3959, 1976.
107. **Duke, C. C., Letham, D. S., Parker, C. W., MacLeod, J. K., and Summons, R. E.**, The complex O-glucosylzeatin derivatives formed in *Populus* species, *Phytochemistry*, 18, 819, 1979.
108. **Duke, C. C., MacLeod, J. K., Summons, R. E., Letham, D. S., and Parker, C. W.**, Lupinic acid and O-β-D-glucopyranosylzeatin from *Lupinus angustifolium, Aust. J. Chem.*, 31, 1291, 1978.
109. **Letham, D. S., Parker, C. W., Duke, C. C., Summons, R. E., and MacLeod, J. K.**, O-Glucosylzeatin and related compounds — a new group of cytokinin metabolites, *Ann. Bot.*, 40, 261, 1976.
110. **Fujii, T. and Ogawa, N.**, The absolute configuration of (−)-dihydrozeatin, *Tetrahedron Lett.*, 3075, 1972.
111. **Mensuali, S. A. and Lorenzi, R.**, Cytokinins in immature seeds of *Phaseolus coccineus* L., *Z. Pflanzenphysiol.*, 108, 343, 1982.
112. **Leonard, N. J. and Fujii, T.**, The synthesis of compounds possessing kinetin activity. The use of a blocking group at the 9-position of adenine for the biosynthesis of 1-substituted adenines, *Proc. Natl. Acad. Sci. U.S.A.*, 51, 73, 1964.
113. **Helgeson, J. P. and Leonard, N. J.**, Cytokinins: identification of compounds isolated from *Corynebacterium fascians, Proc. Natl. Acad. Sci. U.S.A.*, 56, 60, 1966.
114. **Klämbt, D., Thies, G., and Skoog, F.**, Isolation of cytokinins from *Corynebacterium fascians, Proc. Natl. Acad. Sci. U.S.A.*, 56, 52, 1966.
115. **Scott, I. M., Browning, G., and Eagles, J.**, Ribosylzeatin and zeatin in tobacco crown gall tissue, *Planta*, 147, 269, 1980.
116. **Tanaka, Y., Abe, H., Uchiyama, M., Taya, Y., and Nishimura, S.**, Isopentenyladenine from *Dictyostelium discoideum, Phytochemistry*, 17, 543, 1978.
117. **Letham, D. S. and Wettenhall, R. E. H.**, Transfer RNA and cytokinins, in *The Ribonucleic Acids*, Stewart, P. R. and Letham, D. S., Eds., Springer-Verlag, New York, 1977, 129.
118. **Whenham, R. J. and Fraser, R. S. S.**, Does tobacco mosaic virus RNA contain cytokinins? *Virology*, 118, 263, 1982.
119. **Letham, D. S. and Palni, L. M. S.**, The biosynthesis and metabolism of cytokinins, *Ann. Rev. Plant Physiol.*, 34, 163, 1983.
120. **Abe, H., Uchiyama, M., Tanaka, Y., and Saito, H.**, Structure of discadenine, a spore germination inhibitor from the cellular slime mold, *Dictyostelium discoideum, Tetrahedron Lett.*, 3807, 1976.
121. **Tsoupras, G., Luu, B., and Hoffmann, J. A.**, A cytokinin (isopentenyl-adenosyl-mononucleotide) linked to ecdycone in newly laid eggs of *Locusta migratoria, Science*, 220, 507, 1983.
122. **Van Staden, J. and Drewes, S. E.**, 6-(2,3,4-Trihydroxy-3-methylbutylamino)-purine, a cytokinin with high biological activity, *Phytochemistry*, 21, 1783, 1982.
123. **Skoog, G., Hamzi, H. Q., Szewykowska, A. W., Leonard, N. J., Carraway, K. L., Fujii, T., Helegeson, J. P., and Leoeppky, R. N.**, Cytokinins — structure-activity relations, *Phytochemistry*, 6, 1169, 1967.
124. **Kuhnle, J. A., Fuller, G., Corse, J., and Mackey, B. E.**, Antisenescent activity of natural cytokinins, *Physiol. Plant.*, 41, 14, 1977.
125. **Andonova, T., Lilov, D., and Kotseva, E.**, Identification of free cytokinins in grape vine inflorescences, *Fiziol. Rast.*, 8, 23, 1982.
126. **Horgan, R.**, Nature and distribution of cytokinins, *Philos, Trans. R. Soc. London, Ser. B*: 284, 439, 1978.
127. **Skoog, F. and Schmitz, R. Y.**, Biochemistry and physiology of cytokinins, in *Biochemical Action of Hormones*, Vol. 6, Litwack, G., Ed., Academic Press, New York, 1979, 335.

128. **Bearder, J. R.**, Plant hormones and other growth substances. Their background, structures and occurrence, in *Hormonal Regulation of Development* I, *Molecular Aspects of Plant Hormones, Encyclopedia of Plant Physiology*, New Series, Vol. 9, MacMillan, J., Ed., Springer-Verlag, Berlin, 1980, 9.
129. **Horgan, R.**, Analytical procedures for cytokinins, in *Isolation of Plant Growth Substances*, Hillman, J. R., Ed., Cambridge University Press, Cambridge, 1978, 97.
130. **Parker, C. W. and Letham, D. S.**, Regulators of cell division in plant tissue. XVI. Metabolism of zeatin by radish cotyledons and hypocotyls, *Planta*, 114, 199, 1973.
131. **Bieleski, R. L.**, The problem of halting enzyme action when extracting plant tissue, *Anal. Biochem.*, 9, 431, 1964.
132. **Brown, E. G.**, Acid-soluble nucleotides of mature pea seeds, *Biochem. J.*, 85, 633, 1962.
133. **Laloue, M., Terrine, C., and Gawer, M.**, Cytokinins formation of the nucleoside-5'-triphosphate in tobacco and *Acer* cells, *FEBS Lett.*, 46, 45, 1974.
134. **Letham, D. S.**, Regulators of cell division in plant tissues. XXI. Distribution coefficients for cytokinins, *Planta*, 118, 361, 1974.
135. **Miller, C. O.**, Ribosyl-*trans*-zeatin, a major cytokinin produced by crown gall tumor tissue, *Proc. Natl. Acad. Sci. U.S.A.*, 71, 334, 1974.
136. **Yoshida, R. and Oritani, T.**, Cytokinin glucoside in roots of the rice plant, *Plant Cell Physiol.*, 13, 337, 1972.
137. **Armstrong, D. J., Burrows, W. J., Evans, P. K., and Skoog, F.**, Isolation of cytokinins from tRNA, *Biochem. Biophys. Res. Commun.*, 37, 451, 1969.
138. **Biddington, N. L. and Thomas, T. H.**, Effect of pH on the elution of cytokinins from polyvinylpyrrolidone column, *J. Chromatogr.*, 121, 107, 1976.
139. **Kannangara, T., Durley, R. C., and Simpson, G. M.**, High-performance liquid chromatographic analysis of cytokinis in *Sorghum bicolor* leaves, *Physiol. Plant.*, 44, 295, 1978.
140. **Stahly, E. A. and Buchanan, D. A.**, High-performance liquid chromatographic procedure for separation and quantification of zeatin and zeatin riboside from pears, peaches and apples, *J. Chromatogr.*, 235, 453, 1982.
141. **Laloue, M., Terrine, C., and Guern, J.**, Cytokinins: metabolism and biological activity of N^6-(Δ^2-isopentenyl)adenosine and N^6-(Δ^2-isopentenyl)adenine in tobacco cells and callus, *Plant Physiol.*, 59, 478, 1977.
142. **Tegley, J. R., Witham, F. H., and Kranuk, M.**, Chromatographic analysis of a cytokinin from tissue cultures of crown-gall, *Plant Physiol.*, 47, 581, 1971.
143. **Nitsch, J. P.**, Natural cytokinins, in *Society of Chemical Industry Monograph No. 31*, Society of Chemical Industry, London, 1968, 111.
144. **Challice, J. S.**, Separation of cytokinins by high pressure liquid chromatography, *Planta*, 122, 203, 1975.
145. **MacDonald, E. M. S., Akiyoshi, D. E., and Morris, R. O.**, Combined high-performance liquid chromatography radioimmunoassay for cytokinins, *J. Chromatogr.*, 214, 101, 1981.
146. **Andersen, R. A. and Kemp, J. R.**, Reversed-phase high-performance liquid chromatography of several plant cell division factors (cytokinins) and their *cis* and *trans* isomers, *J. Chromatogr.*, 172, 509, 1979.
147. **Horgan, R. and Kramers, M. R.**, High-performance liquid chromatography of cytokinins, *J. Chromatogr.*, 173, 263, 1979.
148. **Walker, M. A. and Dumbroff, E. B.**, Use of ion-pair, reversed-phase high-performance liquid chromatography for the analysis of cytokinins, *J. Chromatogr.*, 237, 316, 1982.
149. **Holland, J. A. McKerrell, E. H., Fuell, K. J., and Burrows, W. J.**, Separation of cytokinins by reversed-phase high-performance liquid chromatography, *J. Chromatogr.*, 166, 545, 1978.
150. **Scott, I. M. and Horgan, R.**, High-performance liquid chromatography of cytokinin ribonucleoside 5'-monophosphates, *J. Chromatogr.*, 237, 311, 1982.
151. **Crozier, A. and Zaerr, J. B.**, High-performance steric exclusion chromatography of plant hormones, *J. Chromatogr.*, 198, 57, 1980.
152. **Martin, G. C., Horgan, R., and Scott, I. M.**, High-performance liquid chromatographic analysis of permethylated cytokinins, *J. Chromatogr.*, 219, 167, 1981.
153. **Most, B. H., Williams, J. C., and Parker, K. J.**, Gas chromatography of cytokinins, *J. Chromatogr.*, 38, 136, 1968.
154. **Zelleke, A., Martin, G. C., and Labavitch, J. M.**, Detection of cytokinins using a gas chromatograph equipped with a sensitive nitrogen-phosphorus detector, *J. Am. Soc. Hortic. Sci.*, 105, 50, 1980.
155. **Ludewig, M. Dörffling, K., and König, W. A.**, Electron-capture capillary gas chromatography and mass spectrometry of trifluoroacetylated cytokinins, *J. Chromatogr.*, 243, 93, 1982.
156. **Kemp, T. R. and Andersen, R. A.**, Separation of modified bases and ribonucleosides with cytokinin activity using fused silica capillary gas chromatography, *J. Chromatogr.*, 209, 467, 1981.
157. **Kemp, T. R. Andersen, R. A., Oh, J., and Vaughn, T. H.**, High-resolution gas chromatography of methylated ribonucleosides and hypermodified adenosines. Evaluation of trimethylsilyl derivatization and split and splitless operation modes, *J. Chromatogr.*, 241, 325, 1982.

158. **Kemp, T. R., Andersen, R. A., and Oh, J.,** Cytokinin determination of tRNA by fused-silica capillary gas chromatography and nitrogen-selective detection, *J. Chromatogr.*, 259, 347, 1983.
159. **Whenham, R. J.,** Evaluation of selective detectors for the rapid and sensitive gas chromatographic assay of cytokinins, and application to the analysis of cytokinins in plant extracts, *Planta*, 157, 554, 1983.
160. **Stafford, A. and Corse, J.,** Fused-silica capillary gas chromatography of permethylated cytokinins with flame-ionization and nitrogen-phosphorus detection, *J. Chromatogr.*, 247, 176, 1982.
161. **Yokota, T., Murofushi, N., and Takahashi, N.,** Extraction, purification, and identification, in *Hormonal Regulation of Development* I, *Molecular Aspects of Plant Hormones, Encyclopedia of Plant Physiology*, New Series, Vol. 9, MacMillan, J. Ed., Springer-Verlag, Berlin, 1980, 113.
162. **Horgan, R.,** Modern methods for plant hormone analysis, in *Progress in Phytochemistry*, Vol 7, Reinhold, L., Harborne, J. B., and Swain, T., Eds., Pergamon Press, Oxford, 1981, 137.
163. **Brenner, M. L.,** Modern methods for plant growth substance analysis, *Ann. Rev. Plant Physiol.*, 32, 511, 1981.
164. **Thompson, A. G., Horgan, R., and Heald, J. K.,** A quantitative analysis of cytokinin using single-ion-current monitoring, *Planta*, 124, 207, 1975.
165. **Weiler, E. W.,** Radioimmunoassays for *trans*-zeatin and related cytokinins, *Planta*, 149, 155, 1980.
166. **Weiler, E. W and Spanier, K.** Phytohormones in the formation of crown gall tumors, *Planta*, 153, 326, 1981.
167. **Vold, B. S. and Leonard, N. J.,** Production and characterization of antibodies and establishment of a radioimmunoassay for ribosylzeatin, *Plant Physiol.*, 67, 401, 1981.
168. **Morris, R. O., Akiyoshi, D. E., MacDonald, E. M. S., Morris, J. W., Regier, D. A., and Zaerr, J. B.,** Cytokinin metabolism in relation to tumor induction by *Agrobacterium tumefaciens*, in *Plant Growth Substances 1982, Proc. 11th Int. Conf. Plant Growth Substances*, Wareing, P. F., Ed., Academic Press, London, 1982, 175.
169. **Townsend, L. B., Robins, R. K., Leoppky, R. N., and Leonard, N. J.,** Purine nucleosides. VIII. Reinvestigation of the position of glycosidation in certain synthetic "7"-substituted 6-dimethylaminopurine nucleosides related to puromycin, *J. Am. Chem. Soc.*, 86, 5320, 1964.
170. **Leonard, N. J., Carraway, K. L., and Helgeson, J. P.,** Characterization of N_x, N_y-disubstituted adenines by ultraviolet absorption spectra, *J. Heterocycl. Chem.*, 2, 291, 1965.
171. **Shugar, D. and Fox, J. J.,** Spectrophotometric studies of nucleic acid derivatives and related compounds as a function of pH. I. Pyridmidines, *Biochim. Biophys. Acta*, 9, 199, 1952.
172. **Leonard, N. J. and Deyrup, J. A.,** The chemistry of triacanthine, *J. Am. Chem. Soc.*, 84, 2148, 1962.
173. **Shannon, J. S. and Letham, D. S.,** Regulators of cell division in plant tissues. IV. The mass spectra of cytokinins and other 6-aminopurines, *N. Z. J. Sci.*, 9, 833, 1966.
174. **Shaw, S. J., Desiderio, D. M., Tsuboyama, K., and McCloskey, J. A.,** Mass spectrometry of nucleic acid components. Analogs of adenosine, *J. Am. Chem. Soc.*, 92, 2510, 1970.
175. **Elion, G. E., Burgi, E., and Hitching, G.,** Studies on condensed pyrimidine systems. IX. The synthesis of some 6-substituted purines, *J. Am. Chem. Soc.*, 74, 411, 1952.
176. **Schmitz, R. Y., Skoog, F., Hecht, S. M., and Bock, R. M.,** Comparison of cytokinin activities of naturally occurring ribonucleosides and corresponding bases, *Phytochemistry*, 11, 1603, 1972.
177. **Verman, H. J., Schmitz, R. Y., Skoog, F., Playtis, A. J., Frihart, C. R., and Leonard, N. J.,** Synthesis of 2-methylthio-*cis*- and *trans*-ribosylzeatin and their isolation from *Pisum t*-RNA, *Phytochemistry*, 13, 31, 1974.
178. **Ikehara, M., Ohtsuka, E., Ono, E., and Imamura, K., III.,** Interaction between synthetic adenosine triphosphate analogues and actomyosin, *Biochem. Biophys. Acta*, 100, 471, 1965.
179. **Daly, J. W. and Christensen, B. E.,** Purines. VI. The preparation of certain 6-substituted- and 6,9-disubstituted purines, *J. Org. Chem.*, 21, 177, 1950.
180. **Shaw, G., Smallwood, B. M., and Wilson, D. V.,** Purines, pyrimidines, and imidazoles. XXIV. Synthesis of zeatin, a naturally occurring adenine derivative with plant cell-division-promoting activity, and its 9-β-D-ribofuranoside, *J. Chem. Soc. C.*, 921, 1966.
181. **Rathey, R. S. and English, J.,** Formation of cyclopropane derivatives from 4-bromocrotonic esters, *J. Org. Chem.*, 25, 2213, 1960.
182. **Müller, E., Metzger, H., and Fries, D.,** Über Nitroso-verbindungen. II. Miteil.: Reduktion geminaler Chloro-nitroso-Verbindungen Oximen, *Chem. Ber.*, 87, 1449, 1954.
183. **Leonard, N. J., Playtis, A. J., Skoog, F., and Schmitz, R. Y.,** A stereoselective synthesis of *cis*-zeatin, *J. Am. Chem. Soc.*, 93, 3056, 1971.
184. **Playtis, A. J. and Leonard, N. J.,** The synthesis of ribosyl-*cis*-zeatin and thin-layer chromatographic separation of the *cis* and *trans* isomers of ribosylzeatin, *Biochem. Biophys. Res. Commun.*, 45, 1, 1971.
185. **Sato, T.,** *Synthetic Procedures in Nucleic Acid Chemistry*, Vol. 1, Zorbach, W. W. and Tipson, R. S., Eds., Wiley Interscience, New York, 1968, 264.
186. **Leonard, N. J. and Fujii, T.,** The synthesis of compounds possessing kinetin activity. The use of a blocking group at the 9-position of adenine for the synthesis of 1-substituted adenines, *Proc. Natl. Acad. Sci. U.S.A.*, 51, 73, 1961.

187. **Robins, M. J., Hall, R. H., and Thedford, R.,** N^6-(Δ^2-isopentenyl)adenosine. A component of the transfer ribonucleic acid of yeast and of mammalian tissue, methods of isolation, and characterization, *Biochemistry*, 6, 1837, 1967.
188. **Hecht, S. M., Helgeson, J. P. and Fujii, T.,** in *Synthetic Procedures in Nucleic Acid Chemistry*, Vol. 1, Zorbach, W. W. and Tipson, R. S., Eds., Wiley Interscience, New York, 1968, 8.
189. **Grim, W. A. H. and Leonard, N. J.,** Synthesis of the "minor nucleotide: N^6-(γ,γ-dimethylallyl)adenoside 5'-phosphate and relative rates of rearrangement of 1 to N^6-dimethylallyl compounds for base, nucleoside, and nucleotide, *Biochemistry*, 6, 3625, 1967.
190. **Schmitz, R. Y., Skoog, F., Vincze, A., Walker, G. C., Kirkegaard, L. H., and Leonard, N. J.,** Comparison of cytokinin activities of the base ribonucleoside and 5'- and cyclic-3', 5'-monophosphate ribonucleotides of N^6-isopentenyl-, N^6-benzyl- or 8-bromo-adenine, *Phytochemistry*, 14, 1479, 1975.
191. **Leonard, N. J., Achmatowicz, S., Loeppky, R. N., Carraway, K. L., Grimm, W. A., Szweykowska, A., Hamzi, H. Q., and Skoog, F.,** Development of cytokinin activity by rearrangement of 1-substituted adenines to 6-substituted aminopurines: inactivation by N^6, 1-cyclization, *Proc. Natl. Acad. Sci. U.S.A.*, 56, 709, 1966.
192. **Hecht, S. M., Leonard, N. J., Schmitz, R. Y., and Skoog, F.,** Cytokinins: synthesis and growth-promoting activity of 2-substituted compounds, in the N^6-isopentenyladenine and zeatin series, *Phytochemistry*, 9, 173, 1970.
193. **Damman, L. G., Leonard, N. J., Schmitz, R. Y., and Skoog, F.,** Cytokinins: synthesis of 2- and 8- and 2,8-substituted 6-(3-methyl-2-butenylamino)purines and their relative activities in promoting cell growth, *Phytochemistry*, 13, 329, 1974 (and references therein).
194. **Robins, R. K.,** in *Heterocyclic Compounds*, Vol. 8, Elderfield, R. C., Ed., Wiley, New York, 1967, 281.
195. **Theiler, J. B., Leonard, N. J., Schmitz, R. Y., and Skoog, F.,** Photoaffinity-labeled cytokinins. Synthesis and biological activity, *Plant Physiol.*, 58, 803, 1976.
196. **Mornet, R., Theiler, J. B., Leonard, N. J., Schmitz, R. Y., Moore, F. H., III, and Skoog, F.,** Active cytokinins. Photo-affinity labelling agent to detect binding, *Plant Physiol.*, 64, 600, 1979.
197. **Sussman, M. R. and Kende, H.,** The synthesis and biological properties of 8-azido-N^6-benzyladenine, a potential photo-affinity reagent for cytokinin, *Planta*, 137, 91, 1977.
198. **Robins, R. K.,** Potential purine antagonists. XV. Preparation of some 6,8-disubstituted purines, *J. Am. Chem. Soc.*, 80, 6671, 1958.
199. **McDonald, J. J., Leonard, N. J., Schmitz, R. Y., and Skoog, F.,** Cytokinins: synthesis and biological activity of ureidopurines, *Phytochemistry*, 10, 1429, 1971.
200. **Moore, F. H., III,** A cytokinin-binding protein from wheat germ. Isolation by affinity chromatography and properties, *Plant Physiol.*, 64, 594, 1979.
201. **Berridge, M. V., Ralph, R. K., and Letham, D. S.,** The binding of kinetin to plant ribosomes, *Biochem. J.*, 119, 75, 1970.
202. **Letham, D. S. and Young, H.,** The synthesis of radioisotopically labelled zeatin, *Phytochemistry*, 10, 2077, 1971.
203. **Elliott, D. C. and Murray, A. W.,** A quantitative limit for cytokinin incorporation into transfer ribonucleic acid by soyabean callus tissue, *Biochem. J.*, 130, 1157, 1972.
204. **Fox, J. E., Sood, C. K., Buckwalter, B., and McChesney, J. D.,** The metabolism and biological activity of a 9-substituted cytokinin, *Plant Physiol.*, 47, 275, 1971.
205. **Walker, G. C., Leonard, N. J., Armstrong, D. J., Murai, N., and Skoog, F.,** The mode of incorporation of 6-benzylaminopruine into tobacco callus transfer ribonucleic acid. A double labeling determination, *Plant Physiol.*, 54, 737, 1974.
206. **Sussman, M. R. and Firu, R.,** The synthesis of a radioactive cytokinin with high specific activity, *Phytochemistry*, 15, 153, 1976.
207. **Fox, J. E., Erion, J. L., and McChesney, J. D.,** A new intermediate in the synthesis of a tritiated cytokinin with high specific activity, *Phytochemistry*, 18, 1055, 1979.
208. **Bruce, M. F. and Zwar, J. A.,** Cytokinin activity of some substituted ureas and thioureas, *Proc. R. Soc. London Ser. B:*, 165, 245, 1966.
209. **Takahashi, S., Shudo, K., Okamoto, T., Yamada, K., and Isogai, Y.,** Cytokinin activity of N-phenyl-N'-(4-pyridyl)urea derivatives, *Phytochemistry*, 17, 1201, 1978.
210. **Okamoto, T., Shudo, K., Takahashi, E., Kawachi, E., and Isogai, Y.,** 4-Pyridylureas are surprisingly potent cytokinins. The structure-activity relationship, *Chem. Pharm. Bull.*, 29, 3748, 1981.
211. **Hong, C. I., Cheda, G. B., Dutta, S. P., O'Grady-Curtis, A., and Tritsch, G. L.,** Synthesis and biological activity of naturally occurring 6-ureidopurines and their nucleosides, *J. Med. Chem.*, 16, 139, 1973.
212. **Trigoe, Y., Akiyama, M., Hirobe, M., Okamoto, T., and Isogai, Y.,** Cytokinin activity of azaindene, azanaphthalene, naphthalene, and indole derivatives, *Phytochemistry*, 11, 1623, 1972.
213. **Rogozinska, J. H., Kroon, C., and Alemink, C. A.,** Influence of alterations in the purine ring on biological activity of cytokinins, *Phytochemistry*, 12, 2087, 1973.

214. **Sugiyama, T., Kitamura, E., Kubokawa, S., Kobayashi, S., Hashizume, T., and Matsubara, S.,** *Phytochemistry,* 14, 2539, 1975.
215. **Weiss, R., Robins, R. K., and Noell, C. W.,** Potential purine antagonists. XXIII. Synthesis of some 7-substituted amino-v-triazolo(d)pyrimidines, *J. Org. Chem.,* 25, 765, 1960.
216. **De Roose, K. B. and Salemink, C. A.,** Deazapurine derivatives. V. A new synthesis of 1- and 3-deazaadenine and related compounds, *Rec. Trav. Chim. Phys-Bas.* 88, 1263, 1969.
217. **Robins, R. K., Holum, L. B., and Furcht, F. W.,** Potential purine antagonists. V. Synthesis of some 3-methyl-5, 7-substituted pyrazolo(4,3-d)pyrimidines, *J. Org. Chem.,* 21, 833, 1956.
218. **Hecht, S. M., Bock, R. M., Schmitz, R. Y., Skoog, F., Leonard, N. J., and Occolowitz, J. L.,** Question of the ribosyl moiety in the promotion of callus growth by exogenously added cytokinins, *Biochemistry,* 10, 4224, 1971.
219. **Skoog, F., Schmitz, R. Y., Hecht, S. M., and Frye, R. B.,** Anticytokinin activity of substituted pyrrolo(2,3-d)pyrimidines, *Proc. Natl. Acad. Sci. U.S.A.,* 72, 3508, 1975.
220. **Iwamura, H., Masuda, N., Koshimizu, K., and Matsubara, S.,** Cytokinin-agonistic and antagonistic activities of 4-substituted-2-methylpyrrolo(2,3-d)pyrimidines, 7-deaza analogs of cytokinin-active adenine derivatives, *Phytochemistry,* 18, 217, 1979.
221. **Iwamura, H., Murakami, S., Koga, J., Matsubara, S., and Koshimizu, K.,** Quantitative analysis of anticytokinin activity of 4-substituted-2-methylthiopyrido(2,3-d)pyrimidines, *Phytochemistry,* 18, 217, 1979.
222. **Uchiyama, M. M. and Abe, H.,** A synthesis of ($\pm$)-discadenine, *Agric. Biol. Chem.,* 41, 1549, 1977.
223. **Sembdner, G., Gross, D., Liebusch, H.-W., and Schneider, G.,** Biosynthesis and metabolism of plant hormones, in *Hormonal Regulation of Development I, Molecular Aspects of Plant Hormones, Encyclopedia of Plant Physiology,* New Series, Vol. 9, MacMillan, J., Ed., Springer-Verlag, Berlin, 1980, 281.
224. **Klemen, F. and Klämbt, D.,** Half life of sRNA from primary roots of *Zea mays,* A contribution to the cytokinin production, *Physiol. Plant.,* 31, 186, 1974.
225. **Maasz, H. and Klämbt, D.,** On the biogenesis of cytokinins in roots of *Phaseolus vulgaris, Planta,* 151, 353, 1981.
226. **Palni, L. M. S. and Horgan, R.,** Cytokinins in transfer RNA of normal and crown-gall tissue of *Vinca rosea, Planta,* 159, 178, 1983.
227. **Borek, E., Baliga, B. S., Gehrke, C. W., Kuo, C. W., Belman, S., Troll, W., and Waalkes, T. P.,** High turnover rate of transfer RNA in tumor tissue, *Cancer Res.,* 37, 3362, 1977.
228. **Beutelmann, P.,** Untersuchungen zur Biosynthese eines Cytokinins in Calluszellen von Lanbmoossporophyten, *Planta,* 112, 181, 1973.
229. **Chen, C.-M. and Eckert, R. L.,** Evidence for the biosynthesis of transfer RNA-free cytokinin, *FEBS Lett.,* 64, 429, 1976.
230. **Nishinari, N. and Syono, K.,** Biosynthesis of cytokinins by tobacco cell cultures, *Plant Cell Physiol.,* 21, 1143, 1980.
231. **Stuchbury, T., Palni, L. M., Horgan, R., and Wareing, P. F.,** The biosynthesis of cytokinins in crown-gall tissue of *Vinca rosea, Planta,* 147, 97, 1979.
232. **Burrows, W. J.,** Evidence in support of biosynthesis *de novo* of free cytokinins, *Planta,* 138, 53, 1978.
233. **Short, K. C. and Torrey, J. G.,** Cytokinins in seedling roots of pea, *Plant Physiol.,* 49, 155, 1972.
234. **Hall, R. H.,** Cytokinins as a probe of developmental processes, *Ann. Rev. Plant Physiol.,* 24, 415, 1973.
235. **Taya, Y., Tanaka, Y., and Nishimura, S.,** 5'-AMP is a direct precursor of cytokinin in *Dictyostelium discoideum, Nature (London),* 271, 545, 1978.
236. **Chen, C.-M. and Melitz, D. K.,** Cytokinin biosynthesis in a cell-free system from cytokinin-autotrophic tobacco tissue cultures, *FEBS Lett.,* 107, 15, 1979.
237. **Nishinari, N. and Syono, K.,** Cell-free biosynthesis of cytokinins in cultured tobacco cells, *Z. Pflanzenphysiol.,* 99, 383, 1980.
238. **Chen, C.-M.,** Cytokinin biosynthesis in cell-free systems, in *Plant Growth Substances 1982: Proc. 11th Int. Conf. Plant Growth Substances,* Wareing, P. F., Ed., Academic Press, London, 1982, 155.
239. **Palni, L. M. S. and Horgan, R.,** Cytokinin biosynthesis in crown gall tissue of *Vinca rosea:* metabolism of isopentenyladenine, *Phytochemistry,* 22, 1597, 1983.
240. **Barnes, M. F., Tien, C. L., and Gray, J. S.,** Biosynthesis of cytokinins by potato cell cultures, *Phytochemistry,* 19, 409, 1980.
241. **Letham, D. S., Tao, G. Q., and Parker, C. W.,** An overview of cytokinin metabolism, in *Plant Growth Substances 1982: Proc. 11th Int. Conf. Plant Growth Substances,* Wareing, P. F., Ed., Academic Press, London, 1982, 143.
242. **Miura, G. A. and Miller, C. O.,** 6-(γ,γ-dimethylallylamino)purine as a precursor of zeatin, *Plant Physiol.,* 44, 372, 1969.
243. **Miura, G. and Hall, R. H.,** *trans*-Ribosylzeatin: its biosynthesis in *Zea mays* endosperm and the mycorrhizal fungus, *Rhizopogon roseolus, Plant Physiol.,* 51, 563, 1973.

244. **Parker, C. W., Letham, D. S., Gollnow, B. I., Summons, R. E., Duke, C. C., and MacLeod, J. K.,** Regulators of cell division in plant tissues. XXV. Metabolism of zeatin by lupin seedlings, *Planta,* 142, 239, 1978.
245. **Entsch, B., Letham, D. S., Parker, C. W., Summons, R. E., and Gollnow, B. I.,** Metabolites of cytokinins, in *Plant Growth Substances 1979: Proc. 10th Int. Conf. Plant Growth Substances,* Skoog, F., Ed., Springer-Verlag, Berlin, 1980, 109.
246. **Wareing, P. F., Horgan, R., Henson, I. E., and Davis, W.,** Cytokinin relations in the whole plant, in *Plant Growth Regulation, Proc. 9th Int. Conf. Plant Growth Substances,* Pilet, P. E., Ed., Springer-Verlag, Berlin, 1977, 147.
247. **Henson, I. E.,** Types, formation, and metabolism of cytokinins in leaves of *Alnus glutinosa* (L.) Gaertn., *J. Exp. Bot.,* 29, 935, 1978.
248. **McGaw, B. A. and Horgan, R.,** Cytokinin oxidase from *Zea mays* kernels and *Vinca rosea* crown-gall tissue, *Planta,* 159, 30, 1983.
249. **Terrine, C. and Laloue, M.,** Kinetics of N^6-(Δ^2-isopentenyl)adenosine degradation in tobacco cells. Evidence of a regulatory mechanism under the control of cytokinins, *Plant Physiol.,* 65, 1090, 1980.
250. **Paces, V., Werstiuk, E., and Hall, R. H.,** Conversion of N^6-(Δ^2-isopentenyl)adenine to adenosine by enzyme activity in tobacco tissue, *Plant Physiol.,* 48, 775, 1971.
251. **Whitty, C. C. and Hall, R. H.,** A cytokinin oxidase in *Zea mays, Can. J. Biochem.,* 52, 789, 1974.
252. **Brownlee, B. G., Hall, R. H., and Whitty, C. D.,** 3-Methyl-2-butenal: an enzymic degradation of the cytokinin N^6-(Δ^2-isopentenyl)adenine, *Can. J. Biochem.,* 53, 37, 1975.
253. **McGaw, B. A. and Horgan, R.,** Cytokinin catabolism and cytokinin oxidase, *Phytochemistry,* 22, 1103, 1983.
254. **Parker, C. W. and Letham, D. S.,** Regulators of cell division in plant tissues. XVIII. Metabolism of zeatin in *Zea mays* seedlings, *Planta,* 115, 337, 1974.
255. **Palmer, M. V., Scott, I. M., and Horgan, R.,** Cytokinin metabolism in *Phaseolus vulgaris* L. II. Comparative metabolism of exogenous cytokinins by detached leaves, *Plant Sci. Lett.,* 22, 187, 1981.
256. **Horgan, R., Palni, L. M. S., Scott, I. M., and McGaw, B. A.,** Cytokinin biosynthesis and metabolism in *Vinca rosea* crown gall tissue, in *Metabolism and Molecular Activities of Cytokinins,* Guern, J. and Peaud-Lenoël, C., Eds., Springer-Verlag, Berlin, 1981, 56.
257. **Parker, C. W., Letham, D. S., Wilson, M. M., Jenkins, I. D., MacLeod, J. K., and Summons, R. E.,** The identity of two new cytokinin metabolites, *Ann. Bot.,* 39, 375, 1975.
258. **Henson, I. E. and Wheeler, C. T.,** Metabolism of [8-^{14}C]zeatin in root nodules of *Alnus glutinosa* L. Gaertn., *J. Exp. Bot.,* 106, 1087, 1977.
259. **Palmer, M. V., Horgan, R., and Wareing, P. F.,** Cytokinin metabolism in *Phaseolus vulgaris* L. I Variation in cytokinin levels in leaves of decapitated plants in relation to lateral bud outgrowth, *J. Exp. Bot.,* 32, 1231, 1981.
260. **Letham, D. S., Palni, L. M. S., Tao, G.-Q., Gollnow, B. I., and Bates, C.,** Regulators of cell division in plant tissues. XXIX. The activities of cytokinin glucosides and alanine conjugates in cytokinin bioassays, *J. Plant Growth Reg.,* 2, 103, 1983.
261. **Gordon, M. E., Letham, D. S., and Parker, C. W.,** The metabolism and translocation of zeatin in intact radish seedlings, *Ann. Bot.,* 38, 809, 1974.
262. **Parker, C. W., Wilson, M. M., Letham, D. S., Cowley, D. E., and MacLeod, J. K.,** The glucosylation of cytokinins, *Biochem. Biophys. Res. Commun.,* 55, 1370, 1973.
263. **Entsch, B. and Letham, D. S.,** Enzymic glucosylation of the cytokinin, 6-benzylaminopurine, *Plant Sci. Lett.,* 14, 205, 1979.
264. **Entsch, B., Parker, G. W., Letham, D. S., and Summons, R. E.,** Preparation and characterization using HPLC of an enzyme forming glucosides of cytokinins, *Biochim. Biophys. Acta,* 570, 124, 1979.
265. **Murakoshi, I., Ikegami, F., Ookawa, N., Haginiwa, J., and Letham, D. S.,** Enzymic synthesis of lupinic acid, a novel metabolite of zeatin in higher plants, *Chem. Pharm. Bull. Tokyo,* 25, 520, 1977.
266. **Entsch, B., Parker, C. W., and Letham, D. S.,** An enzyme from lupin seeds forming alanine derivatives of cytokinins, *Phytochemistry,* 22, 375, 1983.
267. **Tzou, D., Galson, E. C., and Sondheimer, E.,** The effects and metabolism of zeatin in dormant and nondormant ash embryos, *Plant Physiol.,* 51, 894, 1973.
268. **Sivadjian, A., Sadorge, P., Gawer, M., Terrine, C., and Guern, J.,** Enzymic synthesis of some N^6-substituted adenine nucleosides, *Physiol. Vég.,* 7, 31, 1969.
269. **Doree, M. and Terrine, C.,** Enzymatic synthesis of ribonucleoside-5'-phosphate from some N^6-substituted adenosines, *Phytochemistry,* 12, 1017, 1973.
270. **Chen, C.-M. and Kristopeit, S. M.,** Metabolism of cytokinins. Dephosphorylation of cytokinin ribonucleotide by 5'-nucleotidases from wheat germ cytosol, *Plant Physiol.,* 67, 494, 1981.
271. **Chen, C.-M. and Kristopeit, S. M.,** Deribosylation of cytokinin ribonucleoside by adenosine nucleosidase from wheat germ cells, *Plant Physiol.,* 68, 1020, 1981.

272. **Chen, C.-M., Melitz, D.K., and Clough, F. W.**, Metabolism of cytokinins. Phosphoribosylation of cytokinin bases by adenine phosphoribosyltransferase from wheat germ, *Arch. Biochem. Biophys.*, 214, 634, 1982.
273. **Laloue, M. and Pethe, C.**, Dynamics of cytokinin metabolism in tobacco cells, in *Plant Growth Substances 1982: Proc. 11th Int. Conf. Plant Growth Substances*, Wareing, P. F., Ed., Academic Press, London, 1982, 185.
274. **Matsubara, S.**, Structure-activity relationships of cytokinins, *Phytochemistry*, 19, 2239, 1980.
275. **Leonard, N. J., Hecht, S. M., Skoog, F., and Schmitz, R. Y.**, Cytokinins: synthesis, mass spectra, and biological activity of compounds related to zeatin, *Proc. Natl. Acad. Sci. U.S.A.*, 63, 175, 1964.
276. **Dekhuijzen, H. M. and Overeem, J. C.**, Cytokinin activity of N^6, $O^{2'}$-dibutyryl cyclic AMP and N^6-butyryladenine, *Phytochemistry*, 11, 1669, 1972.
277. **Martin, J. H., Fox, J. E., and McChesney, J. D.**, Synthesis and cytokinin activity of some N^6-acylaminopurines, *Phytochemistry*, 12, 749, 1973.
278. **Ariga, T., Masumura, M., and Kuraishi, S.**, Synthesis of kinetin-analogues, II., *Bull. Chem. Soc. Jpn.*, 32, 883, 1959.
279. **Iwamura, H., Fujita, T., Koyama, S., Koshimizu, K., and Kumazawa, Z.**, Quantitative structure-activity relationship of cytokinin-active adenine and urea derivatives, *Phytochemistry*, 19, 1309, 1980.
280. **Matsubara, S., Sugiyama, T., and Hashizume, T.**, Cytokinin activity of benzoylaminodeazapurines, pentanoylaminodeazapurines and their corresponding purine analogs in five bioassays, *Physiol. Plant.*, 42, 114, 1978.
281. **Hecht, S. M., Bock R. M., Schmitz, R. Y., Skoog, F., and Leonard, N. J.**, Cytokinins: development of a potent antagonist, *Proc. Natl. Acad. Sci. U.S.A.*, 68, 2608, 1971.
282. **Skoog, F., Schmitz, R. Y., Bock R. M., and Hecht, S. M.**, Cytokinin antagonists: synthesis and physiological effects of 7-substituted 3-methylpyrazolo(4,3-d)pyrimidines, *Phytochemistry*, 12, 25, 1973.
283. **Iwamura, H., Ito, T., Kumazawa, Z., and Ogawa, Y.**, Synthesis and anticytokinin activity of 4-substituted-7-(β-D-ribofuranosyl)pyrrolo(2,3-d)pyrimidines, *Phytochemistry*, 14, 2317, 1975.
284. **Iwamura, H., Kurimoto, M., Matsubara, S., Ogawa, Y., Koshimizu, K., and Kumazawa, Z.**, Effect of 4-substituted pyrrolo(2,3-d)-pyrimidines on the tobacco callus growth, *Agric. Biol. Chem.*, 40, 1885, 1976.
285. **Hecht, S. M.**, Probing the cytokinin receptor site, in *Plant Growth Substances 1979: Proc. 10th Int. Conf. Plant Growth Substances*, Skoog, F., Ed., Springer-Verlag, Berlin, 1979, 144.
286. **Iwamura, H., Masuda, N., Koshimizu, K., and Matsubara, S.**, Quantitative aspects of the receptor binding of cytokinin agonists and antagonists, *J. Med. Chem.*, 26, 838, 1983.
287. **Okamoto, T., Shudo, K., Takahashi, S., Kawachi, E., amd Isogai, Y.**, 4-Pyridylureas are surprisingly potent cytokinins. The structure-activity relationship, *Chem. Pharm. Bull.*, 29, 3748, 1981.

Chapter 5

ABSCISIC ACID

Nobuhiro Hirai

TABLE OF CONTENTS

I.	History and Occurrence		202
	A.	History	202
	B.	Occurrence	204
II.	Chemistry		204
	A.	Isolation and Detection	204
		1. Isolation	204
		a. Extraction	204
		b. Partition	205
		c. Paper and Thin Layer Chromatographies	206
		d. Column Chromatography	207
		2. Detection for Identification and Quantification	207
		a. HPLC	207
		b. GC	207
		c. Immunoassay	208
		d. Others	209
	B.	Structural Determination	209
		1. Melting Point and Color Reaction	209
		2. UV, Spectrometry, Optical Rotatory Dispersion (ORD), and Circular Dichroism (CD)	210
		3. Infrared (IR) and Nuclear Magnetic Resonance (NMR) Spectrometries	210
		4. Mass Spectrometry (MS)	210
	C.	Synthesis	214
		1. Racemic ABA	214
		2. Absolute Configuration Determination of ABA by Synthesis	215
		3. Phaseic Acid	220
		4. Dihydrophaseic and *Epi*-Dihrophaseic Acids	222
		5. ABA-β-GEs and DPA-β-GEs	223
III.	Biosynthesis and Metabolism		223
	A.	Biosynthesis	223
		1. Stereochemical Origin of the Carbon and Hydrogen Atoms	223
		a. Hydrogens at C-2 and 5′	223
		b. Hydrogens at C-4, 5, and 3′	225
		c. Methyl Groups at C-3, 2′, and 6′	225
		2. Direct Biosynthetic Pathway and Potent Precursors	225
		3. Carotenoid Pathway	227
		4. Biosynthesis in Fungi	228
	B.	Metabolites of ABA	232
		1. *t*-ABA	232
		2. Conjugates of ABA	232
		3. 6′-HydroxymethylABA (**96**) and Phaseic Acid (PA), (**53**)	234

    4. Conjugates of 6'-HydroxymethylABA and Phaseic Acid.........235
    5. Epimeric Dihydrophaseic Acids and a Conjugate of Dihydrophaseic Acid 235
    6. Other ABA-Related Metabolites 236
  C. Metabolic Pathway 236

IV. Biological Activity..238
  A. Exogenous ABA 238
  B. Structural Requirements for Activity 239
  C. Problems in Structure-Activity Relationship Studies....................241

References..241

I. HISTORY AND OCCURRENCE

A. History

The history of the discovery of abscisic acid is suggestive in considering the process of developmental research of biologically active substances. It resembles little tributaries joining to form a big river.

The research on abscisic acid was started by several groups whose work was not apparently related initially. After many turns and twists, they finally converged on abscisic acid, which was found to be a plant hormone with important roles in plant physiology.

The discovery of auxin suggested that plant growth was regulated by low molecular weight compounds, and researchers attempted to explain all physiological phenomena by changes of auxin levels in tissues. However, at the same time they also noticed the presence of substances interfering with *Avena* coleoptile growth, the standard bioassay for auxin.

The pioneering work on growth inhibitors was carried out by Hemberg. Using the *Avena* coleoptile growth bioassay, he found (1949) that potato peels contained growth inhibitors at high levels, and that the levels of the inhibitors decreased during conditions which broke dormancy.[2] He further showed the presence of a similar inhibitor in buds of *Fraxinus excelsior,* and found a correlation between the inhibitor levels and the degree of bud dormancy.[3] Based on these results, he pointed out the possibility that plant growth was regulated by levels of both auxin and inhibitor.

In 1952, Bennet-Clark et al.[4] applied paper chromatography (PC) to the analysis of plant growth substances in plant extracts, and introduced a bioassay for monitoring active substances. A paper chromatogram developed with a mixture of 2-propanol, ammonia, and water was cut into ten equal strips, and activity of their water extracts on *Avena* coleoptile growth was tested. In addition to growth promotion by auxin, growth inhibition was observed at an R_f range between 0.6 and 0.7. Subsequently this growth inhibitor was shown to be widely distributed in various plant species, and changes of the inhibitor levels in response to environmental conditions were observed. These also suggested the physiological importance of the growth inhibitor, as Hemberg pointed out, and it was named "inhibitor β".[5,6]

Bud primordia of woody plants such as Betulaceae and Aceraceae are formed at the leaf bases in July, and after completion of bud formation in September, buds enter into dormancy, which is broken the next spring. Phillips and Wareing[7] investigated the changes of promotor and inhibitor levels in terminal buds and leaves of *Acer pseudoplatanus* with wheat coleoptile growth and lettuce seed germination tests. The levels of the inhibitor in terminal buds were high in October, fell in winter, and reached a minimum in June. The changes of the inhibitor levels in leaves were opposite to those in buds, suggesting that the inhibitor formed in leaves in summer moved to terminal buds at the beginning of autumn. They supposed that the seasonal changes were related to photoperiodism, and examined the inhibitor levels in *Acer* seedlings grown under long or short day conditions.[8] The wheat coleoptile growth test showed that the inhibitor levels in the short day treated seedlings were significantly higher than those in the long day treated seedlings. Furthermore, the inhibitor was shown to be present in leaves of *Betula pubescens*. When the crude inhibitor was applied to *Betula* leaves being grown under long day photoperiodism, the leaf growth was inhibited and the terminal buds became dormant.[9] Wareing named the inhibitor "dormin" after the ability to induce dormancy.

Wisterias bloom from the base of flower clusters to the apex, but only the first one or two flowers fructify and the others abort. If all flowers fructified, competition for nutrition would occur and all seeds could not mature. Clustered flowers of *Lupinus luteus* show a similar behavior of abortion. Van Steveninck[10] observed in 1957 that *Lupinus* flowers infected by pea mosaic virus before flowering did not abort. He investigated the effect of fruits on the abscission of flowers, and found that the abscission of upper flowers was accelerated by lower immature seeds. When the lower immature seeds were removed, the upper flowers did not fall. This result suggested that an abscission-accelerating substance was formed in immature seeds. The abscission-accelerator occurring in the fruit pods was acidic, and caused growth inhibition of wheat coleoptiles.[11] Rothwell and Wain[12] purified the abscission-accelerator to obtain a crystalline solid with a melting point at 145 to 152°C, but did not succeed in determining the structure.

The isolation and elucidation of the structure of the abscission-accelerator were successfully carried out by Addicott's group, which studied the abscission of immature cotton fruits. Several-day-old cotton fruits easily abscise, and the maximum percentage of abscissing fruits extends to 65%. Addicott et al.[13] found that the bases of immature fruits contained a substance inhibiting the curvature response of *Avena* coleoptiles to auxin, and confirmed that the substance had a growth-inhibitory activity on *Avena* coleoptiles. The inhibitor was an ether-soluble acidic compound, and was present at the highest level in 6-day-old immature fruits. For the isolation of the abscission-accelerator, not the misleading growth inhibition test, the cotton explant test was employed as the monitoring bioassay.[13] The explant consisted of excised cotyledonary nodes of 14-day-old cotton seedlings. The samples were applied to the stumps of cotyledonary petioles, and the numbers of abscissed petioles were counted after applying a force of 5 grams to the end of each petiole at 24-hr intervals. Monitoring the abscission activity with this method, Ohkuma et al.[14] succeeded in the isolation of a highly active substance in a yield of 9 mg from 225 kg of immature cotton fruits, and called it "abscisin II". In 1965, the plane structure (1) of abscisin II was finally elucidated from spectral data of UV, IR, PMR, and MS,[15] and shortly after was verified by synthesis.[16] The absolute configuration at C-1' was determined to be *S* in 1973.

Cornforth et al.[17] purified dormin from *Acer* leaves and found that dormin was the same substance as abscisin II. Further, the flower abscission-accelerator of *Lupinus* was identified as abscisin II, and the major component of inhibitor β was considered to be abscisin II.[18-20] Thus, abscisin II or dormin was shown to be widely distributed in plants and to play important roles in different physiological phenomena.

The unification of the name was discussed at the 6th International Conference on Plant Growth Substances. Considering the priority of the research on abscisin II, the new name abscisic acid and the abbreviation ABA were adopted by the researchers.[21,22] Now the use of the name abscisic acid with its abbreviation has been fully established.

More than 20 years have passed since abscisic acid was isolated. During these years knowledge of the chemistry and physiology of abscisic acid has been accumulated, and a sensitive quantification method developed. Its metabolism and biological activity have been elucidated to a certain degree, but we are still ignorant about its biosynthetic route, mechanism of action, and physiological roles.

(1) ABA

(2) Lunularic acid

B. Occurrence

ABA has been detected in various higher plants including Spermatophyta and Pteridophyta. It is likely that there is no vascular plant which does not contain ABA. The occurrence of ABA extends to the mosses.[23] ABA has not been found from the investigated liverworts, algae, or bacteria. Instead, lunularic acid (2) as a plant growth inhibitor occurs in liverworts and algae, and in these species lunularic acid may play an ABA-like physiological role.[23-25]

A recent remarkable finding is that fungi are able to produce ABA, as gibberellin is produced by *Gibberella fujikuroi*. ABA production by *Penicillium italicum* collected from infected oranges was suggested by Rudnicki et al.,[26] but it was likely that the ABA in the fungus was derived from the orange peel. The first report on fungal ABA production appeared in 1977. During the screening of the genus *Cercospora* for secondary metabolites, Assante et al.[27] found that a strain of *C. rosicola* Passerini, a leaf spot fungus of roses, produced (+)-ABA at a high yield of 6 mg/100 mℓ medium when grown on a potato agar medium.

Since then, two other species of fungi have been shown to produce (+)-ABA. The ABA-producing ability of *Cercospora cruenta* IFO 6164 in shaking culture was reported by Oritani et al.[28] in 1982. At the same time Marumo et al.,[29] who were studying spore germination stimulators, found that (+)-ABA is produced by *Botrytis cinerea*, which causes gray mold disease on grapes, strawberries, and other cultivated plants.

Interestingly the ABA yields of *B. cinerea* are affected by light.[29] While *B. cinerea* grown under a blacklight blue lamp produces ABA in a yield of 0.2 mg/100 mℓ medium, blue light irradiation during the culture enhances the ABA production more than tenfold. Mosses seemed to be the lowest class of plants which have ABA as a self growth-regulating hormone. However, considering the effect of light on ABA production with Borecka's observation that ABA stimulated spore germination of *B. cinerea*,[30] ABA appears to have some physiological role in the life cycle of *B. cinerea*.

Whether the fungal ABA is involved in the infection processes in the host plants is unknown, but the fungal ABA could affect the physiological state of the host plants as a toxin to cause disease symptoms.

The species of ABA-producing fungi known so far are restricted to phytopathogenic ones, thus there is a possibility that the ABA biosynthesis gene of plants might have been transferred into fungi in a process of co-evolution between host plants and parasites, rather than the ABA-producing fungi being the evolutionary origin of higher plants. This may be supported by the discontinuity between the fungi and mosses, and the presence of lunularic acid in the liverworts and algae.

After the isolation of ABA, many ABA related compounds such as ABA conjugates and ABA metabolites have been found from natural sources. Their names and abbreviations are summarized in Table 1.

II. CHEMISTRY

A. Isolation and Detection

1. Isolation

a. Extraction

ABA levels rapidly change in response to drought. If quantification of ABA is the goal, plant materials should be frozen and lyophilyzed as soon as they are prepared. Strong acid

Table 1
ABBREVIATIONS

Abbreviation	Trivial name
ABA	Abscisic acid
t-ABA	2-*trans*-Abscisic acid
ABA-β-GEs	Abscisic acid-1-*O*-β-glucosyl ester
ABA-α(β)-GEt	Abscisic acid-1'-*O*-α(β)-glucoside
ABA-β-MEs	Abscisic acid-1-*O*-β-maltosyl ester
HMG-HOABA	β-Hydroxy-β-methylglutarylhydroxy-abscisic acid
PA	Phaseic acid
PA-β-GEs	Phaseic acid-1-*O*-β-glucosyl ester
PA-β-GEt	Phaseic acid-1'-*O*-β-glucoside
DPA	Dihydrophaseic acid
DPA-β-GEt	Dihydrophaseic acid-4'-*O*-β-glucoside
epi-DPA	*epi*-Dihydrophaseic acid

and basic conditions and heating should be avoided during extraction and isolation. Certain conjugated metabolites decompose in such conditions and give stable unconjugated metabolites, which will mislead researchers. Furthermore, extraction should be carried out in the dark or in dim light or else the 2-*cis* double bond of ABA will isomerize to *trans* and give a mixture of ABA and *t*-ABA.[31] Such isomerization occurs in PA, DPA, and *epi*-DPA.

A number of researchers have used aqueous methanol to extract ABA and related compounds from plant materials. Absolute methanol is a good solvent for extraction because enzymes in plant tissues appear not to function in methanol.[32] ABA and related compounds are quite soluble in methanol. The only fault of methanol is that methanolysis or methylation may occur during extraction to form methyl esters. ABA-β-GEs in avocado fruits decomposes during extraction with basic methanol to yield the methyl ester of ABA as an artifact.[33] Acetone or acidic methanol is used to eliminate the possibility of methanolysis. However, acidic methanol extraction may cause methylation of carboxyl groups; for example, the carboxyl group of the acyl moiety in HMG-HOABA is easily methylated with acidic methanol to form the monomethyl ester of HMG-HOABA. Therefore, the use of acetone has been recommended. The ABA metabolites known so far do not possess a visinal dihydroxyl group, so the possibility that acidic acetone may cause a formation of acetonide is presumably excluded. In addition, ethanol and mixtures of solvents have been used for the extraction of ABA.

b. Partition

As ABA is a weak acid, the ordinary procedures for solvent partitioning of acidic compounds can be used. After concentration of the methanol extract under vacuum, the insoluble material including chlorophylls is removed by centrifugation or filtration. Standing the concentrate overnight at low temperature is effective in precipitating the insoluble material. The aqueous concentrate is adjusted to pH 3 and extracted with ether or ethyl acetate at least three times. If a large amount of lipophilic substances is contained in the aqueous concentrate, extraction with hexane may be performed before the ether or ethyl acetate extraction. To obtain only acidic substances, the ether or ethyl acetate phase is extracted with 5% sodium bicarbonate. Neutral substances such as xanthoxin remain in the organic phase. The aqueous phase is adjusted to pH 3 and extracted with ether or ethyl acetate. ABA and some PA, DPA, and *epi*-DPA are contained in the organic phase. Extraction of the aqueous phase with water-saturated *n*-butanol takes up PA, DPA, *epi*-DPA, and the residual ABA. Some of the conjugated metabolites are extracted with the *n*-butanol phase, but conjugated metabolites almost all remain in the aqueous phase. The extraction procedure

Table 2
PARTITION COEFFICIENTS (K_d) FOR ABA WHEN PARTITIONED BETWEEN EQUAL VOLUMES OF ORGANIC SOLVENTS & WATER AT 4 pH VALUES[34]

Organic solvent	Dielectric constant	pH of aqueous phase			
		2.5	5.0	7.0	9.0
Hexane	1.9	93.0	55.9	82.0	72.0
Diethyl ether	4.3	0.302	2.13	6.61	62.9
Chloroform	4.81	0.330	16.3	165	382
Ethylacetate	6.0	0.100	0.736	4.32	138
Methylenechloride	9.09	1.05	6.88	35.1	123
1-Butanol	17.8	0.198	0.186	0.625	126
Water saturated 1-butanol	—	0.019	0.295	0.419	2.14

K_d = Concentration in aqueous phase/concentration in organic phase.

mentioned above is a general method, and other extraction methods are applicable for ABA and related compounds. In partitioning, any solvent cannot completely extract compounds. Chia et al.[34] have reported partition coefficients for ABA which is partitioned between an organic solvent and water adjusted to different pH values. The data, as shown in Table 2, are helpful in knowing the effectiveness of the applied partitioning, but the actual partition coefficients appear to be lower than the ideal values when many impurities are present. The quantification of ABA and its metabolites at the same time seems to be difficult. The error caused by solvent partitioning is unavoidable. It may be necessary to choose the most appropriate partitioning procedure for each compound. Use of an internal standard has been recommended for the quantification of the desired compounds. t-ABA, ($\pm$)-ABA, radioactive ABA, and deuterated ABA have been used as the internal standards for ABA.[31,35-37] Such kinds of internal standards will be applicable for related compounds.

c. Paper and Thin Layer Chromatographies

PC was the major technique in plant hormone research until the 1960s. This technique has been replaced by thin layer chromatography (TLC) because of the time-consuming procedure and the poor separation efficiency.

TLC is widely used to detect ABA and related compounds in fractions purified by column chromatography. If a compound is identified as a spot on TLC, it can be monitored by TLC in the course of purification without using a bioassay. Preparative TLC is effective for final purification, and for preliminary purification for gas chromatography (GC) and high performance liquid chromatography (HPLC).

Silica gel plates containing a fluorescent indicator are usually used. ABA and related compounds have a strong UV absorption moiety and can be observed as quenched spots on TLC under UV_{254} light. When sprayed with ethanolic sulfuric acid followed by heating, the ABA spot gives a bright yellow color at the beginning of heating and yellow-green fluorescence under UV_{365} light.

Various solvent systems have been employed in the development of TLC. Pure ABA, PA, and DPA show R_f values of 0.7, 0.5, and 0.15, respectively, when developed with a mixture of toluene-ethyl acetate-acetic acid (25:15:2) which is a common solvent system for the unconjugated metabolites.[38] The conjugated metabolites are also detected with silica gel TLC. HMG-HOABA has an R_f value of 0.28 in toluene-ethyl acetate-acetic acid (7:12:1), and ABA-β-GEs show R_f values of 0.9 in n-propanol-ethyl acetate-water (3:2:1), and 0.39 in chloroform-methanol-water (75:22:3).[39,40] DPA-β-GEt is observed at R_f 0.28 in n-propanol-ethyl acetate-water (4:15:1). We should bear in mind that R_f values are affected by the applied amounts of compounds, the presence of impurities, and the moisture content of the silica gel.

d. Column Chromatography

Column chromatography is used for isolation, but is not suited to preliminary purification for quantification due to the poor reproducability of elution positions and recovery. Chromatography packings and techniques which are used in the purification of other plant hormones are applicable to ABA and related compounds.

Charcoal, polyvinylpyrrolidone (PVP), and Amberlite® XAD excel in selective removal of impurities such as phenols, and they are used at the first purification step.[31,40-43] These packings are especially suitable for the purification of the polar conjugated metabolites since the elution can be carried out with aqueous solvents.

The chromatographic behavior of ABA and related compounds in silica gel is similar to that on silica gel TLC.

Normal phase partition column chromatography is effective, especially for the purification of HMG-HOABA.[39] Silica gel, Celite®, and Sephadex® LH-20 imbibing 0.5 M formic acid or 1 M phosphate buffer adjusted to pH 5.5 are employed for the carrier and stationary phases, and the mobile phase is a mixture of n-hexane-ethyl acetate or toluene-n-butanol.[39,44-46] Reversed-phase partition column chromatography using ODS is performed under pressure to purify ABA and related compounds including conjugated metabolites at the final step.

Sephadex® LH-20 has characteristic properties which are effective for purification in some cases.[42] Reeve and Crozier[47] have adopted porous polystyrene beads for the separation of plant hormones. ABA is eluted later than gibberellins (GAs), auxin, and cytokinins. In gel permeation, not only steric exclusion, but adsorption and partition effects influence the chromatographic behavior of compounds. Sandberg et al.[48] have reported that combined PVP/Sephadex® LH-20 column eluted at pH 4.5 gives a good separation of plant hormones including ABA.

Ion exchangers have not been widely used for the purification of ABA, but may be applicable to the purification of polar conjugated metabolites.

2. Detection for Identification and Quantification

a. HPLC

A UV_{254} monitor is used for detection. The lower limit of detection of ABA, PA, DPA, and *epi*-DPA is 50 ng. Before the analysis, crude samples should be purified by solvent partitioning and preparative TLC or by a short column such as the C_{18} Sep-pak cartridge.

As shown in Figure 1, ABA, *t*-ABA, PA, DPA, and *epi*-DPA can be separated by a reversed-phase column eluted with aqueous methanol containing 0.1% acetic acid.[49,50] Gradient elution is widely used. A reversed-phase column is also effective for the identification of the polar conjugated metabolites. ABA, *t*-ABA, and PA are separated by an adsorption type column, but DPA and *epi*-DPA are not eluted with a mixture of acetonitrile and chloroform due to their high polarity.[34] Derivatization such as methylation and/or acetylation is necessary for the analysis of DPA and *epi*-DPA by an adsorption type column. Ion exchanger columns can also be applied.[51]

The problem in HPLC analysis is the low specificity of the UV detector. The identification by HPLC is accompanied with some risk that a wrong compound with the same t_R as that of the desired compound may be mistaken for it. To ensure the identification, a combination with other identification method is recommended.

b. GC

Methyl esters of ABA and unconjugated metabolites are detected by an electron capture

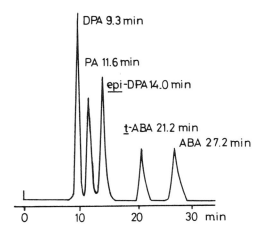

FIGURE 1. HPLC separation of ABA metabolites. Column: μ-Bondapak C_{18}; solvent: 35% methanol containing 0.1% acetic acid; flow rate: 0.6 mℓ/min; detection: UV_{254}.

detector (ECD) with high selectivity and sensitivity.[52] Most of the impurities detected by a flame ionization detector (FID) are not detected by ECD. The detection limit of an ECD is on the picogram order, and there is linearity between 10 pg and 50 ng.[34] GC-ECD is, therefore, one of the most reliable methods for the identification and the quantification of ABA and unconjugated metabolites. Zeevaart and Milborrow[38] have reported the t_R of the volatile derivatives of ABA, PA, DPA, *epi*-DPA, and their *trans* isomers when analyzed with XE-60, SE-30, and OV-17 columns.

Gas chromatography/mass spectromety (GC/MS) is a powerful method for identification, but the samples must be cleaned up so that the peaks of the desired compounds are clearly observed without overlapping with impurity peaks in GC-FID. GC/MS operated in selective ion monitoring (SIM) is comparable to GC-ECD in high selectivity and sensitivity. An example of quantification of *t*-ABA will be described later.

c. Immunoassay

Immunoassay for ABA has rapidly developed since 1979 and has been confirmed as the most sensitive and selective detection method for ABA. Preparation of antigen and antiserum is a time-consuming process, but the advantage of immunoassay is that a number of crude samples without preliminary purification is semiautomatically tested in a short time with high accuracy. Use of immunoassay is most highly recommended for quantification of ABA and related compounds. The bioassay for quantification will be replaced by immunoassay in the future.

Bovine serum albumin (BSA), human serum albumin (HSA), and hemocyanin have been used for carrier proteins to be conjugated with a hapten ABA.[53,55] There are two ways of conjugation, as shown in Figure 2.[56] Antigen conjugated to C-1 of ABA through an amide bond is used for total ABA determination; antigen conjugated to C-4' of ABA through a hydrazone linkage is used for free ABA determination. The difference between the total and free ABA contents corresponds to the content of conjugated ABA.

A rabbit is immunized by repeated injection of the antigen, and 8 weeks after the antiserum is collected containing the antibody specifically binding ABA in vitro to form a complex. While the antiserum prepared from (±)-ABA-C-1-BSA binds (−)-ABA rather than (+)-ABA, the antisera from (+)-ABA-C-1-HSA and (±)-ABA-C-4'-BSA show high affinity to (+)-ABA, and only the latter shows cross-reaction with (−)-ABA.[53,56] When quantified with these antisera, the detection limit is 30 to 50 pg. Mertense et al.[57] have reported the

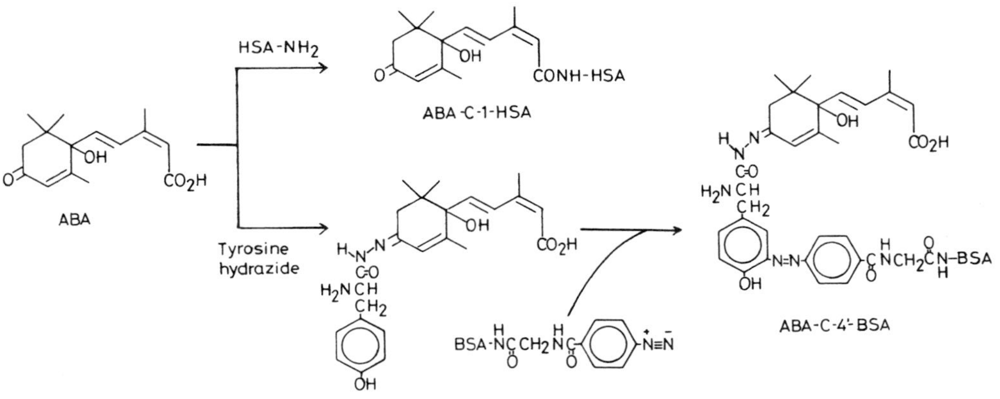

FIGURE 2. Synthesis of ABA-serum albumin conjugates, ABA-C-1-HSA and ABA-C-4'-BSA.[56]

production of monoclonal antibody which exhibits high affinity and specificity to (+)-ABA. In the radioimmunoassay using this monoclonal antibody, the detection limit is 4 pg of (+)-ABA, which allows direct quantification of ABA in raw plant extracts.

Three methods for complex determination have been reported. Fuchs and Mayak[55] used conjugates of ABA with bacteriophage T4 which are completely inactivated by the antibody. Free ABA inhibits this inactivation, and the surviving active phages are observed as plaques on agar plates containing the host *E. coli*. The sensitivity of this bioimmunoassay is not high. Radioimmunoassay for ABA has been introduced by Weiler and Walton et al.[53,54,56] In the radioimmunoassay, [^{3}H]-ABA is used instead of the ABA-linked bacteriophage in the bioimmunoassay. After incubation of the antiserum with [^{3}H]-ABA and free ABA, the radioactivity bound to the antibody is counted. The radioactivity percentage is in inverse proportion to the amount of the added free ABA. Daie and Wyse[58] have developed an enzyme-linked immunosorbent assay (ELISA) based on competitive binding between free and alkaline phosphatase-labeled ABA with antibodies. The bound enzyme activity is measured using the hydrolysis of *p*-nitrophenylphosphate.

Immunoassay will be applied to the metabolites of ABA and is expected to contribute to understanding the relationships between physiological phenomena and changes of ABA levels by accurate quantification.

d. Others

Radioassay for ABA has been reported by Whenham and Fraser.[59] This assay utilizes the radioactivity of ^{14}C as a marker and that of ^{3}H as the internal standard. After methylation of ABA in crude plant extracts with ^{14}C-diazomethane, the labeled ABA is purified by TLC, and the radioactivity counted. The ABA concentration is corrected by measuring the recovery of the internal standard [^{3}H]-ABA.

The intense ORD of ABA can be applied to quantification, but the presence of other optically active and UV absorbing substances will interfere, and the preliminary purification of plant extract is needed.

B. Structural Determination

This section deals with the physicochemical and the spectral properties for identifying ABA and related compounds. The precise spectral data of each compound should be looked up in the corresponding reference. The absolute configuration was determined by synthetic studies and discussed in Section II.C.2.

1. Melting Point and Color Reaction

While the synthetic racemate of ABA melts at 191°C,[16] natural (+)-ABA recrystallized

(3) (4)

from a mixture of *n*-hexane and benzene shows a melting point at 160 to 161°C.[14] Usually natural (+)-ABA shows a slightly lower melting point (159 to 161°C) than that of synthetic (+)-ABA.[60] The methyl esters of PA and DPA melt at 157 to 158°C and 138 to 139°C, respectively.[49]

As mentioned before, ABA forms a yellow-green fluorescent compound after sulfuric acid spray followed by heating on silica gel.[61] The compound, however, has not been identified. When ABA is treated with a mixture of formic acid and hydrochloric acid, several neutral compounds are formed, one of which has been shown to be an unsaturated γ-lactone (3).[62] Further treatment of the lactone with alcoholic alkali gives the violet-red color caused by the zwitterion (4). The violet-red color fades rapidly as the lactone is hydrolyzed.

2. UV Spectrometry, Optical Rotatory Dispersion (ORD), and Circular Dichroism (CD)

ABA shows an intense UV absorption due to the enone, α, β and γ,δ-unsaturated carboxyl groups. The absorption maximum of the methyl ester is observed at 264 to 265 nm (ε 20,900).[17] The UV spectrum of ABA is affected by pH because of the dissociation of the carboxyl group. While in acidic ethanol ABA exhibits an absorption maximum at 262 nm with a shoulder at 240 nm; in basic ethanol the maximum occurs at 245 nm.[63] The methyl esters of PA, DPA, and *epi*-DPA show maxima at 265, 268, and 268 nm, respectively.[49] Their molar extinctions are comparable to that of ABA.

ABA molecule has an asymmetric carbon at C-1', and two chromophores of the dienecarboxyl and enone groups, which give ABA its strong optical activity. In acidic ethanol, the $[\alpha]_D$ is +430°, and an intense positive Cotton effect with extreme at 289 nm (+24,000°) and 246 nm (−6,900°) is observed.[35] In its CD spectrum, ABA shows extreme at 262 nm ($\Delta\epsilon$ +39.5), 230 nm ($\Delta\epsilon$ −34), and 318 nm ($\Delta\epsilon$ −2.5).[35] These spectra also change in response to pH. PA, which does not have an enone group, shows a negative plain curve in the ORD.[64] The $[\alpha]_D$ values of DPA and *epi*-DPA are too small to measure.

3. Infrared (IR) and Nuclear Magnetic Resonance (NMR) Spectrometries

The IR spectrum of ABA has been demonstrated by Hashimoto et al.[65] and discussed by MacMillan and Pryce[66] and Ohkuma et al.[15] When measured by the KBr pellet method, the absorptions by the hydroxyl and the enone groups occur at 3405 and 1650/cm, respectively. The broad band at 3405 to 2300/cm and the bands at 1674, 1623, and 1600/cm are assigned to the dienecarboxyl group, which has been supported by the comparison with the IR spectrum of *cis, trans*-β-methylsorbic acid.[15] PA shows the absorption band of the carboxyl group at 1718/cm, and the other band positions are similar to those of ABA. The IR spectrum of PA has been interpreted by MacMillan and Pryce.[66] The steric relationships between the 4'- and 1'-hydroxyl groups of DPA and *epi*-DPA have been studied by Milborrow.[67]

The PMR spectrum of ABA has been reported by Ohkuma et al.[15] The characteristic point in the PMR is that the proton at C-4 is observed at δ7.93 ppm due to the deshielding effect of the carboxyl group. The proton at C-4 of *t*-ABA occurs at δ6.47 ppm. Milborrow[67] has summarized the PMR spectra of PA, DPA and *epi*-DPA. CMR of ABA obtained from a fungus has been reported by Assante et al.[27]

4. Mass Spectrometry (MS)

The publications on electron impact (EI) MS data of ABA and related compounds have been summarized by Dörffling and Tietz.[68] Gray et al.[69] have precisely studied the frag-

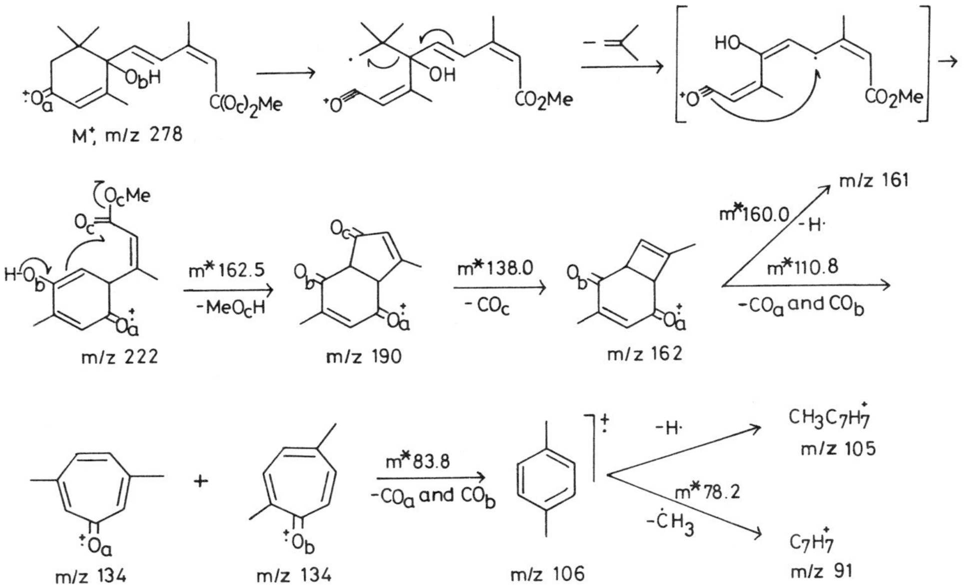

FIGURE 3. Major fragmentation pathway of ABA methyl ester.[69]

mentation pattern of the methyl ester of ABA in EIMS using isotopically labeled and metastable analysis, and defined the fragmentation sequence as shown in Figure 3.

EIMS generally needs derivatization of samples such as methylation for carboxyl and acetylation or trimethylsilylation for hydroxyl groups before measurements to increase the volatility. Derivatization has two disadvantages: it is time consuming and compounds such as the glucosyl ester of ABA decompose by derivatizing treatments.

The ionization method for MS has recently made rapid progress, and chemical ionization (CI), field desorption (FD), and fast atom bombardment (FAB) have quickly spread among mass spectroscopists. These ionization methods can ionize even highly polar and nonvolatile compounds without derivatization.

CIMS has been shown for a number of polyfunctional compounds, such as glucosides and saccharides, to provide definite molecular weights and valuable structural information on molecular weights of aglycon and sugar moieties.[70,71]

Netting et al.[37] have shown that in CIMS, methyl ester of ABA gives the base peak at m/z 261, corresponding to the water-lost ion, whereas methyl ester of t-ABA is at m/z 279, corresponding to the quasimolecular ion.

Takeda et al.[72] have applied CIMS not only to ABA but also to related compounds including conjugated metabolites. The CIMS data with iso-butane, ammonia, and deuterated ammonia as reagent gases for ABA and the unconjugated metabolites are summarized in Table 3. Although these compounds uniformly fail to give prominent molecular ions in the EIMS without chemical derivatization, intense quasimolecular ions ([QM]$^+$; e.g., [MH]$^+$, [M·NH$_4$]$^+$, [M·ND$_4$]$^+$) are clearly observed under iso-butane, ammonia, and deuterated ammonia modes of ionization. In the former two cases, the observed major fragment ions can easily be interpreted by sequential losses of water and/or carbon dioxide from quasimolecular ions. The water-lost ion peaks are present in the CIMS with both iso-butane and ammonia. These ions seem to form by the loss of the 1′-hydroxyl group from the parent ions. The [MH−2H$_2$O]$^+$ and [M·NH$_4$−2H$_2$O]$^+$ are characteristic of DPA, which has two hydroxyl groups in the molecule. The expulsion of carbon dioxide from the quasimolecular ions ([MH−CO$_2$]$^+$ and [M·NH$_4$−CO$_2$]$^+$) is also observed in ABA and unconjugated metabolites. Those diagnostic ions (including quasimolecular ions) are well defined by using a shift technique with deu-

Table 3
DIAGNOSTIC IONS FROM THE CIMS OF ABA AND ITS UNCONJUGATED METABOLITES[72]

		ABA	PA	DPA	epi-DPA
iso-C_4H_{10}	$[MH]^+$	265(100)	281(18)	283(26)	283(30)
	$[MH-H_2O]^+$	247(56)	263(100)	265(100)	265(100)
	$[MH-2H_2O]^+$	—	—	247(18)	247(10)
	$[MH-CO_2]^+$	221(19)	237(6)	239(14)	239(7)
	$[MH-CO_2-H_2O]^+$	203(5)	—	221(20)	221(8)
NH_3	$[M \cdot NH_4]^+$	282(100)	298(100)	300(100)	300(100)
	$[M \cdot NH_4-H_2O]^+$	264(7)	280(20)	282(32)	282(40)
	$[M \cdot NH_4-2H_2O]^+$	—	—	264(5)	264(6)
	$[M \cdot NH_4-CO_2]^+$	238(8)	254(38)	256(6)	256(12)
	$[M \cdot NH_4-CO_2-H_2O]^+$	—	—	238(5)	238(3)
	$[MH]^+$	265(6)	—	283(10)	283(20)
	$[MH-H_2O]^+$	—	—	265(27)	265(60)
	$[MH-2H_2O]^+$	—	—	247(5)	247(4)
	$[MH-CO_2]^+$	221(4)	—	239(5)	239(5)
	$[MH-CO_2-H_2O]^+$	203(3)	—	221(20)	221(8)
No. of active H (n)		2	2	3	3
ND_3	$[d_nM \cdot ND_4]^+$	288(100)	304(100)	307(100)	307(100)
	$[d_nM \cdot ND_4-D_2O]^+$	268(8)	284(9)	287(16)	287(33)
	$[d_nM \cdot ND_4-2D_2O]^+$	—	—	267(12)	267(37)
	$[d_nM \cdot ND_4-CO_2]^+$	244(4)	260(4)	263(5)	263(6)
	$[d_nM \cdot ND_4-CO_2-D_2O]^+$	—	—	243(5)	243(3)
	$[d_nMD]^+$	268(8)	—	287(16)	287(33)
	$[d_nMD-D_2O]^+$	—	—	267(12)	267(37)
	$[d_nMD-2D_2O]^+$	—	—	247(*)	247(*)
	$[d_nMD-CO_2]^+$	224(2)	—	243(3)	243(3)
	$[d_nMD-CO_2-D_2O]^+$	204(*)	—	223(8)	223(4)

Note: Relative intensities are indicated in parentheses. Ions marked by an asterisk were ambiguously observed.

terated ammonia as a reagent gas, since active hydrogens are completely exchanged for deuteriums. The quasimolecular ions and several diagnostic fragment ions with the ammonia CI mode shift to appropriate m/z values corresponding to deuterated ion species.

In CIMS, ABA-β-GEs, DPA-β-GEt, and HMG-HOABA show several fragment ions derived from aglycone or sugar or acyl moieties in addition to quasimolecular ions. For these conjugated metabolites, however, desorption chemical ionization (DCI) can give better mass spectra than CI.[73] The apparatus of DCI consists of a coiled tungsten wire and a chemical ionization chamber. Compounds on the heated tungsten coil are ionized by collision with an ionized reagent gas and desorbed from the coil surface. DCI-MS (ammonia and deuterated ammonia) of ABA-β-GEs, DPA-β-GEt, and HMG-HOABA are shown in Figure 4. The quasimolecular ion $[M \cdot NH_4]^+$ of ABA-β-GEs is clearly observed at m/z 444 as a base peak with 170° for the source temperature. The intensity of fragment ions increases with source temperature. The fragment ions at m/z 282, 198, and 180 observed at 230°C are assigned to aglycone-derived ion $[A \cdot OH+NH_4]^+$, sugar-derived ions $[S \cdot OH+NH_4]^+$, and $[S \cdot NH_3]^+$, respectively. These assignments are confirmed by using deuterated ammonia. The details of fragmentation are also shown in Figure 4. DPA-β-GEt also shows a quasimolecular ion $[M \cdot NH_4]^+$ as a base peak at 170°. The fragment ion at m/z 418 is characteristic of DPA-β-GEt and corresponds with $[M \cdot NH_4-CO_2]^+$. This decarboxylated ion is not observed in DCI-MS of esters like ABA-β-GEs. A high source temperature gives fragment ions at 300, 282, and 180 due to $[A \cdot OH+NH_4]^+$, $[A \cdot NH_3]^+$, and $[S \cdot NH_3]^+$, respectively. Although HMG-HOABA gives a weak quasimolecular ion in CIMS, an intense quasimol-

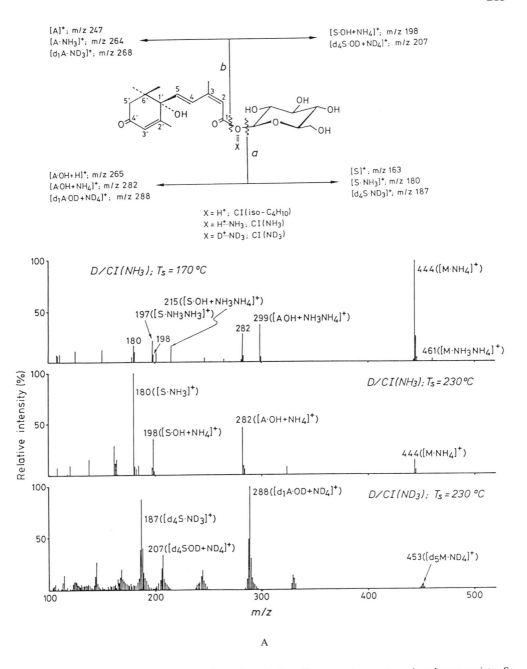

FIGURE 4 (A—C). DCI-MS of ABA conjugated metabolites (Ts: source temperature, A: aglycone moiety, S: sugar moiety). (A) ABA-β-GEs; (B) DPA-β-GEt; (C) HMG-HOABA.

ecular ion occurs at m/z 442 in DCI-MS. At 230°C source temperature, 6′-hydroxymethyl ABA-derived ions [A·OH + NH$_4$]$^+$ and [A·NH$_3$]$^+$, and the HMG-derived ion [Acyl·NH$_3$]$^+$ are observed at m/z 298, 280, and 162, respectively. The details of the fragmentation are shown in Figure 4C.

Thus, CI- and DCI-MS provide strong means by which to identify ABA and related compounds including conjugated metabolites. If CI- or DCI-MS combined with micro HPLC (LC/CIMS) becomes practical, they will replace GC/MS.

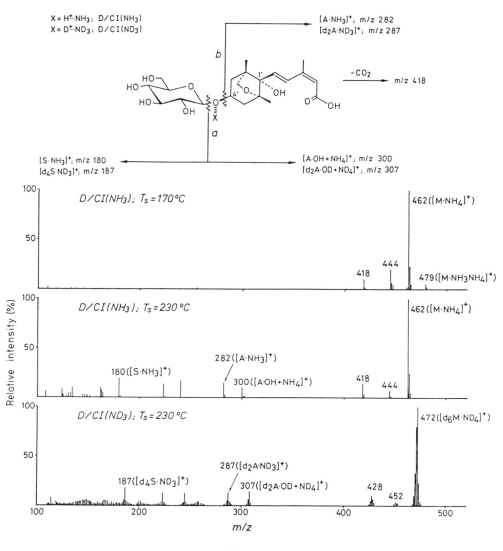

FIGURE 4B.

C. Synthesis

1. Racemic ABA

ABA is a first monocyclic sesquiterpene of the ionone type. The synthesis of ABA, including the stereospecific synthesis, was an intriguing target for organic chemists.

Cornforth et al.[16] reported the synthesis of racemic ABA as soon as Ohkuma presented the structure.[15] They started the synthesis with 3-methyl-5-(2,6,6-trimethylcyclohexa-1,3-dienyl)-2Z,4E-pentadienoic acid (6), which is a synthetic intermediate for vitamin A_2 and is synthesized from β-ionone (5) and methyl bromoacetate, as shown in Figure 5A. Irradiation of (6) with visible light in an atmosphere of oxygen in the presence of eosin as a photosensitizer readily gave crystalline epidioxide (7). ABA and its *trans* isomer were formed through rearrangement with alkaline treatment of (7) at a yield of 7%. This synthetic route is suitable for small scale preparation in the laboratory.

A three-step synthesis, as shown in Figure 5B, was developed by Roberts et al.[74] Dehydrovomifoliol (9) obtained from α-ionone (8) by *t*-butylchromate oxidation was coupled with carboethoxymethylenetriphenylphosphorane to yield equal amounts of ethyl esters of

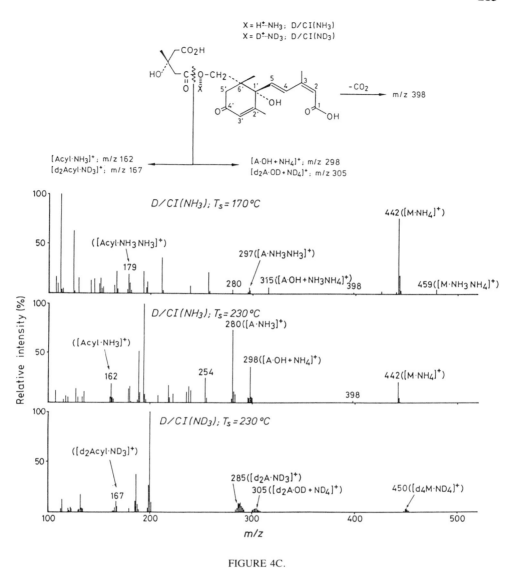

FIGURE 4C.

ABA and its *trans* isomer. After saponification ABA was readily obtained by slow crystallization from ether. This synthesis is easy, but several by-products are formed and the total yield, 11%, is low.

A third method (Figure 5C), developed by the Hofmann-La Roche Co.,[75] is superior in the geometric selectivity of double bonds to other methods. The intermediate (**12**) was prepared from the adduct of (**10**) and (**11**), and converted to the unsaturated aldehyde (**14**) via hydrogenation of the alkyn (**13**) with Lindlar catalyst. In this hydrogenation of a triple bond, a *cis* double bond was not formed. Finally, oxidation of (**14**) with silver oxide gave (±)-ABA.

2. Absolute Configuration Determination of ABA by Synthesis

There are many biologically active compounds which are optically active, and generally unnatural enantiomers are inactive or less active than the natural ones. The absolute structure elucidation at C-1' of ABA and its synthesis were important to correctly evaluate the relationship between the structure and the biological activity. In addition, the stereochemistry

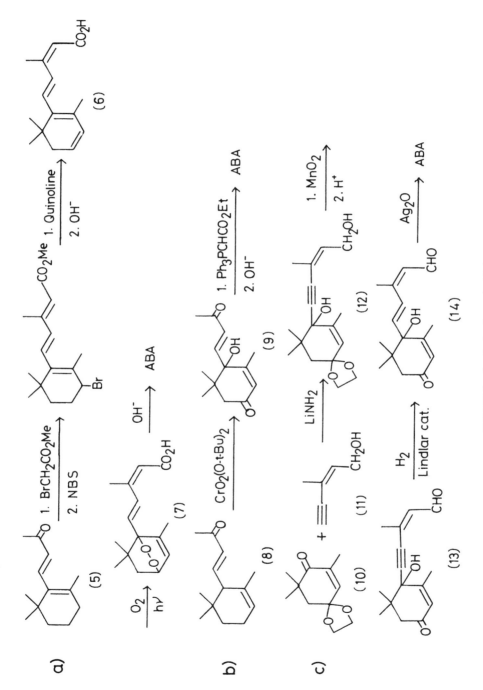

FIGURE 5. Synthesis of racemic ABA.

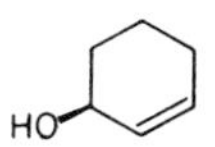

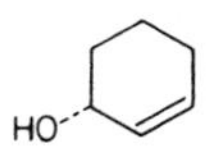

(15) (16)

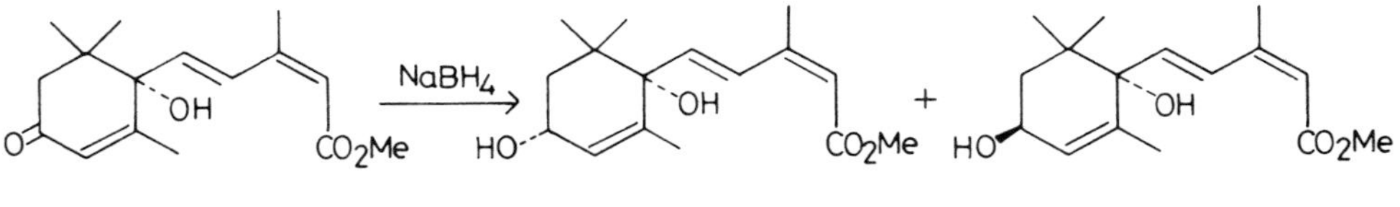

(17) (18) cis-DiolABA methyl ester (19) trans-DiolABA methyl ester

FIGURE 6. Reduction of ABA methyl ester.

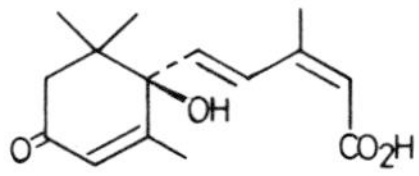

(20) (-)-ABA

of ABA was an intriguing problem in physical chemistry since ABA has an intense positive Cotton effect in the ORD.

Cornforth et al.[76] deduced the absolute structure of ABA in 1967 by applying Mill's rule that 2-cyclohexanol **(15)** is more levoratatory than its enantiomer **(16).**[77] Synthetic (±)-ABA was optically resolved by fractional crystallization of their brucine salts. The methyl ester **(17)** of the resulting (+)-ABA was reduced with sodium borohydride to yield a mixture of optically active *cis*-diol **(18)** and *trans*-diol**(19),** as shown in Figure 6. The diols were separated by TLC, using markers of the racemic *cis*-diols prepared from the epidioxide **(7)** and identified by MS. The *cis*-diol **(18)** was more levorotatory ($[M]_D$ +390°) than the *trans*-diol **(19)** ($[M]_D$ +990°). According to Mill's rule, the secondary hydroxyl of (+)-ABA should be β-configuration. From this result, they assigned the structure **(20)** to (+)-ABA, and it was accepted for several years.

The work of Taylor and Burden,[78] however, cast doubt on Cornforth's structure from the structure of the degradation product of violaxanthin, whose absolute configuration was confirmed by Weedon based on X-ray analysis and synthetic study of model compounds.[79] Zinc permanganate oxidation of violaxanthin **(21)** gave xanthoxin **(22),** the inhibitory activity of which was comparable to that of (+)-ABA in the wheat coleoptile section, the lettuce hypocotyl, and bean abscission tests. As shown in Figure 7, xanthoxin was converted to 1-aldehydeABA **(23)** by chromium trioxide oxidation, and further oxidation of **(23)** with active manganese dioxide yielded ABA. The ABA showed the same positive Cotton effect in the ORD as that of natural (+)-ABA. Since these reactions from violaxanthin to ABA avoided alternation of the cyclohexane ring, the (+)-ABA must have the same configuration as that of violaxanthin. Hence, this result indicated that the absolute configuration of either violaxanthin or (+)-ABA required revision, and reexamination of the absolute configuration of (+)-ABA became necessary.

In 1972, Oritani and Yamashita[80] synthesized (−)-ABA from (−)-α-ionone **(24)** whose absolute configuration was confirmed by Eugster et al.[81] α-Ionone was oxidized with selenium dioxide in ethanol, as shown in Figure 8. The retention of the configuration at the tertiary hydroxyl group was shown by the following facts: selenium dioxide oxidation in ethanol retains the configuration, and the ORD and CD curves of (−)-1′-hydroxy-α-ionone **(25)** were very similar to those of (−)-α-ionone. (−)-1′-Hydroxy-α-ionone was converted to ethyl abscisate **(26)** by the Wittig reaction followed by *t*-butylchromate oxidation. Although

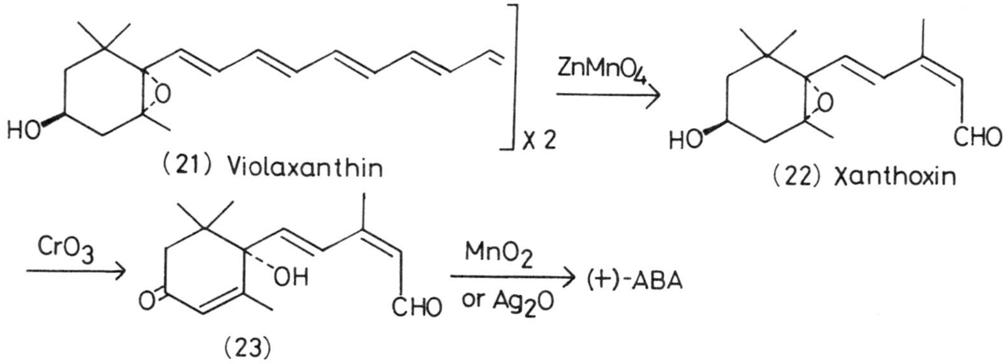

FIGURE 7. Chemical conversion of violaxanthin to (+)-ABA.

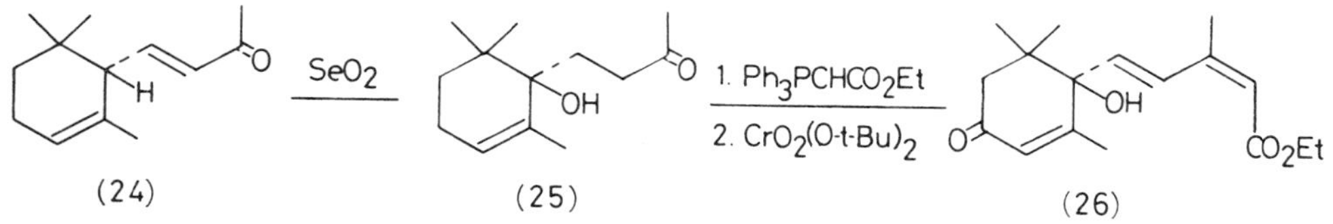

FIGURE 8. Synthesis of (−)-ABA.

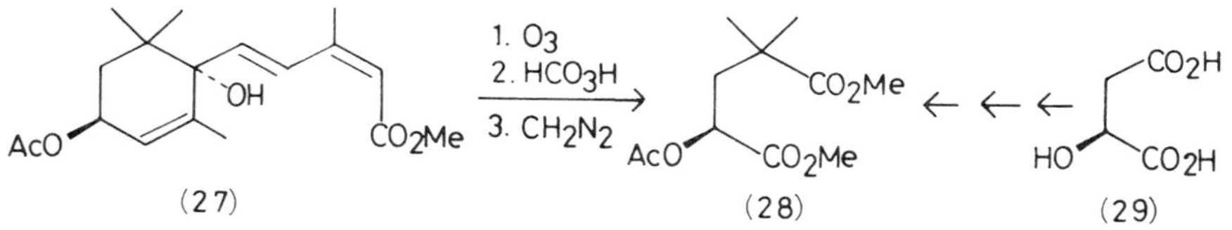

FIGURE 9. Oxidative ozonolysis of (+)-ABA.

the ABA obtained had the same configuration at C-1' as that of the (+)-ABA (structure submitted by Cornforth et al.), it showed a negative Cotton effect in CD. This result indicated that the absolute configuration for (+)-ABA was not (20) but (1).

A few months later, Ryback[82] of Cornforth's group reported the revision of the absolute stereochemistry of (+)-ABA based on another method, as shown in Figure 9. 4'-Acetyl-*trans*-diol (27) derived from (+)-ABA was degraded to (28) by ozonolysis, oxidation with performic acid, and methylation with diazomethane. On the other hand, the acetoxy diester was synthesized by electrolysis from methyl S-2-acetoxy-3-carboxypropionate (29) derived from S-malic acid and methyl dimethylmalonate. Both of the acetoxy diesters showed levorotation, indicating that the absolute configuration of the intermediate, *trans*-diol acetate, was defined as (27), and of (+)-ABA as (1).

Koreeda et al.[83] synthesized (+)-ABA from epidioxide (30) employing HPLC in the resolution of the racemic intermediate. Figure 10 shows the synthetic scheme. Racemic *cis*-diol (31) prepared from (30) was converted into the diastereomeric α-methoxy-α-trifluoromethylphenylacetyl (MTPA) ester. These were successfully separated into (32) and (33) by four recycles in an HPLC column. After hydrolysis of the less polar ester (32) followed by benzoylation and oxidation, they proved the absolute configuration of the benzoate (34) by applying the exciton chirality method. The result showed that the structure of the less polar MTPA ester was (32), and inevitably another MTPA ester had the structure of (33). The MTPA ester (33) was hydrolyzed and oxidized with Jones reagent to give (35), which

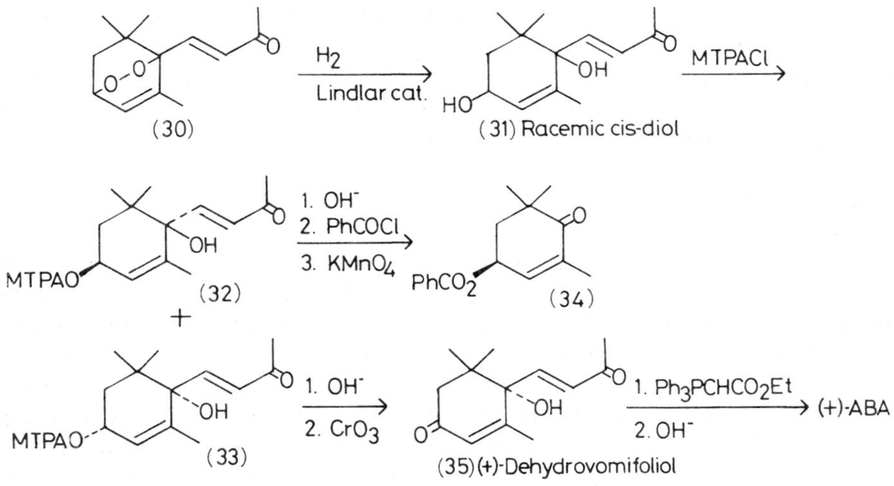

FIGURE 10. Synthesis of (+)-ABA by Koreeda and co-workers.

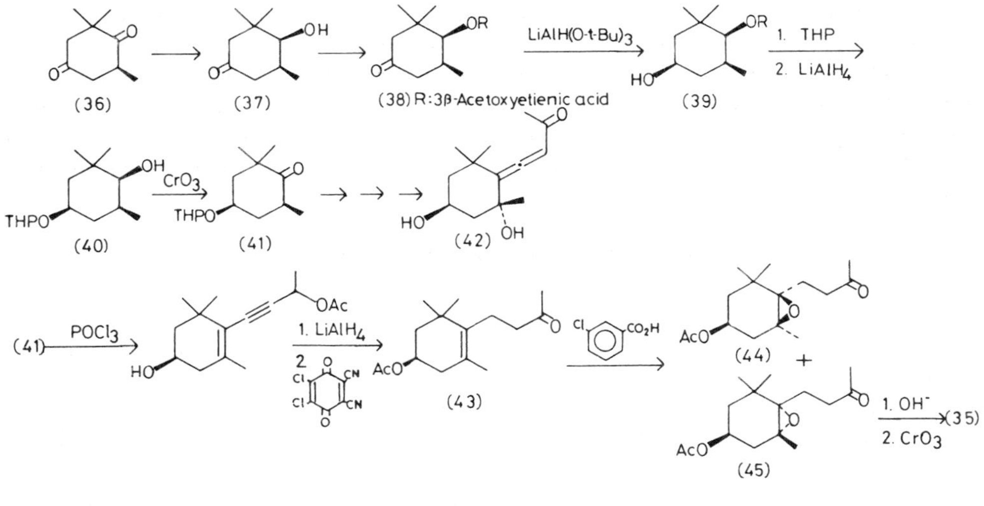

FIGURE 11. Synthesis of (+)-dehydrovomifoliol by Mori.

has been known to be dehydrovomifoliol, isolated from the roots of *Phaseolus vulgaris* by Takasugi et al.[84] The oxidiketone (35) was converted to (+)-ABA by the Wittig reaction, indicating that the structure (1) for (+)-ABA is correct.

Harada[85] directly employed the exciton chirality method to (+)-ABA and defined the configuration as (1) by theoretical calculation.

Mori[86,87] utilized an optically active ketone (41) as a starting compound to synthesize (+)-ABA. The ketone was prepared as shown in Figure 11. The diketone (36) was converted to racemic ketol (37), which was resolved by fractional crystallization of the diastereomers prepared from racemic (37) and 3β-acetoxyetienic acid. The enantiomer (38) was reduced with tri-tertiarybutoxy lithium aluminum hydride to give (39). Reduction of THP ether (40) followed by oxidation afforded the protected ketol (41). The configuration of (41) was deduced from CD comparison of grasshopper ketone (42), synthesized from (41) through several steps, with the natural one whose absolute structure was confirmed by X-ray analysis. The THP ether (41) was converted to 3β-acetoxy-β-ionone (43), which was treated with *m*-

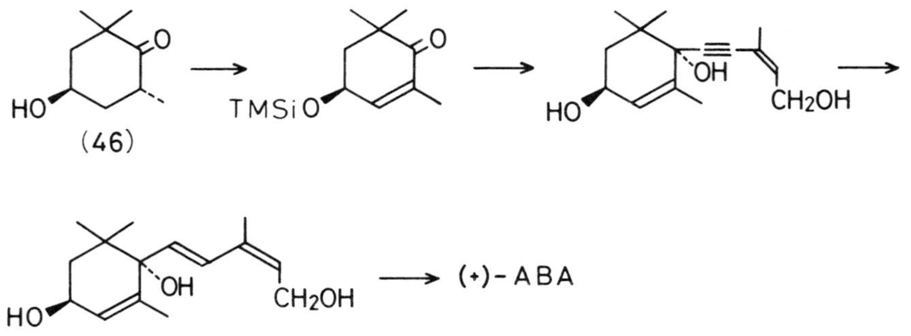

FIGURE 12. Synthesis of (+)-ABA by Kienzle and co-workers.

chloroperbenzoic acid to afford β- and α-epoxide (44) and (45). The major product was assigned as β-epoxide (44) since the steric course of this type of reaction has been known to give the β-epoxide due to the steric hindrance of the quasi-axial methyl group. Therefore, the minor product was α-epoxide (45), which was identical with the degradation product of violaxanthin in mp, IR, PMR, and CD. Hydrolysis of (45) followed by Sarret oxidation yielded (+)-dehydrovomifoliol (35). The configuration of the C-O bond of the epoxy ring is retained in the base-catalyzing opening during the Sarret oxidation. Since (+)-dehydrovomifoliol had been previously converted to (+)-ABA by Koreeda et al.,[83] a link between grasshopper ketone (42) and (+)-ABA was provided by this synthetic work, and the configuration of (+)-ABA was indirectly proved by X-ray analysis.

Kienzle et al.[88] synthesized (+)-ABA and related compounds from one common starting material, 4R,6R-4-hydroxy-2,2,6-trimethylcyclohexane (46), as shown in Figure 12.

3. Phaseic Acid

Phaseic acid (PA, 53) is the first stable metabolite of ABA. The PA molecule has oxabicyclo-(3,2,1)-octane, which is formed by intramolecular Michael addition of the hydroxyl at C-6' to the double bond at C-2'.

Hayase[89] employed dienone ketal (48) to construct this ether bridge, as shown in Figure 13. The dienone ketal (48) was a key compound in the synthesis of trisporic acid and trisporone, and is available in three steps from unsaturated ester ketal (47).[90] Reduction of (48) with lithium aluminum hydride followed by benzoylation gave the dibenzoate (49), which was then converted to the trieneketal (50) by reflux in dimethylaniline. The trieneketal was irradiated with a fluorescent light in the presence of rose bengal. This is the same reaction as the photooxidation used in the synthesis of ABA by Cornforth. Without isolating the epidioxide, the photooxidation product was passed through an alumina column to give the mixture of isomers. The mixture was treated with acetic acid and separated by preparative TLC to give two dienones, (51) and (52). The side chain was constructed by a Wittig reaction with carbomethoxymethylenetriphenylphosphorane. After alkaline hydrolysis of the carboxymethyl and benzoyl groups, acidification of the reaction solution caused Michael addition, and yielded the mixture of cis and trans isomers of PA. Two isomers were separated by preparative TLC to afford natural PA (53) and its trans isomer. Two other isomers, epi-phaseic acid (epi-PA, 54) and its trans isomer, were obtained from (52).

This synthetic work also contributed to the elucidation of the absolute configuration of PA. Milborrow first suggested that the oxymethylene bridge of PA was trans to the tertiary hydroxyl group, since the intramolecular hydrogen bonding between the oxymethylene oxygen and the hydrogen of the tertiary hydroxyl group was not observed in IR spectra measured in carbon disulfide.[91] Hayase and associates, however, revised it by investigating the PMR spectra of the synthetic PA (53) and its epimer (54). They showed that neither (53) nor (54) had intramolecular hydrogen bonding in the IR spectra. In the PMR, the oxymethylene

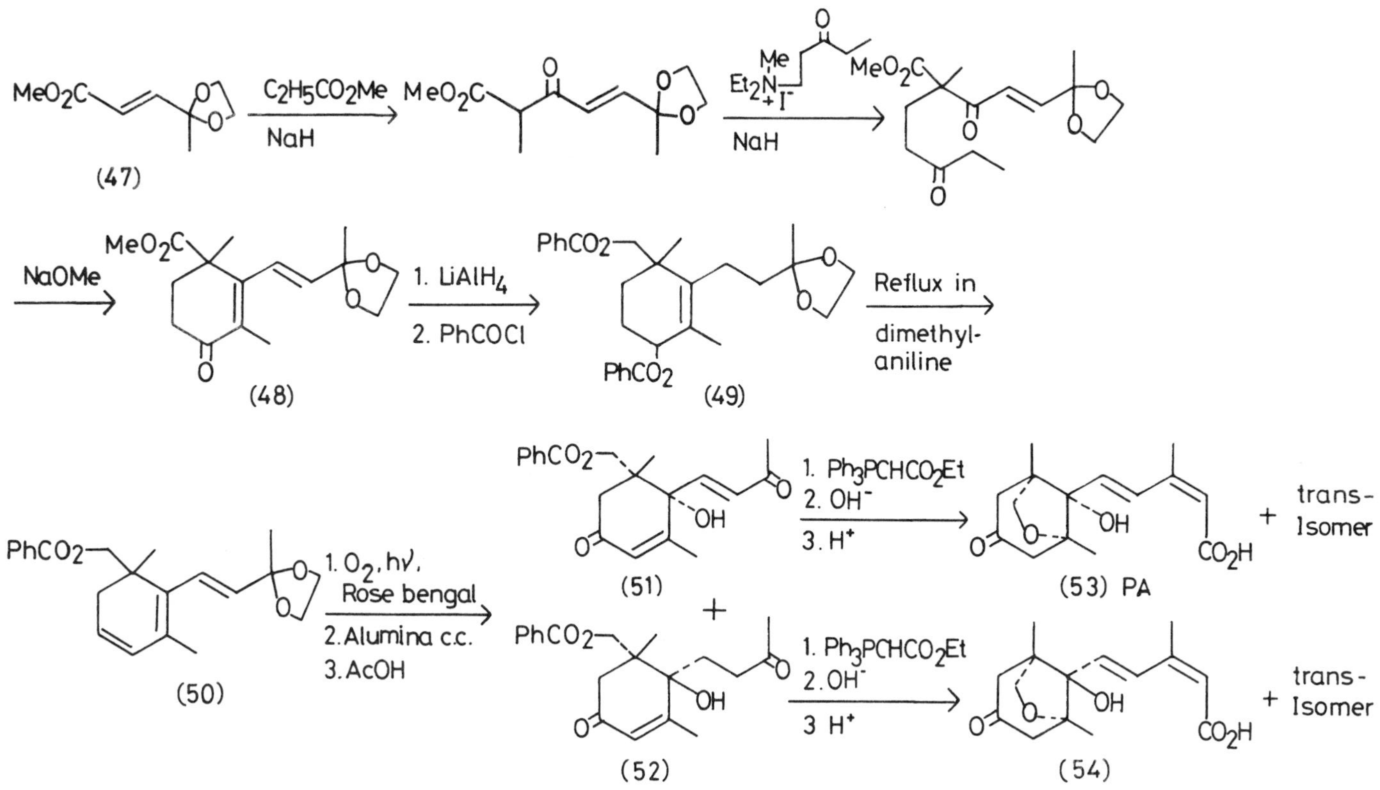

FIGURE 13. Synthesis of PA and *epi*-PA.

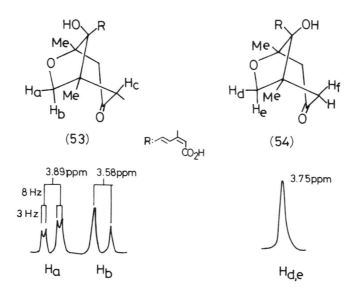

FIGURE 14. Anisotropic effect of tertiary hydroxyl group of PA and *epi*-PA in PMR.[89]

protons of (53) showed an AB-quartet, and the proton H_a at the lower field showed a W-type long-range coupling with proton H_c of the methylene protons at C-5', as shown in Figure 14. If the tertiary hydroxyl group is *cis* to the oxymethylene bridge, this magnetic nonequivalency of the methylene protons can be accounted for by the anisotropic effect of the tertiary hydroxyl oxygen. On the other hand, the signal of the oxymethylene protons of (54) occurred as a singlet of A_2-type, indicating that there is no anisotropic effect. The tertiary hydroxyl group of (54) is, therefore, *trans* to the oxymethylene bridge. The PMR spectra of the reduced compounds, (55) and (56), were obtained to ensure that the methylene protons are not affected by the anisotropic effect of the double bonds of the side chain. The methylene protons of (55) and (56) showed the same signals as those of (53) and (54), respectively. From these results, the absolute configuration of natural PA was assigned as (53). Milborrow[67] confirmed this by the precise investigation of the PMR of methyl esters of PA, DPA, and *epi*-DPA.

4. Dihydrophaseic and Epi-Dihdrophaseic Acids

The methyl ester of dihydrophaseic acid (DPA, 57) and its epimer (*epi*-DPA, 58) are easily prepared by reduction of methyl ester of PA with sodium borohydride in cold methanol.[92] Two methyl esters of DPA and *epi*-DPA can be separated by multiple development on silica gel TLC to give a more polar epimer, methyl ester of DPA, and a less polar epimer, methyl ester of *epi*-DPA.

Milborrow[67] investigated the PMR and IR spectra of methyl esters of PA, DPA, and *epi*-DPA precisely. The IR spectrum of *epi*-DPA showed a hydroxyl absorption whose relative intensity was independent of its concentration in chloroform solution, and in the PMR the signal of the 4'-hydroxyl hydrogen appeared as a doublet split by 4'-hydrogen, indicating the restriction of the 4'-hydroxyl rotation. These structural characters of *epi*-DPA indicated the presence of hydrogen bonding between the 4'-hydroxyl hydrogen and the oxymethylene oxygen, whereas DPA lacked these spectral characteristics in the IR and PMR. Since hydrogen bonding is possible with the 4'-hydroxyl group *cis* to the oxymethylene bridge, the absolute configurations of DPA and *epi*-DPA were assigned at (57) and (58), respectively (see Figure 25).

5. ABA-β-GEs and DPA-β-GEt

The tetraacetate of ABA-β-GEs (59) (see Figure 25) is prepared by the reaction of ABA and α-acetobromoglucose in the presence of triethylamine.[93] Although the selective removal of the acetyl groups was difficult because of the instability of the glucosyl ester linkage, Lehmann and Schütte[94] succeeded by employing a crude esterase of *Helianthus annuus* seeds.

The tetraacetate of DPA-β-GEt methyl ester is synthesized from the methyl ester of DPA and α-acetobromoglucose in the presence of silver nitrate,[49] according to the Koenig-Knorr method.[95] The acetyl and methoxy groups are removed with alkaline treatment to yield DPA-β-GEt (60) (see Figure 25).

III. BIOSYNTHESIS AND METABOLISM

A. Biosynthesis

ABA is a typical sesquiterpene consisting of three isoprene units, which means that ABA is synthesized from (+)-3R-mevalonic acid (MVA). In fact, a feeding experiment showed that natural (+)-3R-MVA was incorporated into ABA by slices of avocado fruit, whereas (−)-3S-MVA was not.[96,97] The stereochemical origin of the carbon and hydrogen atoms of ABA has been confirmed, but the biosynthetic pathway has not been elucidated.

Two alternative biosynthetic pathways have been proposed for ABA in plants. In the first of these, i.e., the direct biosynthetic pathway, it is postulated that a C_{15} precursor is cyclized and then converted into ABA. In the second, the carotenoid pathway, a cleavage of a carotenoid is postulated so that a terminal ring and six carbon atoms of backbone are liberated as a C_{15} moiety for the skeleton of ABA.

1. Stereochemical Origin of the Carbon and Hydrogen Atoms

The stereochemical origin of the hydrogen atoms on the double bonds in ABA has been elucidated using stereochemically labeled MVA with tritium.

a. Hydrogens at C-2 and 5'

If ABA is synthesized from MVA via all *trans*-farnesylpyrophosphate in the direct pathway, as shown in Figure 15, either the pro-4R or 4S hydrogen of MVA is expected to be retained at C-2 and C-5' of ABA. Robinson and Ryback[98] fed [2-^{14}C]-MVA mixed with stereochemically labeled [4R-^{3}H] or [4S-^{3}H]-MVA to avocado fruits. The ABA biosynthesized from the [4R-^{3}H]-MVA showed a ^{14}C:^{3}H ratio close to 3:2, whereas that biosynthesized from the [4S-^{3}H] isomer carried no tritium atom. One of the two tritiums of the former ABA was lost by treatment with a strong base, indicating that the lost tritium had been retained at C-5' of ABA. This result gave further information on the geometry of the double bond formation at C-2 of ABA. It has been known as Cornforth's basic principle for isoprenoid biosynthesis that isopentenylpyrophosphate (IPP) and dimethylallylpyrophosphate (DMAPP) condense in two ways; the elimination of the pro-4S hydrogen from MVA gives a *trans* double bond,[99,100] as in the biosynthesis of squalene, phytoene, and farnesol, and that of the pro-4R hydrogen a *cis* double bond, as in rubber.[101] The retention of the 4R-^{3}H of MVA at C-2 of ABA indicated that the C-2 double bond had been formed in *trans*. Therefore, the C-2 double bond must isomerize from *trans* to *cis* at an early biosynthetic stage because plants cannot isomerize *t*-ABA to ABA.[102]

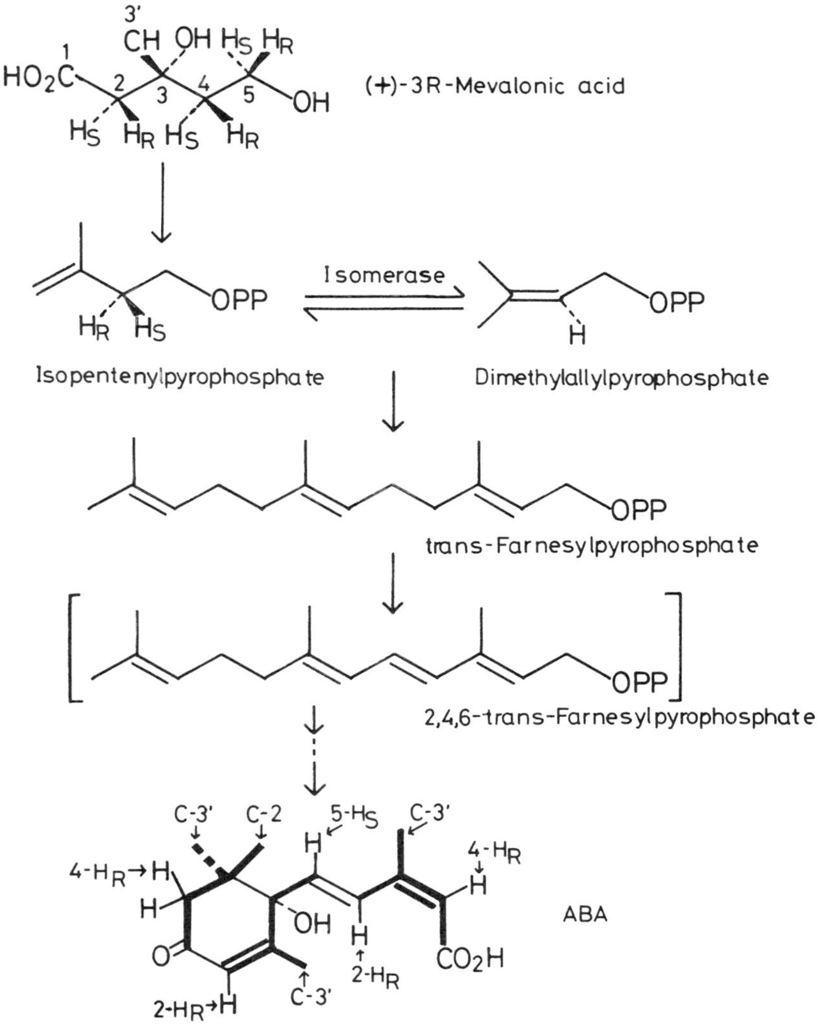

FIGURE 15. Origin of carbon and hydrogen atoms of ABA.

Suga et al.[103] have recently found that in contrast with Cornforth's basic principle, the pro-4S hydrogen of MVA is eliminated to form a *cis* double bond in the biosynthesis of polyprenol in Euphorbiacea. The mechanism of the usual *trans* double bond formation postulates that an allyl residue and an IPP held in prenyltransferase, as shown in Figure 16A, condense to eliminate the pro-2S hydrogen of IPP which corresponds to the pro-4R hydrogen of MVA. Suga and associates[104] have submitted the mirror image structure to the active site of prenyltransferase participating in the biosynthesis of polyprenol in *Mallotus japonica*, as shown in Figure 16b. This finding raises a doubt that the *cis* double bond of ABA might be formed directly by elimination of the pro-4S hydrogen of MVA with the same mechanism as that of the biosynthesis of the *cis* double bonds of polyprenol.

Using a cold trap method, Mallaby[105] showed that [2-^{14}C]-MVA was incorporated into 2,4,6-*trans*-dehydrofarnesol by avocado fruit, but not into 2-*cis*-4,6-*trans*-dehydrofarnesol. Since 2,4,6-*trans*-dehydrofarnesol seems to be a precursor of ABA, this result may support the possibility that the 2-*trans* double bond forms at an early stage and isomerizes to *cis*. However, it has not been confirmed that 2,4,6-*trans*-dehydrofarnesol is metabolized to ABA,

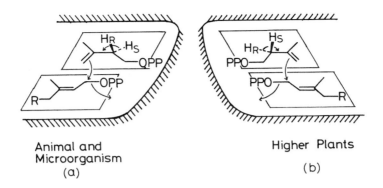

FIGURE 16. Spatial arrangement of the substrate on the enzyme participating in the formation of the *cis*-isoprene units of polyprenols.[104]

and we cannot exclude the possibility that the 2-*cis* double bond in directly formed by the unusual elimination of the pro-4S hydrogen of MVA.

b. Hydrogens at C-4, 5, and 3'

5S-[2-^{14}C,5^3-H]-MVA with a ^{14}C:^{3}H ratio of 3:3 was applied to avocado fruit, and the corresponding ratio in the biosynthesized ABA was 3:1.[106] The position of tritium was determined by chemical treatment to be C-5. Therefore, the origin of the C-5 hydrogen atom of ABA was confirmed to be the pro-5S hydrogen of MVA. In further studies, 2R-[2-^{14}C,2-^{3}H] and 2S-[2-^{14}C,2-^{3}H]-MVA were fed to avocado fruit in the presence of iodoacetamide to inhibit IPP/DMAPP isomerase, causing randomization of the label. The result showed that the pro-2R hydrogen of MVA was retained at C-3' and C-4 of ABA. The retention of the pro-5S and pro-2R hydrogens of MVA at the C-4 *trans* double bond means a *trans* elimination of the pro-5S and pro-2R hydrogens of MVA. This double bond formation is similar to that of carotenoids.

c. Methyl Groups at C-3, 2', and 6'

The origin of the methyl groups of ABA was investigated by Milborrow.[107] ABA biosynthesized from [2-^{14}C, 2-^{3}H]- mevalonolactone by avocado fruit was applied to tomato shoots, and metabolized to PA. The ^{14}C:^{3}H ratio of the isolated PA was similar to that of the applied ABA, suggesting that the unlabeled 6'-methyl was hydroxylated. The hydroxylation has been known to occur at the pro-6'S methyl by the stereochemical study of PA,[67] so it was suggested that the pro-6'S and R-methyl groups of ABA were derived from the C-3' and C-2 methyl groups of MVA, respectively. This was confirmed by the following experiment. Experimental feeding [3'-^{14}C, 3'-^{3}H]-mevalonolactone to avocado fruit showed that the C-3' of MVA was retained at three methyl groups attached to C-3, C-2', and C-6' of ABA.[108]

2. Direct Biosynthetic Pathway and Potent Precursors

Although radioactive ABA can be recovered from plant tissues to which radioactive MVA has been applied, no radioactive intermediate has been identified. The intermediate might be unstable and easily converted to ABA. Another approach to the intermediate employs ABA-derivatives which are supposed to be converted to ABA by oxidation and/or hydroxylation and/or isomerization in plant tissues.

There have been many ABA-derivatives synthesized to compare their biological activities in several bioassay systems with ABA for the study of structure-activity relationship. In these derivatives, certain compounds have been found to be converted to ABA by plants and/or plant tissues.

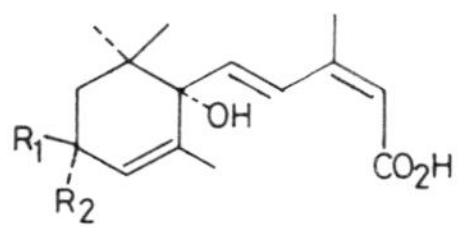

	R_1	R_2	
(61)	H	OH	cis-DiolABA
(62)	OH	H	trans-DiolABA

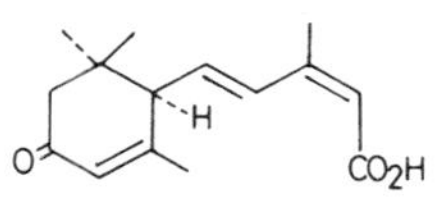

(63) 1'-DeoxyABA

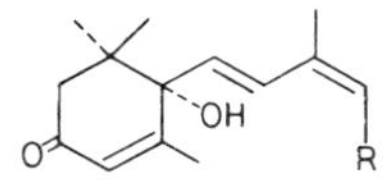

	R	
(64)	CH₂OH	1-AlcoholABA
(65)	CHO	1-AldehydeABA

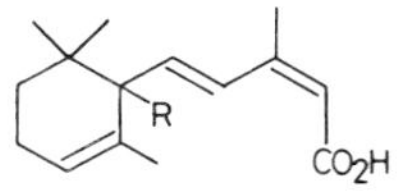

	R
(66)	H
(72)	OH

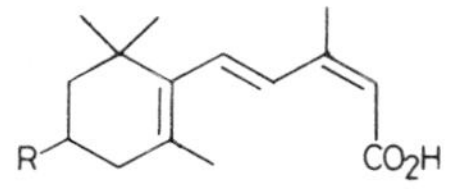

	R
(67)	H
(68)	OH

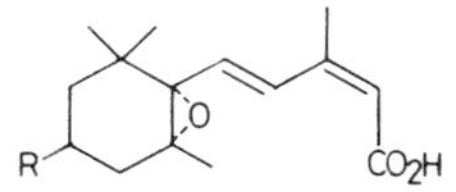

	R
(69)	β-OH
(70)	H

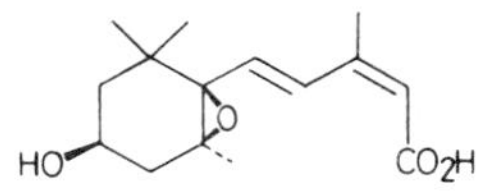

(71)

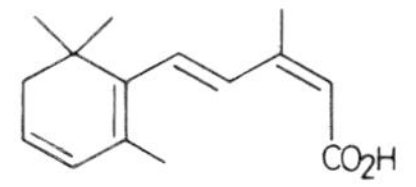

(73)

Reduction of ABA with sodium borohydride gives cis- and trans-diolABA (61) and (62). These compounds are automatically oxidized in an aqueous solution to ABA.[109] Such unstable compounds are the most probable candidates for the precursor of ABA. A cold trap technique was employed to study whether diolABA is an intermediate in the biosynthetic pathway. [2-^{14}C]-MVA was applied to avocado fruit, and after addition of unlabeled diolABA as a cold trap it was recovered. No radioactivity was incorporated into the recovered diolABA. This result showed that neither cis- nor trans-diolABA is involved in the biosynthesis of ABA.

1'-DeoxyABA (63) was synthesized to estimate the contribution of the 1'-hydroxyl to the biological activity,[20] and its methyl ester showed strong inhibitory activity in wheat embryo and soybean seedling assays. This similar activity of 1'-deoxyABA to ABA was interpreted as caused by ABA converted from 1'-deoxyABA by enzymatic hydroxylation in plants. As mentioned later, however, the feeding experiment using α-ionylideneacetic acid showed that 1'-deoxyABA was not metabolized to ABA in plants. It is likely that 1'-deoxyABA itself has inhibitory activity. Although Mallaby[105] showed that 1'-deoxyABA is automatically hydroxylated at C-1' in air to give ABA, the hydroxylation does not seem to occur in plants, and the evidence that 1'-deoxyABA is a precursor of ABA has not been confirmed. The natural occurrence of 1'-deoxyABA in fungi was recently found, and fungi seem to metabolize it to ABA. The fungal biosynthesis of ABA will be described later.

1-AlcoholABA (64) has inhibitory activity in the cress seedling test,[110] but not in the rice seedling test.[111] Cress seedlings seem to be able to oxidize the hydroxyl group to a carboxyl group. Another reduced form of ABA, 1-aldehydeABA (65) has been shown to be easily oxidized to ABA in air.[109] The presence of (64) and (65) in plants has not been examined. Sondheimer et al.[112] found that 1',4'-epidioxide (7) was converted to ABA by a rapid nonenzymatic reaction.

Tomato shoots and seedlings of rice and soybean administered to [2-^{14}C]-2-cis-α-ionylideneacetic acid (66) metabolize it to 1'-deoxyABA and 3'-keto-β-ionylideneacetic acid,

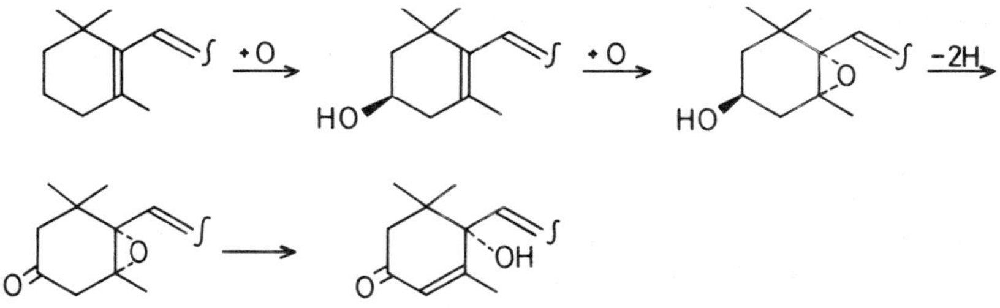

FIGURE 17. Oxidation pathway of xanthoxin.[96]

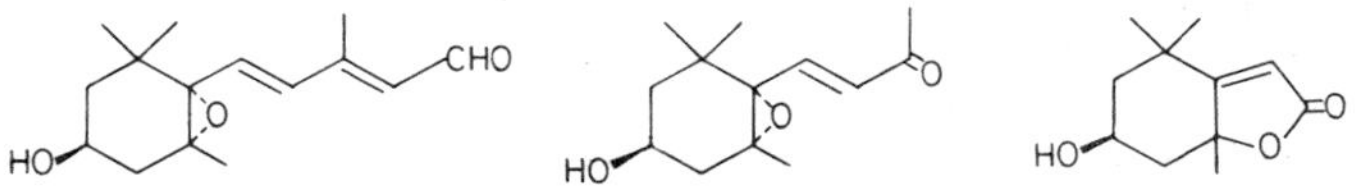

(74)   (75)   (76)

but not to ABA.[113] It has also been shown that seedlings of *Vicia faba*, *Phaseolus vulgaris*, and barley and avocado mesocarp convert *cis*-α-ionylideneacetic acid to 1′-deoxyABA. Oritani and Yamashita[113] showed that 2-*cis*-β-ionylideneacetic acid (67) was metabolized to ABA in tomato shoots and seedlings of rice and soybean. Since 4′-hydroxy-2-*cis*-β-ionylideneacetic acid (68) and (+)-1′S,2′-S,4′S-2-*cis*-xanthoxin acid (69) are rapidly converted into ABA, the conversion pathway of 2-*cis*-β-ionylideneacetic acid to ABA, as shown in Figure 17, has been proposed by Milborrow and Garmston.[109]

(+)-2-*cis*-Epoxy-β-ionylideneacetic acid (70) is also converted into ABA via (+)-1′S,2′S,4′S-2-*cis*-xanthoxin acid by avocado mesocarp.[114] However, (−)-*epi*-1′R,2′R,4′S-2-*cis*-xanthoxin acid (71) formed from (70) is not metabolized to ABA.

1′-Hydroxy-α-ionylideneacetic acid (72) administered to tomato shoots and seedlings of rice, soybean, and oat is metabolized to ABA and dehydro-β-ionylideneacetic acid (73).[113]

3. Carotenoid Pathway

The structural similarity between ABA and the end portion of carotenoid suggests that ABA could be formed from a carotenoid.[15] It has been known that plants grown under bright light contain an inhibitor, and its content is higher than in dark-grown plants.[115] Taylor and Smith[116] found that photooxidation of a mixture of carotenoids yields a neutral substance with inhibitory activity. This inhibitor, termed xanthoxin (22), is shown to be formed with its *trans* isomer (74), butenone (75), and loliolide (76) from violaxanthin (21) by exposure to light.[117] Xanthoxin has high inhibitory activity while its *trans* isomer is less active.

Firn and Friend[118] reported that incubation of violaxanthin with lipoxygenase and linolate afforded xanthoxin, its *trans* isomer, and butenone. This result suggests that xanthoxin could be formed by enzymatic action in vivo without direct involvement of photolysis.

A feeding experiment with synthesized [2-^{14}C]-xanthoxin was carried out by Taylor and Burden.[119] [2-^{14}C]-xanthoxin administered to tomato and dwarf bean was converted to ABA in yields of 10.8 and 7.0%, respectively.

Firn et al.[120] analyzed the ether extract of plants with GC, and showed that xanthoxin is widely distributed in higher plants and certain ferns. The content of 2-*trans*-xanthoxin in plants is higher than that of xanthoxin.

Although these data provide indirect evidence that xanthoxin is a potent precursor of ABA, this pathway has not been confirmed, and most of the evidence on the biosynthesis of ABA favors the direct synthesis route via a C_{15} precursor. When ^{14}C-phytoene, an uncyclized precursor of carotenoids, was applied to avocado fruit with [2-^{3}H]-mevalonolactone, ^{3}H was incorporated into carotenoids and ABA, but no ^{14}C was incorporated into

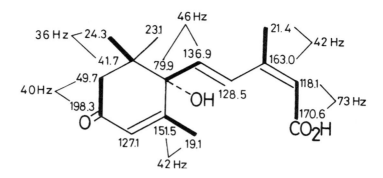

FIGURE 18. Chemical shifts and coupling constants of ABA biosynthesized from [1,2-$^{13}C_2$]-acetate.[121]

ABA.[96] This experiment indicates that carotenoids are not involved in the biosynthesis of ABA in slices of avocado mesocarp tissue.

There remains the possibility that xanthoxin may be an intermediate in the direct biosynthetic route. An experiment with labeled violaxanthin is necessary to study whether ABA is formed from violaxanthin via xanthoxin.

4. Biosynthesis in Fungi

The biosynthesis of ABA in fungi has been intensively investigated since Assante and co-workers[27] reported that ABA is produced by a fungus, *Cercospora rosicola*. Fungi are of great advantage to biosynthetic studies compared with plants. Feeding experiments can be easily carried out by culturing the fungi under controlled conditions. The fungi produce relatively large amounts of ABA, so detection of labeled intermediates should be easier than in plants.

Bennett et al.[121] confirmed that *C. rosicola* biosynthesizes ABA via the normal isoprenoid pathway. Mycelia of *C. rosicola* were cultured in a liquid medium containing [1,2-^{13}C]-sodium acetate, and labeled ABA with ^{13}C was isolated by using HPLC. The protons noise-decoupled CMR of the ABA showed 12 of the 15 carbon resonances were accompanied by satellite doublets due to ^{13}C-^{13}C coupling, while the other three resonances were singlets. The coupling constants and chemical shifts for those carbons are summarized in Figure 18. The values of the coupling constants coincide with those expected for the hybridization states of the carbons involved in coupling: 73 Hz for sp^2-sp^2, 40 to 46 Hz for sp^2-sp^3, and 36 Hz for sp^3-sp^3. This result showed that six intact acetate units and three carbons derived from acetate were incorporated into ABA. This pattern of labeling is in complete agreement with the biosynthesis of ABA via the isoprenoid pathway as in higher plants.

Neill et al.[122] have proposed that 1'-deoxyABA is the immediate precursor of ABA in *C. rosicola*. [2-^3H]-Mevalonolactone was applied to the fungus at the onset of the period of rapid ABA production. After 24 hr, the ether-soluble acidic parts of the medium and mycelium were obtained, and analyzed by HPLC monitored with UV absorption and radioactivity. The chromatogram indicated that in addition to the peak of ABA at t_R 16 min the peak at t_R 21.5 min possessed radioactivity. From the spectral data, the compound corresponding to the latter peak was identified as 1'-deoxyABA. Although it has been believed that 1'-deoxyABA is automatically converted to ABA, they found that 1'-deoxyABA is a stable compound at normal temperature and physiological pH values. Since alkaline hydrolysis of the methyl ester of 1'-deoxyABA does not yield 1'-deoxyABA but its isomer, they have cast a doubt on the purity of 1'-deoxyABA used in biosynthetic studies of ABA in plants. [3-Me-^3H]-1'-deoxyABA fed to a culture of *C. rosicola* was converted into ABA in a good yield of 11% while there was no incorporation of the 2-*trans* isomer of 1'-

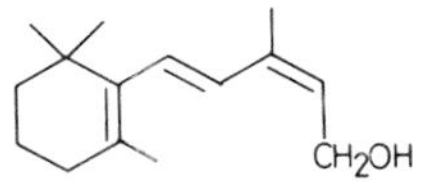

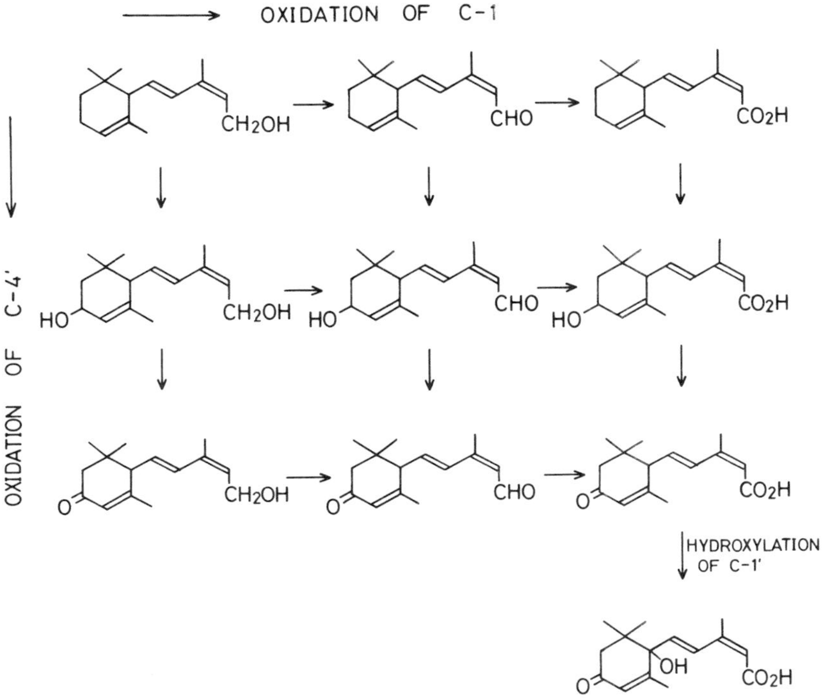

FIGURE 19. Possible biosynthetic pathway of ABA in *C. rosicola*.[123]

deoxyABA. This result strongly suggested that 1′-deoxyABA is the immediate precursor of ABA in *C. rosicola*.

Furthermore, the feeding experiment with [3-Me-^{2}H]-α-ionylideneethanol was carried out since the biosynthetic pathway was supposed to involve successive oxidation of α-ionylidene derivatives.[123] ^{2}H-Labeled α-ionylideneethanol and α-ionylideneacetic acid were applied to *C. rosicola* and after methylation the acidic extracts were analyzed by GC/MS. Deuterium incorporation into 1′-deoxyABA and ABA was quantified by the relative intensity of the deuterated ions. The result showed that both compounds were converted to 1′-deoxyABA and ABA in high yield. By contrast the 2-*trans* isomers of α-ionylideneethanol and α-ionylideneacetic acid were converted to the 2-*trans* isomer of 1′-deoxyABA, but not incorporated into ABA. This suggests that isomerization of the 2,3-double bond precedes or is concomitant with the formation of the first desaturated cyclic intermediate. These results also suggest that the early oxidizing enzymes are rather nonspecific, whereas the final hydroxylase is sensitive to the geometry of the side chain and will not utilize the 2-*trans* isomer of 1′-deoxyABA as a substrate. On the other hand, neither β-ionylideneethanol (**77**) nor β-ionylideneacetic acid (**67**) was incorporated into 1′-deoxyABA or ABA, suggesting that cyclization of farnesol derivatives proceeds via the formation of a 2′-ene intermediate, analogous to ε-ring formation in carotenoid biosynthesis. Based on these results, Neill et al.[123] have proposed a biosynthetic pathway for ABA, as shown in Figure 19. They carried out a similar experiment with *Vicia faba*, and suggested that this biosynthetic pathway is present in higher plants.

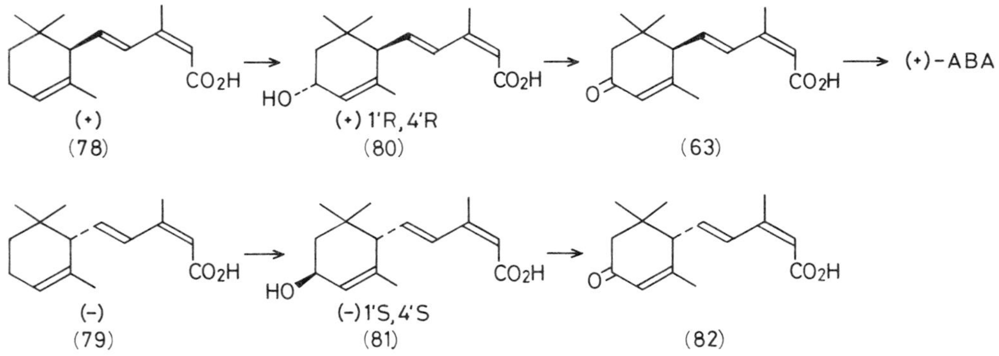

FIGURE 20. Metabolism of α-ionylideneacetic acid by *C. cruenta*.

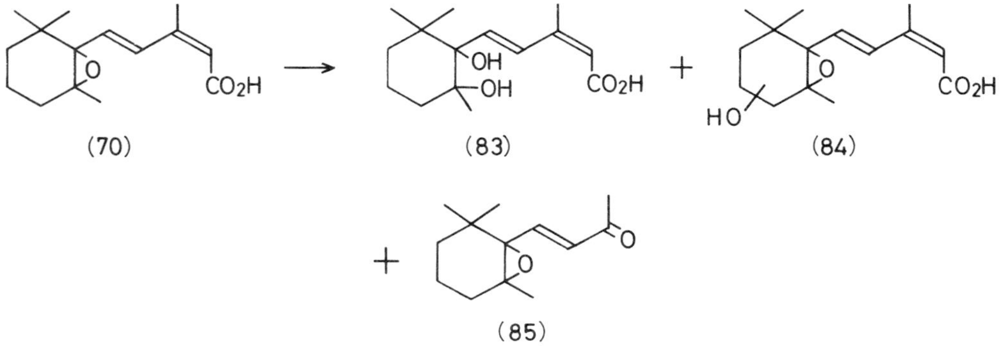

FIGURE 21. Metabolism of epoxy-β-ionylideneacetic acid by *C. cruenta*.

The metabolism of α-ionylideneacetic acid in *Cercospora cruenta* has been precisely studied by Ichimura et al.[124] The sodium salt of (±)-[2-^{14}C]-α-ionylideneacetic acid was administrated to a culture of *C. cruenta*. The radioactivity was incorporated into 4'-hydroxy-α-ionylideneacetic acid, 1'-deoxyABA and ABA. The CMR and PMR of the 4'-hydroxy-α-ionylideneacetic acid indicated that the 4'-hydroxylation of (±)-α-ionylideneacetic acid occurs mainly at the *trans* relation against the 1'-side chain, as shown in Figure 20. Furthermore, the feeding experiment with optically pure (+) and (−)-α-ionylideneacetic acid, **(78)** and **(79)**, showed that (+)-1'R,4'R-4'-hydroxy-α-ionylideneacetic acid **(80)** was converted to 1'-deoxyABA in higher yield than the enantiomer, (−)-1'S,4'S-4'-hydroxy-α-ionylideneacetic acid **(81)**. They have suggested that the enantioselectivity of the microbial oxidation may result from the chirality of the 4'-hydroxyl group rather than the 1'-chirality. (+)-1'-DeoxyABA seems to be converted to ABA faster than (−)-1'-deoxyABA **(82)**, but the enantioselectivity of 1'-hydroxylation has not been confirmed owing to the low incorporation ratio.

The metabolites of (±)-epoxy-β-ionylideneacetic acid **(70)** applied to *C. cruenta* have also been analyzed by Oritani et al.[28] Although (±)-epoxy-β-ionylideneacetic acid has been shown to be converted into ABA by avocado mesocarp, no ABA was detected among the metabolites, which included three compounds identified as 1',2'-dihydroxy-β-ionylideneacetic acid **(83)**, (±)-monohydroxyepoxy acid **(84)**, and (±)-1',2'-epoxy-β-ionone **(85)**, as shown in Figure 21. The presence of (±)-1',2'-epoxy-β-ionone showed that the fungus oxidatively degraded the 1'-double bond of **70** rather than converting it to ABA.

Very recently new ABA-related compounds were found along with 1'-deoxyABA in a culture broth of *C. cruenta*, and identified as γ-ionylideneacetic acid derivatives, **(87)** and

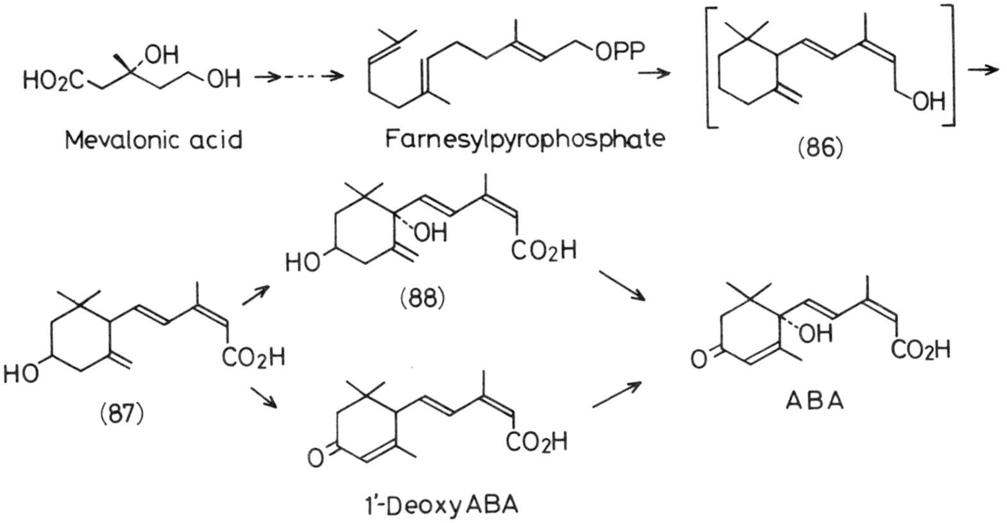

FIGURE 22. Possible biosynthetic pathway of ABA in *C. cruenta*.[126]

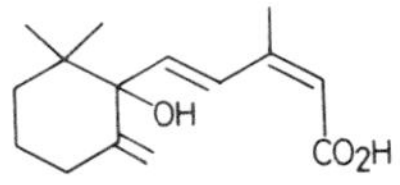

(88).[125,126] A study of the concentrations of these compounds with time showed that (87) accumulated in the early culture period decreased gradually, and (88) and ABA were formed later. Furthermore, when labeled (87) was administered to *C. cruenta*, the radioactivity was incorporated into ABA in a high yield. From these data, Oritani has supposed the presence of γ-ionylideneacetic acid (86) as a first cyclic intermediate, and proposed a new biosynthetic pathway of ABA in *C. cruenta*, as shown in Figure 22. In this scheme, ABA is assumed to be derived from two precursors, (88) and 1'-deoxyABA. The activity of (88) is one tenth that of ABA in the rice seedling growth test, and further, Tamura and Nagao[127] have shown that 1'-hydroxy-γ-ionylideneacetic acid (89) has high activity in the same test. These activities seem to be caused by ABA formed from (88) and (89), suggesting the possibility that γ-ionylideneacetic acid is also a biosynthetic precursor of ABA in plants. This γ-ionylideneacetic acid pathway is under further study.

No biosynthetic studies with another ABA-producing fungus, *Botrytis cinerea*, have been reported.

It is expected that mutants of the ABA-producing fungi will provide a powerful tool in the biosynthetic study of ABA, as that of gibberellins (GAs) was intensively advanced using mutants of *Gibberella fujikuroi*. However, there is no report on mutants of the ABA-producing fungi so far, probably due to the difficulty in selection of useful mutants. ABA is not an essential compound for the fungi, which means that there is no specific medium for the selective growth of mutants. Therefore, mutants selected at random must be cultured and analyzed by bioassays, radioimmunoassay, or HPLC. New methods to detect ABA directly on the medium might be necessary for the effective selection of mutants.

As suggested by Neill, it is likely that fungi and plants have the same biosynthetic pathway of ABA. If the biosynthetic pathway of ABA in fungi is elucidated, it will be applied to plants.

(90) t-ABA

B. Metabolites of ABA

In catabolism, ABA is converted to more polar compounds by oxidation, hydroxylation, isomerization, and conjugation with sugars or carboxylic acids. Although the detection and isolation of the metabolites are more difficult than those of ABA due to their low biological activity and high polarity, the metabolites of ABA and the metabolic pathway have been elucidated by labeling studies and employment of HPLC and spectrometries on the micro scale.

1. t-ABA

ABA isolated from plants is normally accompanied by t-ABA **(90)**. Since ABA easily isomerizes by exposure to light to form t-ABA, it had been regarded as an artifact formed from ABA during extraction and isolation. However, t-ABA is present in plant extracts obtained even under dim light. Milborrow[102] showed that t-ABA is naturally present in the leaves of a wild rose. After addition of (±)-[2-^{14}C]-ABA to an extract of leaves of *Rosa arvensis*, t-ABA was isolated and its optical activity and radioactivity were determined. The content of (+)-t-ABA was 4.1% of that of (+)-ABA, whereas 0.9% of the radioactivity was incorporated into (+)-ABA. This result showed that only 0.9% of (+)-ABA was formed from (+)-ABA during isolation.

t-ABA in roots of *Zea mays* grown in darkness was also quantified by Rivier and Pilet.[128] Hexadeuterated ABA **(98)** (see Figure 24) was prepared from ABA by treatment with alkaline deuterated water, and added to a root extract as an internal standard, and the methylated extract was analyzed by GC/SIM (EI). A prominent ion at m/z 190 was monitored for detection of methyl esters of ABA and t-ABA, and that at m/z 194 for detection of methyl esters of hexadeuterated ABA and its *trans* isomer. Since the ion at m/z 190 possesses the structure **(91)**, the correspondent ion **(92)** of hexadeuterated ABA occurs at m/z 194 by loss of two deuterium atoms. Figure 23 shows the GC/SIM profile of m/z 190 and 194. In the profile, methyl esters of ABA and t-ABA were detected at t_R 8.1 and 8.7 min of m/z 190, respectively, while the methyl ester of hexadeuterated ABA was observed at t_R 8.1 min of m/z 194, but not its *trans* isomer at t_R 8.7 min. This result clearly demonstrated the natural occurrence of t-ABA even in a dark-grown plant. It is likely that t-ABA forms in vivo by an enzymatic reaction rather than by photoisomerization of ABA.

The metabolism of t-ABA was studied by Milborrow.[102] (+)-[2-^{14}C]-t-ABA administered to tomato shoots was metabolized to glucosyl ester 10 times faster than ABA. Since (±)-[2-^{14}C]-t-ABA was not isomerized enzymatically to ABA in tomato shoots, conjugation seems to be only pathway of metabolism for exogenous t-ABA.

2. Conjugates of ABA

In the early 1950s, Van Stevenick[129] reported that the fruit of yellow lupin contained abscission-accelerators. Koshimizu et al.[40] found a highly polar inhibitor along with ABA from immature seeds of yellow lupin. This inhibitor easily decomposed with acid or alkaline treatment to yield ABA and glucose, and was shown to be the β-glucosyl ester of ABA (ABA-β-GEs), **(59)**. This was the first discovery of a conjugated form of ABA.

Glucosyl esterification of ABA in plant tissues was shown by Milborrow.[102] When (±)-[2-^{14}C]-ABA was fed to tomato shoots, three metabolites (A, B, and C) were formed and identified as the methyl ester of ABA, ABA-β-GEs, and 6′-hydroxymethylABA, respectively. Unnatural (−)-ABA was also converted to its glucosyl ester as well as (+)-ABA,

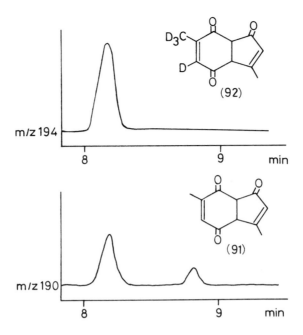

FIGURE 23. SIM chromatogram from maize primary root extract.[128]

but not metabolized to other compounds. The metabolite C was derived only from (+)-ABA, indicating that the enzyme hydroxylating the 6'-methyl group of ABA is specific to (+)-ABA. Since ABA-β-GEs easily yields a methyl ester by methanolysis, the metabolite A appears to be an artifact derived from ABA-β-GEs.

The presence of ABA-β-GEs in plants is suggested by the release of ABA from ABA-free aqueous residues with alkaline hydrolysis. In most plant tissues the amount of released ABA is between 10 and 30% of the content of ABA, but the peels of oranges aged on the tree contain 10 times as much hydrolyzable ABA as free ABA.[130] However, it cannot be concluded that the source of all released ABA is ABA-β-GEs, since other conjugated forms which can release ABA with alkaline hydrolysis are known, as will be mentioned later. The radioimmunoassay recently showed the presence of ABA conjugated at C-1 in leaves of *Phaseolus vulgaris* and *Hyoscyamus niger*,[56] and phloem exudates of *Acer platanoides* and *Robinia pseudacacia*.[131] These conjugated metabolites may be ABA-β-GEs, but have not been definitely identified.

When (±)-[2-^{14}C]-ABA was applied to surface-sterilized apple seeds and incubated for 70 days, radioactivity was incorporated into polar conjugates.[132] The formation of polar conjugates was also observed in tomato plants supplied with (±)-[2-^{14}C]-ABA.[133] Two compounds were isolated from extracts of tomato plants and identified as 1'-O-α-glucoside of ABA (ABA-α-GEt), (93) and 1'-O-β-glucoside of ABA (ABA-β-GEt), (94). It is noteworthy that glucose attaches at the tertiary hydroxyl group with steric hindrance. These glucosides are labile in alkaline solutions and hydrolyzed into ABA and glucose. Even during the methylation with etherial diazomethane, the glucosides decomposed into methyl ester of ABA and glucose. Most importantly, ABA-β-GEt gradually isomerizes to ABA-β-GEs in acidic aqueous methanol via the orthoester formed from the carboxyl group of ABA and the C-1 of a glucosyl residue; ABA-α-GEt is also formed in the course of isomerization.[134] Loveys and Milborrow[134] showed that ABA synthesized during stress is converted into ABA-β-GEt and ABA-β-GEs, whereas ABA-β-GEt is not formed in nonstressed tissues. ABA-β-GEt, like ABA-β-GEs, can be formed from both the (+) and (−) enantiomers of ABA.

Sembdner et al.[135] found β-maltosyl ester of ABA (ABA-β-MEs), (95) (see Figure 25)

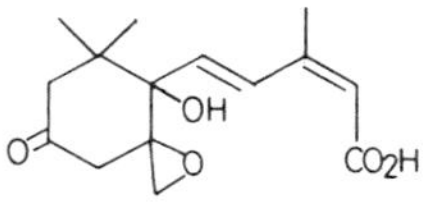

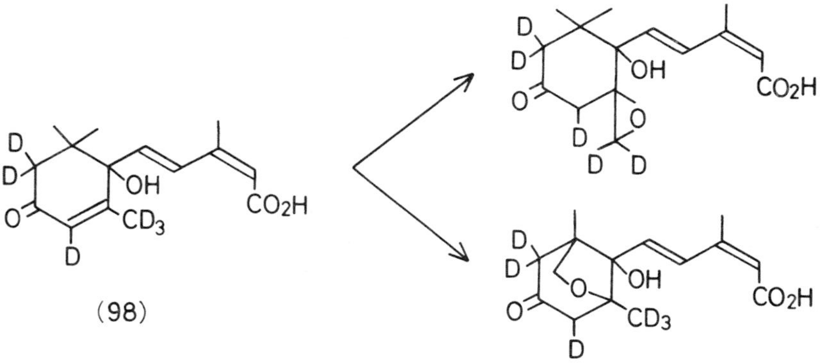

FIGURE 24. Possible structures of PA derived from hexadeuterated ABA.[96]

in both the bark and wood of birch trees (*Betula pubescens*) during the course of study on the role of ABA in dormancy. As ABA-β-MEs is stored in the bark in large amounts, it has been suggested that ABA-β-MEs continues to accumulate over the years without mobilization. The high content of ABA-β-MEs does not change up to the stage when the buds burst and the first leaves unfold. Therefore, ABA-β-MEs seems to be physiologically inactive, and it is unlikely that ABA-β-MEs regulates the bud dormancy of birch trees.

3. 6'-HydroxymethylABA (96) and Phaseic Acid (PA), (53)

As mentioned previously, a compound called "Metabolite C" was formed in tomato plants supplied with (±)-[2-^{14}C]-ABA.[102] Metabolite C showed a positive Cotton effect similar to that of (+)-ABA, but it was easily converted to the methyl ester of PA with a negative plain ORD curve by methylation with diazomethane. PA was isolated previously from seeds of *P. vulgaris* by MacMillan et al.,[136] and its structure had been suggested as an epoxide (97) from the MS data.[137] However, the formation of PA from Metabolite C could not be explained by this structure. Milborrow[64] proposed new structures, (96) and (53), for Metabolite C and PA, respectively, and these structures were elucidated by a feeding experiment. Hexadeuterated ABA (98) with deuteriums at the 2'-methyl, C-3', and C-4' was applied to tomato shoots, and PA was isolated. If (97) is the correct structure for PA, both methyl groups at C-6' remain, and if (53) is correct, one methyl group is lost, as shown in Figure 24. The PMR of the methyl ester of PA clearly showed that one of the 6'-methyl groups was absent, indicating that one methyl group was hydroxylated.[91] The result supported structure (53) for PA, and the structure of Metabolite C was concluded to be 6'-hydroxymethylABA (96) because the conversion of Metabolite C to PA is interpreted as an intramolecular vinylogous addition of the 6'-hydroxymethyl group to the double bond at C-2'. Although subsequent attempts to isolate 6'-hydroxymethylABA were unsuuccessful due to its spontaneous conversion to PA, Gillard and Walton[138] showed evidence for its presence as an intermediate in PA formation. ABA was incubated with an ABA-hydroxylating enzyme obtained from Eastern wild cucumber liquid sperm, and after immediate acetylation the presence of acetate of 6'-hydroxymethylABA was observed on TLC. This result also showed that 6'-hydroxymethylABA can be trapped by acetylation of the 6'-hydroxymethyl group. In fact, an acyl derivative of 6′-hydroxymethylABA has been shown to occur as a stable metabolite in a plant.[39]

Since DPA and *epi*-DPA, reduced forms of PA, occur naturally in plants,[139] it has been confirmed that PA is a natural metabolite of endogenous ABA in plants. However, whether the major endogenous form is PA or 6′-hydroxymethylABA is still unknown. Adesomoju et al.[140] have suggested the natural occurrence of 6′-hydroxymethylABA in immature fruits of *Vigna unguiculata* by GC/MS after methylation and trimethylsilylation, so there is a possibility that some of the PA isolated so far from various plants may be a rearranged product formed from 6′-hydroxymethylABA during extraction.

The absolute configuration at C-6′ of PA has been shown to be *R*, as previously mentioned.[67,89] This means that the pro-*S*-methyl group at C-6′ is stereospecifically hydroxylated in plant tissues. The particulate ABA-hydroxylating enzyme, which seems to be a monooxygenase, showed a requirement for oxygen and NADPH, inhibition by CO_2, and high specificity for (+)-ABA.[138] Such enzymatic studies are expected to contribute to the precise understanding of ABA metabolism.

4. Conjugates of 6′-HydroxymethylABA and Phaseic Acid

A new class of conjugated forms of 6′-hydroxymethylABA was isolated from immature seeds of *R. pseudacacia* by Hirai et al.[39] The structure was shown to be an acyl derivative (**99**) of 6′-hydroxymethylABA, and named β-hydroxy-β-methylglutarylhydroxymethylabscisic acid (HMG-HOABA). HMG-HOABA is easily hydrolyzed by alkali to give HMG and PA rearranged from 6′-hydroxymethylABA.

If HMG-HOABA were derived in vivo by the reaction of 6′-hydroxymethylABA with HMG-CoA, which is a precursor of MVA in terpenoid biosynthesis, the C-3 of the acyl group must have the *S*-configuration. Interestingly, however, the absolute configuration at C-3 was shown to be *R* by selective reduction of the HMG group to *R*-(−)-mevalonolactone with borane.[141] This result suggested that 6′-hydroxymethyl is not acylated by the usual acylation mechanism through HMG-CoA. It is possible that another acylation mehanism is involved in the acylation of 6′-hydroxymethylABA.

The content of HMG-HOABA in *R. pseudacacia* seeds increases after flowering, and 2 months later reaches 7.3 μg/g pod, then rapidly decreases as the seeds mature.[142] Accumulated HMG-HOABA seems to be metabolized to PA after cleavage of the ester linkage during maturation. This seasonal change suggests that HMG-HOABA plays an important role in ABA metabolism in seed maturation of *R. pseudacacia*. HMG-HOABA has not been found in other plant species, so it may be a specific conjugate for *R. pseudacacia*.

1′-*O*-β-Glucoside of PA (PA-β-GEt), (**100**) was isolated from apple seeds and from tomato leaves which were fed with (±)-[2-^{14}C]-ABA.[133] The glucose attaches at the tertiary hydroxyl group at C-1′ of PA. PA-β-GEt is more stable than ABA-α and β-GEt and resists mild alkaline hydrolysis. PA-β-GEt is naturally present in plants.

Applied [^{14}C]-PA was mainly converted to its glucosyl ester, PA-β-GEs (**101**), but not to DPA in *Xanthium strumarium* leaves during rehydration after water stress.[143] The natural occurrence of PA-β-GEs is not known.

5. Epimeric Dihydrophaseic Acids and a Conjugate of Dihydrophaseic Acid

When (±)-[2-^{14}C]-ABA was fed to excised embryonic bean axes, it was rapidly metabolized to two metabolites, M-1, identified as PA, and M-2.[92] The methyl ester of M-2 was oxidized with chromium trioxide to give the methyl ester of PA. From the PMR and MS data, M-2 was assigned as 4′-dihydrophaseic acid (DPA), (**57**). The absolute configuration at C-4′ of DPA and its epimer, *epi*-DPA (**58**) was confirmed by Milborrow.[67]

DPA occurs naturally at a high level in mature seeds of *P. vulgaris*.[139,144] In addition, immature pea seeds,[145] the endosperm of *Marah macrocarpus* (former *Echinocystis macrocarpa*),[146] pear seeds,[147] cowpea,[140] and others have been shown to contain DPA. The

(102)   (103)

presence of *epi*-DPA in *P. vulgaris* seeds was also shown, and its content was 10% of that of DPA.[139] The natural occurrence of *epi*-DPA cast a doubt that *epi*-DPA may be an artifact formed from a conjugate at C-4' by the inversion of the hydroxyl group during extraction. Zeevaart and Milborrow[139] carried out a feeding experiment with *P. vulgaris* shoots fed with [4'-^{18}O]-ABA and determined the content of ^{18}O in isolated DPA and *epi*-DPA by GC/MS. Since both epimers retained ^{18}O equally, it was shown that no loss of oxygen at C-4' occurred during extraction. Furthermore, the existence of an epimerase and microbial reduction of PA were ruled out from feeding experiments with DPA and *epi*-DPA, or under aseptic conditions. Thus, it was concluded that *epi*-DPA is enzymatically derived from ABA and occurs endogenously.

A soluble PA-reducing enzyme has been prepared from *Echinocystis lobata* liquid sperm, and shown to reduce PA to DPA.[138] However, there is no evidence whether the carbonyl group at C-4' of PA is reduced to a hydroxyl group with the opposite orientation by one nonspecific enzyme or by two stereospecific enzymes.

When Setter and Brun[148] carried out metabolic studies with (±)-[2-^{14}C]-ABA applied to soybean leaves and pods, a major proportion of the radioactivity was incorporated into an unidentified metabolite, which accumulated as ^{14}C-incorporation into DPA reached a peak. From an interpretation of MS of its derivatives, it was identified as a glucoside at the C-4' hydroxyl group of DPA, that is, DPA-β-GEt (**60**).[149] The formation of DPA-β-GEt as a major metabolite of ABA was observed in tomato shoots fed with (±)-[2-^{14}C]-ABA and [2-^{14}C]-DPA.[150] Endogenous DPA-β-GEt was isolated from avocado fruits and found in fruits of grape.[49] DPA-β-GEt is a stable conjugate different from other conjugates and appears to occur widely in plants.

6. Other ABA-Related Metabolites

cis-DiolABA (**61**) was found in immature seeds of *V. faba*.[151] This compound has the two possibilities of being a precursor or a metabolite of ABA. However, the conversion of 4'-hydroxyABA to ABA in vivo has not been confirmed, so 4'-hydroxyABA seems to be an occasionally hydroxylated product of ABA. It has been shown that microbial cultures isolated from soil or decaying fruit metabolized ABA to *trans*-diolABA (**62**) with a range of compounds.[132] These results suggest that there is a difference of the stereospecificity of 4'-carbonyl group hydroxylation between plants and microorganisms.

Lehmann et al.[152] isolated a new ABA metabolite from a cell suspension culture of *Nigella damascena* fed with [2-^{14}C]-ABA, and identified as 2'-hydroxymethylABA (**102**). 2'-HydroxymethylABA seems to be an occasional metabolite of exogenous ABA in culture cells.

4'-DesoxyABA (**103**) was detected with PAA and DPA in seedlings of *Pisum sativum* supplied with [1-^{14}C]-ABA at the onset of a recovery period from wilting,[153] but its natural occurrence has not been confirmed.

C. Metabolic Pathway

Metabolism of ABA is now supposed to be as shown in Figure 25, including eight conjugated forms.

ABA is metabolized to 6'-hydroxymethylABA by hydroxylation of the 6'-α-methyl group at a first step, and the resulting 6'-hydroxymethylABA rearranged enzymatically or automatically to PA with a tetrahydrofurane ring. The carbonyl group at C-4' of PA is reduced

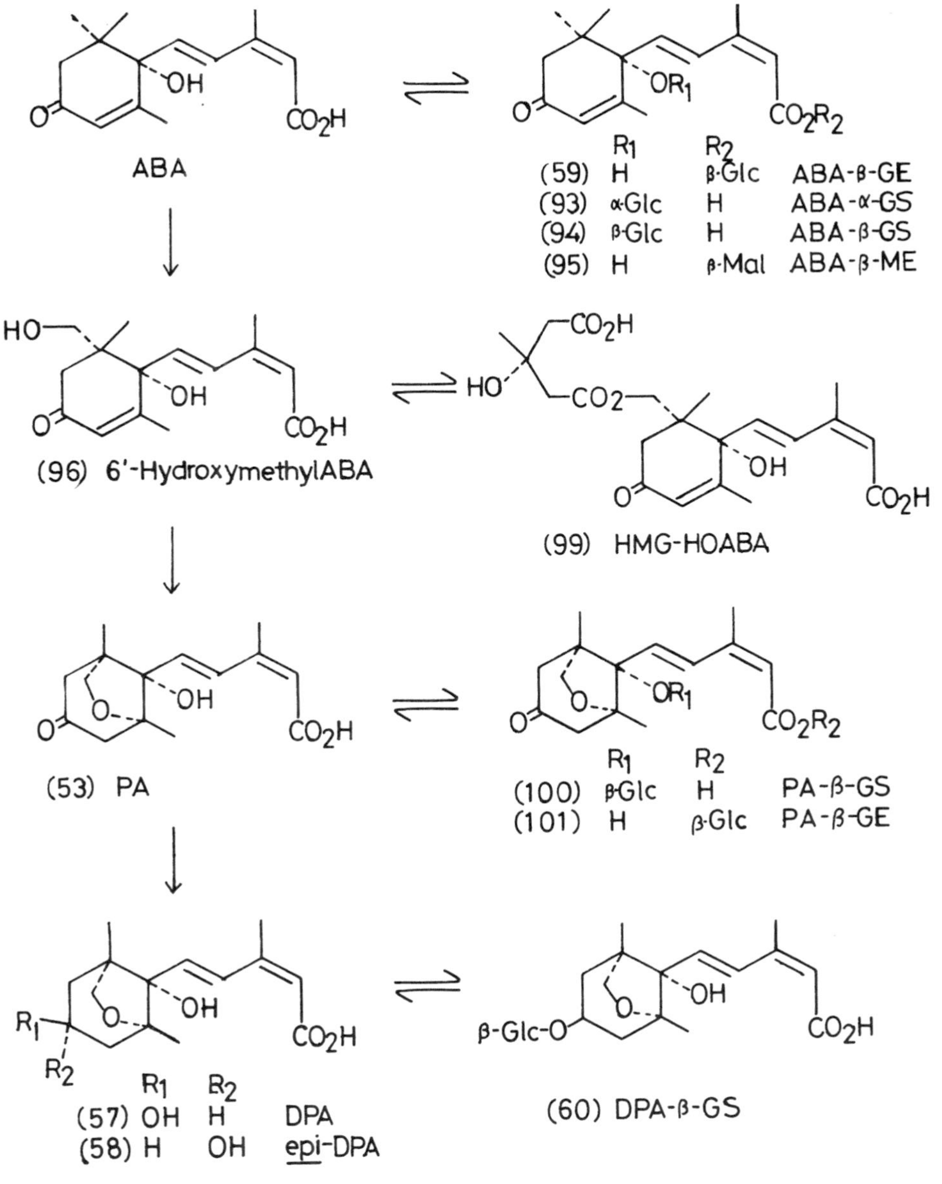

FIGURE 25. Metabolic pathway of ABA.

in two ways to form DPA and *epi*-DPA. This pathway from ABA to DPA and *epi*-DPA seems to be common to a number of plant species. Further metabolites after DPA and *epi*-DPA are unknown. Although Rudnicki and Czapski[154] suggested that the loss of C-1 and C-2 could occur in apple seeds as one of the complete oxidation steps of ABA, it has been shown that such oxidation was caused by microorganisms, not by the apple seeds.[132]

At each metabolic step, those free acids are converted to conjugates as glucosides or glycosyl esters or an acylated form. HMG-HOABA is the only conjugate that does not possess a sugar moiety. A conjugate of *epi*-DPA has not yet been found, probably due to its small amount, but it also appears to form a glucoside as well as DPA. Most of the conjugates have been isolated from only a few of the various plant species. Therefore, each conjugated form cannot be regarded as a metabolite common to all plant species, meaning that forms of conjugation are different between plant species. There is a possibility that new conjugated forms might occur in other plant species which have not been investigated.

Each plant species seems to have its own regulation system for free acid concentration, and whether accumulated free acids are converted to conjugated forms or to more polar free acids depends on the plant species, kinds of organs, and physiological states. Dwarf *P. vulgaris* seedlings convert accumulated ABA to its conjugated form.[155] On the contrary, it has been shown by Dörffling et al.[156] that *P. sativum* seedlings metabolize ABA to PA and DPA rather than to its conjugated form. These results suggest that conjugation occurs in plant tissues where accumulated ABA cannot be rapidly metabolized to more polar inactive free acids because of low activity of the hydroxylation enzyme. This presumption may apply to the conjugation of PA and DPA. It is likely that the conjugates are a temporary storage form and/or a transportation form for removal of free acids from certain cellular sites, and contribute to the regulation of free acid levels in plant tissues.

There is no metabolic study with the ABA-producing fungi fed with radioactive ABA, and the presence of PA and DPA have not been detected in media of the fungi. The fungi appear not to metabolize ABA to other compounds, because metabolism of ABA excreted into medium is assumed to be unnecessary for the fungi.

IV. BIOLOGICAL ACTIVITY

A. Exogenous ABA

ABA exogenously applied to intact plants, certain organs, and tissues shows various biological activities which are closely related to the physiological roles of endogenous ABA. Several activities have been employed in bioassays for the detection and quantification of ABA.

ABA has a high inhibitory effect on the growth of various plants. ($\pm$)-ABA inhibits the growth of wheat coleoptiles at a concentration below $10^{-6}\,M$.[157] Growth of *Hordeum* shoots, bean axes, and second leaf sheaths of rice seedlings are also inhibited by ABA at a concentration between 10^{-7} and $10^{-5}\,M$.[18,158] Furthermore, ABA shows an inhibitory effect on the growth of *Lemna gibba* L.,[159] and its effect is caused at a lower concentration than in other growth inhibitions. Germination of seeds such as cress, lettuce, and radish is inhibited by ABA.[160,161] These inhibitory effects are not specific to ABA. Many other compounds such as phenolic derivatives and pathogenic fungal toxins have been known to cause inhibition of growth and germination. However, while other inhibitors exhibit their effects through toxicity, ABA does not. The inhibitory effect of ABA is characterized by its nontoxicity. Seeds treated with ABA can germinate after removal of the ABA by rinsing and transferring into a new medium containing no ABA. Evaluating the inhibitory effects of compounds, we should distinguish between toxic and nontoxic inhibition so that we do not consider apparent inhibition as an ABA-like effect.

ABA inhibits the α-amylase synthesis induced by GAs in barley aleurone layers.[162] This counteracting ability of ABA against GAs suggests that seed germination is regulated by endogenous levels of ABA and GAs. The recent observation that ABA cannot inhibit the growth of germinated seeds of *Haplopappus gracilis*[163] suggests that inhibition of seed germination by ABA is caused through a different process from that of inhibition of axis elongation.

Abscission of plant organs is accelerated by ABA. When ABA is applied to cut surfaces of cotton or bean explants, abscission of petioles is observed after a few days.[164] ABA is an extremely potent accelerant of abscission. Although GA is also known to accelerate petiole abscission, it is not as potent as ABA, and its mode of action is different from that of ABA.[165]

One of the most important activities of ABA is the ability to cause stomatal closure, and to reduce transpiration.[166] Stomatal closure of *Commelina communis* L. epidermal strips is caused by ABA at a very low concentration range, between 10^{-10} and 10^{-7} M.[167] Such a strong effect seems to reflect the physiologically intrinsic role of ABA in plants. Stomatal closure by ABA is observed also when ABA is applied directly to leaves or to the transpiration stream. Farnesol and certain fatty acids, particularly decanoic and undecanoic acid, can induce stomatal closure in epidermal strips, but these compounds have no effect when applied to the transpiration stream. It has been suggested that these compounds cause irreversible damage to guard cell membranes.[168] Fusicoccin, a fungal toxin produced by *Fusicoccum amygdali*, inhibits stomatal closure induced by ABA, and opens stomata.[169] The effect of fusicoccin seems to be related to transportation of K^+ or H^+ ions in guard cell membranes.[170]

The effects of ABA on flower bud formation are diverse. While ABA inhibits the flower bud formation of long day plants, it promotes that of short day plants such as the morning glory.[171] The promotion effect is considered to be a result of vegetative growth inhibition.

ABA accelerates the senescence of excised leaf discs in all species.[161,171,172] This effect is not observed when ABA is applied to leaves of intact plants. Cytokinin counteracts the senescence promotion induced by ABA.[161]

Moreover, ABA can cause root curvature. When a small agar block containing ABA is placed asymmetrically on the apical cut surface of a root, curvature toward the block-placed side is induced.[173] This shows the basipetal movement of ABA, and is part of the circumstantial evidence supporting the idea that ABA is a geotropism regulator synthesized in the root cap.

In addition to these activities on plants, it has been suggested recently that ABA inhibits the reproduction of insects and tumor growth in a certain strain of mice.[174,175] Although these are not intrinsic activities of ABA as a plant hormone, inhibitory effects on growth of plants, insects, and animals appear to suggest the unique possibility of ABA as a biologically active substance. Effects of ABA in insects and vertebrates have been reviewed by Visscher.[176]

B. Structural Requirements for Activity

The ABA molecule has unsaturated carboxyl, tertiary hydroxyl, and enone groups, and in addition a geminal methyl group and two methyl groups on double bonds. Many analogues have been synthesized so far, and biological activities of derivatives (including natural metabolites) have been tested to confirm which groups are essential for biological activity. Here, the structural requirements for activity are briefly summarized, introducing new results concerning demethyl derivatives. Structure-activity relationships of ABA analogues and metabolites have previously been reviewed by Walton.[177]

The unnatural *R*-(−)-enantiomer (**20**) is highly active, comparable to ABA, in many bioassays, but not in the stomatal closure test.[158,178] This is an exceptional case, because unnatural optical isomers usually show less activity than natural isomers. Milborrow[96] has proposed a hypothetical active site accomodating both enantiomers. In the stomatal closure test, *R*-(+)-α-ionylideneacetic acid (**66,**) possessing the same configuration at C-1′ as ABA, shows higher activity than its enantiomer.[179] There may be two kinds of active sites, one for growth regulation and one for stomatal closure in plant tissues.

The 2-*cis*-4-*trans*-3-methyl-2,4-pentadienoic side chain is essential to maintain activity. Isomerization of the 2-*cis* double bond to 2-*trans*, or lengthening of the side chain reduces activity.[180-183] Shortening of the side chain has given diverse results.[180-181] The activity of 1-aldehydeABA (**65**) and 1-alcoholABA (**64**) is as high as ABA in several bioassays.[110,111,186-188] Whether these compounds are active without conversion to ABA in plants is still unknown, but it is likely that a functional group which can be oxidized to a carboxyl group is necessary. The activity of methyl and glucosyl esters seems to be caused by ABA after hydrolysis.[18,40,189] 3-DemethylABA (**104**) is inactive in the rice seedling growth test,[190] indicating the importance of the 3-methyl group for activity.

Derivatives modified at the 1′-hydroxyl and 4′-carbonyl groups, and the 2′,3′-double

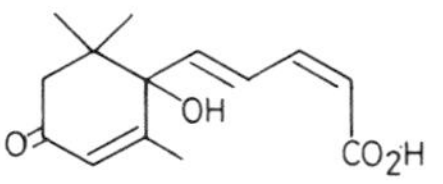

(104) 3-DemethylABA
(105) 2'-DemethylABA 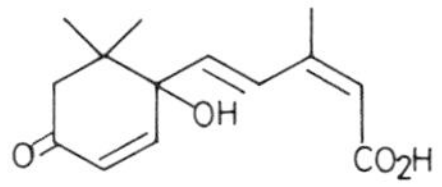
(106) 6'-α-DemethylABA 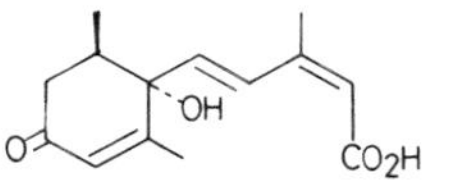
(107) 6'-β-DemethylABA 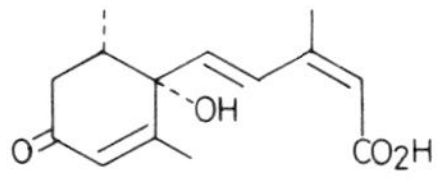
(108) 6',6'-DidemethylABA 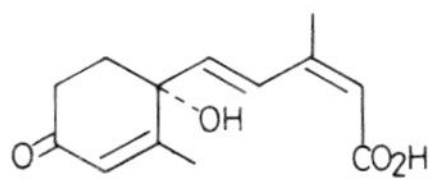
(109) 5'-NorABA 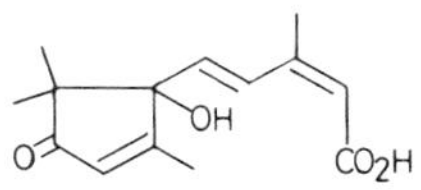

bond also may be converted to ABA by hydroxylation and/or oxidation in plants. Therefore, the necessity of the groups cannot be separately evaluated unless metabolic studies with radioactive derivatives are carried out.

Oritani and Yamashita[191] have suggested that the 1'-hydroxyl group contributes to the reduction of toxicity caused by the carbon skeleton of ABA. While highly active 1'-deoxyABA and 1'-deoxy-4'-deoxoABA cause death of tested plants by wilting, 1'-hydroxy-α-ionylideneacetic acid (72) shows high activity without toxicity. The toxicity of 1'-deoxyABA may suggest that 1'-deoxyABA is not converted to ABA in plants. Saturation of the 2',3'-double bond reduces activity,[192] indicating that the double bond is necessary to maintain activity. The 4'-carbonyl group seems not to be essential for activity. Reduced compounds at the 4'-carbonyl group, diolABA, and 4'-deoxyABA show activity.[127,193,194] The activity has been suggested to be caused by ABA, since these compounds can be converted to ABA through enone formation. A metabolic study with (±)-[2-^{14}C]-1',4'-deoxyABA, however, has shown that no conversion of 1',4'-deoxyABA to ABA occurs in plants.[113]

Oritani et al.[195,196] synthesized demethyl derivatives, (105) to (108), and examined the contribution of methyl groups in the ring system to activity in the rice seedling growth test. While the 2'-demethyl derivative (105) completely loses activity, the 6',6'-didemethyl derivative (108) does not, and the 6'-monodemethyl derivatives, (106) and (107), have activity comparable to ABA. 6'-Demethyl derivatives of 1',2'-epoxyionylideneacetic acid do not completely lose activity.[196,197] These results show that a 2'-methyl group is essential for activity, and that at least the presence of one methyl group at C-6' is sufficient to keep activity. The latter fact further suggests the difference of structural requirements between the active site and metabolic enzyme for ABA. A metabolic study with the 6'-α-demethyl derivative (106), lacking the methyl group to be hydroxylated, may give an interesting result about the stereospecificity of the ABA-hydroxylating enzyme.

Kienzle[198] has synthesized 5'-norABA (109) and showed the possibility that cyclohexenone can be substituted for by cyclopentenone without changing the activity.

All metabolites of ABA are less active than ABA. Since 6'-hydroxymethylABA is easily converted to PA, it is difficult to test its activity. Its acyl derivative, HMG-HOABA, shows as low activity as PA in the rice seedling test, suggesting conversion of HMG-HOABA to PA in plants.[39]

PA possesses activities in the inhibition test of α-amylase production and the stomatal closure test, although not as high as ABA.[199-201] Whether PA acts on the same site as that for ABA has not been confirmed. Sharkey and Raschke[201] have proposed an interesting hypothesis that ABA is a stress messenger initiating stomatal closure while PA accumulated by conversion of ABA maintains closure.

DPA slightly inhibits bean axis growth, but is inactive in other bioassays.[199-201] The activity reduction from PA to DPA is similar to that from ABA to diolABA, suggesting that the 4'-carbonyl group partially contributes to the activity.

C. Problems in Structure-Activity Relationship Studies

Several structural requirements for activity have been deduced from structure-activity relationship studies. At the same time, however, such studies have revealed the difficulty that all ABA-like activities caused by application of compounds cannot be assigned to the compounds themselves. As mentioned in Section III.A, several analogues have been shown to be automatically or enzymatically converted to ABA. To determine the real active principle, metabolic studies of the applied compounds are necessary. Furthermore, structure-activity relationships vary depending on bioassay systems. This seems to be caused by differences of the permeability of compounds, metabolic enzyme activity, and active site structure between plants and plant tissues. Therefore, the activities of compounds tested in different bioassays cannot be evaluated using the same criterion.

The stomatal closure test using epidermal strips of *C. communis*, developed by Ogunkanmi et al.,[167] seems to be most suited to the study on structure-activity relationships because of its high specificity to ABA. The presence of compounds which are inactive in the stomatal closure test but active in other bioassays indicates that the structural requirement of the active site in guard cell is stricter than that of other active sites in plant tissues. Exogenously applied ABA is accumulated in guard cells of *C. communis* epidermis,[202,203] indicating the localization of specific sites binding to ABA which presumably consists of a protein of a carrier, receptor, or metabolic enzyme. Although the stomatal closure test is highly specific to ABA, the conversion of ABA to PA and DPA can occur as in epidermal strips of *C. communis* and *Tulipa gesnerians*.[204] Structure-activity relationships of ABA, therefore, should be discussed in terms of not only mode of action, but also metabolism. Clarification of the substance responsible for specific binding of ABA will contribute to the study on the structure-activity relatonships and structural requirement for activity.

REFERENCES

1. **Addicott, F. T.**, Ed., *Abscisic Acid*, Praeger, New York, 1983.
2. **Hemberg, T.**, Significance of growth-inhibiting substances and auxins for the rest-period of the potato tuber, *Physiol. Plant.*, 2, 24, 1949.
3. **Hemberg, T.**, Growth-inhibiting substances in terminal buds of *Flaxinus*, *Physiol. Plant.*, 2, 37, 1949.
4. **Bennet-Clark, T. A., Tambiah, M. S., and Kefford, N. P.**, Estimation of plant growth substances by partition chromatography, *Nature (London)*, 169, 452, 1952.
5. **Bennet-Clark, T. A. and Kefford, N. P.**, Chromatography of the growth substances in plant extracts, *Nature (London)*, 171, 645, 1953.
6. **Kefford, N. P.**, The growth substances separated from plant extracts by chromatography. II. The coleoptile and root elongation properties of the growth substances in plant extracts, *J. Exp. Bot.*, 6, 245, 1955.
7. **Phillips, I. D. J. and Wareing, P. F.**, Studies in dormancy of sycamore. I. Seasonal changes in the growth-inhibitors in *Acer pseudoplatanus*, *J. Exp. Bot.*, 9, 350, 1958.
8. **Phillips, I. D. J. and Wareing, P. F.**, Effect of photoperiodic conditions on the level of growth inhibitors in *Acer pseudoplatanus*, *Naturwissenschaften*, 45, 317, 1958.
9. **Wareing, P. F., Eagles, C. F., and Robinson, P. M.**, Natural inhibitors as dormancy agents, in *Régulateurs Naturels de la Croissance Végétale*, Nitsch, J. P., Ed., Centre Nat. Rech. Sci., Paris, 1964, 377.
10. **Van Stevenick, R. F. M.**, Influence of pea-mosaic virus on the reproductive capacity of yellow lupine, *Bot. Gaz.*, 119, 63, 1957.

11. **Van Steveninck, R. F. M.**, Factors affecting the abscission of reproductive organs in yellow lupins (*Lupinus luteus* L.). III. Endogenous growth substances in virus-infected and healthy plants and their effect on abscission, *J. Exp. Bot.*, 10, 367, 1959.
12. **Rothwell, K. and Wain, R. L.**, Studies on a growth inhibitor in yellow lupin (*Lupinus luteus* L.), in *Régulateurs Naturels de la Croissance Végétale*, Nitsch, J. P., Ed., Centre Nat. Rech. Sci., Paris, 1964, 363.
13. **Addicott, F. T., Carns, H. R., Lyon, J. L., Smith, O. E., and McMeans, J. L.**, On the physiology of abscisins, in *Régulateurs Naturels de la Croissance Végétale*, Nitsch, J. P., Ed., Centre Nat. Rech. Sci., Paris, 1964, 687.
14. **Ohkuma, K., Lyon, J. L., Addicott, F. T., and Smith, O. E.**, Abscisin II, an abscission-accelerating substance from young cotton fruit, *Science*, 142, 1592, 1963.
15. **Ohkuma, K., Addicott, F. T., Smith, O. E., and Thiessen, W. E.**, The structure of abscisin II, *Tetrahedron Lett.*, 2529, 1965.
16. **Cornforth, J. W., Milborrow, B. V., and Ryback, G.**, Synthesis of (±) abscisin II, *Nature (London)*, 206, 715, 1965.
17. **Cornforth, J. W., Milborrow, B. V., Ryback, G., and Wareing, P. F.**, Chemistry and physiology of "dormins" in sycamore. Identity of sycamore "dormin" with abscisin II, *Nature (London)*, 205, 1269, 1965.
18. **Koshimizu, K., Fukui, H., Mitsui, T., and Ogawa, Y.**, Identity of lupin inhibitor with abscisin II and its biological activity on growth of rice seedlings, *Agric. Biol. Chem.*, 30, 941, 1966.
19. **Cornforth, J. W., Milborrow, B. V., Ryback, G., Rothwell, K., and Wain, R. L.**, Identification of the yellow lupin growth inhibitor as (+)-abscisin II ((+)-dormin), *Nature (London)*, 211, 742, 1966.
20. **Milborrow, B. V.**, Identification and measurement of (+)-abscisic acid in plants, in *Biochemistry and Physiology of Plant Growth Substances: Proc. 6th Int. Conf. Plant Growth Substances*, Wightman, F. and Setterfield, G., Eds., Runge Press, Ottawa, 1968, 1531.
21. **Addicott, F. T., Carns, H. R., and Cornforth, J. W.**, Abscisic acid: a proposal for the redesignation of abscisin II (dormin), in *Biochemistry and Physiology of Plant Growth Substances: Proc. 6th Int. Conf. Plant Growth Substances*, Wightman, F. and Setterfield, G., Eds., Runge Press, Ottawa, 1968, 1527.
22. **Addicott, F. T., Lyon, J. L., Ohkuma, K., Thiessen, W. E., Carns, H. R., Smith, O. E., Cornforth, J. W., Milborrow, B. V., Ryback, G., and Wareing, P. F.**, Abscisic acid: a new name for abscisin II (dormin), *Science*, 159, 1493, 1968.
23. **Pryce, R. J.**, The occurrence of lunularic and abscisic acids in plants, *Phytochemistry*, 11, 1759, 1972.
24. **Pryce, R. J.**, Lunularic acid, a common endogenous growth inhibitor of liverworts, *Planta*, 97, 354, 1971.
25. **Schwabe, W. W. and Valio, I. F. M.**, Growth and dormancy in *Lunularia cruciata* (L.) Dum., *J. Exp. Bot.*, 21, 122, 1970.
26. **Rudnicki, R., Borecka, H., and Pieniazek, J.**, Abscisic acid in *Penicillium italicum*, *Planta*, 86, 195, 1969.
27. **Assante, G., Merlini, L., and Nasini, G.**, (+)-Abscisic acid, a metabolite of the fungus *Cercosspora rosicola*, *Experientia*, 33, 1556, 1977.
28. **Oritani, T., Ichimura, M., and Yamashita, K.**, The metabolism of analogs of abscisic acid in *Cercospora cruenta*, *Agric. Biol. Chem.*, 46, 1959, 1982.
29. **Marumo, S., Katayama, M., Komori, E., Ozaki, Y., Natsume, M., and Kondo, S.**, Microbial production of abscisic acid by *Botrytis cinerea*, *Agric. Biol. Chem.*, 46, 1967, 1982.
30. **Borecka, H. and Pieniazek, J.**, Stimulatory effect of abscisic acid on spore germination of *Gloeosporium album* Osterw. and *Botrytis cinerea* Pers., *Bull. Acad. Pol. Sci. Cl. 5:*, 16, 657, 1968.
31. **Lenton, J. R., Perry, V. M., and Saunders, P. F.**, The identification and quantitative analysis of abscisic acid in plant extracts by gas-liquid chromatography, *Planta*, 96, 271, 1971.
32. **Nitsch, J. P.**, Method for the investigation of natural auxins and growth inhibitors, in *The Chemistry and Mode of Action of Plant Growth Substances*, Wain, R. L. and Wightman, F., Eds., Butterworths, London, 1956, 3.
33. **Milborrow, B. V. and Mallaby, R.**, Occurrence of methyl (+)-abscisate as an artefact of extraction, *J. Exp. Bot.*, 94, 741, 1975.
34. **Chia, A. J., Brenner, L., and Brun, W. A.**, Rapid separation and quantification of abscisic acid from plant tissues using high-performance liquid chromatography, *Plant Physiol.*, 59, 821, 1977.
35. **Milborrow, B. V.**, The identification of (+)-abscisin II in plants and measurement of its concentration, *Planta*, 76, 93, 1967.
36. **Rivier, L., Milon, H., and Pilet, P.-E.**, Gas chromatographic-mass spectrometric determinations of abscisic acid levels in the cap and the apex of maize roots, *Planta*, 134, 23, 1977.
37. **Netting, A. G., Milborrow, B. V., and Duffield, A. M.**, Determination of abscisic acid in *Eucalyptus haemastoma* leaves using gas chromatography/mass spectrometry and deuterated internal standards, *Phytochemistry*, 21, 385, 1982.

38. **Zeevaart, J. A. D. and Milborrow, B. V.**, Metabolism of abscisic acid and the occurrence of epidihydrophaseic acid in *Phaseolus vulgaris, Phytochemistry,* 15, 493, 1976.
39. **Hirai, N., Fukui, H., and Koshimizu, K.**, A novel abscisic acid metabolite from seeds of *Robinia pseudacacia, Phytochemistry,* 17, 1625, 1978.
40. **Koshimizu, K., Inui, M., Fukui, H., and Mitsui, T.**, Isolation of (+)-abscisyl-β-D-glucopyranoside from immature fruit of *Lupinus luteus, Agric. Biol. Chem.,* 32, 789, 1968.
41. **Glenn, J. L., Kuo, C. C., Durley, R. C., and Pharis, R. P.**, Use of insoluble polyvinylpyrrolidone for purification of plant extracts and chromatography of plant hormones, *Phytochemistry,* 11, 345, 1972.
42. **Anderson, B., Haggstrom, N., and Anderson, K.**, Identification of abscisic acid in shoots of *Picea sylvestris* by combined gas chromatography-mass spectrometry. A versatile method for clean-up and quantification, *J. Chromatogr.,* 157, 303, 1978.
43. **Anderson, B. and Anderson, K.**, Use of Amberlite XAD-7 as a concentrator column in the analysis of endogenous plant growth hormones, *J. Chromatogr.,* 242, 353, 1982.
44. **Most, B. H.**, Abscisic acid in immature apical tissue of sugar cane and in leaves of plants subjected to drought, *Planta,* 101, 67, 1971.
45. **Durley, R. C., Crozier, C. A., Pharis, R. P., and McLaughlin, G. E.**, Chromatography of 33 gibberellins on a gradient eluted silica gel partition column, *Phytochemistry,* 11, 3029, 1972.
46. **MacMillan, J. and Wels, C. M.**, Partition chromatography of gibberellins and related diterpenes on columns of Sephadex LH-20, *J. Chromatogr.,* 87, 271, 1973.
47. **Reeve, D. R. and Crozier, A.**, Purification of plant hormone extracts by gel permeation chromatography, *Phytochemistry,* 15, 791, 1976.
48. **Sandberg, G., Dunberg, A., and Ode'n, P.-C.**, Chromatography of acid phytohormones on columns of Sephadex LH-20 and insoluble poly-*N*-vinylpyrrolidone, and application to the analysis of conifer extracts, *Physiol. Plant.,* 53, 219, 1981.
49. **Hirai, N. and Koshimizu, K.**, A new conjugate of dihydrophaseic acid from avocado fruit, *Agric. Biol. Chem.,* 47, 365, 1983.
50. **Durley, R. C., Kannangara, T., and Simpson, G. M.**, Leaf analysis for abscisic, phaseic and 3-indolylacetic acids by high-performance liquid chromatography, *J. Chromatogr.,* 236, 181, 1982.
51. **Quebedeaux, B., Sweetser, P. B., and Rowell, J. C.**, Abscisic acid levels in soybean reproductive structures during development, *Plant Physiol.,* 58, 363, 1976.
52. **Seeley, S. D. and Powell, L. E.**, Electron capture-gas chromatography for sensitive assay of abscisic acid, *Anal. Biochem.,* 35, 530, 1970.
53. **Weiler, E. W.**, Radioimmunoassay for the determination of free and conjugated abscisic acid, *Planta,* 144, 255, 1979.
54. **Walton, D., Dashek, W., and Galson, E.**, A radioimmunoassay for abscisic acid, *Planta,* 146, 139, 1979.
55. **Fuchs, Y. and Mayak, S.**, Detection and quantitative determination of abscisic acid by immunological assay, *Planta,* 103, 117, 1972.
56. **Weiler, E. W.**, Radioimmunoassays for the differential and direct analysis of free and conjugated abscisic acid in plant extracts, *Planta,* 148, 262, 1980.
57. **Mertense, R., Deus-Neuman, B., and Weiler, E. W.**, Monoclonal antibodies for the detection and quantitation of the endogenous plant growth regulator, abscisic acid, *FEBS Lett.,* 160, 269, 1983.
58. **Daie, J. and Wyse, R.**, Adaptation of the enzyme-linked immunosorbent assay (ELISA) to the quantitative analysis of abscisic acid, *Anal. Biochem.,* 119, 365, 1982.
59. **Whenham, R. J. and Fraser, R. S. S.**, A rapid and simple radioassay for abscisic acid using ^{14}C-diazomethane, *J. Exp. Bot.,* 32, 1223, 1981.
60. **Fukui, H., Koshimizu, K., Usuda, S., and Yamazaki, Y.**, Isolation of plant growth regulators from seeds of *Cucurbita pepo* L., *Agric. Biol. Chem.,* 41, 175, 1977.
61. **Antoszewski, R. and Rudnicki, R.**, Spectrofluometric method for quantitative determination of abscisic acid on thin-layer chromatograms, *Anal. Biochem.,* 32, 233, 1969.
62. **Mallaby, R. and Ryback, G.**, Chemistry of a colour test for abscisic acid, *J. Chem. Soc. Perkin Trans. 2,* 919, 1972.
63. **Cornforth, J. W., Milborrow, B. V., and Ryback, G.**, Identification and estimation of (+)-abscisin II (''dormin'') in plant extracts by spectropolarimetry, *Nature (London),* 210, 627, 1966.
64. **Milborrow, B. V.**, Identification of ''Metabolite C'' from abscisic acid and a new structure for phaseic acid, *Chem. Commun.,* 966, 1969.
65. **Hashimoto, T., Ikai, T., and Tamura, S.**, Isolation of (+)-abscisin II from dormant aerial tubers of *Dioscorea batatas, Planta,* 78, 89, 1968.
66. **MacMillan, J. and Pryce, R. J.**, Plant hormones-IX. Phaseic acid, a relative of abscisic acid from seed of *Phaseolus multiflorus*. Possible structures, *Tetrahedron,* 25, 5893, 1969.
67. **Milborrow, B. V.**, The absolute configuration of phaseic and dihydrophaseic acids, *Phytochemistry,* 14, 1045, 1975.

68. **Dörffling, K. and Tietz, D.**, Methods for the detection and estimation of abscisic acid and related compounds, in *Abscisic Acid,* Addicott, F. T., Ed., Praeger, New York, 1983, 23.
69. **Gray, R. T., Mallaby, R., Ryback, G., and Williams, V. P.**, Mass spectra of methyl abscisate and isotopically labelled analogues, *J. Chem. Soc. Perkin Trans. 2,* 919, 1974.
70. **Takeda, N., Harada, K., Suzuki, M., Tatematsu, A., and Sakata, I.**, Structural characterization of underivatized menthyl glucosides using chemical ionization mass spectrometry, *Biomed. Mass Spectrom.,* 10, 608, 1983.
71. **Takeda, N., Harada, K., Suzuki, M., and Tatematsu, A.**, Chemical ionization mass spectrometry of macrolide antibiotics, *Biomed. Mass Spectrom.,* 8, 332, 1981.
72. **Takeda, N., Harada, K., Suzuki, M., Tatematsu, A., Hirai, N., and Koshimizu, K.**, Structural characterization of abscisic acid and related metabolites by chemical ionization mass spectrometry, *Agric. Biol. Chem.,* 48, 685, 1984.
73. **Takeda, N. and Hirai, N.**, Unpublished data, 1983.
74. **Roberts, D. L., Heckman, R. A., Hege, B. P., and Bellin, S. A.**, Synthesis of (RS)-abscisic acid, *J. Org. Chem.,* 33, 3566, 1968.
75. **Weedon, B. C. L., Mayer, H., and Schmieter, U.**, S. African Patent 68 00. 621, 1969, *Chem. Abstr.,* 70, 11496g, 1969.
76. **Cornforth, J. W., Draber, W., Milborrow, B. V., and Ryback, G.**, Absolute stereochemistry of (+)-abscisin II, *Chem. Commun.,* 114, 1967.
77. **Mills, J. A.**, Correlations between monocyclic and polycyclic unsaturated compounds from molecular rotation differences, *J. Chem. Soc.,* 4976, 1952.
78. **Taylor, H. F. and Burden, R. S.**, Xanthoxin, a recently discovered plant growth inhibitor, *Proc. R. Soc. London Ser. B:,* 180, 317, 1972.
79. **Deville, T. E., Hurthouse, M. B., Russel, S. W., and Weedon, B. L. C.**, Absolute configuration of carotenoids, *Chem. Commun.,* 1311, 1969.
80. **Oritani, T. and Yamashita, K.**, Synthesis of optical active abscisic acid and its analogs, *Tetrahedron Lett.,* 2521, 1972.
81. **Eugster, C. H., Buchecker, R., and Tscharner, C. H.**, Determination of the chirality of the α-cyclogeraniols, α-ionones, γ-ionones, α-carotenes, ε-carotenes, and related compounds by chemical linking reactions, *Helv. Chim. Acta,* 52, 1729, 1969.
82. **Ryback, G.**, Revision of the absolute stereochemistry of (+)-abscisic acid, *Chem. Commun.,* 1190, 1972.
83. **Koreeda, M., Weiss, G., and Nakanishi, K.**, Absolute configuration of natural (+)-abscisic acid, *J. Am. Chem. Soc.,* 95, 239, 1973.
84. **Takasugi, M., Anetani, M., Katsui, N., and Masamune, T.**, The occurrence of vomifoliol, dehydrovomifoliol and dihydrophaseic acid in the roots of kidney bean *(Phaseolus vulgaris), Chem. Lett.,* 245, 1973.
85. **Harada, N.**, Absolute configuration of (+)-*trans*-abscisic acid as determined by a quantitative application of the exciton chirality method, *J. Am. Chem. Soc.,* 95, 240, 1973.
86. **Mori, K.**, Synthesis of the optically active grasshopper ketone, *Tetrahedron Lett.,* 723, 1973.
87. **Mori, K.**, Synthesis of the optically active dehydrovomifoliol. A synthetic proof of the absolute configuration of (+)-abscisic acid, *Tetrahedron Lett.,* 2635, 1973.
88. **Kienzle, F., Mayer, H., Minder, R. E., and Thommen, H.**, Synthesis of optically active, natural carotenoids and structural related compounds. III. Synthesis of (+)-abscisic acid, (−)-xanthoxin, (−)-loliolid, (−)-actinioliolid and (−)-dihydroactinioliolid, *Helv. Chim. Acta,* 61, 2616, 1978.
89. **Hayase, Y.**, Studies on the Syntheses of Naturally Occurring Compounds, Utilizing Photoreaction, Ph.D. thesis, Osaka City University, Osaka, 1974, chap. 2.
90. **Isoe, S., Hayase, Y., and Sakan, T.**, Sexual hormones of mucorals. The synthesis of methyl trisporate B and C, *Tetrahedron Lett.,* 3691, 1974.
91. **Milborrow, B. V.**, Abscisic acid, in *Aspects of Terpenoid Chemistry and Biochemistry,* Goodwin, T. W., Ed., Academic Press, London, 1971, chap. 5.
92. **Tinelli, E. T., Sondheimer, E., Walton, D. C., Gaskin, P., and MacMillan, J.**, Metabolites of 2-^{14}C-abscisic acid, *Tetrahedron Lett.,* 139, 1973.
93. **Lehmann, H. and Schütte, H. R.**, The preparation of O-substituted sugars esters, *J. Pract. Chem.,* 319, 117, 1977.
94. **Lehmann, H., Miersch, O., and Schütte, H. R.**, Synthesis of abscisiyl-β-D-glucopyranoside, *Z. Chem.,* 15, 443, 1975.
95. **Koenig, W. and Knorr, E.**, Ueber einige Derivate des Traubenzuckers und der Galactose, *Chem. Ber.,* 34, 957, 1901.
96. **Milborrow, B. V.**, Abscisic acid, in *Phytohormones and Related Compounds: A Comprehensive Treatise,* Vol. 1, Letham, D. S., Goodwin, P. B., and Higgins, T. J. V., Eds., Elsevier, Amsterdam, 1978, chap. 6.

97. **Noddle, R. C. and Robinson, D. R.**, Biosynthesis of abscisic acid: incorporation of radioactivity from [2-^{14}C]-mevalonic acid by intact fruits, *Biochem. J.*, 112, 547, 1969.
98. **Robinson, D. R. and Ryback, G.**, Incorporation of tritium from [4R-4-^{3}H]-mevalonate into abscisic acid, *Biochem. J.*, 113, 895, 1969.
99. **Cornforth, J. W., Cornforth, R. H., Donninger, C., and Popják, G.**, Studies on the biosynthesis of cholesterol. XIX. Steric course of hydrogen eliminations and of C-C bond formations in squalene biosynthesis, *Proc. R. Soc. London Ser. B:*, 163, 492, 1965.
100. **Goodwin, T. W. and Williams, R. J. H.**, The stereochemistry of phytoene biosynthesis, *Proc. R. Soc. London Ser. B:*, 163, 515, 1965.
101. **Archer, B. L., Barnard, D., and Cockbain, E. G.**, The stereochemistry of rubber biosynthesis, *Proc. R. Soc. London Ser. B:*, 163, 519, 1965.
102. **Milborrow, B. V.**, The metabolism of abscisic acid, *J. Exp. Bot.*, 21, 17, 1970.
103. **Suga, T., Hirata, T., Aoki, T., and Shishibori, T.**, Biosynthesis of polyprenols in higher plants. The elimination of the pro-4S-hydrogen atom of mevalonic acid during the formation of their (Z)-isoprene chain, *J. Am. Chem. Soc.*, 105, 6178, 1983.
104. **Suga, T., Hirata, T., Aoki, T., Okamura, M., Ohnishi, Y., Kataoka, T., and Saragai, Y.**, Biosynthesis of polyprenols in higher plants. The stereochemistry in the biological formation of the Z-isoprene chain, Abstr., 26th Symp. Chemistry of Natural Products, Kyoto, Jpn., 1983, 275.
105. **Mallaby, R.**, Studies on the Chemical and Biological Synthesis of Abscisic Acid, Ph.D. thesis, University of London, 1974.
106. **Milborrow, B. V.**, Stereochemical aspects of the formation of double bonds in abscisic acid, *Biochem. J.*, 128, 1135, 1972.
107. **Milborrow, B. V.**, The stereochemistry of cyclization in abscisic acid, *Phytochemistry*, 14, 123, 1975.
108. **Milborrow, B. V.**, The origin of the methyl groups of abscisic acid, *Phytochemistry*, 14, 2403, 1975.
109. **Milborrow, B. V. and Garmston, M.**, Formation of (−)-1′,2′-epi-2-*cis*-xanthoxin acid from a precursor of abscisic acid, *Phytochemistry*, 12, 1597, 1973.
110. **Oritani, T. and Yamashita, K.**, Studies on abscisic acid. IV. Syntheses of compounds structurally related to abscisic acid, *Agric. Biol. Chem.*, 34, 1184, 1970.
111. **Taylor, H. F. and Burden, R. S.**, Xanthoxin, A recently discovered plant growth inhibitor, *Proc. R. Soc. London Ser. B:*, 180, 317, 1972.
112. **Sondheimer, E., Michniewicz, B. M., and Powell, L. E.**, Biological and chemical properties of the epidioxide isomer of abscisic acid and its rearrangement products, *Plant Physiol.*, 44, 205, 1969.
113. **Oritani, T. and Yamashita, K.**, Synthesis and metabolism of abscisic acid analogs, Abstr., Annu. Mtg. Soc. Chemical *Regulation of Plants*, Japan, 1979, 9.
114. **Milborrow, B. V. and Noddle, R. C.**, Conversion of 5-(1,2-epoxy-2,6,6-trimethylcyclohexyl)-3-methylpenta-*cis*-2-*trans*-dienoic acid into abscisic acid, *Biochem. J.*, 119, 727, 1970.
115. **Masuda, Y.**, Effect of light on a growth inhibitor in wheat roots, *Physiol. Plant.*, 15. 780, 1962.
116. **Taylor, H. F. and Smith, T. A.**, Production of plant growth inhibitors from xanthophylls: a possible source of dormin, *Nature (London)*, 215, 1513, 1967.
117. **Taylor, H. F. and Burden, R. S.**, Identification of plant growth inhibitors produced by photolysis of violaxanthin, *Phytochemistry*, 9, 2217, 1970.
118. **Firn, R. D. and Friend, J.**, Enzymatic production of the plant growth inhibitor, xanthoxin, *Planta*, 103, 263, 1972.
119. **Taylor, H. F. and Burden, R. S.**, Preparation and metabolism of [2-^{14}C]-*cis*, *trans*-xanthoxin, *J. Exp. Bot.*, 24, 873, 1973.
120. **Firn, R. D., Burden, R. S., and Taylor, H. F.**, The detection and estimation of the growth inhibitor xanthoxin in plants, *Planta*, 102, 115, 1972.
121. **Bennett, R. D., Norman, S. M., and Maier, V. P.**, Biosynthesis of abscisic acid from [1,2-^{13}C$_2$]acetate in *Cercospora rosicola*, *Phytochemistry*, 20, 2343, 1981.
122. **Neill, S. J., Horgan, R., Walton, D. C., and Lee, T. S.**, The biosynthesis of abscisic acid in *Cercospora rosicola*, *Phytochemistry*, 21, 61, 1982.
123. **Neill, S. J., Horgan, R., Walton, D. C., and Griffin, D.**, Biosynthesis of abscisic acid, in *Plant Growth Substances 1982: Proc. 11th Int. Conf. Plant Growth Substances*, Wareing, P. F., Ed., Academic Press, London, 1982, 315.
124. **Ichimura, M., Oritani, T., and Yamashita, K.**, The metabolism of (2Z,4E)-α-ionylideneacetic acid in *Cercospora cruenta*, a fungus producing (+)-abscisic acid, *Agric. Biol. Chem.*, 47, 1895, 1983.
125. **Oritani, T., Ichimura, M., and Yamashita, K.**, A novel abscisic acid analog, (+)-(2Z,4E)-5-(1′,4′-dihydroxy-6′,6′-dimethyl-2′-methylenecyclohexyl)-3-methyl-2,4-pentadienoic acid, from *Cercospora cruenta*, *Agric. Biol. Chem.*, 48, 1677, 1984.
126. **Oritani, T. and Yamashita, K.**, *Chemical Regulation of Plants*, 19, 17, 1984.
127. **Tamura, S. and Nagao, M.**, Syntheses and biological activities of new plant growth inhibitors structurally related to abscisic acid. I., *Agric. Biol. Chem.*, 34, 1393, 1970.

128. **Rivier, L. and Pilet, P.-E.**, Quantification of abscisic acid and its 2-trans isomer in plant tissues using a stable isotope dilution technique, in *Stable Isotope*, Schmidt, H. L., Forstel, H., and Heinzinger, K., Eds., Elsevier, Amsterdam, 1982, 535.
129. **Van Steveninck, R. F. M.**, Abscission-accelerators in lupins (*Lupinus luteus* L.), *Nature (London)*, 183, 1246, 1959.
130. **Goldschmidt, E. E., Goren, R. R., EvenChen, Z., and Bitter, S.**, Increase in free and bound abscisic acid during natural and ethylene-induced senescence of citrus fruit peel, *Plant Physiol.*, 51, 879, 1973.
131. **Weiler, E. W. and Ziegler, H.**, Determination of phytohormones in phloem exudate from tree species by radioimmunoassay, *Planta*, 152, 168, 1981.
132. **Milborrow, B. V. and Vaughan, G.**, The long term metabolism of ($\pm$)-[2-^{14}C]-abscisic acid by apple seeds, *J. Exp. Bot.*, 30, 983, 1979.
133. **Milborrow, B. V.** Regulation of abscisic acid metabolism in *Plant Growth Substances 1979: Proc. 10th Int. Conf. Plant Growth Substances*, Skoog, F., Ed., Springer-Verlag, Berlin, 1980, 262.
134. **Loveys, B. R. and Milborrow, B. V.**, Isolation and characterization of 1'-O-abscisic acid-β-D-glucopyranoside from vegetative tomato tissue, *Aust. J. Plant Physiol.*, 8, 571, 1981.
135. **Sembdner, G., Dathe, W., Kefeli, V. I., and Kutacek, M.**, Abscisic acid and other naturally occurring plant growth inhibitors, in *Plant Growth Substances 1979: Proc. 10tth Int. Conf. Plant Growth Substances*, Skoog, F., Ed., Springer-Verlag, Berlin, 1980, 254.
136. **MacMillan, J., Seaton, J. C., and Suter, P. F.**, Plant hormones-I. Isolation of gibberellin A_1 and gibberellin A_5 from *Phaseolus multiflorus*, *Tetrahedron*, 11, 60, 1960.
137. **MacMillan, J. and Pryce, R. J.**, Phaseic acid. A putative relative of abscisic acid, from seed of *Phaseolus multiflorus*, *Chem. Comm.*, 124, 1968.
138. **Gillard, D. F. and Walton, D. C.**, Abscisic acid metabolism by a cell-free preparation from *Echinocystis lobata* liquid sperm, *Plant Physiol.*, 58, 790, 1976.
139. **Zeevaart, J. A. D. and Milborrow, B. V.**, Metabolism of abscisic acid and the occurrence of *epi*-dihydrophaseic acid in *Phaseolus vulgaris*, *Phytochemistry*, 15, 493, 1976.
140. **Adesomoju, A. A., Okogun, J. I., Ekong, D. E., and Gaskin, P.**, GC-MS identification of abscisic acid and abscisic acid metabolites in seed of *Vigna unguiculata*, *Phytochemistry*, 19, 223, 1980.
141. **Hirai, N. and Koshimizu, K.**, Chirality of the acyl group of β-hydroxy-β-methylglutarylhydroxyabscisic acid, *Phytochemistry*, 20, 1867, 1981.
142. **Hirai, N.**, Unpublished data, 1979.
143. **Zeevaart, J. A. D. and Boyer, G. L.**, Metabolism of abscisic acid in *Xanthium strumarium* and *Ricinus communis*, in *Plant Growth Substances 1982: Proc. 11th Int. Conf. Plant Growth Substances*, Wareing, P. F., Ed., Academic Press, London, 1982, 335.
144. **Walton, D. C., Dorn, B., and Fey, J.**, The isolation of an abscisic acid metabolite, 4'-dihydrophaseic acid, from nonimbibed *Phaseolus vulgaris* seed, *Planta*, 112, 87, 1973.
145. **Frydman, V. M., Gaskin, P., and MacMillan, J.**, Qualitative and quantitative analyses of gibberellins throughout seed maturation in *Pisum sativum*, *Planta*, 118, 123, 1974.
146. **Beeley, L. J., Gaskin, P., and MacMillan, J.**, Gibberellin A_{43} and other terpenes in endosperm of *Echinocystis macrocarpa*, *Phytochemistry*, 14, 779, 1975.
147. **Martin, G. C., Dennis, F. G., Jr., Gaskin, P., and MacMillan, J.**, Identification of gibberellins A_{17}, A_{25}, A_{45}, abscisic acid, phaseic acid, and dihydrophaseic acid in seeds of *Pyrus communis*, *Phytochemistry*, 16, 605, 1977.
148. **Setter, T. L. and Brun, W. A.**, Abscisic acid translocation and metabolism in soybeans following depodding and petiole girdling treatments, *Plant Physiol.*, 67, 774, 1981.
149. **Setter, T. L., Brenner, M. L., Brun, W. A., and Krick, T. P.**, Identification of a dihydrophaseic acid aldopyranoside from soybean tissue, *Plant Physiol.*, 68, 93, 1981.
150. **Milborrow, B. V. and Vaughan, G. T.**, Characterization of dihydrophaseic acid, 4'-O-β-glucopyranoside as a major metabolite of abscisic acid, *Aust. J. Plant Physiol.*, 9, 361, 1982.
151. **Dathe, W. and Sembdner, G.**, Isolation of 4'-dihydroabscisic acid from immature seeds of *Vicia faba*, *Phytochemistry*, 21, 1798, 1982.
152. **Lehmann, H., Press, A., and Schmidt, J.**, A novel abscisic acid metabolite from cell suspension cultures of *Nigella damascena*, *Phytochemistry*, 22, 1277, 1983.
153. **Tietz, D., Dörffling, K., Wöhrle, D., Erxlebenm, I., and Lieman, F.**, Identification by combined gas chromatography-mass spectrometry of phaseic acid and dihydrophaseic acid and characterization of further abscisic acid metabolites in pea seedlings, *Planta*, 147, 168, 1979.
154. **Rudnicki, R. and Czapski, J.**, Uptake and degradation of 1-carbon-14-labelled-abscisic acid by apple seeds during stratification, *Ann. Bot.*, 38, 189, 1974.
155. **Hiron, R. W. P. and Wright, S. T. C.**, The role of endogenous abscisic acid in the response of plants to stress, *J. Exp. Bot.*, 24, 769, 1973.
156. **Dörffling, K., Sonka, B., and Tietz, D.**, Variation and metabolism of abscisic acid in pea seedlings during and after water stress, *Planta*, 121, 57, 1974.

157. **Zeevaart, J. A. D.**, (+)-Abscisic acid content of spinach in relation to photoperiod and water stress, *Plant Physiol.*, 48, 86, 1971.
158. **Sondheimer, E., Galson, E. C., Chang, Y. P., and Walton, D. C.**, Asymmetry its importance to the action and metabolism of abscisic acid, *Science*, 174, 829, 1971.
159. **Tillberg, E.**, An abscisic acid-like substance in dry and soaked *Phaseolus vulgaris* seeds determined by the Lemna growth bioassay, *Physiol. Plant.*, 34, 192, 1975.
160. **Franssen, J. M., Knegt, E., and Bruinsma, J.**, Note on the biological activities of xanthoxin-like substances, *Z. Pflanzenphysiol.*, 94, 155, 1979.
161. **Aspinall, D., Paleg, L. G., and Addicott, F. T.**, Abscisin II and some hormone-regulated plant responses, *Aust. J. Biol. Sci.*, 20, 869, 1967.
162. **Chrispeels, M. J. and Varner, J. E.**, Inhibition of gibberellic acid induced formation of α-amylase by abscisin II, *Nature (London)*, 212, 1066, 1966.
163. **Galli, M. G., Miracca, P., and Sparvoli, E.**, Lack of inhibiting effects of abscisic acid on seeds of *Haplopappus gracilis* (Nutt.) Gray pregerminated in water for short times, *J. Exp. Bot.*, 31, 763, 1980.
164. **Osborne, D. J.**, Hormonal mechanism regulating senescence and abscission, in *Biochemistry and Physiology of Plant Growth Substances: Proc. 6th Int. Conf. Plant Growth Substances*, Wightman, F. and Setterfield, G., Eds., Runge Press, Ottawa, 1968, 815.
165. **Bornman, C. H., Spurr, A. R., and Addicott, F. T.**, Abscisin, auxin, and gibberellin effects on the developmental aspects of abscission in cotton *(Gosypium hirsutum)*, *Am. J. Bot.*, 54, 125, 1967.
166. **Mittelheuser, C. J. and Van Steveninck, R. F. M.**, Stomatal closure and inhibition of transpiration induced by (RS)-abscisic acid, *Nature (London)*, 221, 281, 1969.
167. **Ogunkanmi, A. B., Tucker, D. J., and Mansfield, T. A.**, An improved bio-assay for abscisic acid and other antitranspirants, *New Phytol.*, 72, 277, 1973.
168. **Willmer, C. M., Don, R., and Parker, W.**, Levels of short-chain fatty acids and of abscisic acid in water-stressed and nonstressed leaves and their effects on stomata in epidermal strips and excised leaves, *Planta*, 139, 281, 1978.
169. **Tucker, D. J. and Mansfield, T. A.**, A simple bioassay for detecting "antitranspirant" activity of naturally occurring compounds such as abscisic acid, *Planta*, 98, 157, 1971.
170. **Travis, A. J. and Mansfield, T. A.**, Reversal of the CO_2-response of stomata by fusicoccin, *New Phytol.*, 83, 607, 1979.
171. **El-Antably, H. M. M., Wareing, P. F., and Hillman, J.**, Some physiological responses to D. L. abscisin (dormin), *Planta*, 73, 74, 1967.
172. **Spivastava, B. I. S.**, Acceleration of senescence and of the increase of chromatin-associated nucleases in excised barley leaves by abscisin II and its reversal by kinetin, *Biochim. Biophys. Acta*, 169, 534, 1968.
173. **Pilet, P.-E.**, Abscisic acid as a root growth inhibitor: physiological analysis, *Planta*, 122, 299, 1975.
174. **Visscher, S. N.**, Regulation of grasshopper fecundity, longevity and egg viability by plant growth hormones, *Experientia*, 36, 130, 1980.
175. **Livingston, B. W.-C.**, Abscisic tablets and process, U.S. Patent 3,958,025, 1976.
176. **Visscher, S. N.**, Effects of abscisic acid in animal growth and reproduction, in *Abscisic Acid*, Addicott, F. T., Ed., Praeger, New York, 1983, 553.
177. **Walton, D. C.**, Structure-activity relationships of abscisic acid analogs and metabolites, in *Abscisic Acid*, Addicott, F. T., Ed., Praeger, New York, 1983, 113.
178. **Cummins, W. R. and Sondheimer, E.**, Activity of the asymmetric isomers of abscisic acid in a rapid bioassay, *Planta*, 111, 365, 1973.
179. **Yamashita, K., Nagano, E., and Oritani, T.**, Optical resolution and biological activities of α-ionylideneacetic acid, *Agric. Biol. Chem.*, 44, 1441, 1980.
180. **Kriedman, P. E., Loveys, B. R., Fuller, G. L., and Leopold, A. C.**, Abscisic acid and stomatal regulation, *Plant Physiol.*, 49, 842, 1972.
181. **Sondheimer, E. and Walton, D. C.**, Structure-activity correlations with compounds related to abscisic acid, *Plant Physiol.*, 45, 244, 1970.
182. **Oritani, T. and Yamashita, K.**, Studies on abscisic acid. I. Synthesis of keto-ionylideneacetic acids, *Agric. Biol. Chem.*, 34, 108, 1970.
183. **Oritani, T. and Yamashita, K.**, Studies on abscisic acid. V. The epoxidation products of methyl dehydro-β-ionylideneacetates, *Agric. Biol. Chem.*, 34, 1821, 1970.
184. **Oritani, T. and Yamashita, K.**, Studies on abscisic acid. II. The oxidation products of methyl α- and β-cyclocitrylideneacetates and methyl α-cyclogeranate, *Agric. Biol. Chem.*, 34, 198, 1970.
185. **Coke, L. B., Stuart, K. L., and Whittle, Y. G.**, Further effects of vomifoliol on stomatal aperture and on the germination of lettuce and the growth of cucumber seedlings, *Planta*, 122, 307, 1975.
186. **Raschke, K., Firn, R. D., and Pierce, M.**, Stomatal closure in response to xanthoxin and abscisic acid, *Planta*, 125, 149, 1975.
187. **McWha, J. A., Philipson, J. J., Hillman, J. R., and Wilkins, M. B.**, Molecular requirements for abscisic acid activity in two bioassay systems, *Planta*, 109, 327, 1973.

188. **Orton, P. J. and Mansfield, T. A.**, The activity of abscisic acid analogs as inhibitors of stomatal opening, *Planta,* 121, 263, 1974.
189. **Jones, R. J. and Mansfield, T. A.**, Effects of abscisic acid and its ester on stomatal aperture and the transpiration ratio, *Physiol. Plant,* 26, 321, 1972.
190. **Yamashita, K., Watanabe, T., Watanabe, M., and Oritani, T.**, Synthesis and biological activity of methyl 3-demethylabscisate and its related analogs, *Agric. Biol. Chem.,* 46, 3069, 1982.
191. **Oritani, T. and Yamashita, K.**, Chemistry of abscisic acid, *Kagaku to Seibutsu,* 13, 351, 1975.
192. **Oritani, T. and Yamashita, K.**, Synthesis and biological activity of ($\pm$)-2′,3′-dihydroabscisic acid, *Agric. Biol. Chem.,* 46, 817, 1982.
193. **Walton, D. C. and Sondheimer, E.**, Activity and metabolism of ^{14}C-($\pm$)-abscisic acid derivatives, *Plant Physiol.,* 49,, 290, 1972.
194. **Mousseron-Canet, M., Mani, J.-C., Durand, B., Nitsch, J., Dormand, J., and Bonnafous, J.-C.**, Analogues de l'acide abscisique ($\pm$) hormone de dormance, *C. R. Acad. Sci, Paris,* 270, 1936, 1970.
195. **Nagano, E., Oritani, T., and Yamashita, K.**, Synthesis and physiological activity of methyl 6′,6′-didemethyl abscisate and new chiral analogs, *Agric. Biol. Chem.,* 44, 2095, 1980.
196. **Nanjo, M., Oritani, T., and Yamashita, K.**, Synthesis and physiological activity of monodemethyl abscisic acids and methyl 5-(1′,6′-epoxy-2′,2′-dimethylcyclohexyl)-3-methyl-(2Z,4E)-2,4-pentadienoate, *Agric. Biol. Chem.,* 41, 1711, 1977.
197. **Adachi, T., Oritani, T., and Yamashita, K.**, Synthesis of ($\pm$)-methyl 5-(1′,2′-epoxy-2′, 6′-dimethyl-cyclohexyl)-3-methyl-(2Z,4E)-2,4-pentadienoate, *Agric. Biol. Chem.,* 39, 1681, 1975.
198. **Kienzle, F.**, Notiz zur Syntheses von rac. *cis-* und *trans-*2-Nor-abscisinsäure, *Helv. Chim. Acta,* 62, 155, 1979.
199. **Ho, T.-h.D.**, On the mode of action of abscisic acid in barley aleurone layer cells, *Plant Physiol.,* 63(Suppl.), 79, 1979.
200. **Dashek, W. V., Singh, B. N., and Walton, D. C.**, Abscisic acid localization and metabolism in barley aleurone layers, *Plant Physiol.,* 64, 43, 1979.
201. **Sharkey, T. D. and Raschke, K.**, Effects of phaseic acid and dihydrophaseic acid on stomatal and the photosynthetic apparatus, *Plant Physiol.,* 65, 291, 1980.
202. **Itai, C., Weyers, J. D. B., Hillman, J. R., Meidner, H., and Willmer, C.**, Abscisic acid and guard cells of *Commelina communis* L., *Nature (London),* 271, 652, 1978.
203. **Itai, C. and Meidner, H.**, Effect of abscisic acid on solute transport in epidermal tissue, *Nature (London),* 271, 653, 1978.
204. **Singh, B. N., Galson, E., Dashek, W., and Walton, D. C.**, Abscisic acid levels and metabolism in the leaf epidermal tissue of *Tulipa gesneriana* L. and *Commelina communis* L., *Planta,* 146, 135, 1979.

Chapter 6

ETHYLENE

Hidemasa Imaseki

TABLE OF CONTENTS

I. Occurrence 250

II. Chemistry 251
    A. Isolation 251
    B. Characterization 252

III. Biosynthesis and Metabolism 252
    A. Biosynthesis 252
    B. Metabolism 257

IV. Biological Activity 257

References 259

I. OCCURRENCE

Ethylene is the simplest form of organic compounds that have ever been recognized as growth regulators of plants. Like other growth regulators known as plant hormones, ethylene is produced in a minute amount by tissues and organs of almost all plant species, but in some cases in large amounts. They include bacteria,[1,2] fungi,[3-5] algae,[6,7] and various parts of higher plants.[7,8] A literature search reveals that although ethylene is produced in both unicellular and multicellular plants, its physiological action appears to be confined to multicellular plants, and there has been no report on physiological significance of ethylene on unicellular bacteria or algae.

A long history of ethylene research in plant science is described in detail by Burg,[7] Abeles,[8] and briefly by Osborne,[9] and this will not be reproduced here. It should be mentioned, however, that recognition of ethylene as a plant growth regulator originates from examinations of various unusual phenomena such as early shedding of street trees,[10] unusual growth pattern of etiolated peas in laboratory,[11] "sleep" of carnation flowers in a greenhouse,[12] earlier flowering of pineapples treated with smoke,[13] yellowing of orange fruits by the gaseous combustion products of kerosene stoves,[14] and so forth.

Use of sensitive gas chromatography (GC) enables us to measure accurately amounts of ethylene of less than several picomoles by a simple injection of a gas sample containing ethylene,[15,16] and has established characteristic occurrence of ethylene in plants; the rate of ethylene production by plant tissues greatly varies with changes in the physiological state of the tissues, and the rise and fall of the production rate are fairly rapid.[17-23] Although a small portion of ethylene produced by or supplied to plant tissues is metabolized,[24] or bound to cellular components,[25,26] plant tissues in general have no large capacity to bind, trap, or metabolize the gas to influence the rate of release from tissues. Most ethylene produced in tissues is released to the ambient atmosphere depending on the concentration gradient between the inside and outside of the tissues.[27] Therefore, the cellular concentration of ethylene, which should be an important factor for its biological action, is primarily controlled by the rate of production. Measurement of the rate of release from tissues generally gives sufficient information on the rate of production in the tissues, thus, the relative changes in cellular concentrations.

The rate of ethylene production per fresh weight is relatively greater in younger tissues such as apical portions of stem and lateral buds,[18,27] decreases as tissues become mature, and again increases greatly when tissues enter the senescent stage.[29-31] Auxin promotes ethylene production in vegetative tissues,[32-34] but not in fruit tissues,[35] and a high rate in young tissues is attributed to their relative higher concentration of auxin.[18,24] This is indeed the case in photo- or geostimulated stems.[36] According to the classical Went-Cholodny theory, tropic responses result from differential elongation due to unequal distribution of auxin between both sides of the stimulated tissue. Abeles and Rubinstein,[36] using bean stem, showed that the lower side of the horizontally placed stem which should contain more auxin[37] produced more ethylene than the upper side of the tissue which should contain less auxin.

Great increases in the ethylene production rate in senescent tissues are normally observed with leaves,[31,38,39] flowers,[29,30,39] and fruits.[7,8,38,40-42] Light irradiation changes the rate of ethylene production either to increase, as in sorghum seedlings,[43] cultured rose cells,[44] and germinating cucumber seeds,[45] or to decrease, as in bean hooks,[46] etiolated pea seedlings,[20] and rice coleoptiles.[47] Red light irradiation causes opening of plumules and hook in parallel with the decrease in the production rate; the opening is inhibited or reversed by a low concentration of ethylene applied to the irradiated pea plants, implying that the unique morphology of young seedlings of dicots germinated in the dark or in soil is maintained by the gas produced in those tissues.[48] Changes by red light of ethylene production are partially controlled by the phytochrome system, as the red-far red reversibility of the production rate is observed.[20,46,47]

Stress of various kinds imposed on plant tissues greatly stimulates the rate of ethylene production. Injury caused by pathogenic microorganisms[49-51] and by local application of toxic chemicals,[52,53] wounds made by cutting,[22,23] water stress,[54-56] or even physical stress such as contact of tissues to obstacles or bending[19,57-59] causes a transient burst of ethylene production by the affected cells. In the cases of injury or wound, live cells located near the cells destroyed or killed by injury or wound are activated to produce ethylene.[51,52]

Thus, occurrence of ethylene in the plant kingdom has been firmly established, but it is not the simple occurrence but changes in the production rate and in the sensitivity of tissues to the gas that are of physiological significance.

II. CHEMISTRY

As is clear from the chemical structure of ethylene, it is a symmetric molecule having one double bond. The biological activity of ethylene appears to relate to its unsaturated bond which is attached to a terminal carbon atom. A weak, but definite ethylene-like activity is found in propylene, acetylene, and carbon monooxide (all of which have common structural characteristics mentioned above), but not in ethane and carbon dioxide which lack either one of the above two structural characteristics.[60,61] A great reduction in biological activity of substituted ethylene, propylene, and monochloroethylene has been attributed to a steric or ionic hindrance by the substituted group to a possible binding site of cell.[62]

A. Isolation

Since ethylene is a gas which is released out of the producing tissue and since ethylene contained in air or other gases can be directly analyzed by a gas chromatograph equipped with a hydrogen flame ionization detector,[15,16] isolation of the gas is not necessary in most cases. A central problem, however, is to collect the gas quantitatively. When the production rate of ethylene in tissue is relatively high (usually more than 5 pmol/g fresh weight per hour), tissue (whole organs or excised tissue sections) is placed in a gas-tight vessel (test tubes, sample vials, or flasks sealed with vaccine stoppers or silicone rubber stoppers) for an hour or so, then a portion of the headspace gas withdrawn with a gas-tight syringe is directly injected into a gas chromatograph. An advantage of this method is that repeated measurement with the same tissue samples is possible and changes with time of the production rate are easily determined. In order to minimize possible influence of changes in the ambient air composition during the prolonged incubation in the sealed vessel, flushing of the incubation vessels with fresh air is made after each measurement.

If, however, the ethylene production rate of tissue is very small or if a whole intact plant must be used, the incubation period of the tissue must be long to accumulate ethylene high enough for direct measuremeent of the gas. This will lead to changes in composition of the ambient air by respiration, which in turn will affect the rate of ethylene production. To overcome this problem, the continuous flow-through incubation method has been frequently used.[63] Bassi and Spencer[63] developed an all glass-metal cuvette which eliminates all possible sources of artifactual ethylene formation, and can accommodate an intact plant through which ethylene-free air is continuously introduced to maintain a constant composition of the ambient air. Bradford and Dilley[62] designed a flow-through vessel which allowed separate measurement of ethylene produced by the shoot and root systems of a plant. As the concentration of ethylene in the outflow air is usually below the detection limit of GC, for measurement the air-flow must be stopped for 1 or 2 hr to accumulate ethylene in the ambient air,[62] or ethylene in the outflow air must be collected by a small column of silica gel cooled in dry ice-acetone slurry.[64,66] To obtain ethylene-free air, Eastwell et al.[67] found that a small column containing 50% (w/w) cupric oxide powder on Chromosorb P, 30% (w/w) cupric oxide, 0.3% (w/w) ferric oxide in Kieselguhr pellets, or 5% platinum on asbestos heated at 800 or 650°C efficiently removed all the hydrocarbons contained in air without significant changes in concentrations of carbon dioxide and oxygen in air.

"Extraction" of air contained in intercellular space of plant tissues by the evacuation method has been reported by Beyer and Morgan,[68] and has been successfully employed to determine concentration of ethylene in the intercellular space.[68,69]

B. Characterization

Practically speaking, ethylene contained in an air sample can be identified by its retention time of GC with an alumina[15] or Porapak[71,72] column. These columns efficiently separate neutral hydrocarbons of low molecular weight at a low temperature (lower than 80°C). Retention times for methane, ethylene, and ethane, however, are very close to each other on an alumina column, which has been most widely used; for more accurate characterization several simple separation procedures must be taken. Neutral hydrocarbons are relatively insoluble in dilute alkali and acid solution, and unsaturated hydrocarbons are quantitatively absorbed by mercuric perchlorate in acidic condition.[73] Ethylene absorbed by the mercuric perchlorate is quantitatively recovered by addition of excess chloride ions.[73] Therefore, if several milliliters of a gas sample taken in a syringe are successively washed with 0.1 N NaOH, 0.1 N HCl, and 0.25 M mercuric perchlorate solution, and if after each washing, a small fraction of the gas sample is analyzed by GC, ethylene disappears from the gas sample only after the last washing. The mercuric perchlorate solution is transferred to a small graduated test tube. The tube is sealed with a silicone stopper, cooled in an ice-bath, evacuated through a hypodermic needle inserted through the stopper, and 4 M LiCl solution is introduced into the tube until the gas space becomes 1.0 to 1.5 mℓ. Clean air is then introduced to the atmospheric pressure and the gas phase in the tube is taken to a syringe and analyzed by GC. Ethylene should now be present in the gas sample. The entire procedure can be completed within $1/2$ hr.

III. BIOSYNTHESIS AND METABOLISM

A. Biosynthesis

Feeding of various radioactive materials to ethylene-producing plant tissues has not been successful in obtaining a definite pathway of ethylene biosynthesis.[74-76] However, during the course of studies to establish the in vitro ethylene-producing systems, nonenzymatic model systems producing ethylene were elaborated by Lieberman and Mapson,[77] and by Abeles and Rubinstein.[78] Extensive studies on the model systems revealed that methionine was a precursor of ethylene in the systems,[79,80] and radioactive methionine supplied to apple fruit tissue was found to be converted to ethylene.[81,82] Subsequently, methionine was indeed proved to be a good precursor of ethylene in a number of plant tissues including fruits,[82-84] auxin-treated tissues,[82,85] and flowers.[83,86] During conversion of methionine to ethylene by plant tissues, the C-3 and C-4 of methionine are converted to ethylene[81,82] with the methylthio group being retained in the tissue;[82] the C-1 is released as carbon dioxide[81,82] as occurred in the model systems.[79,80] The C-2 atom is converted to formate.[87] Methionine is not an only source of ethylene in plants; ethylene produced by *Penicillium digitatum*, the citrus fungus, grown in the static culture has been shown to originate from α-ketoglutarate or glutamate,[88] but the fungus grown in the shake culture produces ethylene from methionine.[89] Other minor sources of the gas are described by Abeles.[8]

Although methionine has been conclusively shown as a major precursor of ethylene produced in higher plant tissues, the in vitro ethylene-producing system has never been established because the ethylene-producing activity is very labile and completely lost upon disruption of tissues.[77] Consequently, characterization of the ethylene-producing system has been extensively made at the tissue level, mainly with climacteric fruit slices or plugs[81,84,90-93] and auxin-treated stem sections of etiolated pea[82,94] and mung bean[95-98] seedlings. Slices of apple fruits at the climacteric stage produce ethylene at an almost constant rate for more than several hours, if slices are incubated without buffers or with buffers containing 0.2 M mannitol or KCl.[99] In contrast, stem sections of pea and mung bean seedlings produce little ethylene, but when the tissue sections are incubated with indole-3-acetic acid (IAA), the rate of ethylene production starts to increase dramatically after about a 1 hr lag period.[36,94,95,97] The action of IAA to increase the ethylene production rate appears at 1 μM,[36,94,95] and the increase in the production rate is dependent upon concentrations of IAA up to about 1 mM.[97] The high rate of ethylene production in IAA-treated tissue sections is maintained only in the continued presence of IAA.[94,95] Only compounds having the auxin activity act in a similar way.[36]

The properties of both ethylene-producing systems are similar except one feature; the apple system is not inhibited by inhibitors of protein and RNA synthesis[99] whereas the auxin-induced system is inhibited.[36,94,95] The ethylene-producing activity in both systems is inhibited by aminoethoxyvinylglycine (AVG),[92,100] aminooxyacetic acid,[101] Co^{++} ion,[102] anaerobiosis,[74,99] surface active substances,[93,103] uncouplers of oxidative phosphorylation,[104] and L-canaline.[105] Since development of auxin-induced ethylene production is inhibited by inhibitors of protein and RNA synthesis,[36,94,95] auxin has been thought to induce *de novo* synthesis of protein(s) involved in ethylene biosynthesis.[94,95] The protein has been shown to be rapidly inactivating within cells with its half-life being about 30 min,[94,95] because the high ethylene-producing activity of the IAA-treated stem sections rapidly decays if further protein synthesis is inhibited by adding cycloheximide to the incubation medium.

Ethylene production in apple fruit has high affinity to oxygen similar to the respiratory cytochrome system,[74,99] and is inhibited by 2,4-dinitrophenol (DNP) and arsenate; inhibition by arsenate is reversed by phosphate.[99] A large portion of methionine supplied is converted to S-adenosylmethionine (SAM).[82] Based on these observations, Burg[99] suggested that the high energy phosphate bond was involved in ethylene biosynthesis from methionine and that SAM might be a possible intermediate. Murr and Yang[104] reported that ethylene production in both apple fruit and IAA-treated mung bean stems is severely inhibited by DNP and L-canaline, an inhibitor of pyridoxal phosphate (PLP)-linked enzymes, and they suggested involvement of ATP and PLP in ethylene biosynthesis. To explain these results, Murr and Yang[104] also proposed SAM as a possible intermediate in the conversion of methionine to ethylene and predicted that 5′-methylthioadenosine (MTA) would be formed in parallel with ethylene synthesis. Subsequently, formation of MTA and 5-methylthioribose (MTR) was demonstrated only in the climacteric (ethylene producing) apple fruit tissue, but not in the preclimacteric (ethylene nonproducing) apple and the climacteric tissue treated with AVG.[105] MTR is formed from MTA in apple fruit tissue and incorporation of methylthio group of MTR back to methionine is also demonstrated.[105,106] Thus, Adams and Yang[105] proposed a cyclic pathway of ethylene biosynthesis. Recycling of MTR derived from MTA into methionine is also shown in ripe tomato pericarp tissue.[107]

Ethylene production involves an aerobic reaction.[7,99] When apple fruit tissue which has been placed in a nitrogen atmosphere for some time is transferred to air, the tissue immediately restarts ethylene production at a rate higher than the control tissue which has not been placed in nitrogen. This phenomenon was interpreted to be that a presumptive precursor of ethylene accumulated within the tissue during the anaerobic period and the precursor rapidly converted to ethylene on supply of oxygen.[108] The step requiring oxygen has been assumed to locate between methionine and ethylene,[84] probably at the final stage of biosynthesis.[109]

Climacteric apple fruit tissue fed with U-^{14}C-methionine and placed in nitrogen atmosphere accumulates a radioactive metabolite which disappears on transfer of the tissue from nitrogen to air concomitant with the restart of ethylene production.[110] Formation of the metabolite is accompanied by formation of MTR and is inhibited by AVG, an inhibitor of ethylene production.[110] The metabolic behavior of the metabolite accounted for the properties of an

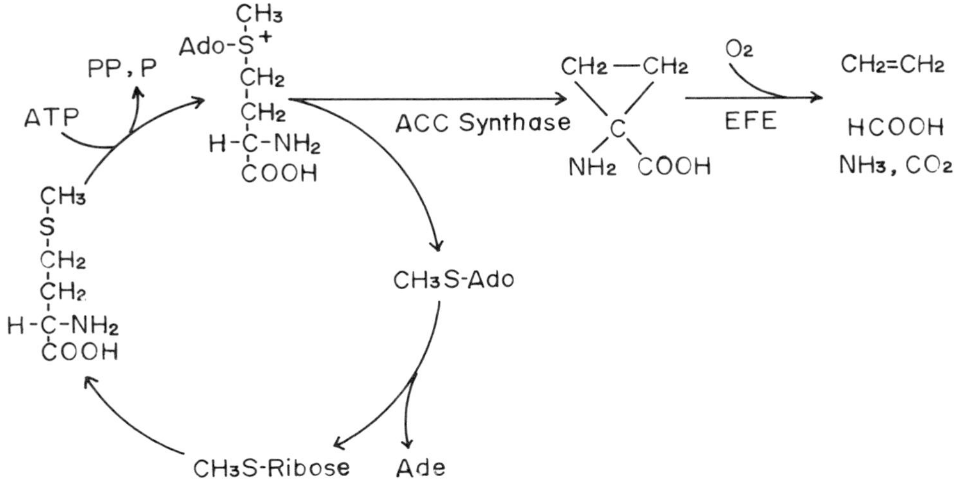

FIGURE 1. Biosynthetic pathway of ethylene from methionine with a cyclic methionine metabolism.[105,106,110]

ethylene precursor and the metabolite was identified as 1-aminocyclopropane-1-carboxylic acid (ACC).[110] ACC supplied to apple fruit tissue is efficiently converted to ethylene and the conversion is the enzymatic reaction requiring oxygen.[110] ACC was originally found in perry pear and cider apple by Burroughs,[111] and also in cowberries by Vähätalo and Virtanen[112] as a nonprotein amino acid.

Lürssen et al.[113] independently found that soybean leaf discs and other several plant tissues produced a large amount of ethylene when the discs were supplied with synthetic ACC and characteristics of the ACC-dependent ethylene formation were similar to those of endogenous ethylene production by several criteria. They proposed that ACC derived from SAM in plant tissues served as an endogenous precursor of ethylene. Subsequently, conversion of ACC to ethylene has been shown in a number of plant tissues.[114] In the auxin-treated mung bean stem sections, a methionine metabolite which accumulated only after treatment with IAA at concentrations to induce sizable ethylene production was again identified as ACC.[115] Conversion of ACC to ethylene occurs by the control tissue without auxin treatment, indicating that auxin stimulates induction of the enzyme involved in ACC synthesis and the enzyme catalyzing ACC conversion to ethylene, an ethylene-forming enzyme (EFE), is a constitutive enzyme.[115-117] The ability of plant tissues to produce ACC is always related to the ability of the tissues to produce ethylene.[116,120-123] The pathway of ethylene biosynthesis is illustrated in Figure 1.

Enzymatic formation of ACC from SAM was demonstrated by a crude extract prepared from ripe tomato fruits.[118,123] The enzyme, ACC synthase, is soluble, requires PLP for its reaction, and uses SAM and to a lesser extent S-adenosylethionine as substrates.[118,123] It was also found that AVG competitively inhibited the ACC synthase reaction.[118,123] Formation of ACC from SAM is a unique reaction involving elimination of α-hydrogen concerted with γ-elimination of methylthioadenosine of SAM.[110,124] The molecular weight of ACC synthase of tomato fruit is estimated to be about 57,000 by gel filtration,[125] but only limited information on properties of the enzyme is available, as the enzyme has not been purified.[118,124,125]

The ACC synthase activity in tomato fruit increases as the fruit ripen to produce ethylene,[118,124] and that in etiolated pea[116] and mung bean stem sections[117,120] also increases after auxin treatment, whereas formation of SAM from methionine is not affected by auxin.[115] Cytokinin which stimulates and abscisic acid (ABA) which inhibits auxin-induced ethylene production enhances and reduces, respectively, the endogenous ACC synthase activity.[120] Ethylene, the end product of the biosynthetic pathway, partially suppresses formation of the enzyme in mung bean stem sections.[126] In mung bean stem sections placed under the various hormonal conditions, the endogenous content of ACC, the endogenous activity of ACC synthase, and the rate of ethylene production change almost in parallel to each other, indicating that auxin stimulates inductive formation of ACC synthase which increases the endogenous content of ACC and in turn the rate of ethylene production.[115,120] Tissue wounding, which is also known to increase ethylene production, stimulates the ACC synthase activity in wounded tomato[127] and squash[128] tissue. Development of the ability to synthesize ACC in wounded citrus peel discs is suppressed while the discs are incubated with exogenous ethylene.[129]

The half-life of ACC synthase activity within tissues is extremely short, being 25 min in auxin-treated mung bean stems[126] and 30 to 40 min in tomato fruit.[127] These findings clearly indicate that the labile protein which has been assumed to be produced by auxin action and to be essential to the ethylene producing system is ACC synthase itself.[126,130] All the results obtained so far with higher plant tissues indicate that ACC synthase is a rate-limiting enzyme of ethylene biosynthesis in most cases.

The EFE activity, which is widely distributed in plant tissues, has been shown only at the tissue level; the exogenous ACC-dependent ethylene forming activity. The EFE activity is inhibited by uncouplers of oxidative phosphorylation,[131,132] Co^{++} ion,[117] free radical scavengers,[132,133] at temperatures higher than 35°C,[131,133] and by perturbation of the membrane integrity of the cell.[113,133] Temperature dependence of the ethylene-producing activity of apple and tomato fruits shows a discontinuous two-phase change in the Arrhenius plots at 10 to 12°C, suggesting that the phase transition of the membrane lipids affects ethylene production.[93] Auxin-induced ethylene production is reversibly inhibited by low concentrations of detergents such as lecithin and Tween®20.[93,103] Osmotic shock applied to mung bean stem sections inhibits α-aminoisobutyric acid transport of the treated tissue indicating that the membrane integrity is impaired by the osmotic shock treatment and auxin-induced ethylene production of the sections is inhibited by the treatment to a similar degree to the amino acid transport.[98] High concentrations of Ca^{++} and Mg^{++} added to apple fruit slices suppress both a decline in the ethylene-producing activity and an increase in proton leakage from the tissue, which occurs during the prolonged incubation in the absence of the two metal ions.[134] Those results together with the complete loss of the EFE activity on cell disruption indicate that integrity of the cellular membrane system is essential for ethylene biosynthesis, and the EFE activity shows all of those characteristics. Although there is no direct evidence that EFE localizes in the cellular membranes, EFE has been strongly inferred to be associated with the cellular membranes.[133]

As the in vitro system representing the EFE activity has not been established, the enzymatic mechanism of ACC conversion to ethylene is not known. However, by analogy to the chemical oxidation of substituted cyclopropylamines to ethylene, a reaction scheme via nitrenium ion formed by oxidation of ACC has been proposed.[135] More recently, a hypothesis which explains the unique properties of the EFE activity was proposed.[136] The hypothesis presumes that EFE localizes in the cytoplasmic membranes and ACC oxidation couples with a transmembrane, electrogenic proton flux.[136]

Stereospecific interaction of ACC with EFE has been assessed by Hoffman et al.[137] using 2-substituted ACC, 1-amino-2-ethylcyclopropane-1-carboxylic acid (AEC) isomers. Although two methylene groups of ACC are equivalent, if a substituted group is introduced to one methylene hydrogen, the substituted molecule creates two asymmetric carbon atoms, thus producing four stereoisomers (Figure 2). Among the four stereoisomers of AEC, apple fruit tissue, cantaloupe fruit tissue, and IAA-treated mung bean stem sections convert only (1R,2S)-isomer to 1-butene.[137] Because the activity to convert (1R,2S)-AEC to 1-butene parallels with the activity of ACC conversion to ethylene under the various conditions, and is inhibited by anaerobiosis and Co^{++}. As the AEC isomer inhibits ACC conversion to ethylene and vice versa, Hoffman et al.[137] concluded that the same enzyme catalyzed both

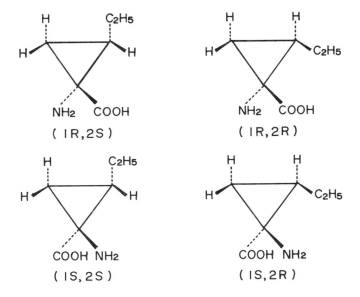

FIGURE 2. Four stereoisomers of 1-amino-2-ethylcyclopropane-1-carboxylic acid.

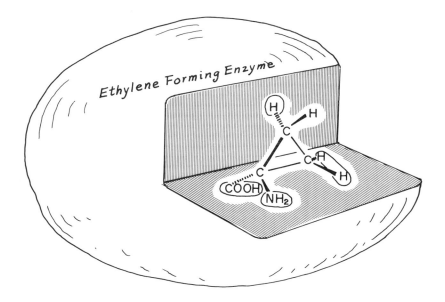

FIGURE 3. A possible interaction of ACC with an ethylene-forming enzyme proposed by Hoffman et al.[138]

reactions; thus ACC is stereospecifically converted to ethylene by EFE. To explain these observations, they assume that ACC interacts with EFE at four points: carboxyl, amino, pro(S)-methylene group, and pro(R)-hydrogen of pro(S)-methylene group (Figure 3).

In vitro systems to convert ACC to ethylene have been reported in enzyme preparations extracted from etiolated peas,[138,139] and *Citrus unshiu* peel.[140] These preparations require oxygen, Mn^{++}, and low molecular soluble materials of the tissue or a monophenol as cofactors, and the reaction rate is not saturated in terms of ACC concentrations.[138,139] The reaction is very similar to the oxidative reaction of peroxidase represented by IAA oxidase.[141]

Peroxidase is an abundant enzyme in plant extracts, and indeed the preparation from *Citrus unshiu* contains IAA oxidase activity.[140] Pea stem extracts converts all four stereoisomers of AEC, indicating that the in vitro system is not operating in vivo.[142] A microsomal fraction prepared from pea stems is reported to produce ethylene from ACC without adding any other cofactors, but the in vivo function of the system is questioned.[143]

B. Metabolism

Results of earlier works on metabolism of ethylene in plant tissues are not definitive, because purity of ^{14}C-ethylene used in those experiments was not seriously considered.[24] Curious metabolic behavior of labeled ethylene regenerated from mercuric perchlorate complex[144] was attributed to labeled degradation products which were metabolized by the tissues.[24,144,145] Highly purified ^{14}C-ethylene prepared by two-step GC produces, during storage in the gaseous state, labeled degradation products which are soluble in an alkaline solution.[144] Using the ultra-high pure ^{14}C-ethylene, Beyer[24] demonstrated that ethylene was oxidized to CO_2 and also metabolized to nonvolatile materials by aseptic pea seedlings, although oxidation and incorporation into tissue were less than 2% of the supplied ethylene.[24] The rate of oxidation to CO_2 by pea seedlings is greater than the rate of incorporation into tissue, and both oxidation and incorporation are reduced by lowering the oxygen tension and by separating the seedling into cotyledons and shoot-root axis before the application of ^{14}C-ethylene.[24] An elevated concentration of CO_2 (5%) reduces only oxidation to CO_2, but not incorporation into tissue.[24] Metabolism of ethylene has been observed in many other plant tissues including carnation flower,[146] morning glory flower,[147] cotton abscission zone,[148] and tomato fruit.[149]

Ethylene oxidation to CO_2 and conversion to nonvolatile materials, however, are independent of each other.[149] As cited above,[24] oxidation is inhibited by high CO_2 concentration, but incorporation into tissue is not. In contrast, Ag^+, which inhibits ethylene action,[150] reduces incorporation into tissue, but not oxidation to CO_2.[151] Changes in the two metabolic activities of ethylene during petal expansion of carnation flower,[146] abscission of cotton leaves,[148] and ripening of tomato fruit[149] are markedly different. The rate of oxidation to CO_2 is greatest in dark-green tomato fruit, decreases as the fruit ripens to the orange-red stage, then increases again at the red stage, whereas incorporation into tissue gradually increases during the ripening process.[149] A common feature of both processes of ethylene metabolism is that the rate of ethylene metabolism increases when tissues become responsive to ethylene.[148,149] The metabolic rates of ethylene in the abscission zone of young cotton leaves are very low, but if leaf blades are removed, the abscission zone increases the rate of ethylene oxidation in parallel with the responsiveness to ethylene.[148] Although ethylene oxidation to CO_2 and incorporation into tissue apparently occur independently, ethylene action always accompanies ethylene metabolism and inhibition of either one of the two processes inhibits ethylene action.[149] Thus, Beyer and Bloomstrom[149] propose that both ethylene oxidation to CO_2 and conversion to nonvolatile materials couple with the initial event of ethylene action (Figure 4).

The nonvolatile metabolites of ethylene contain both neutral and basic materials,[152] and the two neutral metabolites have been identified as ethyleneglycol and its glucoside.[153] Ethylene oxide is presumed to be an intermediate,[149] and ^{14}C-ethylene supplied to cotyledons of a specific cultivar of *Vicia faba,* cv. Aquadulce and their enzyme extracts is converted to ethylene oxide and no further oxidation to CO_2 is observed.[154,155]

IV. BIOLOGICAL ACTIVITY

Following the discovery that ethylene has various morphogenetic activity on many plants, a number of gaseous substances were examined for their biological activity and it was found that ethylene is most active.[60,61] Based on the detailed comparison of chemical structures of

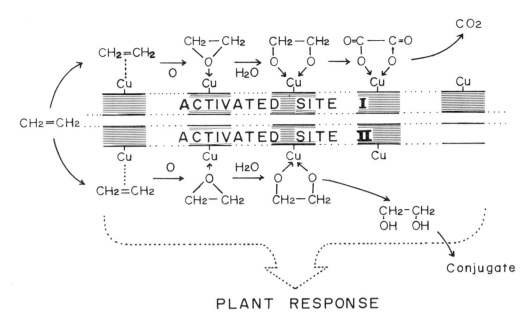

FIGURE 4. Possible mechanisms of ethylene metabolism and their relation to ethylene action proposed by Beyer.[150]

different gases and their biological activity to retard elongation growth of etiolated pea stem sections, Burg and Burg[61] postulated that the biological activity is associated with the following characteristics:

1. Only unsaturated aliphatic compounds are active.
2. Biological activity is inversely related to molecular size; the smaller the molecular size, the more active.
3. Substitutions which lower electron density in the double bond reduce biological activity, thus chloroethylene is less active than propylene.
4. The unsaturated bond must be attached to a terminal carbon atom, thus methylacetylene ($CH_3C\equiv C$) is active, but CO_2 ($O=C=O$) and acetonitrile ($CH_3C\equiv N$) are not.
5. The terminal carbon atom must not be positively charged, thus carbon monoxide which has a resonance form, $\ominus C=O \oplus$, is active while formaldehyde ($H_2C^{\delta+} = O^{\delta-}$) is not.

In a substituted ethylene, its biological activity is a compromise between the size of the substituted group and the ability of the substituted group to reduce electron density in the double bond.[61]

Biological activities of those unsaturated aliphatic compounds are closely related to the stability of their complexes with silver; expression of ethylene activity requires oxygen which is known to bind with metal-containing proteins, and none of the inactive substances has ability to form the metal complex.[61] These considerations led Burg and Burg[61] to suggest that the possible receptor of ethylene may be a metal-containing component.

That binding of growth regulators with cellular receptors is a necessary step for expression of their biological activity is generally accepted, but no direct evidence to show the presence of such receptors in plants has so far been presented, although presence of binding proteins has been shown.[156] If binding of ethylene with cellular components occurs, the interaction can be through covalent, ionic, hydrogen, coordinate, or van der Waals bonding, and depending upon the nature of the bonding, either break of carbon-hydrogen bond or exchange of hydrogen of ethylene with cellular hydrogen is expected to occur. As the bond energy of carbon and hydrogen is significantly different from that of carbon and deuterium, if break of carbon-hydrogen bond is involved in ethylene action, biological activity of fully deuterated ethylene (C_2D_4) may be different from that of ethylene (C_2H_4). However, deuterated ethylene showed the same degree of biological activity as ethylene on growth of pea seedlings and detectable amounts of hydrogen-exchanged species (O_2D_3H, $C_2D_2H_2$) are not formed from deuterated ethylene (C_2D_4) applied to the plant after the ethylene effect was evident.[157,158]

The presence of cellular binding components for ethylene has been reported in tobacco leaves,[25] cultured tobacco cells,[25] mung bean seedlings,[159] and *Phaseolus vulgaris* cotyledons.[26,160] Binding of ^{14}C-ethylene to the components is replaceable by excess of unlabeled ethylene,[25,26] is saturable in terms of ethylene concentrations,[26] and does not occur after heat treatment of the components.[25,26] The binding components are localized in the 12,000 to 100,000 g pellets in mung bean[159] and in the endoplasmic reticulum and protein body membranes in *Phaseolus vulgaris* cotyledons.[161] These components are solubilized by treatment of the particulate fractions with Triton®X100.[159,162] Whether the particulate binding components serve as receptor of ethylene in expressing physiological effects is presently not known.

A subject related to ethylene metabolism is the metabolism of ACC. Conversion of ACC to ethylene is a small portion of ACC produced in or supplied to plant tissue, and the major portion is recently shown to be malonylated to form *N*-malonyl ACC.[163-165]

REFERENCES

1. **Freebairn, H. I. and Buddenhagen, I. W.**, Ethylene production by *Pseudomonas solanacearum, Nature (London),* 202, 313, 1964.
2. **Primrose, S. B. and Dilworth, M. J.**, Ethylene production by bacteria, *J. Gen. Microbiol.,* 93, 177, 1976.
3. **Miller, E. V., Winston, J. R., and Fisher, D. F.**, Production of epinasty by emanations from normal and decaying fruits and from *Penicillium digitatum, J. Agric. Res.,* 660, 269, 1940.
4. **Biale, J. B.**, Effect of emanations from several species of fungi on respiration and color development of citrus fruits, *Science,* 91, 458, 1940.
5. **Ilag, L. and Curtis, R. W.**, Production of ethylene by fungi, *Science,* 159, 1357, 1968.
6. **Watanabe, T. and Kondo, N.**, Ethylene evolution in marine algae and a proteinaceous inhibitor of ethylene biosynthesis from red algae, *Plant Cell Physiol.,* 17, 1159, 1975.
7. **Burg, S. P.**, The physiology of ethylene formation, *Annu. Rev. Plant Physiol.,* 13, 265, 1962.
8. **Abeles, F. B.**, *Ethylene in Plant Biology,* Academic Press, New York, 1973, chap. 3.
9. **Osborne, D. J.**, Ethylene, in *Phytohormone and Related Compounds: A Comprehensive Treatise,* Vol. 1, Letham, D. S., Goodwin, P. B., and Higgins, T. J. V., Eds., Elsevier/North-Holland, Amsterdam, 1978, 265.
10. **Girardin, J. P. L.**, Einfluss des Leuchtgases auf die Promnaden und Strassenbläume, *Jahresber. Agrikult.,* 7, 199, 1864.
11. **Neljubow, D.**, Ueber die horizontale Nutation der Stengel von *Pisum sativum* und einiger änderen Pflanzen, *Bot. Centralbl.,* 10, 128, 1901.
12. **Crocker, W. and Knight, L. I.**, Effect of illuminating gas and ethylene upon flowering carnations, *Bot. Gaz.,* 46, 259, 1908.
13. **Rodriguez, A. G.**, Influence of smoke and ethylene on the fruiting of the pineapple (*Ananas sativus* Shult), *J. Dept. Agric. P. R.,* 16, 5, 1932.
14. **Denny, F. E.**, Hastening the coloration of lemons, *J. Agr. Res.,* 27, 757, 1924.
15. **Burg, S. P.**, Vapour phase chromatography, in *Modern Methods of Plant Analysis,* Vol. 5, Springer-Verlag, Heidelberg, 1962, 97.
16. **Meigh, D. F., Norris, K. H., Craft, C. C., and Lieberman, M.**, Ethylene production by tomato and apple fruits, *Nature (London),* 186, 902, 1960.
17. **Burg, S. P. and Burg, E. A.**, Role of ethylene in fruit ripening, *Plant Physiol.,* 37, 179, 1962.
18. **Burg, S. P. and Burg, E. A.**, Ethylene formation in pea seedlings; its relation to the inhibition of bud growth caused by indole-3-acetic acid, *Plant Physiol.,* 43, 1069, 1968.

19. **Goeschl, J. D., Rappaport, L., and Pratt, H. K.,** Ethylene as a factor regulating the growth of pea epicotyls subjected to physical stress, *Plant Physiol.,* 41, 877, 1966.
20. **Goeschl, J. D., Pratt, H. K., and Bonner, B. A.,** An effect of light on the production of ethylene and the growth of the plumular portion of etiolated pea seedlings, *Plant Physiol.,* 42, 1077, 1967.
21. **Jackson, M. B. and Osborne, D. J.,** Ethylene, the natural regulator of leaf abscission, *Nature (London),* 225, 1019, 1970.
22. **Saltveit, M. E., Jr. and Dilley, D. R.,** Rapidly induced wound ethylene from excised segments of etiolated *Pisum sativum* L. cv. Alaska. I. Characterization of the response, *Plant Physiol.,* 61, 447, 1978.
23. **Saltveit, M. E., Jr. and Dilley, D. R.,** Rapidly induced wound ethylene from excised segments of etiolated *Pisum sativum* L. cv. Alaska. II. Oxygen and temperature dependency, *Plant Physiol.,* 61, 675, 1978.
24. **Beyer, E. M., Jr.,** $^{14}C_2H_4$: its incorporation and metabolism by pea seedlings under aseptic conditions, *Plant Physiol.,* 56, 273, 1975.
25. **Sisler, E. C.,** Measurement of ethylene binding in plant tissue, *Plant Physiol.,* 64, 538, 1979.
26. **Bengochea, T., Dodds, J. H., Evans, D. E., Jerie, P. H., Niepel, B., Shaari, A. R., and Hall, M. A.,** Studies on ethylene binding by cell-free preparations from cotyledons of *Phaseolus vulgaris* L. I. Separation and characterization, *Planta,* 148, 397, 1980.
27. **Burg, S. P. and Burg, E. A.,** Fruit storage at subatmospheric storage, *Science,* 153, 314, 1966.
28. **Burg, S. P.,** Ethylene, plant senescence and abscission, *Plant Physiol.,* 43, 1503, 1968.
29. **Kende, H. and Baumgarter, B.,** Regulation of aging in flowers of *Ipomoea tricolor* by ethylene, *Planta,* 116, 279, 1974.
30. **Mayak, S., Halevy, A. H., and Katz, M.,** Correlative changes in phytohormones in relation to senescence processes in rose petals, *Physiol. Plant.,* 27, 1, 1972.
31. **Aharoni, N., Lieberman, M., and Sisler, H. D.,** Patterns of ethylene production in senesciing leaves, *Plant Physiol.,* 64, 796, 1979.
32. **Morgan, P. W. and Hall, W. C.,** Effect of 2,4-dichlorophenoxyacetic acid on the production of ethylene by cotton and grain sorghum, *Physiol. Plant.,* 15, 420, 1962.
33. **Morgan, P. W. and Hall, W. C.,** Accelerated release of ethylene by cotton following application of indoel-3-acetic acid, *Nature (London),* 201, 99, 1964.
34. **Abeles, F. B.,** Auxin stimulation of ethylene production, *Plant Physiol.,* 41, 585, 1966.
35. **Lieberman, M., Baker, J. E., and Sloger, M.,** Influence of plant hormones on ethylene production in apple, tomato, and avocado slices during maturation and senescence, *Plant Physiol.,* 60, 214, 1977.
36. **Abeles, F. B. and Rubinstein, B.,** Regulation of ethylene production and leaf abscission by auxin, *Plant Physiol.,* 39, 963, 1964.
37. **Kang, B. G. and Burg, S. P.,** Ethylene action in lateral auxin transport in tropic response, leaf epinasty, and horizontal mutation, in *Plant Growth Substances 1973: Proc. 8th Int. Conf. Plant Growth Substances,* Hirokawa Publishing, Tokyo, 1974, 1090.
38. **McGlasson, W. B., Poovaiah, B. W., and Dostal, H. C.,** Ethylene production and respiration in aging leaf segments and in disks of fruit tissue of normal and mutant tomatoes, *Plant Physiol.,* 56, 347, 1975.
39. **Aharoni, N. and Liegerman, M.,** Ethylene as a regulator of senescence in tobacco leaf discs, *Plant Physiol.,* 64, 801, 1979.
40. **Kende, H. and Hansen, A. D.,** Relationship between ethylene evolution and senescence in morning glory flower tissue, *Plant Physiol.,* 57, 523, 1976.
41. **Rasmussen, G. K.,** Cellulase activity, endogenous abscisic acid and ethylene in four citrus cultivars during maturation, *Plant Physiol.,* 56, 765, 1975.
42. **Sawamura, M., Knegt, E., and Bruinsma, J.,** Levels of endogenous ethylene, carbon dioxide, and soluble pectin, and activities of pectin methylesterase and polygalacturonase in ripening tomato fruits, *Plant Cell Physiol.,* 19, 1061, 1978.
43. **Craker, L. E., Abeles, F. B., and Shrosphire, W., Jr.,** Light stimulated ethylene production in sorghum, *Plant Physiol.,* 51, 1082, 1973.
44. **LaRue, T. A. G. and Gamborg, O. L.,** Ethylene production by plant cell cultures, *Plant Physiol.,* 48, 394, 1971.
45. **Saltveit, M. E., Jr. and Pharr, D. M.,** Light-stimulated ethylene production by germinating cucumber seeds, *J. Am. Soc. Hort. Sci.,* 105, 364, 1980.
46. **Kang, B. G., Yocum, C. S., Burg, S. P., and Ray, P. M.,** Ethylene and carbon dioxide: mediation of hypocotyl hook opening response, *Science,* 156, 958, 1967.
47. **Imaseki, H., Pjon, C. J., and Furuya, M.,** Phytochrome action in Oryza sativa L. IV. Red and far red reversible effect on the production of ethylene in excised coleoptiles, *Plant Physiol.,* 48, 241, 1971.
48. **Goeschl, J. D. and Pratt, H. K.,** Regulatory roles of ethylene in the etiolated growth habit of *Pisum sativum,* in *Biochemistry and Physiology of Plant Growth Substances: Proc. 6th Int. Conf. Plant Growth Substances,* Wightman, F. and Setterfield, G., Eds., Runge Press, Ottawa, 1968, 1229.
49. **Williamson, C. E.,** Ethylene, a metabolic product of diseased or injured plants, *Phytopathology,* 40, 203, 1950.

50. **Stahmann, M. A., Clare, B. G., and Woodburg, W.**, Increased disease resistance and enzyme activity induced by ethylene and ethylene production by black rot infected sweet potato tissue, *Plant Physiol.*, 41, 1505, 1966.
51. **Imaseki, H., Teranishi, T., and Uritani, I.**, Production of ethylene by sweet potato roots infected by black rot fungus, *Plant Cell Physiol.*, 9, 769, 1968.
52. **Imaseki, H., Stahmann, M. A., and Uritani, I.**, Production of ethylene by injured sweet potato root tissue, *Plant Cell Physiol.*, 9, 757, 1968.
53. **Cooper, W. C. and Henry, W.**, Abscission chemicals in relation to citrus fruit harvest, *Agric. Food Chem.*, 19, 559, 1971.
54. **McMichael, B. L., Jordan, W. R., and Powell, R. D.**, An effect of water stress on ethylene production by intact cotton petioles, *Plant Physiol.*, 49, 658, 1972.
55. **Wright, S. T. C.**, The relationship between leaf water potential (ψ_{leaf}) and the levels of abscisic acid and ethylene in excised wheat leaves, *Planta*, 134, 183, 1977.
56. **Apelbaum, A. and Yang, S. F.**, Biosynthesis of stress ethylene induced by water deficit, *Plant Physiol.*, 68, 594, 1981.
57. **Jaffe, M. J.**, Physiological studies on pea tendrils. VII. Evaluation of a technique for the assymmetrical application of ethylene, *Plant Physiol.*, 46, 631, 1970.
58. **Hiraki, Y. and Ota, Y.**, The relationship between growth inhibition and ethylene production by mechanical stimulation in *Lilium longiflorum*, *Plant Cell Physiol.*, 16, 185, 1975.
59. **Robitaille, H. A. and Leopold, C. A.**, Ethylene and the regulation of apple stem growth under stress, *Physiol. Plant.*, 32, 301, 1974.
60. **Knight, L. I. O., Rose, C., and Crocker, W.**, Effect of various gases and vapors upon etiolated seedlings of the sweet pea, *Science*, 31, 635, 1910.
61. **Burg, S. P. and Burg, E. A.**, Molecular requirements for the biological activity of ethylene, *Plant Physiol.*, 42, 144, 1967.
62. **Bradford, K. J. and Dilley, D. R.**, Effects of root anaerobiosis on ethylene production, epinasty, and growth of tomato plants, *Plant Physiol.*, 61, 506, 1978.
63. **Bassi, P. K. and Spencer, M. S.**, A cuvette design for measurement of ethylene production and carbon dioxide exchange by intact shoots under controlled environmental conditions, *Plant Physiol.*, 64, 488, 1979.
64. **Gibson, M. S. and Young, R. E.**, Acetate and other carboxylic acids as precursors of ethylene, *Nature (London)*, 210, 529, 1966.
65. **Stitt, F. and Tomimatsu, Y.**, Removal and recovery of traces of ethylene in air by silica gel, *Anal. Chem.*, 25, 181, 1953.
66. **Stinson, R. A. and Spencer, M. S.**, Alanine as an ethylene precursor. Investigations towards preparation, and properties, of a soluble enzyme from a sub-cellular particulate fraction of bean cotyledons, *Plant Physiol.*, 44, 1217, 1969.
67. **Eastwell, K. C., Bassi, P. K., and Spencer, M. S.**, Comparison and evaluation of methods for the removal of ethylene and other hydrocarbons from air for biological studies, *Plant Physiol.*, 62, 723, 1978.
68. **Beyer, E. M., Jr. and Morgan, P. W.**, A method for determining the concentration of ethylene in the gas phase of vegetative plant tissues, *Plant Physiol.*, 46, 352, 1970.
69. **Kawase, M.**, Causes of centrifugal root promotion, *Physiol. Plant.*, 25, 64, 1971.
70. **Burg, S. P. and Burg, E. A.**, The interaction between auxin and ethylene and its role in plant growth, *Proc. Natl. Acad. Sci. U.S.A.*, 55, 262, 1966.
71. **Galliard, T. and Grey, T. C.**, A rapid method for the determination of ethylene in the presence of other volatiles in natural products, *J. Chromatogr.*, 41, 442, 1969.
72. **Muir, R. M. and Richter, E. W.**, The measurement of ethylene from plant tissues and its relation to auxin effect, in *Plant Growth Substances 1970: Proc. 7th Int. Conf. Plant Growth Substances*, Carr. J., Ed., Springer-Verlag, Berlin, 1970, 518.
73. **Young, R. E., Pratt, H. K., and Biale, J. B.**, Manometric determination of low concentration of ethylene, *Anal. Chem.*, 24, 551, 1952.
74. **Burg, S. P. and Thimann, K. V.**, Studies on the ethylene production of apple tissues, *Plant Physiol.*, 35, 24, 1960.
75. **Burg, S. P. and Thimann, K. V.**, The conversion of glucose-C^{14} to ethylene by apple tissue, *Arch. Biochem. Biophys.*, 95, 450, 1961.
76. **Burg, S. P. and Burg, E. A.**, Biosynthesis of ethylene, *Nature (London)*, 203, 869, 1964.
77. **Lieberman, M. and Mapson, L. W.**, Genesis and biogenesis of ethylene, *Nature (London)*, 204, 343, 1964.
78. **Abeles, F. B. and Rubinstein, B.**, Cell-free ethylene evolution from etiolated pea seedlings, *Biochim. Biophys. Acta*, 93, 675, 1964.
79. **Lieberman, M., Kunishi, A. T., Mapson, L. W., and Wardale, D. A.**, Ethylene production from methionine, *Biochem. J.*, 97, 449, 1965.

80. **Yang, S. F., Ku, H. S., and Pratt, H. K.**, Photochemical production of ethylene from methionine and its analogues in the presence of flavin mononucleotide, *J. Biol. Chem.*, 242, 5273, 1967.
81. **Lieberman, M., Kunishi, A., Mapson, L. W., and Wardale, D. A.**, Stimulation of ethylene production in apple tissue slices by methionine, *Plant Physiol.*, 41, 376, 1966.
82. **Burg, S. P. and Clagett, C. O.**, Conversion of methionine to ethylene in vegetative tissue and fruits, *Biochem. Biophys. Res. Commun.*, 27, 125, 1967.
83. **Mapson, L. W., March, J. F., Rhodes, M. J. C., and Wooltorton, L. S. C.**, A comparative study of the activity of methionine or linolenic acid to act as precursors of ethylene in plant tissues, *Biochem. J.*, 117, 473, 1970.
84. **Baur, A. H., Yang, S. F., and Pratt, H. K.**, Ethylene biosynthesis in fruit tissues, *Plant Physiol.*, 47, 696, 1971.
85. **Sakai, S. and Imaseki, H.**, Ethylene biosynthesis: methionine as an *in vivo* precursor of ethylene in auxin-treated mung bean hypocotyl segments, *Planta*, 105, 165, 1972.
86. **Konze, J. R. and Kende, H.**, Interactions of methionine and selenomethionine with methionine adenosyltransferase and ethylene generating systems, *Plant Physiol.*, 63, 507, 1979.
87. **Sierbert, K. J. and Clagett, C. O.**, Formic acid from carbon 2 of methionine in ethylene production in apple tissue, *Plant Physiol.*, 44(Suppl.), 30, 1969.
88. **Chou, T. W. and Yang, S. F.**, The biogenesis of ethylene in *Penicillium digitatum*, *Arch. Biochem. Biophys.*, 157, 73, 1973.
89. **Chalutz, E., Lieberman, M., and Sisler, H. D.**, Methionine-induced ethylene production by *Penicillium digitatum*, *Plant Physiol.*, 60, 402, 1977.
90. **Baur, A. H. and Yang, S. F.**, Precursors of ethylene, *Plant Physiol.*, 44, 189, 1969.
91. **Baur, A. H. and Yang, S. F.**, Methionine metabolism in apple tissue in relation to ethylene biosynthesis, *Phytochemistry*, 11, 3207, 1972.
92. **Baker, J. E., Lieberman, M., and Anderson, J. D.**, Inhibition of ethylene production in fruit slices by a rhizobitoxin analog and free radical scavengers, *Plant Physiol.*, 61, 886, 1975.
93. **Mattoo, A. K., Baker, J. E., Chalutz, E., and Lieberman, M.**, Effect of temperature on the ethylene-synthesizing systems in apple, tomato and *Penicillium digitatum*, *Plant Cell Physiol.*, 18, 715, 1977.
94. **Kang, B. G., Newcomb, W., and Burg, S. P.**, Mechanism of auxin-induced ethylene production, *Plant Physiol.*, 47, 504, 1971.
95. **Sakai, S. and Imaseki, H.**, Auxin-induced ethylene production by mung bean hypocotyl segments, *Plant Cell Physiol.*, 12, 349, 1971.
96. **Lau, O. L. and Yang, S. F.**, Mechanism of a synergistic effect of kinetin on auxin-induced ethylene production: suppression of auxin conjugation, *Plant Physiol.*, 51, 1011, 1973.
97. **Imaseki, H., Kondo, K., and Watanabe, A.**, Mechanism of cytokinin action on auxin-induced ethylene production, *Plant Cell Physiol.*, 16, 707, 1975.
98. **Imaseki, H. and Watanabe, A.**, Inhibition of ethylene production by osmotic shock. Further evidence for control of ethylene production by membrane, *Plant Cell Physiol.*, 19, 345, 1978.
99. **Burg, S. P.**, Ethylene in plant growth, *Proc. Natl. Acad. Sci. U.S.A.*, 70, 591, 1973.
100. **Owens, L. D., Lieberman, M., and Kunishi, A.**, Inhibition of ethylene production by rhizobitoxine, *Plant Physiol.*, 48, 1, 1971.
101. **Amrhein, N. and Wenker, D.**, Novel inhibitors of ethylene production in higher plants, *Plant Cell Physiol.*, 20, 1635, 1979.
102. **Lau, O. L. and Yang, S. F.**, Inhibition of ethylene production by cobalt ion, *Plant Physiol.*, 58, 114, 1976.
103. **Odawara, S., Watanabe, A., and Imaseki, H.**, Involvement of cellular membranes in regulation of ethylene production, *Plant Cell Physiol.*, 18, 569, 1977.
104. **Murr, D. P. and Yang, S. F.**, Inhibition of in vivo conversion of methionine to ethylene by L-canaline and 2,4-dinitrophenol, *Plant Physiol.*, 55, 79, 1975.
105. **Adams, D. O. and Yang, S. F.**, Methionine metabolism in apple tissue, *Plant Physiol.*, 60, 889, 1977.
106. **Murr, D. P. and Yang, S. F.**, Conversion of 5'-methylthioadenosine to methionine by apple tissue, *Phytochemistry*, 14, 1291, 1975.
107. **Wang, S. Y., Adams, D. O., and Lieberman, M.**, Recycling of 5'-methylthio-adenosine-ribose carbon atoms into methionine in tomato tissue in relation to ethylene production, *Plant Physiol.*, 70, 117, 1982.
108. **Burg, S. P. and Thimann, K. V.**, The physiology of ethylene formation in apples, *Proc. Natl. Acad. Sci. U.S.A.*, 45, 335, 1959.
109. **Imaseki, H., Watanabe, A., and Odawara, S.**, Role of oxygen in auxin-induced ethylene production, *Plant Cell Physiol.*, 18, 569, 1977.
110. **Adams, D. O. and Yang, S. F.**, Ethylene biosynthesis: 1-aminocyclopropane-1-carboxylic acid as an intermediate in the conversion of methionine to ethylene, *Proc. Natl. Acad. Sci. U.S.A.*, 76, 170, 1979.
111. **Burroughs, L. F.**, 1-Aminocyclopropane-1-carboxylic acid: a new amino acid in perry pears and cider apples, *Nature (London)*, 179, 360, 1957.

112. **Vähätalo, M. L. and Virtanen, A. I.**, A new cyclic α-aminocarboxylic acid in berries of cowberry, *Acta Chem. Scand.*, 11, 741, 1957.
113. **Lürssen, K., Nauman, K., and Schroder, R.**, 1-Aminocyclopropane-1-carboxylic acid-an intermediate of the ethylene biosynthesis in higher plants, *Z. Pflanzenphysiol.*, 92, 285, 1979.
114. **Cameron, A. C., Fenton, C. A. L., Yu, Y., Adams, D. O., and Yang, S. F.**, Increased production of ethylene by plant tissues treated with 1-aminocyclopropane-1-carboxylic acid, *HortScience*, 14, 178, 1979.
115. **Yoshii, H., Watanabe, A., and Imaseki, H.**, Biosynthesis of auxin-induced ethylene in mung bean hypocotyls, *Plant Cell Physiol.*, 21, 279, 1980.
116. **Jones, J. F. and Kende, H.**, Auxin-induced ethylene biosynthesis in subapical stem sections of etiolated seedlings of *Pisum sativum* L., *Planta*, 146, 649, 1979.
117. **Yu, Y. B. and Yang, S. F.**, Auxin-induced ethylene production and its inhibition by aminoethoxyvinylglycine and cobalt ion, *Plant Physiol.*, 64, 1074, 1979.
118. **Boller, T., Herner, R. C., and Kende, H.**, Assay for and enzymatic formation of an ethylene precursor, 1-aminocyclopropane-1-carboxylic acid, *Planta*, 145, 293, 1979.
119. **Boller, T. and Kende, H.**, Regulation of wound ethylene synthesis in plants, *Nature (London)*, 286, 259, 1980.
120. **Yoshii, H. and Imaseki, H.**, Biosynthesis of auxin-induced ethylene. Effects of indole-3-acetic acid, benzyladenine and abscisic acid on 1-aminocyclopropane-1-carboxylic acid (ACC) and ACC synthase, *Plant Cell Physiol.*, 22, 369, 1981.
121. **Apelbaum, A. and Yang, S. F.**, Biosynthesis of stress ethylene induced by water deficit, *Plant Physiol.*, 68, 594, 1981.
122. **Bradford, K. J. and Yang, S. F.**, Xylem transport of 1-aminocyclopropane-1-carboxylic acid, an ethylene precursor, in waterlogged tomato plants, *Plant Physiol.*, 65, 322, 1980.
123. **Yu, Y. B. and Yang, S. F.**, Biosynthesis of wound ethylene, *Plant Physiol.*, 66, 281, 1980.
124. **Yu, Y. B., Adams, D. O., and Yang, S. F.**, 1-Aminocyclopropanecarboxylate synthase, a keyenzyme in ethylene biosynthesis, *Arch. Biochem. Biophys.*, 198, 280, 1979.
125. **Acaster, M. A. and Kende, H.**, Properties and partial purification of 1-aminocyclopropane-1-carboxylate synthase, *Plant Physiol.*, 72, 139, 1983.
126. **Yoshii, H. and Imaseki, H.**, Regulation of auxin-induced ethylene biosynthesis. Repression of inducible formation of 1-aminocyclopropane-1-carboxylate synthase by ethylene, *Plant Cell Physiol.*, 23, 639, 1982.
127. **Kende, H. and Boller, T.**, Wound ethylene and 1-aminocyclopropane-1-carboxylate synthase in ripening tomato fruit, *Planta*, 151, 476, 1981.
128. **Hyodo, H., Tanaka, K., and Watanabe, K.**, Wound-induced ethylene production and 1-aminocyclopropane-1-carboxylic acid synthase in mesocarp tissue of winter squash fruit, *Plant Cell Physiol.*, 24, 963, 1983.
129. **Riov, J. and Yang, S. F.**, Autoinhibition of ethylene production in citrus peel discs. Suppression of 1-aminocyclopropane-1-carboxylic acid synthesis, *Plant Physiol.*, 69, 687, 1982.
130. **Imaseki, H., Yoshii, H., and Todaka, I.**, Regulation of auxin-induced ethylene biosynthesis in plants, in *Plant Growth Substances 1972: Proc. 11th Int. Conf. Plant Growth Substances*, Wareing, P. F., Ed., Academic Press, London, 1982, 259.
131. **Yu, Y. B., Adams, D. O., and Yang, S. F.**, Inhibition of ethylene production by 2,4-dinitrophenol and high temperature, *Plant Physiol.*, 66, 286, 1980.
132. **Apelbaum, A., Wang, S. Y., Burgoon, A. C., Baker, J. E., and Lieberman, M.**, Inhibition of the conversion of 1-aminocyclopropane-1-carboxylic acid to ethylene by structural analogs, inhibitors of electron transfer, uncouplers of oxidative phosphorylation, and free radical scavengers, *Plant Physiol.*, 67, 74, 1981.
133. **Apelbaum, A., Burgoon, A. C., Anderson, J. D., Solomos, T., and Lieberman, M.**, Some characteristics of the system converting 1-aminocyclopropane-1-carboxylic acid to ethylene, *Plant Physiol.*, 67, 80, 1981.
134. **Lieberman, M. and Wang, S. Y.**, Influence of calcium and magnesium on ethylene production by apple tissue slices, *Plant Physiol.*, 69, 1150, 1982.
135. **Yang, S. F. and Adams, D. O.**, Biosynthesis of ethylene, in *The Biochemistry of Plants*, Vol. 4, Stumpf, P. K. and Conn, E. E., Eds., Academic Press, London, 1980, 163
136. **John, P.**, The coupling of ethylene bioynthesis to a transmembrane, electrogenic proton flux, *FEBS Lett.*, 152, 141, 1983.
137. **Hoffman, N. E., Yang, S. F., Ichihara, A., and Sakamura, S.**, Stereo-specific conversion of 1-aminocyclopropanecarboxylic acid to ethylene by plant tissues. Conversion of stereoisomers of 1-amino-2-ethyl-cyclopropanecarboxylic acid to 1-butene, *Plant Physiol.*, 70, 195, 1982.
138. **Konze, J. R. and Kende, H**, Ethylene formation from 1-aminocyclopropane-1-carboxylic acid in homogenates of etiolated pea seedlings, *Planta*, 146, 292, 1979.
139. **Konze, J. R. and Kwiatkowski, G. M. K.**, Enzymatic ethylene formation from 1-aminocyclopropane-1-carboxylic acid by manganese, a protein fraction and a cofactor of etiolated pea shoots, *Planta*, 151, 320, 1981.
140. **Shimokawa, K.**, An ethylene-forming enzyme in *Citrus unshiu* fruits, *Phytochemistry*, 22, 1903, 1983.

141. **Ray, P. M.,** Destruction of auxin, *Annu. Rev. Plant Physiol.,* 9, 81, 1958.
142. **Yang, S. F., Hoffman, N. E., McKeon, T., Riov, H., Kao, C. H., and Yung, K. H.,** Mechanism and regulation of ethylene biosynthesis, in *Plant Growth Substances 1982: Proc. 11th Int. Conf. Plant Growth Substances,* Wareing, P F., Ed., Academic Press, London, 1982, 239.
143. **McRae, D. G., Baker, J. E., and Thompson, J. E.,** Evidence for involvement of superoxide radical in the conversion of 1-aminocyclopropane-1-carboxylic acid to ethylene by pea microsomal membranes, *Plant Cell Physiol.,* 23, 275, 1982.
144. **Hall, W. C., Miller, C. S., and Herrero, F. A.,** Studies with ^{14}C-labelled ethylene, in *Plant Growth Regulation,* Iowa State University Press, Ames, 1961, 751.
145. **Beyer, E. M., Jr.** $^{14}C_2H_4$: its purification for biological studies, *Plant Physiol.,* 55, 845, 1975.
146. **Beyer, E. M., Jr.** $^{14}C_2H_4$: its incorporation and oxidation to $^{14}CO_2$ by cut carnations, *Plant Physiol,* 60, 203, 1977.
147. **Beyer, E. M., Jr. and Sundin, O.,** $^{14}C_2H_4$ metabolism in morning glory flowers, *Plant Physiol.,* 61, 896, 1978.
148. **Beyer, E. M., Jr.** ^{14}C-Ethylene metabolism during leaf abscission in cotton, *Plant Physiol.,* 64, 971, 1979.
149. **Beyer, E. M., Jr. and Bloomstrom, D. C.,** Ethylene metabolism and possible physiological role in plants, in *Plant Growth Substances 1979: Proc. 10th Int. Conf. Plant Growth Substances,* Skoog, F., Ed., Springer-Verlag, Berlin, 1980, 208.
150. **Beyer, E. M., Jr.** A potent inhibitor of ethylene action in plants, *Plant Physiol.,* 58, 268, 1976.
151. **Beyer, E. M., Jr.** Effect of silver ion, carbon dioxide, and oxygen on ethylene action and metabolism, *Plant Physiol.,* 63, 169, 1979.
152. **Giaquinta, R. and Beyer, E. M., Jr.** $^{14}C_2H_4$: distribution of ^{14}C-labeled tissue metabolites in pea seedlings, *Plant Cell Physiol.,* 18, 141, 1977.
153. **Blomstrom, D. C. and Beyer, E. M., Jr.** Plants metabolise ethylene to ethylene glycol, *Nature (London),* 283, 66, 1980.
154. **Jerie, P. H. and Hall, M. A.,** The identification of ethylene oxide as a major metabolite of ethylene in *Vicia faba* L., *Proc. R. Soc. London Ser. B:,* 200, 87, 197.
155. **Dodds, J. H, Musa, S. K., Jerie, P. H., and Hall, M. A.,** Metabolism of ethylene to ethylene oxide by cell-free preparations from *Vicia faba* L., *Plant Sci. Lett.,* 17, 109, 1979.
156. **Kende, H. and Gardner, G.,** Hormone binding in plants, *Annu. Rev. Plant Physiol.,* 27, 207, 1976.
157. **Beyer, E. M., Jr.** Mechanism of ethylene action. Biological activity of deuterated ethylene and evidence against isotopic exchange and *cis-trans* isomerization, *Plant Physiol.,* 49, 672, 1972.
158. **Abeles, F. B., Ruth, J. M., Forrence, L. E., and Leather, G. R.,** Mechanism of hormone action. Use of deuterated ethylene to measure isotopic exchange with plant material and biological effects of deuterated ethylene, *Plant Physiol.,* 49, 669, 1972.
159. **Sisler, E. C.,** Partial purification of an ethylene-binding component from plant tissue, *Plant Physiol.,* 66, 404, 1980.
160. **Bengochea, T., Acaster, M. A., Dodds, J. H., Evans, D. E., Jerie, J. H., and Hall, M. A.,** Studies on ethylene binding by cell free preparations from cotyledons of *Phaseolus vulgaris* L. II. Effects of structural analogues of ethylene and inhibitors, *Planta,* 148, 407, 1980.
161. **Evans, D. E., Dodds, J. H., Lloyd, P. C., apGwynn, I., and Hall, M. A.,** A study of the subcellular localisation of an ethylene binding site in developing cotyledons of *Phaseolus vulgaris* L. by high resolution autoradiography, *Planta,* 154, 48, 1982.
162. **Hall, M. A., Cairns, A. J., Evans, D. E., Smith, A.R., Smith, P. G., Taylor, J. E., and Thomas, C. J. R.,** Binding sites for ethylene, in *Plant Growth Substances 1982: Proc. 11th Int. Conf. Plant Growth Substances,* Wareing, P. F., Ed., Springer-Verlag, Berlin, 1982, 375.
163. **Amrhein, N., Schneebeck, D., Skorupka, H., Tophof, S., and Stockigt, J.,** Identification of a major metabolite of the ethylene precursor 1-aminocyclopropane-1-carboxylic acid in higher plants, *Naturwissenschaften,* 68, 619, 1981.
164. **Hoffman, N. E, Yang, S. F., and McKeon, T.,** Identification of 1-(malonyl-amino)cyclopropane-1-carboxylic acid as a major conjugate of 1-aminocyclopropane-1-carboxylic acid, and ethylene precursor in higher plants, 104, 765, 1982.
165. **Hoffman, N. E., Fu, J. R., and Yang, S. F.,** Identification and metabolsim of 1-(malonylamino)cyclopropane-1-carboxylic acid in germinating peanut seeds, *Plant Physiol.,* 71, 197, 1983.

INDEX

A

ABA, see Abscisic acid
t-ABA, see 2-trans-Abscisic acid
ABA-β-GEs, see Abscisic acid-1-O-β-glucosyl ester
ABA-α-GEt, see Abscisic acid-1'-O-α-glucoside
ABA-β-GEt, see Abscisic acid-1'-O-β-glucoside
Abbreviations, 205
1-Abietic acid, 117
Abscisic acid (ABA), 37, 201—248, 254
   abbreviations, 205
   absolute configuration determination by synthesis, 215—220
   1-alcohol, 226, 239
   1-aldehyde, 217, 226, 239
   biological activity, 238—241
   biosynthesis, 223—231
   carotenoid pathway, 227—228
   chemistry, 204—223
   CIMS, 211—212
   circular dichroism, 210
   CMR, 210
   color reaction, 209—210
   column chromatography, 207
   3-demethyl, 239
   1'-deoxy, 226, 228—230, 240
   4'-desoxy, 236
   1',4'-cis-diol, 217, 226, 236
   1',4'-trans-diol, 217, 226, 236
   direct biosynthetic pathway, 225—227
   EIMS, 210—211
   exogenous, 238—239
   extraction, 204—205
   GC, 207—208
   growth inhibition, 238
   hexadeuterated, 232, 234
   history, 202—203
   HPLC, 207—208
   hydrogens, 223—225
   hydroxylating enzymes, 235
   6'-hydroxymethyl, see 6'-Hydroxymethyl abscisic acid
   identification, 207—209
   immunoassay, 208—209
   infrared spectrometry, 210
   isolation, 204—207
   mass spectrometry, 210—215
   melting point, 209—210
   metabolic pathway, 236—238
   metabolites, 232—236
   methyl groups at C-3, 2', and 6', 225
   5'-nor, 240
   nuclear magnetic resonance, 210
   occurrence, 204
   ORD, 209—210
   paper chromatography, 206—207
   partition, 205—206
   PMR, 210
   precursors, 225—227
   quantification, 207—209
   racemic, 214—215
   radioassay, 209
   R-($-$)-enantiomer, 239
   R_f value, 206—207
   stereochemical origin of carbon and hydrogen atoms, 223—225
   structural determination, 209—215
   structural requirements for activity, 239—241
   structure-activity relationship, 239—241
   synthesis, 214—223
   thin layer chromatography, 206—207
   UV spectrometry, 210
2-trans-Abscisic acid (t-ABA), 205, 208, 210, 211, 232
Abscisic acid-1'-O-α-glucoside (ABA-α-GEt), 223
Abscisic acid-1'-O-β-glucoside (ABA-β-GEt), 223
Abscisic acid-1-O-β-glucosyl ester (ABA-β-GEs), 232
   DCI/MS, 213—215
   isolation, 205
   R_f value, 206—207
   synthesis, 223
Abscisic acid-related metabolites, 236
Abscission, 238
Absidia ramosa, 12
ACC, see 1-Aminocyclopropane-1-carboxylic acid
Acer pseudoplatanus, 156
Acetonitrile, 258
Acetylation, 99
Acetylene, 251
Acid degradation products, gibberellins, 97
Acid fast auxin, 15—16
Acid hydrolysis of GAs-GE, 102
Acid treatment, 99
Actinidia chinensis, 156
N^6-Acyladenines, 179
S-Adenosylmethionine (SAM), 253—255
AEC, see 1-Amino-2-ethylcyclopropane-1-carboxylic acid
AG$^+$, 257
Algae, 250
Alkali flame ionization detector (AFID), 29
Alkaline hydrolysis of GAs-GE, 102
N^6-n-Alkyladenines, 179
Allogibberic acid, 96—97, 103
All trans-farnesylpyrophosphate, 223
Altheae rosea, 67
3-(3-Amino-3-carboxylpropyl)iP, 157
1-Aminocyclopropane-1-carboxylic acid (ACC), 254—257, 259
Aminoethoxyvinylglycine (AVG), 253—254
1-Amino-2-ethylcyclopropane-1-carboxylic acid (AEC), 255—256
3-Aminomethylindole, 23
Aminoxyacetic acid, 253
AMO-1618, 122, 126, 129

α-Amylase, 135, 238
α-Amylase mRNA, 139
Anaerobiosis, 253, 255
Ancymidol, 122
Anemia phylltidis, 63
Anion exchange resins, 161
Antheridiogen, 63
Antheridium formation, 137
Antheridium-inducing activity of gibberellin methyl esters, 135
Anti-auxins, 10, 19, 23—26
Antibody, 86
Anticytokinins, 170, 183, 184
Antigen, preparation of, 84
Antisera, 85
Apple, see also *Malus,* 252—254
5-*O*-β-L-Arabinopyranosyl-DL-(indole-3-acetyl)-*myo*-inositol, 22
5-*O*-β-L-Arabinopyranosyl-2-*O*-(indole-3-acetyl)-*myo*-inositol, 22
Arsenate, 253
Ascorbigen A, 19, 23
Assay plants for gibberellins, 135
Atisagibberellin A_{12}, 126
Atisagibberellin A_{14}, 126
ATP, 253
Auxin a, 12—14
Auxin a lactone, 12—14
Auxin b, 12, 14
Auxin b lactone, 13
Auxin-induced ethylene production, 253—254
Auxins, see also specific topics, 9—56
   analysis of, 27—39
   biological activity, 47
   biosynthesis of, 41—45
   bound, 47
   definition, 10
   ethylene production, 250, 252, 254
   GC analysis, 27—33
   GC/MS analysis, 27—33
   GC/SIM analysis of, 30, 32
   history, 10—16
   metabolism of, 44—47
   natural, 15—23
   synthesis of, 39—41
   synthetic, 10
Avena coleoptile bioassay, 14
Avena coleoptile section test, 23
Avena coleoptile straight growth test, 18
Avena coleoptile test, 13
Avena curvature test, 11—12, 16, 43
Avena root growth test, 13
Avena sativa, 10—11, 67
AVG, see Aminoethoxyvinylglycine

B

Bacteria, 250
Baeyer-Villiger type oxidation of aldehyde group on C-10, 124

BAP, see 6-Benzylamino-purine
Barley, see also *Hordeum,* 23
Barley aleurone layers, 139
Beckman Ultrasphere ODS column, 37
1,2-Benzisothiazol-3-ylacetic acid (BIA), 10
N^6-Benzyl adenine analogues, cytokinins, 157
6-Benzylamino-purine (BAP), 157
Beta vulgaris, 156
BIA, see 1,2-Benzisothiazol-3-ylacetic acid
Binding components for ethylene, 259
Binding proteins, 258
Bioassay systems for gibberellins, 135
Biological activity, see also specific topics
   auxins, 47
   cytokinins, 178—189
   DPA, 241
   epi-DPA, 241
   ethylene, 251, 257—259
   GAs, 135
   HMG-HOABA, 240
   PA, 240—241
Biosynthesis, see also specific topics
   abscisic acid, 223—231
   cytokinins, 173—175
   ethylene, 252—257
   fungi, 228—231
   GAs, 119—135
      Gibberella fujikuroi, 121—126
      higher plants, 126—135
   IAA, 41—45
   isoprenoid, 223
Biosynthetic conversion of gibberellins, 125—126
Biosynthetic pathway
   abscisic acid, 225—227
   from MVA to *ent*-kaurene, 121, 123
Bis(trimethylsilyl)acetamide (BSA), 28
Bis(trimethylsilyl)trifluoroacetamide (BSTFA), 28—29
μ-Bondapak C_{18}, 37, 38
Botrytis cinerea, 204, 231
Bound auxins, 47
Brassica, 18—21, 23
Brassica napus, 138
Brassica oleracea, 20—21, 23, 156
Brassica pekinensis, 17, 20
Bruguiera gymnoriza, 68
Bryophyllum diageremontianum, 68
BSA, see Bis(trimethylsilyl)acetamide
BSTFA, see Bis(trimethylsilyl)trifluoroacetamide
1-Butene, 255
Butenone, 227

C

[^{14}C]-labeled IAA, 41
C-20, 124, 125
C_{20}-GAs, 94, 101
Ca^{++}, 255
Cabbage, see *Brassica*
Calonyction aculeatum, 63, 66

Calystegia soldanella, 66
L-Canaline, 253
Canavalia gladiata, 63, 66
Cantaloupe, 67, 255
Carbon atoms, abscisic acid, 223—225
Carbon dioxide, 251, 257—258
Carbon monoxide, 251, 258
3-*O*-Carboxymethylviridicatin, 26
Carnation, 257
Carotenoid pathway, abscisic acid, 227—228
Cassia fistula, 66
Catalytic reduction, 99
Catanea spp., 21, 156
CD, see Circular dichroism
Cell differentiation, 154
Cell division, 154
Cell-free extracts, soluble fraction of, 129
Cell-free systems, 121, 128, 131, 133
Cellular receptors, 258
Cercospora cruenta, 204, 230, 231
Cercospora rosicola, 204, 228, 229
Chamaecyparis lawsoniana, 36
Characterization
    cytokinins, 161—164
    ethylene, 252
    GAs, see also Gibberellins, 68—95
Charcoal adsorption chromatography of gibberellins, 71—72
Chemical cleavage of conjugated gibberellins, 76—77
Chemical ionization mass spectrometry (CIMS), 165, 211—212
Chemistry
    abscisic acid, see also Abscisic acid, 204—223
    cytokinins, see also Cytokinins, 161—174
    ethylene, 251—252
    GAs, see also Gibberellins, 68—120
Chinese cabbage, 17
α-Chloro-β-(3-chloro-*o*-tolyl)propionitrile, 10
Chloroethylene, 258
4-Chloroindole-3-acetic acid (4-Cl-IAA), 14, 16—17, 20
4-Chloroindole-3-acetyl-L-aspartic acid monomethyl ester, 19
4-Chlorophenoxyisobutyric acid (PCIB), 25
4-Chlorotryptophan, 19, 22
Cichorium intybus, 156
CIMS, see Chemical ionization mass spectrometry
cis-Cinnamic acid, 10
trans-Cinnamic acid, 23
Circular dichroism (CD), 102, 210
Ciridicatin, 26
Citrus, 255
Citrus auxin, 16, 18
Citrus limon, 67
Citrus sinensis, 67
Citrus unshiu, 18, 62, 65, 67
    ethylene production, 256—257
    natural auxins, 20
4-Cl-IAA, see 4-Chloroindole-3-acetic acid
4-Cl-IAA-Me, see Methyl 4-chloroindole-3-acetate

Climacteric apple, 253
CMR, see ^{13}C-Nuclear magnetic resonance
Co^{++} ion, 253, 255
Coconut milk, 154
Cocos nucifera, 156
Cold-requiring plants, 138
Collidine treatment, 99
Color reaction, abscisic acid, 209—210
Column chromatography, abscisic acid, 207
Commelina communis L., 238
Conforth's basic principle, 223
Conjugated abscisic acid, 232—234
Conjugated dihydrophaseic acid, 235—236
Conjugated *epi*-dihydrophaseic acid, 236
Conjugated gibberellins, 64—68
    charcoal adsorption chromatography, 72
    chemical cleavage, 76—77
    enzymatical cleavage, 76—77
    enzymatic hydrolysis, 102
    enzyme hydrolysis, 77
    hydrolysis rates, 77
    isolation examples, 90—91, 94—95
    reversed phase HPLC, 86
    silicic acid adsorption chromatography, 73
    solvent fractionation, 69
    structural determination, 102
    structures, 64
    synthesis, 119
    thin layer chromatography, 81
Conjugated 6'-hydroxymethyl abscisic acid, 235
Conjugated indole-3-acetic acid, 19
Conjugated phaseic acid, 235
Conjugates, see specific types
Copalol, 121
Copalyl pyrophosphate, 121, 128
Corn kernels, 67, 156—157
    GC/MS analysis of auxins, 31—32
    IAA levels, 33
    IAA metabolization, 37
    isolation of indole-3-acetic acid, 14
    isolation of zeatin, 154
    natural auxins, 20, 22
Cornmeal, 14, 20
Corylus avellana, 68
Corynebacterium fascians, 157
Cotton, 33, 67, 156, 257
Countercurrent distribution of gibberellins, 71—72
Cowberries, 254
Cross-reactivity of anti-GA-antisera, 86, 91—92
Cucumber, 18, 21, 38, 67
Cucumis melo, 67, 255
Cucumus sativus, 18, 21, 38, 67
Cucurbita maxima, 63, 67, 129—130
Cucurbita pepo, 63, 67, 89
Cultured tobacco cells, 259
Cyclamen, 138
12,16-Cyclogibberellin A_9, 126
12,16-Cyclogibberellin A_{12}, 126
Cycloheximide, 253
Cytisus scoparius, 63, 65—66, 102
Cytokinin glucosides, 170

Cytokinin receptor, 186
Cytokinins, 153—199, 254
   N^6-benzyl adenine analogues, 157
   biological activity, 178—189
   biosynthesis, 173—175
   cell differentiation, 154
   cell division, 154
   characterization, 161—164
   chemical ionization mass spectrometry, 165
   chemistry, 161—174
   extraction, 161
   functions, 154
   GC, 163
   GC/MS, 163—164
   GC/SIM, 164
   HPLC, 155, 162
   identification, 163—164
   infrared spectra, 165
   isolation, 155, 161—164
   N^6-isoprenoid adenine analogues, 155
   linked to ecdysone, 157
   mass spectrometry, 155, 164—165
   metabolism, 173, 175—178
   mode of action, 189
   naturally occurring in higher plants, 156—160
   nulcear magnetic resonance, 165
   occurrence, 154—161
   other spectroscopy, 165—166
   permethylation, 163
   photoaffinity-labeled, 168
   purification, 161—163
   radioimmunoassay, 164
   radioisotopically labeled, 169
   separation, 161—163
   structural determination, 164—166
   structure-activity relationship, 179
   synthesis, 166—174
   trifluoroacetylated, 163
   trimethylsilylated, 163
   UV spectroscopy, 164
Cytoplasmic membranes, 255

D

2,4-D, see 2,4-Dichlorophenoxyacetic acid
Datura innoxia, 156, 163
Daucus carota, 68, 156
Day neutral plants, 138
DCI, see Desorption chemical ionization
DEAE-cellulose, 161
Deaza analogues, 181
Decarboxylation hypothesis, 125
1-Decene-1,10-dicarboxylic acid, 154
1-Decylimidazole, 122, 126, 129
Degradation products of gibberellins, 114—115
Degradation studies of $GA_{1,3}$, 96—97
2,4,6-*trans*-Dehydrofarnesol, 224
Dehydrogenation of $GA_{1,3}$, 96
3-Deoxygibberellin C-Me, 115
Desorption chemical ionization (DCI), 212—215

Desthioglucobrassicin, 44
Detergents, 255
Deuterated ethylene (C_2D_4), 259
Deuterium-labeled 4-Cl-IAA-Me, 17
2,4,5,6,7-Deuterium-labeled indole-3-acetic acid, 41
4,5,6,7-Deuterium-labeled indole-3-acetic acid, 41
Di-*p*-bromobenzoate of GA_3, 97
2,4-Dichloroanisole, 23
2,4-Dichlorophenoxyacetic acid (2,4-D), 10
2,6-Dichlorophenoxyacetic acid, 24
Dictyostelium discoideum, 157
Digitalis purpurea, 138
Dihydro-GA_{31}, 131
Dihydrolupinic acid, 157, 160
Dihydrophaseic acid (DPA), 205
   biological activity, 241
   CIMS, 212
   GC, 208
   HPLC, 208
   infrared spectrometry, 210
   isolation, 205
   melting point, 210
   metabolic pathway of ABA, 237
   PMR, 210
   R_f value, 206—207
   synthesis, 222—223
   UV spectrometry, 210
epi-Dihydrophaseic acid (*epi*-DPA), 235—236
   biological activity, 241
   CIMS, 212
   GC, 208
   HPLC, 208
   infrared spectrometry, 210
   isolation, 205
   melting point, 210
   metabolic pathway of ABA, 237
   PMR, 210
   R_f value, 206—207
   synthesis, 222—223
   UV spectrometry, 210
Dihydrophaseic acid-4'-*O*-β-glucoside (DPA-β-GEt), 236
   DCI/MS, 213—215
   R_f value, 206—207
   synthesis, 223
ent-6α,7α-Dihydroxy-13-acetoxykaurenoic acid, 126
2,6-Dihydroxycinchoninic acid, 47
ent-6α,7α-Dihydroxykaurenoic acid, 121, 130
ent-7α,13-Dihydroxykaurenoic acid, 126
ent-3α,7β-Dihydroxykaurenolide, 116
ent-7α,13-Dihydroxykaurenolide, 126
Dihydrozeatin, 157, 159
Dihydrozeatin-9-glucoside, 157
Dihydrozeatin *O*-glucoside, 159
Dihydrozeatin riboside, 157, 159
Dimethylallylpyrophosphate (DMAPP), 223
1,7-Dimethylfluorene, 96
Dimorphic crystals of gibberellins, 103
2,4-Dinitrophenol (DNP), 253
Dioxindole-3-acetic acid derivatives, 47
N,N'-Diphenylureas, 169, 185

Discadenine, 157, 173
Distribution coefficients of gibberellins, 71—72
DMAPP, see Dimethylallylpyrophosphate
DNP, see 2,4-Dinitrophenol
Dolichos lablab, 65—66, 156
Dormancy break, effect of gibberellins on, 138
Dormin, 203
Douglas fir, 32
DPA, see Dihydrophaseic acid
epi-DPA, see *epi*-Dihydrophaseic acid
DPA-β-GEt, see Dihydrophaseic acid-4'-*O*-β-glucoside
DPX-1840, 26
Dwarf mutants, 135

E

ECD, see Electron capture detector
Ecdysone, 157
Ehmann's spray reagent, 17
Ehrlich's reagent, 17
EIMS, see Electron impact mass spectrometry
Electrogenic proton flux, 255
Electron capture detector (ECD), 30, 33, 207—208
Electron impact mass spectrometry (EIMS), 210—211
Eleocharis tuberosa, 156
Endoplasmic reticulum, 259
Enhydra fluctuans, 68
Enmein, 117
Enzymatical cleavage of conjugated gibberellins, 76—77
Enzymatic hydrolysis, conjugated gibberellins, 102
Enzyme hydrolysis, 77
Enzyme immunoassay, gibberellins, 84
Epigibberic acid, 114, 116—117
Ethane, 251, 252
Ethyl chlorogenate, 16
Ethylene, 249—264
    activity, 258
    auxin-induced production, 253—254
    auxin promoting production of, 250
    binding components, 259
    binding proteins, 258
    biological activity, 251, 257—259
    biosynthesis, 252—257
    biosynthetic pathway from methionine, 254
    characterization, 252
    chemistry, 251—252
    evacuation method, 252
    extraction, 252
    flow-through incubation, 251
    fruit tissues, 250
    inhibitors, 253
    initial event of action, 257—258
    intercellular space, 252
    in vitro producing systems, 252
    isolation, 251—252
    light irradiation, 250
    metabolism, 257—258
    nonenzymatic model systems, 252
    occurrence, 250—251
    oxidation to CO_2, 257
    precursor to, 252, 254
    rate of oxidation, 257
    rate of production, 250
    receptor of, 258
    red light, 250
    responsiveness to, 257
    retention times, 252
    senescent tissues, 250
    sensitivity of tissues, 251
    stress, 251
    vegetative tissues, 250
^{14}C-Ethylene, 257, 259
Ethylene-forming enzyme (EFE), 254—256
Ethylene-free air, 251
Ethyleneglycol, 257
Ethyleneglycol glucoside, 257
Ethylene oxide, 257
Ethylene-producing system, 252
Evacuation method, ethylene, 252
Exogenous abscisic acid, 238—239
Extraction
    abscisic acid, 204—205
    cytokinins, 161
    ethylene, 252
    GAs, 69—71

F

FID, see Flame ionization detector
Fisher indole synthesis, 39
Flame ionization detector (FID), 29
Flowering, effect of gibberellins on, 137—138
Flowers, ethylene production, 252
Flow-through incubation, 251
Formaldehyde, 258
Fragmentation patterns of gibberellins, 107
Fragment ions, 111, 113, 114
Free gibberellins, 60—61, 64—65
    charcoal adsorption chromatography, 72
    chemical evidence, 97—98
    degradation studies, 96—97
    identification by GC/MS analysis, 98—102
    isolation examples, 88—90, 92—93
    silicic acid adsorption chromatography, 72—73
    solvent fractionation, 69
    spectrometric methods, 97—98
    structural determination, 96—102
    structure of, 60—61, 65
Free radical mechanism, 130—131
Free radical scavengers, 255
Fruits, ethylene production, 252
Fruit-setting factor, 14
Fruit tissues, ethylene production, 250
Fucus sp., 63
Fujenal, 121
Fungal gibberellins, 58—62
Fungi, 250

biosynthesis of abscisic acid, 228—231
Fusarium moniliforme, 58
Fusicoccin, 239

G

GA, see Gibberellins
GA_1, 97, 119
$GA_{1,3}$, 96—97
GA_3, 96, 97, 117, 119—120
GA_4, 98, 117
GA_7, 126
GA_8-Me, 119
GA_{12}-7-alcohol, 124
GA_{12}-7-aldehyde, 115—116
    conversion of *ent*-kaurene to, 121, 123—124
    conversion to CA_{19}-GAs, 130—132
    formation from mevalonic acid, 128—130
    formation of C_{19}-GAs from, via C_{20}-GAs, 124—125
    3β-hydroxylation, 125
    metabolism of, 133
    synthetic route, 117
GA_{13}-7-aldehyde, 131
GA_{13}-Me, 98
GA_{14} aldehyde, 116
GA_{14}-7-aldehyde, 125
GA_{19}, 119
GA_{25}-anhydride, 131
GA_{35}, 103
GA_{35}-GEt, 103
C_{19}-GA, 94, 100, 130—132
C_{20}-GA, 124—125, 131
GA-GE, 102, 138
GA-GE-TMSi, 114, 115
GA-GEt, 137
GA-GEt-Me-TMSi, 112—113
GA-Me, 108—109
GA-Me-TMSi, 109—111
GA-TMSi-TMSi, 111
5-*O*-L-Galactopyranosyl-2-*O*-(indole-3-acetyl)-*myo*-inositol, 22
Gas chromatography (GC)
    abscisic acid, 207—208
    auxins, 27—33
    cytokinins, 163
    DPA, 209
    epi-DPA, 208
    GAs, 80—83, 87—90
    phaseic acid, 208
    retention times, 87—89
Gas chromatography/mass spectrometry (GC/MS)
    auxins, 27—33
    cytokinins, 163—164
    GAs, 63, 80—83, 87—90, 110
    new, 98—102
Gas chromatography/SIM
    auxins, 30, 32
    cytokinins, 164
    GAs, 63, 80—83, 87—90, 102

plant hormone research, 41
GC, see Gas chromatography
GC/MS, see Gas chromatography/mass spectrometry
Gel permeation columns for gibberellins, 79
Geraniol, 121
Geranylgeraniol, 121
Geranylgeranyl pyrophosphate, 128
trans-Geranylgeranyl pyrophosphate, 121
Germination, effect of gibberellins on, 138
Gibbane, 92
Gibberella fujikoroi, 45, 58—59, 62, 119, 130
    abscisic acid, 204
    ACC 917 strain, 125
    B1-41a mutant, 125—126, 128, 129
    biosynthesis of GAs in, see also Biosynthesis, 121—126
    metabolism of nonfungal GAs by, 126
    mutant strains of, 120
    R-9 mutant, 125
ent-Gibberellan-7-al-19-oic acid, 121
ent-Gibberellane, 91, 121
Gibberellenic acid, 70
Gibberellethione, 132
Gibberellin C, 96, 115, 117
Gibberellin C methyl ester, 118
Gibberellins (GAs), see also GA; specific topics, 57—151
    assay plants for, 135
    bioassay systems for, 135
    biological activities of, 135
    biosynthesis, 119—135
    biosynthetic conversion of, 125—126
    characterization, see also other subtopics hereunder, 68—95
    charcoal adsorption chromatography, 71—72
    chemistry, see also other subtopics hereunder, 68—120
    CMR, 104
    conjugated, see Conjugated gibberellins
    countercurrent distribution, 71—72
    degradation products, 114—115
    dimorphic crystals, 103
    discovery, 59—68
    distribution coefficients, 71—72
    enzyme immunoassay, 84
    extraction, 69—71
    fragmentation patterns, 107
    free, see Free gibberellins
    fungal, 58—62
    GC, 80—83, 87—90
    GC/MS, 63, 80—83, 87—90, 110
    GC/SIM, 63, 80—83, 87—90, 102
    higher plants, 63, 65, 68
    HPLC, 77—80, 82—87
    hydrated crystals, 103
    hydroxylation, 125
    3β-hydroxylation, 125
    immunoassay, 83—92
    intact plants, conversion in, 130
    isolation, see also other subtopics hereunder, 68—95

isomerization, 69—70
isotope-labeled, 114
mass chromatograms, 82—83, 89
mass chromatography, 82—83
mass spectrometric studies, 107
metabolism, 126, 131
methoxycoumaryl esters, 87
methyl ester, 135
methyl esterification, 81—82
^{13}C-NMR (CMR), 98
^{1}H-NMR (PMR), 98, 105—106
nomenclature, 91
olefinic double bond, introduction of, 125
paper chromatography, 77—78
partial synthesis, 114—117
partition chromatography, 73—76
partition coefficients, 71—72
physiology, 132—139
plant, 59, 61—68
PMR, 98, 105—106
radioimmunoassay, 84—86
relative activities of, 136
shoot elongation, 132
silicic acid adsorption chromatography, 72—73
solvent fractionation, 69—71
structural determination, 91—120
synthesis, 114—120
thin layer chromatography, 77, 79—81
total synthesis, 114, 117—120
trimethylsilylation, 82
trivial nomenclature, 92
X-ray analysis, 97
Gibberene, 96
Gibberethione, 102, 119, 135
Gibberic acid, 96—97, 103, 114, 116
Gibberone, 114, 116
Gleditschia triacanthos, 21
Glucobrassicin, 19, 23, 44
O-β-D-Glucopyranosylzeatin, 171
O-Glucosyl-dihydrozeatin, 157
O-Glucosyldihydrozeatin riboside, 157, 160
O-Glucosylzeatin, 159
O-Glucosyl-zeatin riboside, 155
Glutamate, 252
Glycine max, 33, 156, 254
Gossypium hirsutum, 33, 67, 156, 257
Gramine, 23
Gymnaster, 138

H

Helianthus annuus, 15, 18, 42, 63, 68, 156
  natural auxins, 21
Heptafluorobutyric anhydride (HFBA), 28, 30
Hetero-auxin, 12
HFBA, see Heptafluorobutyric anhydride
Higher plant gibberellins, 63, 65, 68
Higher plants, 119
  biosynthesis of GAs, see also Biosynthesis, 126—135
  ethylene, 250
  in vitro ethylene-producing systems, 252
  naturally occurring cytokinins in, 156—160
HMG-HOABA, see β-Hydroxy-β-methylglutarylhydroxyabscisic acid
Hordeum distichon, 130
Hordeum vulgare, 23, 67, 156
Horseradish peroxidase, 47
HPLC
  abscisic acid, 207—208
  cytokinins, 155, 162
  dihydrophaseic acid, 208
  epi-DPA, 208
  GAs, 77—80, 82—87
  gel permeation columns, 79
  IAA, 33—38
  indolecarboxylic acids, 34, 35
  ion-pair, 34
  nonderivatized GAs, 78
  phaseic acid, 208
  radioactive GAs and GA precursors, 82
  reversed phase, 83—84, 86—87
  reversed phase columns, 79—80
Human urine, 12, 20
Humulus lupulus, 68, 156
Hydrated crystals of gibberellins, 103
Hydrogenase activity, activation by gibberellins, 139
Hydrogenation of GA$_3$, 96
Hydrogen atoms, abscisic acid, 223—225
Hydrogenolysis of GA$_3$, 96
Hydrolysis rates, 77
ent-7α-Hydroxyatis-16-en-19-oic acid, 126, 129
6-(*O*-Hydroxybenzylamino)purine, 158
6-(*O*-Hydroxybenzylamino)-9-β-D-ribofuranosyl-purine, 157
5-Hydroxydioxindole-3-acetic acid, 47
5-Hydroxyindole-3-acetic acid (5-OH-IAA), 20, 47
ent-7α-Hydroxykaurenoic acid, 121, 128—130
ent-15α-Hydroxykaurenoic acid, 126, 129
ent-7α-Hydroxykaurenolide, 130
ent-7β-Hydroxykaurenolide, 116
Hydroxylating enzymes, 235
Hydroxylation, 125
3β-Hydroxylation, 125
6'-Hydroxymethyl abscisic acids, 232, 234—236, 240
6-(4-Hydroxy-3-methyl-2-*trans*-butenylamino)purine, 154
L-2-[6-(4-Hydroxy-3-methyl-2-*trans*-butenyl-amino)purin-9-yl]alanine, 155
β-Hydroxy-β-methylglutarylhydroxyabscisic acid (HMG-HOABA), 205—207, 213—215, 240
3-Hydroxymethyloxindole, 45—46
5-Hydroxy-*N*-methyltryptamine, 23
5-Hydroxytryptamine, 17, 23
2-Hydroxyzeatin, 159
Hyoscyamus niger, 138
Hypersil-ODS, 33, 36

I

IAA, see Indole-3-acetic acid
IAald, see Indole-3-acetaldehyde
d_4-IAA, 41
d_5-IAA, 41
^{3}H-IAA, 37
IAA-2,4,5,6,7-d_5, 42
IAA-4,5,6,7-d_4, 42
IAA-aspartate, see Indole-3-acetyl-L-aspartic acid
IAA-[^{14}C]-Asp, 38
IAA-[^{14}C]-Glu, 38
IAA-*myo*-inositol, 19
IAA-*myo*-inositol glycosides, 19
IAA-Me, see Methyl indole-3-acetate
IAA oxidase, 45, 256—257
IAA peroxidase, 45
IACRA, see Indole-3-acrylic acid
IAM, see Indole-3-acetamide
IAN, see Indole-3-acetonitrile
IEt, see Indole-3-ethanol
Immunoassay
  abscisic acid, 208—209
  GAs, 83—92
Immunogen, preparation of, 84, 90
Impatiens balsamina, 138
Indole-3-acetaldehyde (IAald), 21, 41—43
[^{14}C]-Indole-3-acetaldehyde, 45
Indoleacetaldehyde dehydrogenase, 43
Indole-3-acetaldoxime, 44
Indole-3-acetamide (IAM), 17, 20
Indole-3-acetate, 29
Indole-3-acetic acid (IAA), see also IAA, 10, 20
  biosynthesis of, 41—45
  ethylene production, 253—254
  hetero-auxin, 12
  high molecular weight, 19
  HPLC analysis of, 33—38
  isolation, 14
  isotopically labeled, 41
  metabolism of, 44—47
  oxidation metabolites of, 47
  paper chromatography, 14
  preparation of derivatives for GC analysis, 28
  radioimmunoassay of, 38—39
  separation by LiChrosorb RP-8 column, 36
  synthesis of, 39—41
  synthetic methods of, 40—41
  TMSi derivatives of, 28, 31
Indoleacetonitrile, 44
Indole-3-acetonitrile (IAN), 17, 20
Indole-3-acetyl-L-aspartic acid (IAA-aspartate), 19, 21
(Indole-3-acetyl)-glucan, 22
4-*O*-(Indole-3-acetyl)-D-glucopyranose, 22
5-*O*-(Indole-3-acetyl)-D-glucopyranose, 22
Indole-3-acetyl-*myo*-inositol, 22
1-DL-(Indole-3-acetyl)-*myo*-inositol, 22
2-*O*(Indole-3-acetyl)-*myo*-inositol, 22
Indole-3-acrylate, 29
Indole-3-acrylic acid (IACRA), 18, 20
Indole-3-butyrate, 29
Indole-3-butyric acid, 23
Indole-3-carboxyaldehyde (indole-3-CHO), 18, 21
Indole-2-carboxylate, 29
Indole-3-carboxylate, 29
Indole-5-carboxylate, 29
Indolecarboxylic acids, 34, 35
Indole-3-carboxylic acid, 18, 21
Indole compounds, retention values of, 29
Indole dimers, 47
Indole-3-ethanol (IEt), 18, 21, 44
Indole-3-glyoxylate, 29
Indole-3-lactate, 29
Indole-3-lactic acid, 23
Indolenine epoxide, 45—46
Indolenine hydroperoxide, 45—46
Indole-1-propionate, 29
Indole-3-propionate, 29
Indole-3-propionic acid, 23
Indole-3-pyruvic acid, 41—43
Indolepyruvic acid decarboxylase, 43
Indole trimers, 47
Indolo-α-pyrone spectrofluorometric method, 30
Infrared (IR) spectrometry
  abscisic acid, 210
  cytokinins, 165
  DPA, 210
  epi-DPA, 210
  PA, 210
  physicochemical properties of GAs, 103
Inhibitor β, 202
Injury, 251
Intact plants, conversion of gibberellins in, 130, 131
Intercellular space, ethylene in, 252
Inumakilactone A, 27
Ion exchange resins, 161
Ion-pair chromatography, 36
Ion-pair HPLC, 34
α-Ionylideneacetic acid, 231
α-Ionylideneethanol, 229
Ipomea batatas, 66, 156
Ipomea pes-caprae, 66
IPP, see Isopentenylpyrophosphate
Isatis japonica, 23
Isatis tinctoria, 23, 45
Isoden trichocarpus, 117
Isolation
  ABA, 204—207
  cytokinins, 155, 161—164
  DPA, 205
  epi-DPA, 205
  ethylene, 251—252
  GAs, see also Gibberellins, 68—95
  HMG-HOABA, 205
  9-β-D-ribofuranosyl zeatin, 154
  zeatin, 154
  zeatin ribotide, 154
Isomerization of gibberellins, 69—70
Isopentenyladenine, 167, 181
N^6-(Δ^2-Isopentenyl)-adenine, 160

N^6-(Δ^2-Isopentenyl)-adenosine, 160
Isopentenylpyrophosphate (IPP), 223
N^6-Isoprenoid adenine analogues of cytokinins, 155
Isoprenoid biosynthesis, 223
Isotope-labeled gibberellins, 114
Isotopically labeled indole-3-acetic acids, 41

J

Jerusalem artichoke test, 16

K

ent-Kaur-6,16-dienoic acid, 130
ent-Kaurenal, 126, 129
ent-Kaurene, 12, 121, 123—124
ent-Kaurenoic acid, 121, 129
ent-Kaurenol, 126, 129
3-Keto-GA$_3$, 119
α-Ketoglutarate, 252
Kinetin, 154, 166, 181

L

Labdadienol, 121
Lactuca, 138
Lagenaria leucantha, 63, 67
Lathyrus latifolium, 17, 20
Lathyrus maritimus, 17, 20
Lathyrus odoratus, 17, 20
Lathyrus sativus, 17, 20
Lecithin, 255
Lemna gibba L., 238
Lens cultinaris, 17—18, 20
Leucaena leucocephala, 66
LiChrosorb RP-8 column, 36
LiChrosorb RP-18, 34—35
Light irradiation, ethylene production, 250
Locusta migratoria, 157
Loliolide, 227
Long day plants, 138
Lunularic acid, 204
Lupinic acid, 157, 159
L-Lupinic acid, 171
Lupinus augustifolus, 21
Lupinus luteus, 62—63, 66, 156
Lupinus spp., 157
Lycopersicon esculentum, see Tomato
Lygodium japonicum, 63, 68

M

Magnolia sp., 21
Maize root tips, 32
Maleic hydrazide, 23
N-Malonyl ACC, 259
N-Malonyl derivatives of 4-chlorotryptophan, 22

Malonyltryptophan, 23
N-Malonyl-tryptophan, 19, 21
Malus, 138, 252—254
Malus auxin, 15
Malus sylvestris, 63, 67
Mangifera indica, 156
Marah macrocarpus, 63, 67, 128
Mass chromatograms, 82—83, 89
Mass chromatography, 82—83
Mass spectrometry (MS), see also Gas chromatography/mass spectrometry
    cytokinins, 155, 164—165
    GAs, 107—114
Mass spectrometry fragmentation, 108—115
Matricaria, 138
MBHFBA, see N-Methylbis(heptafluorobutyramide)
Melting point, 209—210
Membrane integrity, 255
Membrane lipids, phase transition of, 255
Mercaptopyruvic acid, 119, 132
Mercurialis ambigua, 156
Mercurialis annua, 156
Mercuric perchlorate, 252, 257
Metabolic pathway, abscisic acid, 236—238
Metabolism, see also specific topics
    cytokinins, 173, 175—178
    ethylene, 257—258
    GA$_{12}$-7-aldehyde, 133
    GAs, 131, 134
    IAA, 44—47
    nonfungal GAs by *Gibberella fujikuroi*, 126
Metabolite C, see 6'-Hydroxymethyl abscisic acid
Metabolites of abscisic acid, 232—236
Methane, 252
Methionine, 252—254
U-^{14}C-Methionine, 253
1-Methoxy-indole-3-acetonitrile, 17, 20
4-Methoxyindole-3-acetonitrile, 17—18, 20
3-Methylaminomethylindole, 23
N-Methylbis(heptafluorobutyramide) (MBHFBA), 30
Methyl 4-chloroindole-3-acetate (4-Cl-IAA-Me), 20, 31
3-Methyleneoxindole, 45—46
3-Methyleneoxindole reductase, 45
Methyl esterification of gibberellins, 81—82
1-Methyl-7-hydroxyfluorene, 96—97
Methyl indole-3-acetate (IAA-Me), 17, 20
2-Methylindolo-α-pyrone (2-MIP), 30, 36
2-Methyl-6-(3-methyl-2-butenylamine)-9-β-D-ribofuranosylpurine, 157
3-Methyloxindole, 45—46
5'-Methylthioadenosine (MTA), 253
2-Methylthio-6-(O-hydroxy-benzylamino)-9-β-D-glucofuranosylpurine, 158
5-Methylthioribose (MTR), 253
2-Methylthiozeatin, 159
2-Methylthio-*cis*-zeatin, 155, 159
2-Methylthio zeatin riboside, 155, 159
2-Methylthio-*cis*-zeatin riboside, 155
N-Methyltryptamine, 23

Mevalonate, 121
Mevalonic acid (MVA), 121, 123, 128—130
Mg^{++}, 255
Microdrop method for gibberellin bioassay, 135
MicroPak CH, 34—35
Microsomal enzymes, 129
Microsomal fraction, 130—131, 257
2-MIP, see 2-Methylindolo-α-pyrone
Mirosinase, 45
Mn^{++}, 256
Mode of action, cytokinins, 189
Monochloroethylene, 251
Monomethyl 4-chloroindole-3-acetyl-L-aspartic acid, 21
Monophenol, 256
Morning glory, 257
MS, see Gas chromatography/mass spectrometry; Mass spectrometry
MTA, see 5′-Methylthioadenosine
MTR, see 5-Methylthioribose
Mung bean, 252—255, 259
MVA, see Mevalonic acid

N

NAA, see 1-Naphthaleneacetic acid
NADPH, 129
Nagilactones, 27
1-Naphthaleneacetic acid (NAA), 10
α-Naphthaleneacetic acid, 38
β-Naphthaleneacetic acid, 38
α-(2-Naphthoxy)-propionic acid, 23
Natural anti-auxins, 19, 26—27
Natural auxins, 15, 20—23
   chemistry of, 16—19
   paper chromatography, 15
   related metabolites, 20—23
Naturally occurring cytokinins, see Cytokinins
Neoglucobrassicin, 19, 23
Neurospora crassa, 59, 62
Neutral hydrocarbons, 252
Nicotiana, 138, 259
Nicotiana glauca, 23
Nicotiana langsdorffii, 23
Nicotiana tabacum, 38—39, 68, 156
Nitrilase, 45
NMR, see Nuclear magnetic resonance
Nomenclature of gibberellins, 91
Nonenzymatic model systems, ethylene production, 252
Normal phase partition chromatography of gibberellins, 76
Nuclear magnetic resonance (NMR)
   abscisic acid, 210
   cytokinins, 165
   physicochemical properties of GAs, 104—107
^{13}C-Nuclear magnetic resonance (CMR), 98, 100, 101, 103, 104, 210
^{1}H-Nuclear magnetic resonance (PMR), 98, 105—106, 210

Nucleosil C_{18}, 34

O

Octadecyl silane (ODS), 162
5-OH-IAA, see 5-Hydroxyindole-3-acetic acid
Olefinic double bond of gibberellins, 125
Optical rotatory dispersion (ORD), abscisic acid, 209—210
Oryza sativa, 47, 67, 156, 238
Osmotic shock, 255
Oxidation, 99, 257
Oxindole-3-acetic acid derivatives, 47
Oxygen, ethylene production, 253, 256, 258
Ozonolysis, 99

P

PA, see Phaseic acid
Papaver somniferum, 21
Paper chromatography
   abscisic acid, 206—207
   ethyl chlorogenate, 16
   GAs, 77—78
   natural auxins, 15
   plant growth regulators, 14
Parthenocarpy, effect of gibberellins on, 138—139
Partial synthesis of gibberellins, 114—117
Partisil-10-ODS, 36
Partisil PXS, 37
Partisil-10-SAX, 33
Partition, abscisic acid, 205—206
Partition chromatography of gibberellins, 73—76
Partition coefficients
   abscisic acid, 206
   GAs, 71—72
Paucus carpa, 138
PCIB, see 4-Chlorophenoxyisobutyric acid
Pea, see *Pisum sativum*
Pear, see *Pyrus communis*
Penicillium digitatum, 252
Peracid hypothesis, 125
Perilla, 138
Permethylation, cytokinins, 163
Peroxidase, 256—257
Phalaris canariensis, 10
Pharbitis, 138
Pharbitis nil, 63, 65—66, 79, 88, 90, 102
Pharbitis purpurea, 65—66, 71, 83, 91
Phaseic acid (PA), 205, 234—235
   biological activity, 240—241
   CIMS, 212
   GC, 208
   HPLC, 208
   infrared spectrometry, 210
   isolation, 205
   melting point, 210
   metabolic pathway of ABA, 236
   PMR, 210

R_f value, 206—207
reducing enzyme, 236
synthesis, 220—222
UV spectrometry, 210
Phaseolus coccineus, 61, 63, 65—66, 83, 99, 130
Phaseolus multiflorus, 61
Phaseolus mungo, 18, 20—21
Phaseolus vulgaris, 61, 63, 65—66, 69, 91, 102, 156—157
    cell-free system from immature seeds, 131
    ethylene components, 259
    IAA levels, 33
    seeds, metabolism of GAs in, 134
Phase transition of membrane lipids, 255
Phenylacetic acid, 10
Phosphon D, 122
Photoaffinity-labeled cytokinins, 168
Phototropism, 10
Phyllostachys edulis, 62—63, 67
Physical stress, 251
Physicochemical properties of gibberellins, see also Structural determinations, 103—114
Physiology of gibberellins, 132—139
Phytochrome system, 250
Phytohormones, see also specific topics, 1
Phytotoxins, 4—7
Picea sitchensis, 33, 65, 67, 156
Pimpinella anisum, 68
Pinces attenuata, 65, 67
Pinto bean, see *Phaseolus vulgaris*
Pinus sylvestris, 34, 36
Pisum sativum, 16—17, 19, 42, 63, 66, 130, 252—254, 256—257
    analysis of natural auxins, 31—32
    cell-free system, 130, 133
    natural auxins, 20—22
Plant gibberellins, 59, 61—68
Plant growth inhibitors, 2—7
Plant growth promotors, 1—4
Plant growth promoting effect of gibberellins, 132, 135—137
Plant growth regulators, 1—7, 14
Plant growth retardants, 120—122, 126
Plant hormones, see also specific topics, 1
    auxins, 9—56
    defined, 1, 10
    separation of, 38
PLP, see Pyridoxal phosphate
PMR, see ^{1}H-Nuclear magnetic resonance
Polyvinylpyrrolidone (PVP), 162
Populus robusta, 156
Primaradiene, 121
Primula, 138
Propylene, 251, 258
Protein body, 259
Prunus, 138
Prunus armenica, 67
Prunus cerasus, 67, 156
Prunus persica, 63, 67
Purification, cytokinins, 161—163
N-(Purine-6-yl-carbamoyl)threonine, 157

PVP, see Polyvinylpyrrolidone
N-Pyridyl-*N'*-phenylureas, 189
Pyridoxal phosphate (PLP), 253—254
Pyrus communis, 63, 67, 156, 254
Pyrus malus, 67
2-Pyruvoylaminobenzamide, 27

Q

Quamoclit pennata, 66, 180
Quantification, abscisic acid, 207—209

R

Racemic abscisic acid, 214—215
Radioassay, abscisic acid, 209
Radioimmunoassay (RIA)
    cytokinins, 164
    GAs, 84—86
    IAA, 38—39
Radioisotopically labeled cytokinins, 169
Radish, see *Raphanus*
Radish leaf expansion test, 179
Raphanatin, 155
Raphanus, 138, 179
Raphanus sativus, 156
Rate of oxidation, ethylene, 257
Rate of production, ethylene, 250
RCM-100, 38
Receptor of ethylene, 258
Red light, ethylene production, 250
Reducing enzyme, 236
Relative activities, 136—138
Retention times, 252
Retention values of indole compounds, 29
Reversed phase columns for gibberellins, 79—80
Reversed phase HPLC
    conjugated GAs, 86
    GAs, 83—84, 87
Reversed-phase partition chromatography on Sephadex LH-20 of gibberellins, 76
R_f values, 206—207
Rhizophora mucranata, 68
Rhyzopus suinus, 12, 14, 20
9-β-D-Ribofuranosyl benzylaminopurine 5'-phosphate, 167
9-β-D-Ribofuranosyl isopentenyladenine, 167
9-β-D-Ribofuranosyl isopentenyladenine 5'-phosphate, 167
9-β-D-Ribofuranosyl zeatin, 154, 166
Rice, see *Oryza sativa*
Ricinus communis, 31, 128
Root curvature, 239

S

SAM, see S-Adenosylmethionine
Secale cereale, 67

Senescence, 239
Senescent tissues, ethylene production, 250
Sensitivity of tissues, ethylene, 251
Separation, cytokinins, 161—163
Sephadex LH-20, 161
Sex expression, effect of gibberellins on, 139
Shoot elongation by gibberellins, 132
Short day plants, 138
Silene, 138
Silica gel, 251
Silicic acid adsorption chromatography of gibberellins, 72—73
Solanum tuberosum, 156
Soluble enzyme system, 121
Soluble fraction, 131
Solvent fractionation of gibberellins, 69—71
Sonneralitia aplata, 68
Soybean, see *Glycine max*
Sphaceloma manihoticola, 59, 62
Spinacia, 138
Spinacia oleracea, 68
Squash, see also *Cucurbita*, 255
Stereochemical origin of carbon and hydrogen atoms, abscisic acid, 223—225
Stereoisomers, 257
Stereospecific interaction, 255—256
Steviol, 126
Steviol acetate, 126
Stomatal closure, 238, 240—241
Stress, ethylene production, 251
Structural determination, see also specific topics
   abscisic acid, 209—215
   cytokinins, 164—166
   GAs, 91—120
      conjugated GAs, 102
      free GAs, see also Free gibberellins, 96—102
      IR spectra, 103
      mass spectra, 107—114
      NMR spectra, 104—107
      physicochemical properties, 103—114
      UV spectra, 103
Structural requirements, abscisic acid, 239—241
Structure, see specific chemical
Structure-activity relationship
   abscisic acid, 239—241
   anticytokinins, 184
   cytokinins, 179
   N,N'-diphenylureas, 185
   N-pyridyl-N'-phenylureas, 189
Substance F, 16
2-Substituted ACC, 255
Substituted ethylene, 251
N^6-Substituted-2-methylthioadenine, 166
1-Sulfoglucobarssicin, 19, 23
Sunflower, see *Helianthus annuus*
Surface active substances, 253
Synthesis
   ABA, 214—223
   ABA-β-GEs, 223
   auxins, 39—41
   cytokinins, 166—174
   dihydrophaseic acid, 222—223
   epi-dihydrophaseic acid, 222—223
   DPA-β-GEt, 223
   GAs, see also Gibberellins, 114—120
   phaseic acid, 220—222
Synthetic anti-auxins, 19, 23—27
Synthetic auxins, 10
Synthetic methods of indole-3-acetic acid, 40—41
Synthetic route, 117, 118

T

Terminal carbon atom, 251
Tetraselmis sp., 63
TFA, see Trifluoroacetylation
TFAA, see Trifluoroacetic anhydride
Thin layer chromatography (TLC)
   abscisic acid, 206—207
   GAs, 77, 79—81
TLC, see Thin layer chromatography
Tobacco, see *Nicotiana*
Tobacco callus bioassay, 179, 184
Tomato, 156
   ethylene metabolism, 257
   ethylene production, 253—254
   natural auxins, 20
   seedlings, 21
   shoots, 23
Tosylation, 99
Total synthesis
   GA_1, 119
   GA_3, 117, 119—120
   $(\pm)$-GA_{15}, 117
   GA_8-Me, 119
   GAs, 114, 117—120
Trachylobanic acid, 126, 129
Transpiration, 238
Traumatin, 154
2,4,6-Trichlorophenoxyacetic acid, 24
Trifluoroacetic anhydride (TFAA), 30
Trifluoroacetylation (TFA), 28, 163
ent-6α,7α,13-Trihydroxykaurenoic acid, 126
6-(2,3,4-Trihydroxy-3-methyl-butylamino) purine, 157
N,N,N-Trimethyl-1-methyl-(3′,3′,5′-trimethylcyclohexan-1′-yl)prop-2-enylammonium iodide, 122
N,N,N-Trimethyl-1-methyl(3′,3′,5′-trimethyl-cyclohex-2′-en-1′-yl)prop-2-enylammonium iodide, 122
bis(Trimethylsilyl)acetamide (BSA), 28
Trimethylsilylation
   cytokinins, 163
   GAs, 82
bis(Trimethylsilyl)trifluoroacetamide, 28—29
N,N,N-Trimethyl-3-(3′,3′,5′-trimethylcyclohexyl)-2-propenylammonium iodide, 126
Triticum aestivum, 33, 63, 67
Triton X100, 259
Trivial nomenclature of gibberellins, 92

Tryptamine, 23, 43
Tryptamine oxidase, 43
L-Tryptophan, 41
Tryptophan aminotransferase, 42
Tryptophan decarboxylase, 43
Tryptophol, 23
Tulipa, 138
Tween 20, 255

U

Uncouplers, 253, 255
Unsaturated aliphatic compounds, 258
Unsaturated bond, 251
Unsaturated hydrocarbons, 252
Ureidopurines, 169
UV spectrometry
    abscisic acid, 210
    cytokinins, 164
    DPA, 210
    epi-DPA, 210
    GAs, 103
    PA, 210

V

Varian MCH-10, 37
Vegetative tissues, ethylene production, 250
Vernalization, 138
Vicia faba, 17, 63, 66
    ethylene metabolism, 257
    natural auxins, 20
Vicia sativa, 17, 20
Vigna unguiculata, 66
Vinca rosea, 156, 164—165
Viola tricolor, 138
Violaxanthin, 217, 227

W

Wagner-Meerwein rearrangement, 97
Water stress, 251
Wheat, see *Triticum aestivum*
Wistaria floribunda, 66
Wounding, 255
Wounds, 251

X

Xanthium, 138
Xanthoxin, 217, 227
X-ray analysis, 97, 98, 102

Y

Yeast, 20
Yellow cornmeal, see Cornmeal

Z

Zantedeschia aethiopica, 156—157, 164
Zea mays, see Corn kernels
Zeanic acid, 47
Zeatin (Z), 37
    isolation of, 154
    occurrence, 155
    preparation, 166
    structure, 158
cis-Zeatin, 155, 158, 166
Zeatin-7-glucoside, 155, 158
Zeatin-9-glucoside, 155, 158
Zeatin-*O*-glucoside, 155, 159
Zeatin-riboside, 37, 155, 158
cis-Zeatin riboside, 155
Zeatin ribotide, 154, 158
Zinnia elegans, 138
Zorbax-ODS, 33
Zorbax-SIL, 33